# Neurophysiological Basis of Cerebral Blood Flow Control: An Introduction

# Neurophysiological Basis of Cerebral Blood Flow Control: An Introduction

**Editors:**
**Sima Mraovitch and Richard Sercombe**

John Libbey

LONDON • PARIS • ROME

**British Library Cataloguing in Publication Data**

Neurophysiological Basis of Cerebral Blood Flow Control: An Introduction

    I.  Mraovitch, S.      II.  Sercombe, Richard
    612.825

ISBN: 0 86196 272 9

Published by

**John Libbey & Company Ltd,** 13 Smiths Yard, Summerley Street,
London SW18 4HR, England
**Telephone: +44 (0) 181-947 2777 – Fax: +44 (0) 181-947 2664**
**John Libbey Eurotext Ltd,** 127 rue de la République, 92120 Montrouge, France
**John Libbey - C.I.C. s.r.l.,** via Lazzaro Spallanzani 11, 00161 Rome, Italy

Printed in Hong Kong by Dah Hua Printing Press Co., Ltd.

# Foreword

It is now widely accepted that the central nervous system plays an important role in regulating its own circulation. Indeed, numerous experiments have shown that electrical or chemical stimulation of specific brain regions and their intrinsic neuronal pathways (i.e. pathways whose neurons and their processes are entirely contained within the central nervous system) elicit changes in cerebral blood flow (CBF) and/or metabolism.

Important advances have also been made in the last 10–20 years concerning the definition and roles of the extrinsic innervations (sympathetic, parasympathetic and sensory) in the control of CBF. Our aim in this book was therefore to provide an in-depth treatment of CBF control within the framework of current knowledge in neurobiology.

*Neurophysiological Basis of Cerebral Blood Flow Control: an Introduction* has been conceived as an advanced textbook for all 'students' of the cerebral circulation. It provides a thorough introduction to the subject for undergraduate and graduate students of biological sciences and medicine. However, it is also intended for the specialist, clinician or researcher, since its describes systematically and in review form the current understanding of CBF control. Each chapter contains the essential bibliography, including both reviews and original literature, for those who wish to delve further into the matter, but the authors have avoided an excessive number of citations in the text in order to facilitate the reading. Considerable effort has also been made to provide clear didactic illustrations and tables wherever possible.

Part I concerns the anatomical, pharmacological, biochemical and physiological bases. Our hope is that the reader will find in it most of the background information necessary for understanding Part II, which deals with the methods of study, the current hypotheses on CBF control, and the most important related clinical disorders. It should be noted that we have not attempted to cover in detail all the experimental work on ischaemia, but the mechanisms of stroke are discussed in Chapter 13. This subject has been under intense investigation in recent years with whole symposia being devoted to it, and it would require a monograph to itself.

We have, inevitably, given considerable emphasis to our own research preoccupations, which are intimately related to the issue of neurogenic regulation, because we believe that this field promises many major advances in the next few years. As we go to press, the exciting new messenger molecule nitric oxide is invading the whole subject of CBF control, but the reader will have to wait for this aspect to mature (like a young wine) before it can be properly appreciated and assimilated. Meanwhile, several useful papers and reviews may be recommended for the interested reader[1-5].

Last but not least, we wish to give our fullest praise and thanks to the team of specialists who devoted so much of their time to writing this book, and who helped us take this 'photograph' of present understanding of CBF control.

S. MRAOVITCH          R. SERCOMBE

## References

1. Dawson, D.A. (1994): Nitric oxide and focal ischemia: multiplicity of actions and diverse outcome. *Cerebrovasc. Brain Metab. Rev.* **6,** 299–324.

2. Faraci, F.M., Breese, K.R. & Heistad, D.D. (1993): Nitric oxide contributes to dilatation of cerebral arterioles during seizures. *Am. J. Physiol.* **265,** H2209–H2212.

3. Iadecola, C. (1993): Regulation of the cerebral microcirculation during neural activity: is nitric oxide the missing link? *Trends Neurol. Sci.* **16,** 206–214.

4. Iadecola, C., Pelligrino, D.A., Moskowitz, M.A. & Lassen, N.A. (1994): Nitric oxide synthase inhibition and cerebrovascular regulation. *J. Cereb. Blood Flow Metab.* **14,** 175–192.

5. Rigaud-Monnet, A.S., Pinard, E., Borredon, J. & Seylaz, J. (1994): Blockade of nitric oxide synthesis inhibits hippocampal hyperemia in kainic acid-induced seizures. *J. Cereb. Blood Flow Metab.* **14,** 581–590.

# The Editors

**Dr S. Mraovitch** obtained his Ph.D. at the College of Physicians and Surgeons of Columbia University, New York. He later worked at Cornell University Medical Center (New York) where he began studying the intrinsic neural control of cerebral circulation, still his major subject. Since 1984, he has been a Research Scientist of the CNRS (Centre National de la Recherche Scientifique), Paris.

**Dr R. Sercombe** obtained his Ph.D. at the Institute of Psychiatry (London University). In 1972 he began working in Paris in the field of cerebral circulation, his main subjects being the extrinsic neural control and the role of neurotransmitters. Since 1986 he has been a Research Director of INSERM (Institut National de la Santé et de la Recherche Médicale), Paris.

# Table of Contents

---

## PART I

---

**1 – THE AUTONOMIC NERVOUS SYSTEM**
*David A. Ruggiero and Douglas L. Feinstein*                                            1

    Introduction                                                                    1

    The autonomic nervous system: principles of anatomical and functional organization    3

    Central autonomic regulation                                                    11

    References                                                                       38

**2 – ELEMENTS OF CELLULAR NEUROPHYSIOLOGY**
*Michel Lamarche*                                                                       47

    Resting potential                                                               47

    The action potential                                                            53

    Synaptic transmission                                                           57

    References                                                                       61

**3 – NEUROTRANSMITTER SYSTEMS**
*Thérèse M. Jay*                                                                        63

    Introduction                                                                    63

    Identification criteria for neurotransmission and general scheme                 64

    Catecholaminergic pathways                                                      65

    Cholinergic pathways                                                            69

    Serotonergic pathways                                                           70

    Peptide-containing cell groups and pathways                                     72

    Amino acid neurotransmitters                                                    77

    Cotransmission                                                                  80

    Parasynaptic transmission                                                       81

    References                                                                       82

**4 – RECEPTORS, IONIC CHANNELS AND ION PUMPS**
*Robert C. Miller, Martin Galvan and C. Robin Hiley*                                    87

    Introduction                                                                    87

    Receptors                                                                       87

    Transmitter–receptor interactions                                               90

Equilibrium and kinetic binding studies . . . . . . . . . . . . 94
Ionic channels . . . . . . . . . . . . . . . . . . . . . . . . . . . . . . . 96
Ionic pumps and exchange mechanisms . . . . . . . . . . . 107
References . . . . . . . . . . . . . . . . . . . . . . . . . . . . . . . . . . 108

## 5 – THE ANATOMY OF THE BRAIN VASCULATURE
*K. Hodde and R. Sercombe* . . . . . . . . . . . . . . . . . . . . . . 111
Introduction . . . . . . . . . . . . . . . . . . . . . . . . . . . . . . . . 111
Blood supply to the brain . . . . . . . . . . . . . . . . . . . . . . 112
Venous drainage . . . . . . . . . . . . . . . . . . . . . . . . . . . . . 118
Vascular-meningeal relations . . . . . . . . . . . . . . . . . . . 120
Microcirculatory patterns . . . . . . . . . . . . . . . . . . . . . . 121
The vessel wall . . . . . . . . . . . . . . . . . . . . . . . . . . . . . . 126
The vascular innervation . . . . . . . . . . . . . . . . . . . . . . . 127
Blood–brain barrier and brain volume regulation . . . . 136
References . . . . . . . . . . . . . . . . . . . . . . . . . . . . . . . . . . 138

## 6 – CEREBROVASCULAR SMOOTH MUSCLE AND ENDOTHELIUM
*William J. Pearce and David R. Harder* . . . . . . . . . . . . . 145
Introduction . . . . . . . . . . . . . . . . . . . . . . . . . . . . . . . . 145
The biochemistry of vascular smooth muscle contraction . . . 145
Cerebrovascular electrophysiology . . . . . . . . . . . . . . . 153
Cerebrovascular endothelium . . . . . . . . . . . . . . . . . . . 158
References . . . . . . . . . . . . . . . . . . . . . . . . . . . . . . . . . . 173

## 7 – METABOLISM OF THE CENTRAL NERVOUS SYSTEM
*Astrid Nehlig* . . . . . . . . . . . . . . . . . . . . . . . . . . . . . . . . 177
Introduction . . . . . . . . . . . . . . . . . . . . . . . . . . . . . . . . 177
How does brain get and transform energy? . . . . . . . . . 177
Cerebral energy consumption . . . . . . . . . . . . . . . . . . . 190
Cellular heterogeneity of brain energy metabolism . . . 192
Homoeostasis of cerebral energy metabolism . . . . . . . 194
Conclusion . . . . . . . . . . . . . . . . . . . . . . . . . . . . . . . . . 195
References . . . . . . . . . . . . . . . . . . . . . . . . . . . . . . . . . . 196

# PART II

**8 – METHODS OF INVESTIGATION OF CEREBRAL CIRCULATION AND ENERGY METABOLISM**
*Pierre Lacombe and Mirko Diksic*                                                    199

   Introduction                                                      199
     Differences between the techniques used in man and animals     199
     Variety of variables involved                             199
     Interest of quantification and radioactive tracers        200
   Methods used in animals                                           200
     Generalities                                              200
     Investigation of the cerebral blood circulation           205
     Investigation of cerebral capillary permeability and blood volume   216
     Measuremement of cerebral energy metabolism               219
     Conclusions: how to choose a technique?                   229
   Methods used in man                                               230
     Cerebral blood flow measurement in humans                 230
     Oxygen utilization measurements                           233
     Brain glucose utilization measurement with PET            235
   References                                                         238


**9 – REGULATION OF CEREBRAL BLOOD FLOW: AN OVERVIEW**
*Wolfgang Kuschinsky*                                                               245

   Regulation by blood gases 'Chemical' regulation                   245
   Other aspects of regulation Relation between blood flow and metabolism   251
   Autoregulation Blood flow and perfusion pressure                  255
   Neurogenic regulation                                             258
   References                                                         258


**10 – EFFECTS OF NEUROTRANSMITTERS ON CEREBRAL BLOOD VESSELS AND CEREBRAL CIRCULATION**
*Jan Erik Hardebo and Christer Owman*                                               263

   Introduction                                                      263
   Catecholamines                                                    264
   Serotonin                                                         267
   Acetylcholine                                                     268
   Histamine                                                         269
   Neuropeptides                                                     270

Other peptides with effects on cerebrovascular tone     275

Prostanoids and leukotrienes     276

Purinergic agents     277

Amino acids     280

References     281

## 11 – NEUROGENIC REGULATION OF CEREBRAL BLOOD FLOW: EXTRINSIC NEURAL CONTROL

*Peter J. Goadsby and Richard Sercombe*

    285

Introduction     285

The sympathetic nervous system     286

    Anatomy     286

    Effects of ablation, section or blockade     288

    Effects of sympathetic stimulation     292

    Other influences of sympathetic nerves     298

    Overview: role of the sympathetic nervous system     303

The parasympathetic nervous system     304

    Anatomy     304

    Effect of parasympathetic blockade     306

    Stimulation of the parasympathetic nerves     307

    Physiological significance     309

Trigeminal innervation of the cerebral vessels     309

    Anatomy of the trigeminocerebrovascular system     309

    Effect of blocking the trigeminal system     311

    Trigeminal system stimulation     312

    Possible role     314

References     315

## 12 – NEUROGENIC REGULATION OF CEREBRAL BLOOD FLOW: INTRINSIC NEURAL CONTROL

*Sima Mraovitch*

    323

Introduction     323

Experimental considerations     324

Functional organization of central pathways implicated in the regulation of CBF and metabolism     329

The cerebrovascular and metabolic response patterns elicited by specific neural systems     337

Efferent mechanisms of cerebrovascular responses elicited by intrinsic neural systems     346

Conceptual considerations     350

References     354

**13 – CLINICAL IMPLICATIONS (1): STROKE, SUBARACHNOID HAEMORRHAGE AND EPILEPSY**

*Barbro B. Johansson*    359

  Stroke    359

    Definitions and diagnosis    359

    Risk factors    361

    Effects on brain tissue, mechanisms of neuronal death, post-ischaemic reperfusion    363

    Prevention of stroke. Acute treatment of stroke    366

    Vascular dementia    366

  Subarachnoid haemorrhage    367

    Incidence and diagnosis    367

    Causes and mechanisms    367

  Epilepsy    369

    Background    369

    Clinical epilepsy    369

    Animal models of epilepsy    369

    Local blood flow and glucose comsumption    370

  Comparison of neuronal cell damaged in ischaemia, epileptic seizures and hypoglycaemia    371

    Differences and similarities    371

    Distribution of necrosis    371

References    372

**14 – MIGRAINE, AUTONOMIC DYSFUNCTION AND THE PHYSIOLOGY OF THE CEREBRAL VESSELS**

*Peter J. Goadsby*    375

  Migraine    375

    Definition and classification    375

    Premonitory features    376

    Prodrome of migraine with aura    376

    The headache    378

    Response to therapy and the neuropharmacology of migraine    379

    Hypothesis    380

  Autonomic failure    382

  References    383

**APPENDIX**    385

**INDEX**    391

# Neurophysiological Basis of Cerebral Blood Flow Control: An Introduction

**Part 1**

# Chapter 1

# THE AUTONOMIC NERVOUS SYSTEM

## David A. Ruggiero and Douglas L. Feinstein

*Division of Neurobiology, Department of Neurology and Neuroscience, Cornell University Medical College, New York, NY 10021, USA*

## Introduction

The role of the autonomic (or vegetative) nervous system (ANS) is to preserve the preset norm or homoeostasis: an internal milieu compatible with survival. A stable steady state equilibrium is established by neural networks which orchestrate integrated patterns of visceral response: the quick reflex adjustments to signals, perceived and subliminal, visceral and somatic, generated within and outside of the organism. The ANS has traditionally been equated with the visceral effector system. The ANS is defined by the dual motor outflow to involuntary contractile and secretory tissues: the smooth and cardiac muscles and glands. Because sensory signals from diverse sources – the skin, body wall and viscera – often simultaneously influence autonomic motor function, afferents have not been customarily used in defining the ANS. The ANS is simply a convenient functional term used to designate the motor supply of the viscera as opposed to the motor innervation of skeletal (voluntary) muscle. Langley proposed the term 'ANS' for the sympathetic and 'allied nervous systems of cranial and sacral nerves and for the local nervous system of the gut'. Although visceral organs are capable of an intrinsic reflex control independent of central neural input (the enteric nervous system), the ANS enables different organ systems to interact and the organism to adapt more efficiently to environmental perturbations. In life, the automatic and somatic (voluntary) nervous systems are not mutually exclusive and act in concert, enabling the entire organism to adapt to signals conveyed by somatic and visceral receptors. A clear-cut illustration would be the alterations of autonomic response brought on by pain and exercise, and other movements such as head tilt or the assumption of an upright position. The early anatomists conceived of the ANS as distinct and structurally independent of the brain and spinal cord or the central nervous system (CNS; Figs. 1 & 2). The gross anatomy of the sympathetic component was recognized as two bilaterally symmetrical trunks or linear chains of ganglia. These nerve cords lie along the vertebral column and connect by nerves and plexuses with organs of involuntary control.

The ANS is conventionally subdivided into two systems which act reciprocally and stereotypically either to consume, conserve and restore energy (the parasympathetic nervous system) or to mobilize and expend these stores (the sympathetic nervous system). The two divisions of the ANS are also referred to as the craniosacral and thoraco-lumbar outflow, respectively, based on the origins of their motor 'output' neurons in the brain or spinal cord. Sympathetic and parasympathetic divisions differ from the voluntary nervous system in that each employs two motor neurons to deliver an outflow characterized by slower conduction velocities and longer synaptic delays. However their effectiveness lies in a unique ability to coordinate their activities as a unit and, thus, serve the whole organism. The dominance of one component over another is largely determined by the complex interactions between negative and positive feedback systems in the CNS. In response to an elevation in arterial blood pressure, for example, sympathetic nerve discharge is immediately inhibited and vagal parasympathetic discharge is gradually augmented (an example of a reciprocal feedback circuit). The integrated actions of sympathetic and parasympathetic nerves on organ systems devoted to energy use or restoration can direct the flow of neural impulses (and thereby redistribute cardiac output) synergistically and in accordance with the prevailing requirements for homoeostasis (see review by Koizumi & Brooks, 1972).

In primitive vertebrates, visceral end organs are generally innervated by only one division of the ANS, but with the exception of the stomach where their actions are additive and excitatory (see review by Sarnat & Netsky, 1974). Through evolution, a more sophisticated system was selected in

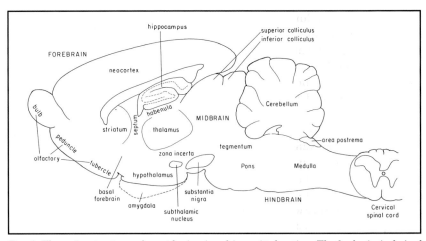

Fig. 1. The major structures of a rat brain viewed in sagittal section. The forebrain is derived from the prosencephalon. The telencephalon comprises the cerebral cortex, white matter and deep subcortical nuclei (e.g. striatum, septum, basal forebrain). The diencephalon gives rise to the epithalamus (pineal gland, habenular nuclei and stria medullaris), thalamus, hypothalamus, subthalamic nucleus and zona incerta. The midbrain (mesencephalon) is divided into the tectum (corpora quadrigemina: superior and inferior colliculi) and the tegmentum and crura cerebri (separated by the substantia nigra). The cerebellum and pons are derived from the metencephalon; the medulla is derived from the myelencephalon. The cerebellum, pons and medulla form the hindbrain (rhombencephalon).

allowing for a balanced antagonism between the two divisions of the ANS. Advanced vertebrates thus have a double-motor innervation of their visceral end organs delivered by sympathetic and parasympathetic motor neurons. Through these dual connections, both divisions of the ANS normally exert balanced antagonistic actions producing either contraction or relaxation. Primitive mechanisms of control, however, were retained as for example in the innervation of salivary *gland cells* where their actions are additive and complementary. The vasomotor innervation of salivary *arterioles*, however, still follows the general rule: sympathetic excitation (depolarization; vasoconstriction) and parasympathetic inhibition (hyperpolarization; vasodilatation).

The neurotransmitters released at the end organ can exert either excitation

or inhibition depending on their site of action. The nature of the response is specified by receptors: specialized membrane-bound protein molecules within the presynaptic or postsynaptic membranes of neurons, muscle fibres or gland cells. Receptor sites normally face outward and are accessible to chemical substances which bind to it (the receptor-binding unit) such as neurotransmitters, pharmacological agents, or circulating hormones (see Chapter 4). When signalled, the binding protein opens a channel (another membrane subunit or ionophore) which allows ions to pass through the plasma membrane (the 'lock and key' hypothesis of receptor-mediated action). Whether an involuntary muscle contracts or relaxes depends on which division predominates. (In this regard, the ANS differs from the skeletal or somatic motor system which always provokes an excitation-contraction response.) Smooth involuntary muscles are unlike striated voluntary muscles in that their plasma membranes are in close apposition and joined by electrical or gap (20 Å) junctions. These channels are formed from protein subunits and characterized by cytoplasmic continuity allowing intercellular exchange of ions and small molecules. When excited, myofilaments of smooth muscle cells contract and the cells as a unit undergo slow-sustained contractions that spread throughout the entire muscle (see Chapter 6).

The CNS plays a key role in determining the dominance of one system over another. It has long been recognized that the ANS, if disconnected from the CNS, can in some ways function automatically; independently of the brain and spinal cord. As will become clear, both systems are intricately connected, and in life, their outputs are the product of their interactions.

## The autonomic nervous system: principles of anatomical and functional organization

The ANS uses two motor neurons (general visceral efferents) to convey signals from the CNS to cardiac and smooth muscles and glands (Figs. 3–6). Preganglionic neurons are smaller than somatic motor neurons and originate in the brain or spinal cord where they form identifiable general visceral efferent cell columns. Preganglionic axons are short or long and finely myelinated and exit from the brain through cranial nerves or the spinal cord through spinal nerves. Preganglionic axons terminate on (are presynaptic to) postganglionic neurons in autonomic ganglia. Postganglionic

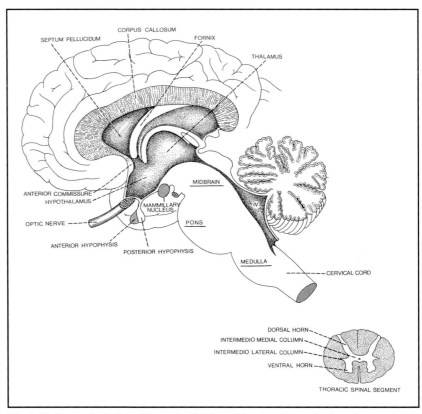

*Fig. 2. Median sagittal view of the major divisions of the adult human brain and a cross-section of a thoracic spinal segment.*

neurons occur in peripheral ganglia which lie in proximity to or are embedded within end organs. These give rise to unmyelinated axons which terminate on the visceral effector cells. Unlike the faster conducting myelinated ($\alpha$ and $\lambda$) somatic motor neurons which directly synapse onto motor end plates within skeletal muscles, postganglionic efferents do not establish well-defined neuromuscular junctions within autonomic target organs (see below). Processes of postganglionic motor neurons travel with visceral nerves and form terminal plexuses associated with the organs they innervate. These networks include the cardiac (heart and great vessels), pulmonary (lungs and tra-

cheobronchial tree), mesenteric (gastrointestinal tract) and pelvic (genitourinary and reproductive systems).

## Sympathetic nervous system

### General organization

The sympathetic nervous system (SNS) is also termed the thoracolumbar division. In mammals, sympathetic preganglionic neurones originate in the spinal cord and are distributed segmentally. Sympathetic neurons are clustered along bilaterally symmetrical columns which extend from the eighth cervical (or first thoracic) spinal segment to the second or third segment of the lumbar spinal cord. Most preganglionic neurons originate in the lateral horns or intermediolateral cell columns (ILC) in the spinal grey matter. [The spinal grey matter contains neuronal somata and axons (unmyelinated and weakly myelinated) and glial cells and blood vessels and is divided into three zones: the dorsal (sensory) horn, the ventral (motor) horn and the intermediate zone where preganglionic neurons of the lateral horn reside (see Fig. 4). These three zones are further divided into groups of neurons arranged in columns. The cell groups are organized along ten laminae (sheets).] Sympathetic motor neurons reside in lamina VII (lamina intermedia) situated between the dorsal and ventral horn. Most are concentrated in the ILC along the lateral aspect of this lamina. Another group of cells is concentrated in its medial aspect (the central autonomic nucleus or intermediomedial cell column (IMC)). Cells in the central zone relay nociceptive (pain) signals from spinal nerves (see review by Vierck *et al.*, 1986) and selectively project to the inferior mesenteric ganglion (Strack *et al.*, 1988). Some preganglionic motor neurons and processes traverse an area of lamina VII intermediate in position between the ILC and IMC (the nucleus intercalatus) or extend into an adjacent locus in the lateral funiculus. The lateral funiculus is mainly composed of fibre bundles (white spinal matter) interposed between the dorsal and ventral horns. Axons of sympathetic preganglionic neurons emerge from the spinal cord through the ventral roots of all thoracic and upper lumbar spinal segments. The spinal segment is defined as a cylindrical portion of the cord connected to the periphery by a pair of spinal

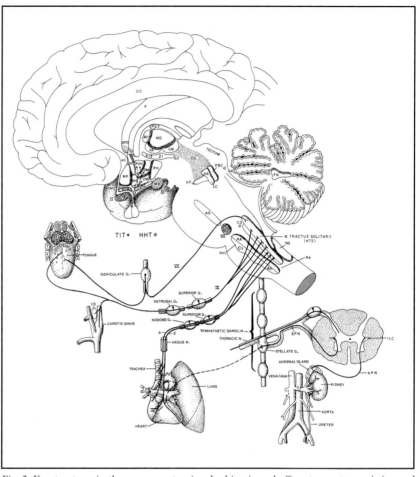

*Fig. 3. Key structures in the nervous system involved in visceral afferent neurotransmission and central autonomic control. The renal sympathetic innervation (not illustrated) arises from paravertebral (T6–L4) and prevertebral ganglia. Renal afferents derive from lower thoracic and upper lumbar dorsal root ganglia.*

nerves. Each spinal nerve is formed of two roots: (1) the dorsal or sensory root formed by pseudounipolar afferent neurons with cell bodies in a dorsal root ganglion located in the intervertebral foramen; (2) the ventral root arising from somatic and visceral motor neurons in the spinal grey.

Sympathetic preganglionic neurons give rise to general visceral efferent axons which travel with somatic fibres and contribute to specific thoracic,

lumbar, and sacral spinal nerves which supply smooth and cardiac muscles and glands (Figs. 4 & 5). These axons are less than 3 μm in diameter and initially travel in the fasciculus proprius (the white matter immediately adjacent to the spinal grey). They ascend or descend intersegmentally in this fasciculus before exiting through the ventral roots (Petras & Faden, 1977). Preganglionic axons terminate on postganglionic multipolar cell bodies in the paravertebral (sympathetic trunk) or prevertebral (abdominal) autonomic ganglia. The sympathetic trunk ganglia are organized in 21 pairs: 3 cervical, 10–12 thoracic, 4 lumbar and 4–5 sacral. They form the sympathetic trunks (chains) which are bilaterally symmetrical and lie along each side (anterolateral) of the vertebral columns. Some axons upon entering a ganglion terminate. Other axons travel a considerable distance along the trunk and end on cells of a ganglion lying below or above the level of entry. Single preganglionic neurons are often related to a large number of postganglionic motor neurons. Their axons may branch and end on multiple peripheral ganglia thus diverting the motor output signal to neurons serving different organs (autonomic integration). The sympathetic trunks extend from the second cervical vertebrae to the coccyx where they merge to form the coccygeal ganglion impar. The segmental arrangement of sympathetic postganglionic cells within specialized ganglia is a phylogenetically more advanced feature of the ANS. This organization is not present in either protochordates or lower vertebrates which emerged before elasmobranches. Primitively, autonomic ganglia lie in the dorsal body wall (e.g. amphioxus) or are scattered in the walls of end organs. The latter architecture was evolutionarily conserved and retained by the parasympathetic nervous system in higher vertebrates (see review by Sarnat & Netsky, 1974).

## Rami communicantes

Sympathetic spinal nerves (T1–L2) join the sympathetic trunk by a small branch of the ventral root, the distal white ramus communicantes (Fig. 4).

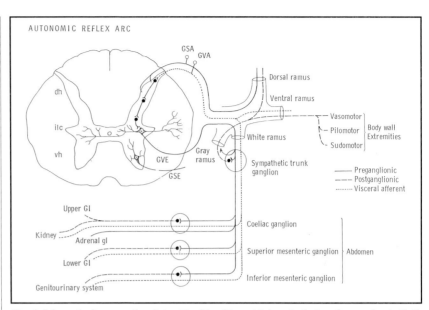

Fig. 4. Schematic diagram of an intraspinal (multisynaptic) sympathetic reflex arc. Sympathetic preganglionic neurons terminate in paravertebral trunk (chain) ganglia or prevertebral (collateral) ganglia. Visceral afferents travel along somatic as well as visceral nerves.

White rami unite spinal nerves to the sympathetic trunk and are composed of myelinated preganglionic fibres and other myelinated and unmyelinated axons. (General visceral afferent fibres, for example, which serve the same thoracic and abdominal visceral organs also travel by way of the white ramus.) Postganglionic cell bodies in the chain ganglia give rise to largely unmyelinated and other finely myelinated axons. These axons are joined to every spinal nerve by the grey rami communicantes and distribute to peripheral blood vessels (vasomotor neurons), sweat glands (sudomotor neurons) and hair follicles (pilomotor neurons). The rami communicantes are composed of visceral efferent as well as afferent axons.

## Neural-vascular plexus

Other postganglionic axons do not rejoin spinal nerves but instead, form plexuses along blood vessels (Fig. 5).

These fibres provide the visceromotor innervation of the head, neck and thorax. The cervical sympathetic ganglia are connected longitudinally by ascending preganglionic axons (white rami). The large superior cervical ganglion connects by the grey rami to neighbouring spinal and cranial nerves. Major branches include the superior cardiac nerve which forms the cardiac plexus (related to the aortic arch), innervating the heart; the internal carotid nerve and plexus innervating the blood vessels of the brain and pupillary dilator muscles and the external carotid plexus distributing to salivary and other glands in the head. Middle and inferior cervical ganglia distribute by grey rami communicantes to adjacent spinal nerves and by the middle and inferior cardiac nerves to the cardiac and pulmonary plexuses. The inferior cervical and first thoracic ganglia combine to form the stellate ganglion.

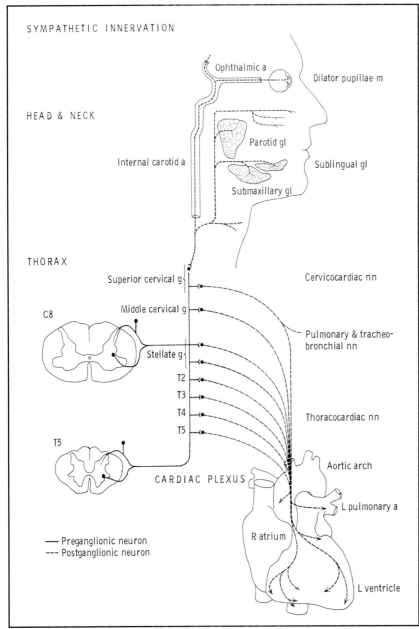

SYMPATHETIC INNERVATION

HEAD & NECK

THORAX

*Fig. 5. Schematic diagram illustrates the sympathetic (paravertebral) postganglionic innervation of key effector structures in the head, neck and thorax. Postganglionic fibres in the thorax contribute to the pulmonary, tracheobroncheal and cardiac nerves. Figure 4 illustrates the distributions of paravertebral postganglionic axons joining spinal nerves (via the grey rami) and the prevertebral sympathetic postganglionic outflow to abdominal and pelvic organs and blood vessels.*

three major branches: the superior, middle and inferior splanchnic nerves. The splanchnic nerves are preganglionic axons which exit from the white rami of the thoracic portion of the sympathetic trunk without synapsing. These fibres, instead, descend through the diaphragm and synapse on 'prevertebral ganglia' which are organized along the aorta. Postganglionic neurons in the prevertebral ganglia innervate effector cells of abdominal and pelvic viscera and reach their targets by perivascular plexuses. Postganglionic neurons in the coeliac, superior and aorticorenal ganglia innervate the abdominal viscera and kidney. Those in the inferior mesenteric ganglia innervate the pelvic viscera including the bladder and reproductive organs by the hypogastric plexus (see DeGroat & Steers, 1990).

## Segmental distribution of sympathetic outflow

Inputs to each sympathetic ganglion derive from preganglionic neurons distributed over multiple segments of the spinal cord. However, each ganglion receives a preferential projection from segments covering one spinal level (C8–T4, head neck and eyes; T1–T5, heart and lungs; upper (T4–T9; T9–T10), and lower (T12) gastrointestinal tract; and T12–L2, pelvic genitourinary organs). The adrenal gland is directly innervated by preganglionic neurons of middle and lower segments of the thoracic spinal cord (Schramm *et al.*, 1975; Rubin and Purves, 1980; Strack *et al.*, 1988). Cholinergic synapses are directly established between sympathetic preganglionic neurons and specialized postganglionic neurons (chromaffin cells) which predominantly synthesize the

The thoracic components of the sympathetic chain ganglia are connected by small spinal nerve branches to aortic and pulmonary plexuses and by

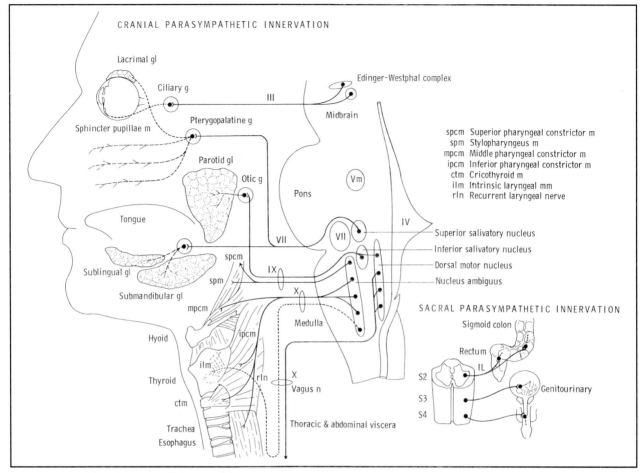

*Fig. 6. Parasympathetic (craniosacral) division of the autonomic nervous system. The cranial division includes the following general visceral efferent cell columns: the Edinger–Westphal nucleus, salivary (lacrimal) nuclei and dorsal motor nucleus. Special visceral efferent or branchiomeric (gill arch) cell columns include the trigeminal (V) and facial (VII) motor nuclei and ambigual motor nuclei of cranial nerves IX–XI. The sacral spinal division is illustrated on the lower right.*

catecholamine, epinephrine and smaller amounts of its precursor, norepinephrine. In lower vertebrates, chromaffin cells lie among sympathetic ganglion cells with which they are synaptically related.

## Parasympathetic nervous system

### General features

The parasympathetic nervous system (PNS) is structurally designed to produce localized responses of short duration (Fig. 6). The PNS is also termed the cranio-sacral system based on the locations of its preganglionic neurons in the brainstem and the sacral spinal cord. Neurons which form the cranial part of the PNS have long been referred to as 'the conserver of bodily resources', and the sacral portion as 'the mechanism of emptying'. These descriptions are based on the functions of the innervated organs. Pre-

ganglionic neurons in the brain issue axons which travel along cranial nerves, whereas those in the sacral spinal cord exit by the second to fourth sacral spinal nerves. In contrast to the sympathetic nervous system, parasympathetic preganglionic motor neurons synapse on a comparatively smaller number of postganglionic neurons. Postganglionic cells of the PNS are located in close proximity to the target organ, in the organ wall or nearby plexuses, unlike sympathetic motor neurons which form ganglia distant from their effector sites. The precision of the PNS in producing a targeted response is possible because the projection to the effector organ is limited to a smaller number of projection sites. Parasympathetic stimulation decreases depolarization of the sinoatrial pacemaker; constricts the coronary arteries and smooth muscles of the tracheobronchial tree. The PNS increases the tone and peristalsis of gastrointestinal smooth muscle with relaxation of the gastrointestinal sphincters, and provides an excitatory input to the detrusor muscle of the urinary bladder and the vasomotor innervation of the reproductive organs. Parasympathetic stimulation also produces global increases in the secretory activities of gland cells in the head, thorax and GI tract and the accessory glands of sexual organs.

## Cranial division

The cranial division of the PNS is illustrated in Fig. 6 and is formed of interrupted columns of neurones extending from the midbrain to the medulla oblongata. These preganglionic neurons are morphologically similar and have fusiform and multipolar shapes, but they relay to functionally distinct effector sites.

### *The dorsal mesencephalic nucleus of Edinger–Westphal*

Preganglionic neurons in the midbrain form the dorsal accessory nucleus of the third (oculomotor) cranial nerve. These cells issue parasympathetic cholinergic process which exit from the brain along the oculomotor nerve and terminate on postganglionic neurons of the ciliary ganglion located behind the eye. Ciliary ganglion (postganglionic) neurons innervate the circular muscle of the iris to produce constriction of the pupil (miosis) in direct response to light. Ciliary postganglionic cell bodies also innervate ciliary muscles which regulate the refractory power of the lens when shifting focus from a distant to a nearby object (accommodation). The pupillary constrictor reflex is mediated by a pathway extending from the retina to the pretectum (a complex heterogeneous nucleus in the dorsal midbrain transitional with the dorsal thalamus). Pretectal cells convey the signal through interneurons to the nucleus of Edinger–Westphal. The accommodation reflex is more complex. The visual stimulus is transmitted by the retina to the lateral geniculate body (the thalamic relay of the visual pathway) and via thalamo-cortical radiations to the cerebral cortex [areas 8 (frontal eye fields) and 19 (occipital cortex)] where the input is refined and relayed to the superior colliculus ('hill'). The superior colliculus (a paired laminated nucleus in the roof (tectum) of the midbrain) projects to the nucleus of Edinger–Westphal. The colliculus also receives direct projections from the retina and is a primitive cortical processing (sensory convergence) centre which integrates the perception of a visual stimulus with signals from the forebrain involved in generating coordinated reflex movements of the head, neck and eyes.

Parasympathetic preganglionic neurons in the lower brainstem are organized in columns and issue general visceral efferent axons distributed by cranial nerves VII, IX and X. Two major groups of motor neurons have traditionally been defined: the lacrimal and salivatory nuclei and the dorsal motor nucleus of the vagus (DMX).

### *Lacrimal and salivatory nuclei*

Lacrimal and salivatory preganglionic neurons form confluent arcs extending diagonally across the parvicellular reticular nucleus (the associative or intermediate relay area of the lower brainstem) (Valverde, 1962; Mehler, 1983). These chains of cholinergic motor neurons are bordered dorsally by the DMX (a general visceral efferent column) and ventrally by the facial nucleus (a special visceral efferent column – see below). Lacrimal and salivatory parasympathetic preganglionic neurons are partially topographically organized. Lacrimal motor neurons are located in the ventral part of the arc and are skewed rostrally and laterally to the facial nucleus. Superior salivatory neurons are intermingled with those of the inferior salivatory nucleus which extend posteriorly within the arc toward the DMX. Because the superior and inferior salivatory neurons are admixed, it has been recommended that these two subdivisions be comprised into one. The terms are imprecise also because these nuclei innervate other tissues as well, including blood vessels, the lacrimal gland and other myoepithelial tissues such as the small glands in the nasal cavity, palate and tongue. The innervation of lacrimal and salivary glands is provided by two cranial nerves (VII and IX).

*Lacrimal (tear) gland.* Lacrimal preganglionic neurons lie lateral and rostral to the facial nucleus (superior salivatory nucleus) and give rise to axons which exit from the brain through the nervous intermedius (Wrisberg's nerve). The intermediate nerve enters a small bony canal in the temporal bone of the skull (the pterygoid canal) and is joined by the deep petrosal nerve composed of postganglionic sympathetic fibres derived from the cervical trunk ganglion. Lacrimal preganglionic neurons synapse on postganglionic cells in the pterygopalatine ganglion which innervate the lacrimal gland, mucous glands in the palate, pharynx and nasal cavity, and blood vessels.

*Salivary glands.* Salivary preganglionic neurons are intermingled in the lateral reticular formation but are distinct from (dorsal and caudal to) lacrimal motor neurons. The superior salivatory cell group also contributes fibres to the nervus intermedius (VII). These axons join another branch of the VIIth nerve (the chorda tympani) before ending in the submandibular ganglion. Postganglionic neurons in this ganglion supply blood vessels and glandular tissues in the submandibular and sublingual glands. When activated (e.g. in response to hunger which precipitates a predatory behaviour or to a maternal-conditioned signal), salivary and lacrimal blood vessels will dilate concomitant with lacrimation and a profuse watery secretion of saliva. The inferior salivatory cell group issues axons which exit the medulla through rootlets of the glossopharyngeal nerve (cranial nerve IX). These fibres join the tympanic nerve and then the inferior petrosal nerve which terminates on secretomotor neurons in the otic ganglion. The otic ganglion lies immediately below the foramen ovale and assumes a position lying between the tensor palati muscle and the mandibular

branch of the trigeminal nerve (cranial nerve V). Postganglionic cells in this ganglion innervate the parotid gland.

### Dorsal motor nucleus of the vagus
The DMX, the dorsal component of the parasympathetic cell columns (Fig. 6) is located in the dorsal medulla and bordered by a cleft (sulcus limitans) indenting the floor of the fourth ventricle. The nucleus tractus solitarii (NTS), the principal sensory relay station for general and special visceral afferents carried by cranial nerves, and the DMX lie dorsal or ventral to the sulcus and are integrated by shared afferents entering into local reflex arcs. Preganglionic motor neurons of the DMX are spatially organized along a heterogeneous cell column running the entire length of the dorsal medulla. The rostral (cephalic) part issues preganglionic axons which exit the brain by the nervus intermedius and distribute by branches of the chorda tympani and greater superficial petrosal nerves. Another group of cells in the DMX exit via rootlets of cranial nerve IX, distribute by the lingual-tonsillar branches of this nerve and synapse on postganglionic neurons which supply arterioles and both mucous and serous glands opening on the posterior surface of the tongue.

The majority of dorsomotor preganglionic neurons reside in caudal parts of the nucleus. These axons leave the medulla through rootlets of the vagus nerve (cranial nerve X) and terminate on ganglia located within or in proximity to their target organs in the thorax and abdomen. The majority of preganglionic neurons at caudal levels of the nucleus project by the sub-

diaphragmatic vagus nerve and are viscerotopically organized. Neurons innervating the stomach predominate and are medial to and rostrocaudally more extensive than those which supply organs innervated by the hepatic and coeliac nerves. A proportionately greater number of neurons in the DMX supplying the GI tract would explain their preponderant influence on GI function (e.g. gastric acid, glucagon and insulin secretion), and the less consistently observed effects of DMX stimulation on arterial blood pressure and heart rate (see Norgren & Smith, 1988). The difficulty in obtaining cardiovascular responses is also probably related to the pattern of distribution of the preganglionic cell bodies. Neurons in the DMX which project to the thorax are more limited in distribution than those which supply the GI tract. A significantly larger number of thoracic-projection cells derive from a dorsal cell group in the subjacent parvicellular reticular formation confluent with DMX and another in the ventrolateral medulla adjacent to the nucleus ambiguus (see below). In fact, a tight correlation exists in rodents between the cells of origin of the cardiac branch of the vagus nerve and physiologically identified sites associated with high responsitivity to low threshold electrical stimulation (Nosaka et al., 1979). This organizational pattern may be species specific. Nonetheless, their different locations suggest that each column of motor neurons might be influenced by afferents arising from functionally distinct sources in the brain. Embryological migration of neurons to different 'niches' in the CNS is not random but genetically pre-programmed to follow a course set by 'radial glial' guides and neuro-

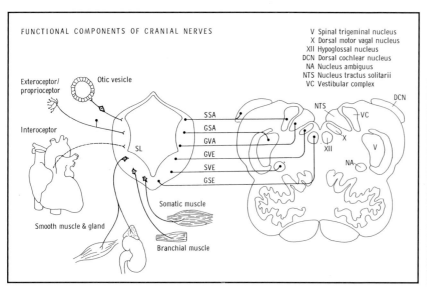

*Fig. 7. Schematic diagram illustrating the functional components of cranial nerves. Topographic relationships of the longitudinally organized cell columns in a human embryo (left) are related to their final positions in the medulla oblongata (right). A longitudinal groove or the sulcus limitans (SL) marks the border between the dorsal sensory (alar) and ventral motor (basal) plates. This developmental plan is shared by both the brainstem and spinal cord. Also shown are spinal precursor cells (neuroblasts) in the basal plate which will differentiate into autonomic preganglionic neurons (e.g. the adrenomedullary projection is also illustrated).*

biotaxic signals. These movements may be analogous to the motions of unicellular organisms toward or away from stimuli (mechanical or chemical) which tend to alter homoeostasis. Neurons of the dorsal (general visceral efferent) cell column are distinguished from a ventral (special visceral efferent) column in the apex of the ventral-lateral medulla: the nucleus ambiguus of Krause.

### Nucleus ambiguus

The embryonic precursors of cells in the DMX are associated with another member of the basal plate: the nucleus ambiguus (NA) (Figs. 6 & 7). The NA is a ventral column of motor neurons in the reticular formation of the ventral-lateral medulla and gives rise to fibres carried by cranial nerves IX and X. The NA was traditionally defined as the origin of the efferent innervation

of the larynx, pharynx and oesophagus. The migrational patterns of neurons belonging to the dorsal and ventral cell columns are species specific. During ontogeny, the movements of these neurons across the lateral medulla are most pronounced in mammals, variable between birds and reptiles and absent in cytostomes through amphibians (Ariëns-Kappers *et al.* 1936; Sarnat & Netsky, 1974). In birds, an incomplete migration of cells between the ventral and dorsal columns may explain why cardioinhibitory responses in pigeons can be obtained from the DMX but not the NA (Schwaber & Cohen, 1978). The NA along with other functionally related cranial nerve cell columns (see below) are categorized as members of the special visceral efferent cell group. These neurons supply striated muscles derived from the branchial

(gill) arches. As illustrated in Fig. 6, the glossopharyngeal (IXth) nerve carries axons from NA motor neurons which innervate the stylopharyngeal muscle, upper pharyngeal constrictor and palatal arch muscles which elevate the pharynx and aid in the swallowing reflex. The vagus (Xth) nerve provides the motor innervation of muscles of the soft palate, the inferior pharyngeal constrictor, oesophageal and intrinsic and extrinsic laryngeal muscles. These parts of the NA contribute to coordinated reflex movements mediating the sequential phases of swallowing, breathing and phonation.

As shown by Bieger & Hopkins (1987), all special visceral efferent lower motor neurons in the NA of rodents are topographically organized in a dorsal (principal) division. This column corresponds to the traditionally defined motor nucleus and is formed of rostral oesophagomotor (compact), intermediate pharyngolaryngomotor (semicompact) and caudal laryngomotor (loose) subdivisions. The general visceral efferent (parasympathetic) components of the NA form the ventral (parvicellular) or 'external' division: a longitudinal column of smaller preganglionic motor neurons in the reticular formation associated with ventral and lateral aspects of the nucleus. These neurons innervate the sinoatrial mode and atria of the heart and unidentified organs supplied by the IXth cranial nerve and superior laryngeal branch of the Xth nerve.

The discharges of general and special visceral efferent motor neurons of the DMX and NA are reflexively coordinated with the outputs of other members of the basal plate. Other functionally related motor neurons are located in the pons (trigeminal motor nucleus), rostral medulla (facial nucleus), spinomedullary junction and upper cervical spinal cord (spinal accessory nucleus). These cells supply other branchiomeric muscles involved in mastication (Vth nerve), facial (mimetic) expression (VIIth nerve) and movements of the head and neck and pectoral girdle by way of the XIth accessory nerve innervation of the sternocleidomastoid and trapezius muscles.

## Sacral division

Members of the sacral division of the PNS are also illustrated in Fig. 6. Sacral parasympathetic (pelvic) nerves derive from preganglionic neurons in the sacral intermediolateral cell column. They terminate on ganglion cells in pelvic plexuses or within organs of the lower GI tract and the genitourinary system.

## Mechanisms of interaction

The mechanisms responsible for most sympathetic and parasympathetic interactions are still poorly understood and take place in the CNS and periphery. In healthy animals, both divisions are normally active and exert antagonistic effects on organs which summate in a non-linear fashion (see review by Levy, 1990). Preganglionic cell bodies of both divisions of the ANS employ acetylcholine (ACh) as their neurotransmitter. Whereas ACh is also used by parasympathetic postganglionic neurons, norepinephrine is the ganglionic neurotransmitter in the sympathetic nervous system. A useful example to illustrate these peripheral mechanisms are the interactions which occur at cardiac neuroeffector sites. Anatomically, sympathetic and parasympathetic postganglionic axons are contiguous in the walls of the atria and ventricles and appear to be enveloped by a common Schwann sheath. Because of their close proximity, the concentrations of ACh and norepinephrine released by each nerve can effectively interact at pre- and post-junctional levels. With an increase in activity of the vagus nerve, ACh is released from parasympathetic postganglionic fibres and inhibits release of norepinephrine from sympathetic nerve endings and vice versa. These are two examples of prejunctional interactions. A third type of prejunctional interaction is the attenuation of the parasympathetic pattern of cardiac response (negative inotropy) by neuropeptide Y. When co-released from sympathetic nerve endings neuropeptide Y will inhibit the release of ACh (Potter, 1985). Postjunctional interactions also occur at the surface of the cardiac effector cell through receptor-mediated changes in adenylate cyclase. These interactions involve second messengers and explain the facilitatory effect of increased sympathetic nerve discharge on parasympathetic inhibition of ventricular contractility (Hollenberg *et al.*, 1965).

## Central autonomic regulation

One of the critical questions in neurophysiology has focused on the mechanisms of differential visceral afferent processing in the CNS (i.e. the central representation of the ANS). Of overriding importance is the functional specificity of neural pathways through which visceral signals are distributed, servocontrolled and integrated in the brain and spinal cord. It is now clear that a segmentally organized hierarchic regulation of the ANS is exerted by neurons involved in the tonic and reflex control of the autonomic outflow including those involved in complex panmodal afferent processing. The latter include neurons which exert intrinsic coordinated reflex control over the cerebral circulation and encode ascending multimodal signals into crude conscious perceptions (e.g. visceral pain, distention), self-awareness (a driving force of goal-related behaviours) and patterns of (motivated and instinctive) emotional expression.

It has long been recognized that autonomic responses are provoked by diverse stimuli conveyed by somatic and visceral afferents, and humoral factors circulating in the blood and other extracellular fluid compartments such as the cerebral ventricles, subarachnoid and Virchow–Robin's spaces and lymphatic system. Visceral stimuli contribute a large percentage of the total afferent inflow to the CNS and are conveyed by the visceral afferent system. The nucleus of the solitary tract (NTS) is located in the dorsal medulla, and associated with a chemoreceptor (circumventricular) organ, the area postrema. The NTS constitutes the principal relay station for all cranial nerve visceral afferent inputs to the CNS. Visceral afferents also derive from the pelvic viscera (DeGroat & Steers, 1990) and all cutaneous and musculoskeletal blood vessels and glands. These afferents are carried by spinal nerves and are first conveyed to the spinal dorsal horn.

## The nucleus tractus solitarii (NTS)

The NTS is the principal nucleus of reception of general and special visceral afferents. As will be described, some visceral afferents are conveyed to somatosensory neurons in the spinal dorsal horn and trigeminal nucleus caudalis (Cervero, 1986). Visceral afferents are conveyed to the NTS by branches of cranial nerves VII (facial), IX (glossopharyngeal) and X (vagus). These nerves transmit visceral sensory signals from the head and neck, thoracic and abdominal cavities. The sensory components of cranial nerves, known as primary or 'first-order' visceral afferents, have their cells of origin in extracerebral ganglia located in the head and neck. These ganglia are homologous to spinal dorsal root ganglia and include the following: (1) the geniculate ganglion of cranial nerve VII (the nervus intermedius) transmits special visceral afferents (gustatory signals) via the chorda tympani and lingual nerve branches from taste buds on the anterior two-thirds of the tongue. The geniculate ganglion also contains the cells of origin of general visceral afferents from the soft palate, tonsil and deep aspects of the face. (2) The petrous (inferior) ganglion of the glossopharyngeal nerve transmits special and general visceral afferents by the lingual-tonsillar and pharyngeal nerves from taste buds along the posterior one-third of the tongue, pharynx, fauces, and palatine tonsil. The carotid sinus nerve conveys general visceral afferents from the carotid sinus and carotid body. (3) The nodose ganglion of the vagus nerve transmits special visceral afferents from taste buds on the epiglottis and general visceral

afferents from the neck (pharynx, larynx and extrathoracic trachea), thoracic cavity (heart and great vessels; intrathoracic trachea, bronchial tree and lungs and the thoracic oesophagus) and the abdominal cavity.

The cell bodies of the VIIth, IXth and Xth cranial nerve ganglia are unipolar. The cell body is ovoid and a developmentally modified bipolar cell also termed a pseudo-unipolar neuron. These cell bodies possess a short process which dichotomizes into two branches: (1) a peripheral or dendritic branch is directed distally in the cranial nerve and provides the sensory endings located in visceral organs, and (2) a central or axonal branch forms the sensory nerve rootlets of the cranial nerve, is directed toward the brain and enters the NTS in the medulla oblongata. In rodents, the superior and inferior ganglia of the IXth and Xth cranial nerves (Altschuler *et al.*, 1989) compose a fused ganglionic mass without distinct demarcation of individual sensory neurons. The individual sensory afferent neurons are crudely organized in a topographic arrangement related to their target organ innervation.

The NTS (the special and general visceral afferent column of the alar plate) is a long, slender nucleus bordered ventromedially by the DMX (the general visceral efferent cell column of the basal plate). The NTS and DMX lie along the dorsal and ventral borders, respectively, of the sulcus limitans, a cleft marking the division between the sensory and motor plates of the developing neural tube. Figure 7 demonstrates the topographic relationship between sensory components of the alar plate and the motor

neurons of the basal plate derived from proliferating neuroepithelial cells. The locations of the embryonic precursors in the primitive neural tube are compared to their final positions in the mature medulla oblongata. The principles governing the developmental organization of the cranial nerve nuclei are identical to those which underlie the organization of the spinal cord. The NTS and DMX are related ventrolaterally to a diagonal laminated strip of reticular formation called the lateral tegmental field (LTF). At middle levels of the medulla, the NTS is bordered dorsally and dorsomedially by the subependymal glial stratum (SGS), a complex layer of large astrocytes and a feltwork of astrocytic processes underlying the ependymal cells which line the floor of the IVth ventricle. This specialized sheet is also composed of microglia and actively dividing immature cells whose progeny are believed to replace cells of the ependymal lining. Collectively, the NTS, DMX and SGS form the alae cinerea or trigonum vagian eminence on the floor of the IVth ventricle. At middle and caudal medullary levels, the NTS is bordered dorsally by the area postrema, a highly-vascularized circumventricular organ lined by non-ciliated ependymal cells and located outside the blood–brain barrier (see reviews by Miselis *et al.*, 1986; Miller & Kucharczyk, 1991). The area postrema (earlier considered to be a chemoreceptor trigger zone or vomiting centre) is composed of parenchymal neurons and astrocytes bearing intricate relationships to its arterial supply. The NTS is separated from the dorsal column nuclei (nucleus gracilis and cuneatus) by the nucleus of the tractus parasolitarius (NPS). The NPS is a cerebellar relay nucleus that conveys afferents from the medial cerebellar nucleus (fastigial nucleus) to the medial accessory olive and the thalamus. There is no direct physiological evidence for a role of the NPS in central autonomic control. At the level of the obex, the caudal bilaterally symmetrical extensions of the NTS merge on both sides of the midline, and form a bridge known as the commissural nucleus (see below). (The obex is the cephalic extension of a dorsomedian sulcus where the central canal opens into the fourth ventricle.) The commissural division of NTS is a composite nucleus lying dorsally to the central canal. At the obex and caudal to this level, a large percentage of primary visceral afferents converge with afferents arising from somatic receptors conveyed by the spinal dorsal horn and its rostral extension, the trigeminal nucleus caudalis.

## Viscerotopic organization of primary visceral afferents within the NTS

Different visceral afferents within the NTS appear to be organized topographically. The central branches of different visceral afferents enter the medulla oblongata and form the tractus solitarius, a longitudinal bundle of fibres surrounded by the sensory relay neurons of the NTS. Afferents travelling within the tractus solitarius enter the NTS and preferentially target discrete subnuclei organized around both sides of the solitary tract. Peripheral afferents to the NTS were mapped in animal studies by transganglionic transport of axonal tracers (e.g. radioactive amino acids which incorporate into proteins; plant lectins) deposited into individual cranial nerve branches or the or-

gans they innervate. The data reveal that special and general visceral afferents of cranial nerves VII, IX and X which subserve gustatory, ingestive, digestive and cardiorespiratory reflexes are distributed differentially within well-circumscribed loci in the NTS. Their distributions in the NTS are organized along its mediolateral and anteroposterior axes. The central representation of different organs and their specialized receptors are compartmentalized in the NTS, and have been described as having a viscerotopic pattern of organization. They exhibit a crude modular organization such that each visceral afferent preferentially targets one locale known as the principal receptive field. Each afferent appears to 'spill over' into an adjacent field representing another visceral modality and termed the secondary receptive field. The distributions of visceral afferents within the NTS appear to reflect two types of structural and functional organization. One type may be organ specific and represented by a precise viscerotopic organization of projections to NTS subnuclei that are devoted to processing organ-specific modalities. The second mode of organization may be integrative (Mifflin & Felder, 1990) whereby afferents from different organs or different receptors within the same organ converge within the solitary complex. Both modes of projection are supported by electrophysiological and anatomical evidence. The confluence of different primary visceral afferents in the NTS might be homologous to the overlap between the central branches of primary spinal root afferents representing different dermatomes (skin areas) and terminating in the dorsal horn of the spinal cord: a mechanism

protecting against complete anesthesia of a dermatome after lesions of a spinal nerve. This plan of organization may allow for interactions between visceral afferent inputs from receptors served by different nerves but differs from the 'strict somatotopic organization' of general somatic afferents from the head and neck ending in the spinal trigeminal nucleus (STN). In the STN, individual neurons are devoted to relaying either ophthalmic, maxillary or mandibular signals carried by the trigeminal (Vth) cranial nerve.

It is thought that overlap between different first-order visceral afferents ending in NTS may be involved in the coordination and patterning of synergistic movements among oropharyngeal, oesophageal and abdominal smooth muscles and glands engaged in feeding, swallowing and digestion and those involved in cardiorespiratory control. As described below, functionally related structures tend to show the strongest and most clearcut overlap of their first order afferents in the NTS. Fibres conveying taste, gastrointestinal and cardiopulmonary afferents are sequentially organized in a rostrocaudal direction within the NTS. The principal subfields of projection of different primary visceral afferents are illustrated in Fig. 8A–C. (Experimental data were obtained in animals and extrapolated to the human.) The purpose of these schematic drawings is to indicate the spatial relationships of each field along the X (medial–lateral) or Y (dorsal–ventral) axes. Because of the limited number of cross-sections drawn through the long axis of the brainstem (Z axis), the reader should reconstruct in three dimensions their

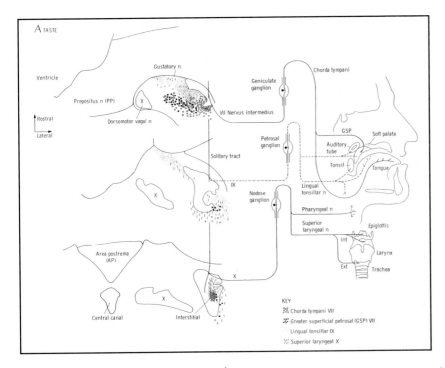

Fig. 8. *Viscerotopic organization of primary visceral afferents in the nucleus of the solitary tract (NTS). Distributions in the human are hypothesized on the basis of data obtained in animal studies.*
*A Taste (gustatory) afferents. B (see p. 16). Gastrointestinal afferents.C (see p. 17). Cardiopulmonary afferents.*

exact anterior–posterior distributions based on the following descriptions.

### Special visceral afferents (Fig. 8A)

The central branches of cranial nerves VII, IX and X convey gustatory afferents from taste buds and terminate within the rostralmost part of the NTS, termed the gustatory nucleus (Hamilton & Norgren, 1984; Sweazy & Bradley, 1986). The anterior-lateral aspect of the NTS is devoted to taste and somatosensory afferent relays from the oral cavity. This part of the NTS exhibits a loose somatotopic organization, i.e. the anterior tongue and palate are located rostrally and the posterior tongue and epiglottis are located caudally. Figure 14A demonstrates the projection fields within the NTS of different gustatory nerve branches of cranial nerves VII, IX and X.

### General visceral afferents

*Upper alimentary and abdominal visceral afferents (Fig. 8B).* The upper alimentary tract. General visceral afferents from the upper alimentary tract travel in cranial nerves IX and X and derive from receptors in the palate, epiglottis, pharynx, larynx and upper oesophagus. Rostrocaudal viscera of the upper alimentary tract tend to be represented along respective rostrocaudal levels of the NTS (Altschuler *et al.*, 1989). Receptors in these regions of the oral pharynx play a role in the sequential stages of swallowing (the buccopharyngeal stages), oesophageal peristalsis and related reflex control of respiration and phonation. Palatal and pharyngeal-wall afferents heavily terminate in the interstitial subnuclei caudally to the principal projections of gustatory afferents. Signals from functionally related structures of the tongue and

upper alimentary tract probably converge on these neurons and mediate synergistic reflex control of swallowing, speech (e.g. velopharyngeal closure) and breathing. These reflex pathways will be described in the next section. A major pharyngeal field occupies dorsal parts of the intermediate subnucleus of the NTS located just medial to the solitary tract. A smaller patch derives from the palate. At rostral levels of the NTS, palatal afferents extend laterally to the solitary tract into the ventrolateral subnucleus.
*Oesophageal afferents* derive from receptors in the tunica muscularis and myenteric ganglia and those scattered within the submucosal and mucosal layers (Clerc & Condamin, 1987) and participate in the reflex control of peristalsis. Oesophageal projections are restricted to an area of the medial NTS termed the nucleus centralis: the origin of a direct and specific reflex projection to the compact division of

the nucleus ambiguus (Ross *et al.*, 1985) (Figs. 8 & 11) Within the nucleus centralis of rodents, oesophageal afferents extend from the obex to a level approximately 800 μm rostrally. Projections from the cervical, thoracic and subdiaphragmatic oesophagus terminate topographically along rostrocaudal levels of the subnucleus centralis (Altschuler *et al.*, 1989).

*Abdominal viscera* are innervated by the subdiaphragmatic portion of the vagus nerve (SDX). The SDX has five major branches and two patterns of projection in the NTS: a gastric field representing the stomach and a hepatic–coeliac field representing the liver, duodenum and pancreas. Afferents from the SDX terminate dorsomedially to fibres derived from the upper alimentary tract and thorax and demonstrate virtually no overlap with those innervating the oesophagus.

*Gastric afferents* end in the dorsomedial NTS in an ovoidal field termed the nucleus gelatinosus. The gastric field is located midway between the solitary tract and the wall of the fourth ventricle and widens at the level of the area postrema. Another smaller projection from the stomach ends in the medial aspect of the nucleus subpostrema. The nucleus subpostrema is enriched in neuroglia and marks the zone of transition with the area postrema. A third projection area lies immediately dorsal to the DMX, and based on a similar pattern of descending projection from the forebrain was termed the arcuate or ventromedial subnucleus (Ruggiero *et al.*, 1987). Gastric nerve afferents to this region of the NTS also might terminate on

dendrites of parasympathetic preganglionic projection neurons of the DMX. Minor gastric afferent fields include contiguous parts of the medial and commissural subnuclei. At caudal levels, fibres cross the midline dorsal to the central canal and enter the area postrema. At rostral levels, another field appears in the periventricular subnucleus of the NTS. Periventricular neurons lie deep to a subependymal glial matrix forming the floor of the IVth ventricle. Gastric afferents do not terminate within lateral subnuclei of the NTS nor do they extend into rostral-lateral aspects of the NTS where taste afferents preferentially terminate (Leslie *et al.*, 1982; Shapiro & Miselis, 1985; Norgren & Smith, 1988; Altschuler *et al.*, 1989). In general, the gastric branches of the SDX appear to concentrate anterior and lateral to afferents derived from the celiac, accessory coeliac and hepatic nerves (see below).

*Coeliac and accessory coeliac* afferent endings are coextensive and concentrated dorsomedially in the medial subnucleus. These afferents lie medial to the nucleus gelatinosus and deep to the tela choroidea, a pial membrane attached to the ependymal cell layer lining the fourth ventricle. Coeliac and accessory coeliac afferents are distributed along the right and left sides of the NTS, respectively, and form a dense band in the nucleus subpostrema, subjacent to the area postrema. These projections are concentrated at the level of the caudal half of the area postrema. Afferents from the *hepatic nerve* which supplies the liver travel along the accessory coeliac nerve and essentially match the distributions of coeliac and accessory coeliac projections. Rostrally to the

area postrema, most hepatic afferents end in the dorsomedial aspect of the medial subnucleus. Their heaviest projection is concentrated on the midline beneath the caudal third of the area postrema.

*General comment.* According to Norgren & Smith (1988), the sensory distribution of SDX afferents within the NTS shows an organ-specific pattern of contiguous projection. Indeed, convergence of different abdominal afferents within the NTS appears to be the rule rather than the exception. Neurally mediated increases in insulin, for example, are obtained by stimulating most component nerves of the SDX (Berthoud *et al.*, 1983). Although each branch of the SDX has a unique target site (i.e. a distinct concentration of primary afferent distribution within NTS), it is the areas of overlap which may be critically involved in sensory integration among different visceral afferents. The absence of clearcut boundaries between the terminal fields of different visceral afferents suggests a redundancy of sensory input to the NTS. Superimpositions of different afferent projections would explain, for example, why large lesions of the solitary complex are required to block the satiating effects of exogenously administered cholecystokinin (Smith *et al.*, 1985) or taste-guided preference behaviours (Vigorito *et al.*, 1987).

*Cardiovascular afferents (Fig. 8C)*
Cardiovascular afferents in the NTS are skewed caudally to special visceral afferents and are concentrated dorsally and laterally to other general visceral afferents of cranial nerves VII–X. Primary cardiovascular afferents derive from chemoreceptors in the ca-

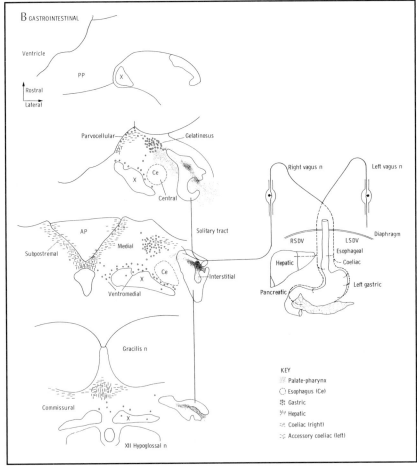

*Fig. 8B. Gastrointestinal afferents.*

rotid and aortic bodies and baroreceptors in the carotid sinus and aortic arch. Afferents from the carotid body and sinus are conveyed to the CNS by the carotid sinus nerve, a branch of the glossopharyngeal nerve. The central projections of cardiovascular afferents are relayed by the IXth and Xth cranial nerves and are topographically separated from other afferent modalities also conveyed by these nerves. The carotid sinus nerve projection (Berger, 1979; Housley *et al.*, 1987) is intermediate in position to the entire sagittally oriented field of glossopharyngeal representation in

the NTS (within approximately 1.0 mm of obex). Two distinct afferent fields are observed: at levels rostral to the obex, carotid sinus afferents are concentrated in the dorsal subnucleus and dorsal part of the intermediate subnucleus. At levels caudal to the obex, these afferents occupy spatially contiguous loci of the commissural nucleus of the vagus and the ventrolateral subnucleus. The projection demonstrates a sharp decrease in innervation density within the NTS, approximately at the level of the obex. This transitional zone appears to represent the division between afferent

projections of baroreceptors and chemoreceptors (see below).

*Chemoreceptor afferents.* The carotid body is an ovoid sensory organ situated at the bifurcation of the common carotid artery and derives its blood supply predominantly from vessels of the external carotid artery (see McDonald & Mitchell, 1975; McDonald, 1981). Microscopically, the carotid body consists of nests or clusters of glomus cells, or type 1 epithelioid cells which receive an afferent innervation from cell bodies in the petrosal ganglion of the glossopharyngeal nerve. In the cat, small neuronal somata in the petrosal ganglia are the predominant cells of origin of carotid body afferents (Claps & Torrealba, 1988) and have patterns of discharge related to arterial chemoreceptors (Donoghue *et al.*, 1984). McDonald & Mitchell (1975) have shown that greater than 95 per cent of the afferent innervation of glomus cells is derived from axons forming reciprocal synapses with the type 1 cells. The carotid body is characterized by a blood flow and oxygen consumption exceeding those of the brain and thyroid gland and is structurally specialized to sense changes in blood flow and temperature (Eyzaguirre *et al.*, 1983). Chemoreceptors in the carotid body also detect changes in blood gases, plasma acidity and osmolarity and play a key role in respiratory and cardiovascular control. When stimulated, chemoreceptor tissues of the carotid body produce action potentials in the glossopharyngeal nerve leading to increases in rate, depth and minute volume of respiration, sympathetic nerve discharge (secretion of catecholamines from the adrenal gland), and cerebrocortical activity. Chemoreceptor-control of the normal breathing

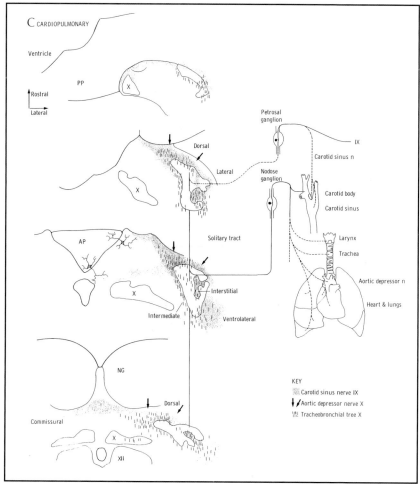

*Fig. 8C. Cardiopulmonary afferents.*

and caudal to the obex. In keeping with anatomical mapping data, few chemoreceptor afferents were activated electrophysiologically from lateral or ventrolateral divisions of the NTS. The latter area is involved in the generation of respiratory rhythms and termed the dorsal respiratory group (von Euler *et al.*, 1973; Richter, 1982). However, chemoreceptor afferents in the NTS might terminate on their dendritic branches. As shown in Fig. 8C, dendrites of neurons in the nucleus subpostrema extend dorsally into the area postrema (a central chemoreceptor) and caudally into the commissural region of NTS which conveys afferents from the carotid body. These neurons may therefore be posited to receive convergent signals from both central and peripheral chemoreceptors.

*Baroreceptor afferents.* Baroreceptors are concentrated in the carotid sinus, common carotid, subclavian and innominate arteries and the arch of the aorta between the brachiocephalic artery and the ligamentum arteriosum (Nonidez, 1937; McDonald & Mitchell, 1975; McDonald, 1983 for references.) An important component of baroreceptor reflex control is exerted by receptors in the carotid sinus and the aortic arch where the innervation density is heaviest. Stimulating baroreceptors produces responses diametrically opposed to the effects of stimulating the carotid body. Nerve endings in the carotid sinus lie in close approximation to a rich network of blood vessels in the adventitia of the carotid sinus wall. In humans, the carotid sinus nerve is divided into a medial and a lateral branch which supply the carotid body and sinus, respectively. In rats, the carotid sinus nerve

rhythm is especially critical during sleep (Sullivan, 1980). Within the NTS of rodents (Ruggiero *et al.*, 1991) primary chemoreceptor afferents are skewed medially and caudally to cardiac and baroreceptor afferents. Following direct injections of lectins into the carotid body, chemoreceptor afferents are anterogradely labelled in the NTS and concentrated caudally to the obex within the commissural nucleus of the vagus. Two subfields are found: one lies in the medial aspect of the medial (commissural) subnucleus; the second forms a thin horizontal strip lining the dorsal border of the nucleus. A similar, although rostrally more extensive distribution of carotid body afferents was described in the cat including moderately dense inputs to dorsal, intermediate and interstitial subnuclei which receive afferents from other extra- and intra-thoracic viscera (Claps & Torrealba, 1988). Chemoreceptor afferent fibres were activated antidromically by stimulating equivalent loci in the NTS (Donoghue *et al.*, 1984) including dorsomedial and medial divisions and the commissural nucleus at the level of

divides into multiple branches. Some traverse the carotid body before ending in the carotid sinus wall, whereas others first join sympathetic nerves. The *carotid sinus* (a dilatation in the carotid bifurcation) is chiefly innervated by the carotid sinus nerve branch of cranial nerve IX; other baroreceptor fibres travel via small vagal branches. In the cat, carotid sinus afferents project bilaterally to the NTS and terminate in areas devoid of chemoreceptor innervation: dorsolateral, intermediate and ventrolateral subnuclei and the lateral extension of the commissural region (Torrealba & Claps, 1988). According to these authors, carotid sinus afferents in the cat are coextensive with those derived from the carotid body, but are more extensive medially and laterally throughout their distribution in the NTS. Electrophysiologically, both myelinated and unmyelinated carotid sinus afferents show a similar pattern of projection to the ipsilateral NTS rostral to the obex, and are activated from dorsal, lateral and commissural divisions of the NTS. In contrast to chemoreceptor afferents, baroreceptor afferents show a preference for lateral aspects of the NTS and appear to avoid ventral and ventrolateral divisions (Donoghue *et al.*, 1984; Lipski & Trzebski, 1975; Lipski *et al.*, 1977). The *aortic depressor nerve* (ADN) arises from the aortic arch and is referred to as a cardiac safety valve responding to and buffering increases in arterial blood pressure by participating in reflex circuits designed to slow the heart rate and decrease arterial resistance. The ADN only carries baroreceptor afferents in the rat and rabbit, whereas in cats it also transmits afferents from chemoreceptors (Neil *et al.*, 1949). ADN afferents in the rat appear to overlap with the carotid sinus component of the carotid sinus nerve within dorsolateral and interstitial subnuclei and contiguous loci in the commissural nucleus. Some fibres also terminate within medial and ventrolateral subnuclei (Ciriello, 1983). Taken collectively, these data demonstrate a partial zonal overlap in the central representation of carotid sinus and aortic depressor nerves in the solitary complex. This region of the NTS plays an important role in the regulation of systemic arterial pressure since lesions produce a sustained arterial hypertension due to removal of baroreceptor inhibition on the sympathetic nervous system. The principal subfield of cardiac afferents in the NTS (not illustrated) is the dorsal subnucleus immediately rostral to the obex (Hellner & von Baumgarten, 1961). According to Kalia & Mesulam (1980), cardiac afferents also terminate in subnuclei receiving afferents from other thoracic and abdominal viscera including the medial, gelatinosus and ventrolateral subnuclei and the area postrema. No cardiac afferents terminate in the interstitial nucleus, an area of pharyngeal and laryngeal representation.

### Respiratory afferents (Fig. 8C)

Afferents involved in respiratory control are derived from the larynx, tracheobronchial tree and lungs. General visceral afferents from these structures are integrated in the NTS and involved in generating the spatiotemporal patterns of activity of respiratory lower motor neurons essential for respiratory rhythmogenesis. The anatomical distributions of primary respiratory afferents within the NTS suggest that neurons receive a common set of converging inputs from different peripheral receptors. The integrated signals contribute to the reflex control of motor neurons engaged in breathing and functionally related viscero-somatic reflexes including phonation, yawning, swallowing and gagging.

*Larynx.* Laryngeal afferents travel by the superior laryngeal branch of the vagus nerve and the pharyngeal branch of the glossopharyngeal nerve. Most laryngeal afferents derive from ganglion cells extending from the caudal pole of the nodose ganglion to the jugular ganglionic cuff. Primary visceral afferents from the larynx (and trachea) terminate in regions of the NTS receiving input from the soft palate and pharynx (Altschuler *et al.*, 1989). Prominent projections end in the interstitial and intermediate subnuclei. Cells in these subnuclei are embedded within or line the medial border of the solitary tract, respectively. In rats, afferents to the interstitial nucleus are crudely topographically related. Laryngeal, pharyngeal and palatal areas of representation occur at caudal, intermediate and rostral levels of the subnucleus.

*Tracheobronchial tree and lungs.* Tracheal afferents travel by the recurrent and superior laryngeal nerves and terminate on stretch receptors on the trachealis muscle and irritant receptors on the tracheal lining. In cats, the central projections of the intrathoracic and extrathoracic trachea overlap in the NTS (Kalia & Mesulam, 1980). Tracheal afferents end predominantly in the ventrolateral subnucleus. *Slowly adapting lung stretch receptor (SAR) afferents* (Kalia & Richter, 1985) are derived from receptors in the tracheobronchial tree and identi-

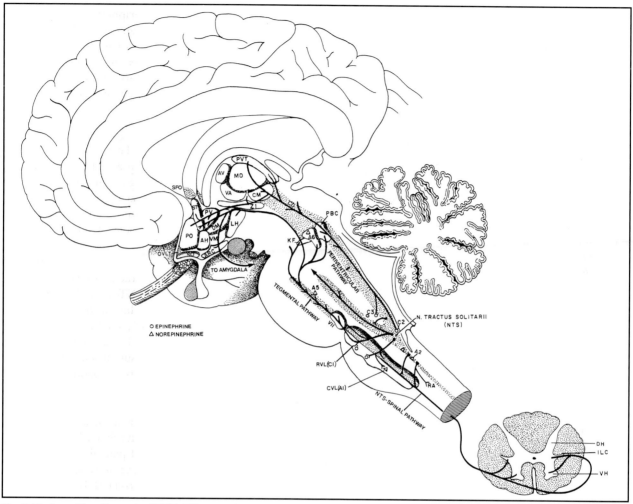

*Fig. 9. Nucleus of the solitary tract (NTS). Efferent projection pathways and terminal fields. Visceral afferents synapse on neurons in the NTS which, in turn, contribute to short reflex circuits in the lower brainstem. Visceral relay neurons in the NTS also give rise to long projections descending to the spinal cord or fibre tracts ascending to the forebrain.*

fied physiologically by increases in discharge during lung inflation and cessation of discharge during noninflation. Upon entering the solitary tract, SAR afferents descend and form an elliptical interrupted longitudinal cell column extending approximately 1.7–2.1 mm rostral to the obex. Terminals in the NTS are restricted to lateral aspects of the nucleus and terminate in the form of en passant boutons concentrated in the ventral and inter-

mediate subnuclei. Scattered projections occur within interstitial and ventrolateral divisions. The anatomical distribution of SAR afferents matches electrophysiologically identified sites in the NTS (Cohen & Feldman, 1984) and is distinct from dorsal-lateral and medial areas receiving inputs from cardiovascular and gastrointestinal receptors. Thus, a modality-specific map of SAR and other visceral afferents appears to exist in the NTS.

*Rapidly adapting pulmonary receptor (RAR) afferents* (Kalia & Richter, 1988) arise from receptors in the tracheobronchial tree and play a role in bronchomotor tone and reflex contractions of airway smooth muscle. A large number of receptors are concentrated in the upper airways and at airway bifurcations. These receptors are associated with airway smooth muscle and lie under the mucosa of cartilaginous portions of the airways. They respond to rapid changes in air flow, transpulmonary pressure (inactivated by pulmonary oedema and congestion) and noxious stimuli, and have also been termed deflation, cough or irritant receptors. The central branches of RAR afferents are concentrated in the dorsal subnucleus at levels rostral to the obex and in the intermediate subnucleus, caudally. In contrast to SAR afferents, they do not project to ventral or ventrolateral subnuclei. A single RAR afferent ascends and descends in the solitary tract and gives off collaterals and preterminal segments to different levels of the nucleus, although the distribution appears to be less extensive than the SAR pathway. Inputs to the dorsal subnucleus overlap pump cells which receive inputs from vagal and laryngeal afferents (Anders & Richter, 1987), and are also coextensive with cardiovascular afferents from the carotid sinus and aortic depressor nerves and the heart (Kalia & Mesulam, 1980).

## Central autonomic reflexes

### Visceral afferent processing in the medulla oblongata

Circuits of the lower brainstem have long been thought of as fixed and ensure a smooth performance of pre-programmed behaviours. Initiated by signals from diverse receptors, these reflexes coordinate tonic and phasic control over autonomic and skeletomotor muscles integrated with the secretomotor activities of glandular tissues. In the medulla oblongata autonomic reflexes are mediated by relatively short connections between the visceral receptor, an integrative limb (intermediary conductor neurons) and the effector organs (Figs. 9–

12). This simple reflex arc is an abstraction. Even the most basic functional unit is far more complex and incorporates closed multiple neuronal chain circuits which are self-sustaining (reverboratory) and self-regulating (degenerative). Components of the reflex circuits exert negative feedback control over all incoming afferents, somatic and visceral, which influence the autonomic outflow.

Medullary visceral reflex arcs are in a unique position to receive signals converging from all segments of the brain and spinal cord. The visceral reflex circuit is diagonally intercalated between two chemoreceptor organs: the area postrema (a dorsal circumventricular organ reciprocally connected with the NTS) and a presumptive ventral chemoreceptor believed to be a neural-glial vascular complex which assumes a close relationship with the subarachnoid space beneath the ventral medullary subpial surface (Loeschcke *et al.*, 1970; Lioy & Trzebski, 1984; Benarroch *et al.*, 1986; Ruggiero *et al.*, 1991). The NTS gives rise to a bidirectional fibre tract (the solitario-reticular or transtegmental pathway) which radiates diagonally toward the ventral medullary surface and bears a close relationship to its target neurons distributed along the subjacent reticular formation and contiguous loci in the spinal grey (Fig. 9). Upon emerging from the NTS linear arrays of axons and terminals extend diagonally across the LTF toward the ventrolateral medulla. These fibres closely invest neurons that lie within columns distributed perpendicularly to the transtegmental trajectory and may, therefore, be posited to influence the associational and reticulomotor limbs of the reflex arcs (see

below). A smooth performance of the brainstem reflex arc might be insured by serial chains of intermediary neurons and associated dendritic arbors (receptive fields) situated between the primary sensory limb and the output neurons (Fig. 10). The sensory, associational and long reticular relay neurons are aligned within oblique laminae in the LTF which allows for a sequential transfer of afferents irradiating diagonally and bidirectionally across the transtegmental fibre tract. Together, the components of the entire projection field may form a single unified system devoted to sensory processing of multiple afferent modalities entering into the tonic and reflex control of the visceral outflow (Reis & Ruggiero, 1990). In the medulla, NTS projections engage in local (short) reflex circuits and longer more circuitous reflex connections with neurons in the reticular formation, periventricular grey and contiguous parts of the spinal grey.

### *Local circuits*

Local intranuclear connections exist between sites in the NTS which relay different primary visceral afferents. Intranuclear pathways are thought to mediate local reflex integration between different sensory modalities. These circuits may explain, for example, why food will elicit salivation and ingestive behaviour or why hypoxia provokes centrally-mediated elevations in arterial blood pressure, heart rate and respiratory drive. The majority of these microcircuits are unknown. However, local connections have been established between sites in the anterior NTS conveying primary gustatory afferents and those in caudal (ventral) subnuclei (Travers, 1988) which engage special visceromotor

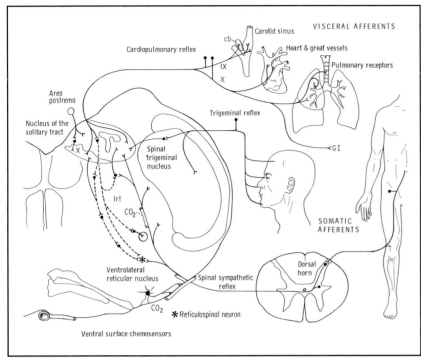

*Fig. 10. Schematic diagram illustrating the basic organizational plan of autonomic reflex arcs within an area of sensory convergence in the reticular formation of the lateral tegmental field. The intermediate reticular zone (Irt) is characterized by collections of neurons organized in longitudinal columns and arranged in radial sheets with primary dendrites aligned diagonally along the solitario-reticular tract. Direct and indirect feedback connections exist between the NTS and reticular formation and are formed by chains of neurons organized in parallel and in series. These neural networks may represent a unified system organized along the solitario-reticular tract and devoted to somato-autonomic reflex control.*

(oromotor) neurons in the nucleus ambiguus, general visceromotor (parasympathetic) neurons in the lateral tegmental field (LTF) (salivatory–lacrimal complex) and viscerosensory reticular relay neurons (Beckman & Whitehead, 1991). By way of these circuits, sites in the NTS involved, for example, in food selection can influence salivation, ingestive and digestive functions. The shortest sensorimotor reflex arc in the medulla is the direct circuit between primary visceral afferents (or secondary relay neurons in NTS) and parasympathetic motor neurons in the DMX (Ross *et al.*, 1985; Rinaman *et al.*, 1989).

*Solitario-reticular (transtegmental) tract* The NTS also issues axons which exit the nucleus and traverse a circumscribed zone of the lateral reticular formation called the intermediate reticular zone (Irt) of the LTF (Figs. 9 & 10). The Irt is present in humans (Arango *et al.*, 1988) as well as other mammals (Jones & Friedman, 1983; Ross *et al.*, 1985). This region of the LTF traversed by the solitario-tegmental tract was long recognized as part of the reticular matrix situated between the dorsal and ventral visceromotor cell groups (DMX and NA), and organized by the prevailing efferent irradiations of cranial nerves

IX, X and XI. The organizational patterns of cell bodies and afferent processes in the LTF likely dictate the direction of information flow along the reflex arcs (King, 1980). Scheibel & Scheibel (1958) recognized the segmental cylindrical organization of afferents entering the reticular formation as aligned with dendrites in planes roughly paralleling the neuraxis. The neurons in the NTS and LTF may also establish a module type organizational pattern similar to the architecture of the cerebral cortex. Cortical afferents terminate within barrel fields (cylindrical columns) which extend vertically to the pial surface, while spreading laterally within laminae running tangentially to the cylinders. The brainstem units are parcelled in an analogous fashion except that they bear an inverse topological relationship to the cortical architecture. Solitario-tegmental and other afferents target modules which are arranged longitudinally, parallel to the neuraxis and perpendicularly to the parallel sheets of sensory afferents. The neurons within each column, their apical/dendrites and associated radial glia run diagonally between the sensory (NTS and area postrema) and reticular limbs of the reflex circuits. This structural arrangement might afford a way by which different sources of afferents might be shared by functionally different columns, while differentially targeting sites along the various branches of individual neurons. The signals would spread laterally yet also maintain a specificity of afferent control over an individual component of the reflex. Afferents extending longitudinally along the cylinders might also synaptically interact with arbors of dendrites running parallel to their

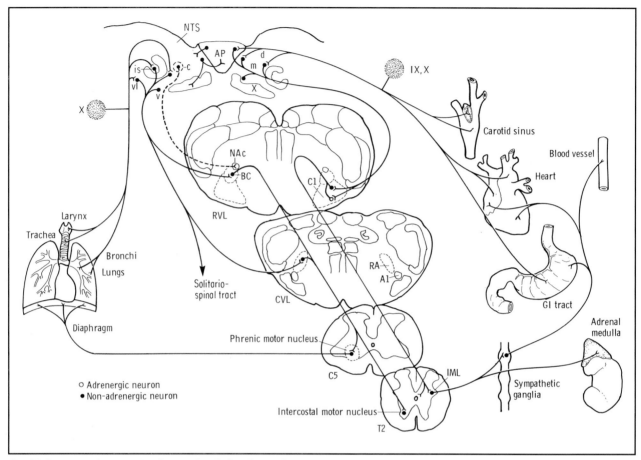

*Fig. 11. Summary of autonomic reflex arcs described in the text. Primary visceral afferents terminate in the nucleus tractus solitarii (NTS) and are organized viscerotopically. NTS projections to the reticular formation and spinal cord are also topographically arranged. The overlap of different visceral afferents within the NTS and both intranuclear and intrareticular 'associational' circuits likely play important roles in autonomic reflex integration. Reciprocating connections between the area postrema (AP) and NTS provide substrates through which substances circulating in the blood (e.g. toxins, hormones) can modulate autonomic reflex function (e.g., the vomiting reflex). The ovoid nucleus centralis (C) – an area of oesophageal representation – is shifted dorsally for illustrative clarity (see text for details; modified after Ruggiero et al., 1991).*

trajectories.

Since different primary visceral afferents are compartmentalized in the NTS, the question remains whether the secondary relay cells preferentially target discrete subfields in the brain and spinal cord. A topographic specificity of NTS projections to the reticular formation and spinal grey is being slowly unravelled (Figs. 11 & 12). The viscerotopic organization of the solitario-tegmental tract has been partially delineated by comparing the distributional patterns of cells in the NTS labelled (backfilled) by depositing retrograde tracers within various sites along the pathway. It is now well established that areal subunits exist within the reticular formation that are characterized by a non-uniform cylindrical distribution of propriobulbar, long reticular relay, ambigual and preganglionic motor neurons. It is also recognized that within each cell column, neurons having different projection fields are spatially segregated (Blessing *et al.*, 1982; Tucker *et al.*, 1987; Ruggiero *et al.*, 1989, 1991; Ellenberger & Feldman, 1990a,b; Ellenberger *et al.*, 1990). Whether this system shares the place and modality – specificity of the cortical receptive fields is unresolved. In general, the data show that neurons distributed along a dorsomedial to ventrolateral gradient in the NTS project to sites extending diagonally in a rostral/ventrolateral to caudal/dorsomedial direction in the lateral medullary reticu-

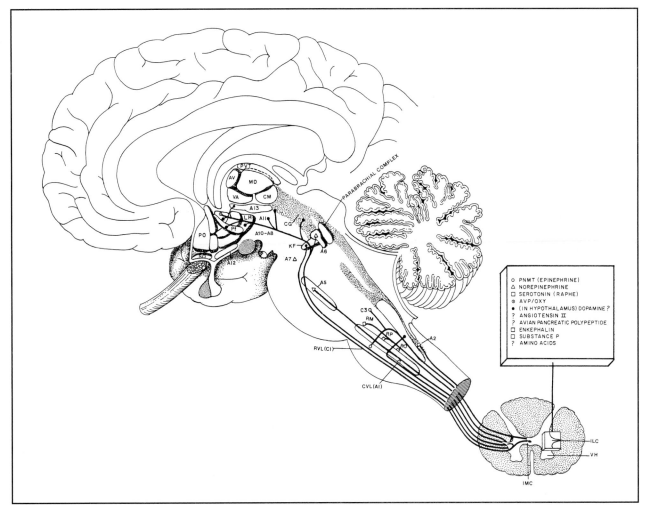

*Fig. 12. Autonomic bulbospinal projections to preganglionic neurons in the thoracic intermediolateral (ILC) and intermediomedial (IMC) cell columns. Some cell groups also project to other spinal laminae (e.g. dorsal horn) or to other segments of the spinal cord. A list of their putative neurotransmitters, or neuromodulators is provided in the inset (see text for details).*

lar formation (Fig. 11). Although the precise 'point to point' connections between the NTS and reticular formation are unknown, the following correlations were observed in our retrograde transport studies. Dorsal and lateral subnuclei of the NTS which relay first order cardiovascular afferents and a neighbouring dorsomedial area of subdiaphragmatic representation in the nucleus subpostrema selectively project to a small cell group in the rostral ventrolateral medulla. Rostral levels of the column correspond to the main body of the rostral ventrolateral reticular nucleus or nucleus RVL. The nucleus RVL is defined by a cluster of neurons (the C1 sympathoexcitatory area) which biosynthesize epinephrine, neuropeptides, and amino acids (Kaneko *et al.*, 1989; Hokfelt *et al.*, 1983; 1985; Ruggiero *et al.*, 1985a,b; 1989). These cells and other chemically unidentified neurons issue direct organized projections to the sympathetic intermediolateral cell columns of the thoracolumbar spinal cord (Dampney *et al.*, 1982; Ross *et al.*, 1981b; 1984a). A small number of fibres descend as far as the sacral lateral horn. The area postrema and another area of chemoreceptor (carotid body) representation in the medial commissural subnucleus of the NTS also selectively project to the C1 area of the nucleus

RVL and may represent anatomical substrates of the chemoreceptor reflex arc (Ross *et al.*, 1985; Ruggiero *et al.*, 1991). Solitario-reticular afferents derived from cells on both sides of the solitary tract extend caudally along this cylinder and terminate on two principal cell populations in the caudal ventrolateral reticular formation (the nucleus CVL): (1) The ventral catecholamine cell groups. A column of adrenergic reticulohypothalamic neurons (caudal Cl area) is situated at rostral and mid-medullary levels of the nucleus CVL and issues collaterals to neurons in the paraventricular hypothalamic (PVN) and median preoptic nuclei involved in volume and electrolyte homoeostasis (Tucker *et al.*, 1987). Noradrenergic reticulohypothalamic neurons (A1 area) at caudal medullary levels also project to magnocellular neurons in the PVN which synthesize and release neurohypophyseal (posterior pituitary) hormones (Blessing *et al.*, 1982; Sawchenko & Swanson, 1982). These columns also derive their inputs from cell groups in NTS which innervate the C1 area of nucleus RVL. (2) The dorsal cell groups. The catecholaminergic subnuclei (the C1 and A1 areas) are bordered dorsally by parallel columns of neurons belonging to the Bötzinger (expiratory) complex and ventral respiratory groups (VRG, Ellenberger *et al.*, 1990). Although partially intermingled with the C1 and A1 neurons, the majority of neurons of the Bötzinger complex and VRG are skewed farther dorsally within the LTF. Many of these cells issue direct projections to respiratory motor neurons in the spinal ventral horn, and are admixed with propriobulbar neurons involved in cranial nerve reflex integration (Holstege & Kuypers,

1977; Holstege *et al.*, 1977; Ruggiero *et al.*, 1982; Mehler, 1983; Ellenberger *et al.*, 1990; Ellenberger & Feldman, 1990a,b). At caudal levels, these cells are also intermingled with retroambigual (e.g. laryngomotor) neurons. The precise topographic projections from the NTS to each of the subsets of neurons are therefore tentative. As shown in Fig. 11, the origins of cells projecting to dorsal parts of the complex derive from sites neighboring ventral and ventrolateral aspects of the solitary tract. These sites predominantly convey afferents from functionally-related receptors in the larynx and pharynx, lungs and tracheobronchial tree.

Cardiorespiratory reflexes appear to be mediated by neurons surrounding the caudal part of the solitary tract. Cells in these loci also project to the spinal cord (von Euler *et al.*, 1973; Loewy & Burton, 1978). The solitario-spinal tract was recently found in rodents to terminate on respiratory lower motor neurons in the midcervical (phrenic motor nucleus) and thoracolumbar ventral horn and on sympathetic preganglionic neurons in the upper thoracic cord (Mtui *et al.*, 1993; Ruggiero *et al.*, 1991). Because neurons in the same locations within the NTS project to the medullary reticular formation (Ross *et al.*, 1985), axons of the solitario-reticular tract might issue collaterals which descend to the spinal cord (Figs. 11 & 12). The topographic organization of projections from the NTS to different subnuclei of the ambiguus complex (Bieger & Hopkins, 1987) has not been extensively investigated. The central part of the NTS (subnucleus centralis) receives an input from the oesophagus and gives rise to a circumscribed projection to

the compact oesophageal-motor division of the nucleus ambiguus (Ross *et al.*, 1985; Ruggiero & Reis, 1988) (Fig. 11). This fibre tract employs as putative neurotransmitters, somatostatin (a hyperpolarizing hypophysiotropic neuropeptide) and enkephalin-like peptides and appears to be involved in the oesophageal phase of swallowing (Cunningham & Sawchenko, 1990).

### *Periventricular fibre system*

The NTS and neighbouring reticular formation also give rise to a long extensive bidirectional fibre pathway which travels along the periventricular grey (Ruggiero *et al.*, 1985b, 1989) (Figs. 9 & 13). The periventricular fibre system is phylogenetically ancient and has long been conceived as relaying protopathic afferents. Behaviours (moods) may be linked physiologically with endocrine activities by virtue of extensive connections with the forebrain and intimate associations between periventricular fibres, neurosecretory neurons and ependymal cells lining the cerebral ventricles. Many neurons lying near the dorsal surface of the lower brainstem are adrenergic and also contain amino acids such as GABA or L-glutamate and acetylcholine (Meeley *et al.*, 1985; Kaneko *et al.*, 1989; Pieribone *et al.*, 1988; Ruggiero *et al.*, 1990a,b). As illustrated in Figs. 9 & 13, neurons in the NTS (C2 area) and the adjacent rostral dorsomedial reticular formation (C3 area) issue long descending projections to sites in the spinal central grey involved in noxious afferent neurotransmission and ascending projections to structures in the dorsal pons, midline-intralaminar thalamus and hypothalamic periventricular grey (Hokfelt *et al.*, 1985; Otake and Ruggiero, 1995). These cell groups, in

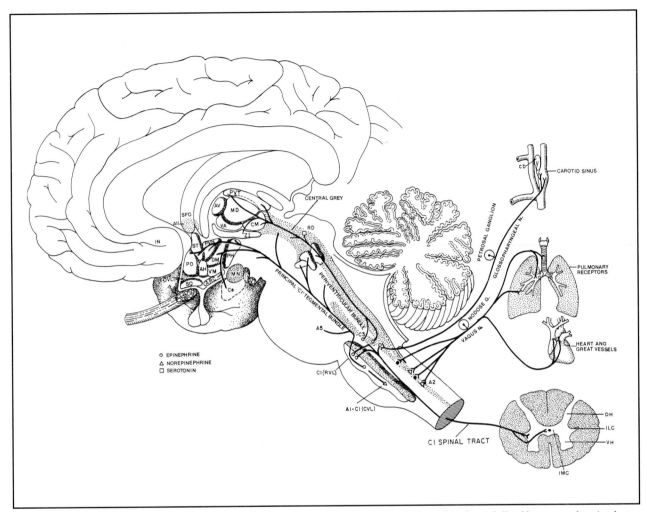

*Fig. 13. Adrenergic efferent pathways in the central nervous system. Adrenergic neurons are located in the medulla oblongata and project by two principal bidirectional fibre pathways. A periventricular fibre system is derived from neurons in the C2 and C3 cell groups. A principal (dorsal) tegmental bundle is derived from adrenergic cell bodies (C1 cell group) in the ventrolateral medulla and contiguous loci in the overlying intermediate reticular zone of the lateral tegmental field. Local catecholaminergic projections to the medullary raphe are not illustrated. Non-catecholaminergic neurons are intermingled with the adrenergic cell bodies and project through similar fibre tracts.*

turn, project to widespread areas of the CNS including the cerebral hemispheres.

Projections to hypothalamic periventricular nuclei may contact hypothalamic neurosecretory cells which exert hypophysiotropic control over the adenohypophysis. Since a large percentage of the total numbers of neurons in the C2 and C3 areas are labelled by deposits of retrograde tracers into different sites along the periventricular grey and midline thalamus it is surmised that these neurons may issue collaterals to different members of the midline (protopathic) system. Moreover, while only modest numbers of neurons in the periventricular grey can be identified at each segment of the brainstem (i.e. on each tissue section), in total they constitute an extensive system which spans the entire length of the midline and neighbouring paramedian reticular formation. The periventricular system has long been considered as of considerable importance to interactions that normally ensue between the neuroendocrine system and the devel-

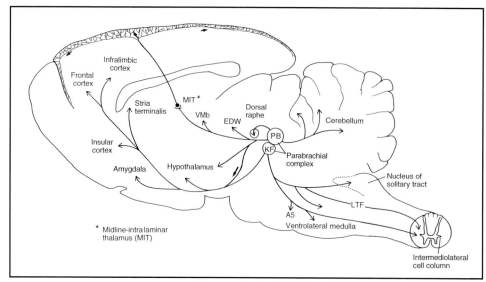

opment of behaviours. As will be described, other chemically-identified neurons (e.g. opioid-peptides) in the pontine and midbrain central grey also issue bidirectional periventricular projections to neurons of the endogenous pain control system, including the midline raphe nuclei. Interestingly, the raphe and central autonomic cell groups are reciprocally connected and may form circuits involved in the afferent regulation of autonomic outflow. Finally, the periventricular system includes sites in the midbrain central grey involved in primitive emotive expressions generated in the forebrain.

Coordinated movements of the muscles of respiration, facial and vocal expression are retained in patients and animals suffering facial paralysis caused by supranuclear lesions of the internal capsule or motor cortex. The central grey and adjacent lateral tegmentum may constitute an efferent system mediating faciovocal and other manifestations of general emotional excitement.

## Visceral afferent processing in the pons and midbrain

### Parabrachial complex (PBC)

The PBC, also called the marginal nucleus, is located in the dorsolateral pons and surrounds the superior cerebellar peduncle (SCP). (The SCP is a thick bundle of axons which bridges the cerebellum and midbrain and extends into the thalamus.) The PBC is the principal pontine relay to the forebrain (Fig. 14) of special and visceral afferents conveyed by the NTS (King, 1980; Ruggiero *et al.*, 1982, 1991; Cechetto *et al.*, 1985; Cechetto, 1987). Cytoarchitecturally, the PBC in humans and other mammals is divided into medial and lateral divisions and the ventrolateral nucleus of Kölliker–Fuse (Fulwiler & Saper, 1984; Block & Estes, 1990). Each division is composed of smaller subgroups which form a mosaic of longitudinally organized cell columns. Parabrachial subnuclei are defined by specific interconnections with the telencephalon, brainstem and spinal cord. The

PBC lies in a key position to engage in somatovisceral integration. Different sensory modalities may be spatially represented within the PBC by virtue of topographically organized afferent projections ascending from the medulla and spinal cord. Since primary afferents are somatotopically distributed within sites in the NTS and spinal dorsal horn which project to the PBC, this organizational pattern might be preserved in the pons.

*Parabrachial afferents*   The distributions of afferents from the NTS and brainstem reticular formation to the PBC are organized topographically. The 'taste' nucleus of the NTS is represented within the anterior one-half of the nucleus and conveys chemoreceptor afferents devoted to gustatory sensibility to the medial parabrachial nucleus and, at rostral levels, the medial halves of both nuclei (Hamilton & Norgren, 1984; Travers, 1988). General visceral afferent pathways to the PBC are also organized in space although the precise viscerotopic plan requires further study. Dorsal-medial divisions of the NTS along with cells neighbouring the DMX and extending into the subjacent reticular formation (dorsomedial LTF) predominantly project to medial loci. Subnuclei of the NTS surrounding the solitary tract and cell columns in the ventrolateral medulla which relay solitario-reticular signals, project to the lateral PBC and the ventrolateral subnucleus of Kölliker–Fuse (King, 1980). Parabrachial projection neurons in the LTF may take part in propriobulbar reflex integration. Afferent cell groups within the LTF are contiguous with projection neurons in laminae V and VII of the spinal cord. These neurons are integrated into sequences of relatively short (local) reflex circuits and afferent feedback loops upon which other neural centres may act. In the reticular formation these neuronal somata and associated glia are organized in columns and, as previously described, a large percentage are aligned along parallel laminae running radially between the NTS and ventrolateral medulla. The associational neurons of the LTF thus lie between the sensory and reticulomotor limbs of autonomic or somato-visceral integrated reflex arcs. In the spinal cord, a similar arrangement of parabrachial projection cells form sheets which extend radially between spinal lamina X (where dorsal root (noxious) afferents terminate) and the lateral horns of lamina VII (where sympathetic or sacral parasympathetic preganglionic motor neurons and spinal interneurons are concentrated). The proximal dendrites of neurons within the reticular formation and spinal grey are aligned transversely along these laminae or extend parasagittally (longitudinally) within the individual cell columns. Since afferents to parabrachial afferent projection cells (or other propriobulbar neurons) align with their proximal dendrites, this organization may dictate the pattern of information transfer within this region of reticular formation (King, 1980). Afferents to the PBC are segmentally organized. Projections to each subnucleus of the PBC arise from multiple segments of the medulla and spinal cord and originate from neurons which occupy similar radial positions along each lamina. This structural plan is believed to be responsible for a transtegmental convergence of afferents (targeting different segments of the medulla or spinal cord) within specific parabrachial subnuclei. The PBC also receives descending projections from areas of the forebrain which differentially encode multimodal signals into emotional behaviours (Saper, 1982; Cechetto & Calaresu, 1983).

### Parabrachial efferents

Parabrachial projection pathways provide neuroanatomical substrates for visceral and somatosensory afferent control of cardiorespiratory, GI and neuroendocrine functions, the sleep–wake cycle and behavioural expression. The three principal ascending parabrachial pathways include: (1) a periventricular system, (2) a dorsal tegmental bundle, and (3) a central tegmental tract (Saper & Loewy, 1980). These fibre tracts are summarized in Fig. 14 and follow a similar course taken by locus ceruleus axons (see next section). The PBC is one of several major brainstem relays of pan-modal sensory information to the forebrain. Parabrachial output neurons directly access the cerebral cortex by way of projections from the medial PBC to a continuous sheet of cortex covering the frontal, infralimbic, granular-insular and septo-olfactory areas. Visceral information is conveyed by the NTS to the cerebral cortex, indirectly, by topographically organized solitariofugal pathways to the PBC and ventral sensory thalamus (see Norgren, 1984; Cechetto, 1987; Cechetto & Saper, 1987 for review and additional references). As will be recalled, special (gustatory) and general (autonomic) visceral afferents are relayed to anterior and posterior parts of the NTS which, in turn, project to medial and lateral divisions of the PBC. In the ventrobasal thalamus, the ventral posterior medial parvicellular thalamic nucleus (VPMpc) conveys special visceral afferents, relayed by the anterior NTS and medial PBC, to the anterior 'dysgranular' insular cortex (the intermediate of three longitudinally organized zones of insular cortex; intercalated between the dorsal granular and ventral agranular insular areas). The ventral posterior lateral parvicellular thalamic nucleus (VPLpc) receives inputs from lateral parts of the PBC which convey general visceral afferents from the posterior half of the NTS. Thalamocortical relay neurons in the VPLpc, in turn, project to the granular insular cortex where units responding to general visceral afferent stimulation are arranged viscerotopically. Parabrachial pathways can also access more widespread areas of the cortex by way of ascending projections to the midline and intralaminar thalamic nuclei (MIT). These projection fields include parts of the centromedian-parafascicular complex (McBride & Sutin, 1976) that have long been known to participate in cortical EEG desynchronization, behavioural arousal and as recently discovered, intrinsic central regulation of cerebral blood flow. Mraovitch *et al.* (1992) discovered that electrical stimulation of the centromedian-parafascicular thalamus, which also projects diffusely upon the cerebral cortex, produced widespread and differential cerebrovasodilatation which in some areas was associated with or, in others, independent of a change in glucose metabolism. Multiple systems appear to mediate regional cerebrovascular responses either through direct neural pathways or by actions of circulating adrenal catecholamines on receptors in circumventricular organ systems (Miselis *et al.*, 1986).

Parabrachial pathways ascending by the dorsal tegmental bundle to the thalamus and through collateral branches to the hypothalamus may constitute a critical component of the reticular activating system. The integrity of each of these structures is essential in maintaining consciousness and such regions have been implicated in mechanisms generating states of arousal and wakefulness (see review by Jones, 1989). Parabrachial projections to subcortical telencephalic nuclei are derived from both divisions and terminate within nuclei selectively involved in autonomic (the ventral amygdalofugal pathway or 'lateral system') or behavioural/neuroendocrine (the stria terminalis-fornix or 'medial system') components of emotional expression. Parabrachial-telencephalic pathways (Nomura *et al.*, 1979; Saper & Loewy, 1980) are characterized by convergence (e.g. both parabrachial divisions project to the central nucleus of the amygdala and its target, the bed nucleus of the stria terminalis) or divergence (speci-

ficity) (e.g. the medial PBC innervates the basolateral amygdala, diagonal band of Broca and magnocellular preoptic area, whereas the lateral division projects to the medial amygdala and median preoptic areas). The functional significance of this organizational scheme is unknown.

The parabrachial-hypothalamic pathway may influence neuroendocrine control by way of a projection from the lateral PBC to the parvicellular neurosecretory system. The lateral PBC, unlike the medial division, projects heavily to the hypophysiotropic nuclei: the mediobasal hypothalamus (arcuate and ventromedial nuclei), periventricular, parvicellular paraventricular and dorsomedial hypothalamic nuclei. The medial and lateral divisions likely influence cells of the magnocellular neurosecretory system (the supraoptic (SON) and paraventricular (PVN) nuclei), indirectly, by way of their connections with the A1 noradrenergic cell group, amygdala and septum, median preoptic nucleus and the subfornical organ. Each of these structures receives a direct input from the PBC and projects to the large magnocellular neurons of the PVN or SON which synthesize the nonapeptides: vasopressin and oxytocin (Miselis *et al.*, 1979). A direct parabrachial pathway to the PVN and lateral hypothalamus may also access neurons with descending projections to the lower brainstem and spinal intermediolateral cell column (Swanson & Mogenson, 1981; Swanson *et al.*, 1980). As reviewed in Ruggiero *et al.* (1984) hypothalamo-spinal (and reticular) projection neurons synthesize, as presumptive neurotransmitters, dopamine, neuropeptides or, as recently demonstrated, an excita-

tory amino acid (Sun & Guyenet, 1986).

Projections from the PBC to the brainstem and spinal cord derive from all divisions including the nucleus of Kölliker–Fuse (Figs. 12 & 14). Descending parabrachial pathways follow the same trajectories taken by those descending from the hypothalamus and the pontine nuclei locus ceruleus and subceruleus. These axons descend through the lateral pontomedullary reticular formation and innervate parasympathetic motor neurons of the nucleus of Edinger–Westphal, salivatory and dorsal motor nuclei, and neurons in the A5 and C1 areas which issue reticulospinal projections to sympathetic and parasympathetic spinal preganglionic neurons. Parabrachial-bulbar fibres terminate monosynaptically in the solitary complex and lateral reticular formation which, in turn, issue direct projections to both autonomic and respiratory motor nuclei of the spinal cord (Ellenberger & Feldman, 1990b; Mifflin & Felder, 1990; Ruggiero *et al.*, 1991; Mtui *et al.*, 1993). Projections to the LTF may also engage neurons of the propriobulbar fibre system which project to cranial nerve motor nuclei. Descending projections to the PBC from the limbic forebrain (e.g. amygdala and lateral hypothalamus) likely orchestrate local reflex control over muscles of mastication and faciovocal expression (Magoun *et al.*, 1937) with a pattern of sympathetic and parasympathetic discharge appropriate to the behavioural programme in progress (see reviews by Koizumi & Brooks, 1972; LeDoux, 1987). Finally, the nucleus of Kölliker–Fuse (n. KF) projects by the central tegmental tract to the lateral preoptic and hypothalamic nuclei and the central nucleus of the amygdala (Saper & Loewy, 1980). The n. KF is characterized by its descending pathway to the ventrolateral medulla (Ruggiero *et al.*, 1989; Van Bockstaele, 1989) and upper thoracic intermediolateral cell column.

The functional significance of the intricate and differentiated wide-ranging projections of the PBC has yet to be unravelled. Interestingly, dramatically different changes in arterial blood pressure can be elicited from different parts of the complex: pressor responses from sites in the lateral PBC and n. KF which have prominent projections to sympathetic vasoconstrictor centres, and depressor responses from an ill-defined area of the caudomedial PBC (Ward, 1988). Perhaps the different outputs of the medial (parabrachial-cortical) and lateral subdivisions may explain why their vasoconstrictor effects on cerebral blood vessels are, respectively, coupled to or independent of cerebral metabolism (Mraovitch *et al.*, 1985).

### Locus coeruleus (LC)

The LC or nucleus pigmentosus pontis is a prominent noradrenergic nucleus in the lateral pontine periventricular grey (A6 area). The LC is unique in that it projects monosynaptically and diffusely to widespread areas of the CNS (Jones & Yang, 1985) and contributes to the central innervation of intraparenchymal blood vessels. Bilateral lesions of all noradrenergic neurons in the LC produce a marked fall-off in the perivascular adrenergic innervation (Edvinsson, 1975). Neurons in other central locations also contribute to the cerebrovascular nerve supply since injections of the neurotoxin, 6-hydroxy-

dopamine, almost completely abolish catecholamine-fluorescent axons associated with most intraparenchymal blood vessels.

The pervasive projections of the LC and their homogeneous discharge properties are consistent with the hypothesis that these neurons act in concert to synaptically gate incoming ascending visceral and somatic panmodal afferents and modulate cerebral and cerebellar-cortical activities. Thus, the LC may influence behavioural states that require arousal or increased vigilance as well as different states of the sleep–wake cycle (Aston-Jones *et al.*, 1986). The NTS (including adrenergic neurons of the 'C2' area) and the spinal dorsal horn are now thought to provide small inputs to the LC. Recent studies suggest that strong inhibitory (state-dependent) inputs to LC may be mediated by neurons in the dorsomedial medulla which may employ the inhibitory amino acid GABA, and those in the rostral ventrolateral medulla which employ adrenaline as a presumptive neurotransmitter (Ennis & Aston-Jones, 1989). The role of state-dependent inhibition of A6 noradrenergic neurons could be related to their decreased responsiveness during behaviours associated with increases in parasympathetic tone: e.g. grooming and consumption and slow-wave sleep.

The efferent projections of the LC were systematically examined by Jones & Yang (1985). LC efferents, although partly shared by those of the PBC and pontomesencephalic reticular formation, are distributed more ubiquitously. In the cerebral cortex, LC fibres ramify extensively within all cortical laminae of the archipalaeo- and neo-cortex. In the subcor-

tical telencephalon, the projections of LC are, again, more widespread, spreading across the entire basal forebrain and involving the basal ganglia. Yet, LC efferents and those of the PBC overlap in the central nucleus of the amygdala, bed nucleus of the stria terminalis and septum. In the hypothalamus, dense projections terminate largely within the lateral hypothalamus (LH) and zona incerta (ZI). Axons of cells in the LH, in turn, issue collateral projections ascending to the basal forebrain, septal complex and amygdala whereas those in the ZI project to sites in the midline-intralaminar complex which project diffusely upon the cerebral cortex. A dorsal branch of the pathway ascending from the LC innervates specific sensory thalamic nuclei and the midline-intralaminar thalamus with especially dense inputs to the parafascicular and anterior thalamic nuclei. Together, these pathways provide parallel routes through which the LC can exert global influence over the cerebral hemispheres. Coeruleal-hypothalamic projections also overlap neurons within the dorsal and posterior hypothalamus with major descending projections to the lower brainstem (e.g. the NTS (Ross *et al.*, 1981a), ventrolateral medulla (Ruggiero *et al.*, 1989; Van Bockstaele *et al.*, 1989) and spinal cord (Hosoya, 1980). Diffuse projections from the LC to parvicellular hypothalamic neurosecretory nuclei may provide neural substrates through which the LC can exert actions on the secretory activity of the anterior lobe of the pituitary gland. Significant connections are also established with the suprachiasmatic nucleus (the terminal nucleus of the retino-hypothalamic tract) through which cyclic changes in environmen-

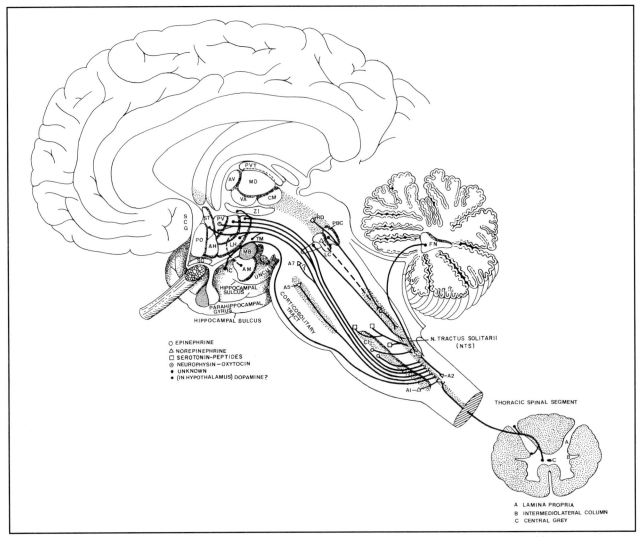

*Fig. 15. Central origins of afferents to the nucleus tractus solitarii and dorsal motor nucleus of the vagus (see text for details)*

tal light can influence metabolic (e.g. electrolyte and volume homoeostasis), reproductive and other behavioural patterns.

Local projections of LC terminate within the dorsal pontomesencephalic raphe which provides a long ascending serotonergic afferent innervation of the forebrain. In the medulla, the principal descending projection ends within sensory (e.g. NTS, Fig. 15), associational (e.g. reticular formation) and motor (cranial nerve motor nuclei) limbs of cranial nerve reflex arcs (Ter Horst *et al.*, 1991). In the spinal cord, efferent fibres terminate predominantly within laminae VII, VIII and X and are sparsely distributed over the dorsal horn (Fig. 16). Coeruleospinal projections densely innervate the sacral parasympathetic cell column (Jones & Yang, 1985).

## Visceral afferent processing in the forebrain

The exact pathways relaying specific unimodal and integrated visceral (viscerosomatic or convergent) signals to the forebrain have only been partially elucidated. It is well recognized that a large part of the afferent projection is relayed through the aforementioned intermediary nuclei intercalated between the first order synapse in the

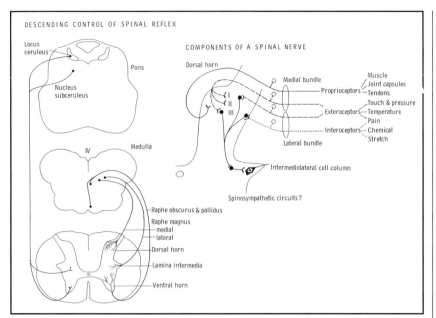

DESCENDING CONTROL OF SPINAL REFLEX

*Fig. 16. Left side: descending supraspinal (monoaminergic) control over the sensory and motor limbs of a typical spinal reflex arc. Right side: schematic representation of the functionally distinct components of a typical spinal nerve. The lateral bundle consists of lightly myelinated axons whereas axons of the medial bundle are heavily myelinated. The schema also provides theoretical circuits through which different peripheral afferent modalities might enter into an autonomic reflex.*

NTS and the forebrain. Anatomical studies have also revealed dense projections derived directly from neurons in the NTS. As indicated above, NTS pathways course by two principal fibre bundles: a periventricular or medial system ascends diffusely through the periventricular grey and the other, a lateral fibre system, travels by way of the pontomedullary reticular formation and medial forebrain bundle (Ricardo & Koh, 1978). The precise contributions of each neural channel have not been demonstrated.

Ascending gustatory projections from the anterior half of NTS are characterized by major species differences. In primates, there is a significant direct projection from the NTS to the taste nucleus of the thalamus (Beckstead *et al.*, 1980). In rodents, gusta-tory pathways form obligate synapses with parabrachial-thalamic neurons but do not directly ascend to the forebrain (Travers, 1988). The posterior NTS clearly gives rise to dense projections to magnocellular neurons in the paraventricular hypothalamic nucleus which synthesize the neurohypophyseal hormones, vasopressin and oxytocin, and nuclei of the parvicellular system (e.g. the dorsomedial and arcuate hypothalamic nuclei) which regulate the secretion of adeno-hypophyseal neurohormones (e.g. ACTH and prolactin). Other ascending pathways engage nuclei which encode polymodal signals entering into the expression of different behaviours such as the consummatory (drinking and feeding) and defence reaction. Major terminal fields include the central nucleus of the amyg-dala, bed nucleus of the stria terminalis, septum and medial preoptic area. As reviewed by Ricardo & Koh (1978), stimulation of peripheral cranial nerves, the NTS or its projection fields tends to induce behavioural expressions, release of hypothalamic or adrenal hormones and either synchronization or desynchronization of the electrocorticogram. Direct structural connections between cardiovascular neurons in the NTS and the forebrain corroborate physiological evidence that electrical stimulation of cardiovascular afferents will elicit short-latency responses of single units in the hypothalamus (Ciriello & Calaresu, 1980).

Recent studies have implicated two cortical areas in central autonomic control: the insular and infralimbic prefrontal regions. Each area shows a close correspondence between physiologically active, cardiovascular and gastrointestinal sites and direct and reciprocating connections between members of the limbic lobe and the ANS. Pyramidal cells in the insular and infralimbic cortex issue dense descending projections to brainstem nuclei (e.g. the NTS and PBC) which convey both general and special visceral afferent impulses to the forebrain. Major cortical projections also descend to the lower brainstem including a region of the lateral medullary reticular formation harbouring propriobulbar neurons and involved in visceral and somatic reflex integration (Holstege & Kuypers, 1977; Holstege *et al.*, 1977; Ruggiero *et al.*, 1982; 1987). The physiological and clinical importance of the insular cortex, where olfactory and gustatory afferents converge, is implicit in the effects of chemical stimulation in raising arterial blood pressure (Ruggiero et al., 1987; Openheimer & Cechetto, 1990; Yasui et al., 1991) and provoking arrhythmias leading to asystole (Oppenheimer et al., 1990). Cechetto & Chen (1990) have identified the lateral hypothalamus as a mandatory relay for cortical-sympathetic responsiveness. Sun (1992) has demonstrated that hypotension evoked by electrical stimulation of the lateral prefrontal cortex is mediated by GABAergic inhibition of vasomotor reticulospinal neurons in the nucleus RVL. Damage to the insular cortex produces electrocardiographic patterns characteristic of stroke (Cechetto et al., 1989). The insular cortex is also extensively wired to the infralimbic prefrontal cortex, dorsal thalamus, hypothalamus and amygdala and, thus, has been implicated in integrating the behavioural and autonomic components of emotional expression (Krushel & van der Kooy, 1988).

The insular cortex is influenced by visceral as well as limbic afferents (e.g. the dorsomedial thalamus and prefrontal cortex) which are segregated along different longitudinally organized cortical strips. This structural arrangement probably reflects a neural substrate through which telencephalic afferents can enter into the generation of an organized adaptive behavioural pattern. Visceral afferents are conveyed to the forebrain by pathways that are organized viscerotopically along their entire trajectory. Primary visceral afferents terminate in the NTS which, in turn, relays to the PBC and, by way of the latter, to visceral thalamocortical relay neurons of the ventral thalamus (Allen et al., 1991). Taste responsive neurons in the insular cortex lie rostral to the

gastric-mechanosensory units and are concentrated anteriorly to units activated by baroreceptor and chemoreceptor stimulation (Cechetto & Saper, 1987). The proximity of visceral and somatic areas of sensory cortical representation (e.g. areas of cardiopulmonary representation in the posterior insular cortex and somatosensory cortical representation of the trunk) is additional evidence that these cortical columns might integrate autonomic and skeletomotor components of emotional expression. Other limbic afferents (e.g. the amygdala and lateral hypothalamus) are spread diffusely across functionally different divisions of insular cortex. These projections are thought to play an integrative role in processing different incoming afferents which would initiate an appropriate physiological response such as the changes in regional vascular resistance and cardiac output associated with the performance of a specific behaviour. A wealth of recent evidence suggests that anatomically segregated subpopulations of cells might mediate the behavioural versus autonomic concomitants of emotions (see LeDoux, 1987; LeDoux et al., 1990). The amygdala is widely recognized as lying in a unique pivotal position, in serving as a convergence centre for multimodal afferents and as an outlet in triggering pre-programmed fixed-action patterns of behaviour and autonomic outflow. Amygdaloid subnuclei, by way of anatomically divergent pathways, may divert these signals to structures mediating different components of a behavioural response including skeletomotor, autonomic, neuroendocrine and affective. Electrical or chemical stimulation of the amygdala (Fernandez de Molina &

Hunsperger, 1962; Hilton & Zbrozyna, 1963; Igic et al., 1970) provoke full expression of the defence reaction which resembles the response to natural provocation (e.g. confrontation). Conversely, destruction of the amygdala diminishes the response to a threatening stimulus (Blanchard & Blanchard, 1972) and abolishes the conditioned potentiation of the acoustic-fear/startle response (Mondlock & Davis, 1985).

The outputs of the insular cortex and amygdala, by virtue of their intimate connections, probably superimpose a hierarchic control over emotional expressions. Pathways mediating autonomic responses from both nuclei appear to constitute a functionally integrated system extending from the cortical mantle to the spinal cord. They share in common similar local connections within the limbic lobe and long descending projections to sensory and associational components of cranial nerve reflex arcs. Sympathetic response patterns provoked by insular cortical and amygdaloid stimulation are blocked by lesions of the lateral hypothalamus (Cechetto & Chen, 1990). The full expression of the cardiovascular pattern of response provoked by stimulating a similar area of the hypothalamus is also disrupted by lesions of sites in the rostral ventrolateral medulla which issue sympathoexcitatory projections to the thoracic spinal cord (Hilton et al., 1983).

The cellular basis for most emotional and volitional concomitants of behaviours is unknown. These include species-specific (parasympathetic-bradycardic versus sympathetic-tachycardic dominant) responses to conditioned aversive stimulation (Kapp *et al.*, 1984); defence versus predatory attack, and 'motivational states' (e.g. hunger, thirst and sex) characterized by a loose correlation between the behavioural response and the observable stimulus. It is well recognized that sites in the brain which are most effective in supporting intracranial self-stimulation (i.e. producing positive reinforcement) are correlated with feedback loops (e.g. between the parabrachial complex and lateral hypothalamus) from which stimulation provokes complex patterns of behaviour and visceral response involved in feeding, drinking and self preservation (acute: defence; long term: reproduction) (Ferssiwi *et al.*, 1987a,b). The precise relationships between neural circuits activated by punishment and reward have yet to be adequately delineated.

## Feedback circuits involved in visceral reflex control

The CNS modulates visceral reflex excitability by exerting functionally specific afferent feedback controls over all three limbs (sensory, associational and reticulomotor) of brainstem visceral reflex arcs. The NTS, a site of visceral-sensory integration, receives peripheral inputs from cranial nerves VII, IX and X and is also richly innervated by anatomically integrated segments of the brain and spinal cord. The principal afferent pathways to the NTS are illustrated in Fig. 15. Projections to the NTS derive from subnuclei of larger traditionally defined nuclei. NTS afferents are segmentally organized within the brainstem and form a functionally unified system of feedback and feed-forward control

circuits. Inputs from different segments of the CNS are distinguished functionally by the level of complexity of their afferent processing centres (e.g. the spinal dorsal horn versus insular-prefrontal cortex). All sources of afferent input to NTS also receive a direct visceral or integrated somatovisceral ascending projection from the NTS, with the exception of the cerebral cortex which requires an obligatory synapse in the lower brainstem (parabrachial complex, locus ceruleus and nucleus raphe dorsalis) or subcortical forebrain (e.g. septum, amygdala and visceral thalamus). Intracellular recording studies indicate that the NTS produces excitatory effects on the PBC with which it is reciprocally connected (Granata & Kitai, 1989). Why multiple afferent cell groups are required to modulate NTS neuronal activity (Mifflin & Felder, 1990) is probably related to the hierarchic control exerted by different segments of the CNS over components of the rigidly wired closed-chain feedback circuits in the lower brainstem.

Afferents to the solitary complex synapse on the primary (first order) sensory neuron, an intermediary interneuron (intercalated between the sensory and motor limb of the reflex loop), an efferent projection neuron or a combination, thereof. The central derivation is heterogeneous (Ross *et al.*, 1981a; van der Kooy *et al.*, 1984; Ruggiero *et al.*, 1987; Mtui *et al.*, 1993; Ter Horst *et al.*, 1991), and comprised of neurochemically distinct perikarya including some capable of co-synthesizing two or more putative neurotransmitters or neuromodulators (van der Kooy *et al.*, 1984; Gray & Magnuson, 1987; Millhorn *et al.*, 1987). Immunocytochemical, biochemical

assay and receptor binding studies have collectively shown that terminals in the NTS employ as potential neurotransmitters either monoamines (serotonin, noradrenaline and adrenaline derived from the lower brainstem and, perhaps, dopamine derived from cell bodies of the A10–A15 areas of the hypothalamus), acetylcholine, neuropeptides (e.g. substance P, galanin, neuropeptide Y) including those which serve dual roles as pituitary neurohormones and hypothalamic releasing or inhibitory factors and, lastly, amino acids such as GABA or L-glutamate. Individual terminal fields within the NTS can be recognized under the light microscope as discrete topographically organized collections of fine punctate processes which represent terminals. Different afferents have unique distributional patterns and can be conceptualized as originating from functionally specific, albeit integrated, subsystems. NTS afferents project to several different sites and have major or minor projection fields which show variable overlap with the viscerotopic distributions of specific first-order afferents. These complicated structural relationships likely provide the neural circuits for divergent coordinated central reflex control over the excitability of different visceral sensory modalities.

### *Closed multiple feedback control circuits*

All components of visceral reflex arcs in the lower brainstem, including neurons involved in the tonic and reflex control of the autonomic outflow, feedback to the NTS. These circuits comprise the servomechanisms (self-correcting error-control systems) of the lower brainstem, and are designed to coordinate changes in reflex excitability of the primary sensory controller (NTS) with the constellations of skeletomotor and visceromotor reflex adjustments to signals continually generated by the internal and external milieu (e.g. those associated with movements (e.g. exercise), pain, and the expression of distinct behaviours). The smooth performance of such goal-directed reflexes is servoregulated by a sequential transfer of afferents to the NTS from collaterals of intermediary neurons organized along its efferent transtegmental fibre tract. These include propriobulbar interneurons within the lateral parvicellular reticular formation, A5, C1 and non-catecholaminergic reticulospinal projection neurons and both C1 and A1 reticulohypothalamic projection neurons (Ciriello & Caverson, 1986; Ruggiero et al., 1991; 1994). Local afferents to the NTS, in part, may derive from axon collaterals of longer projection systems. For example, injections of different fluorescent dyes into the NTS and sites in the cervical ventral horn which harbour respiratory motor neurons (or the thoracic lamina intermedia where sympathetic preganglionic neurons reside) retrogradely label many of the same cell bodies in the brainstem reticular formation (Ruggiero et al., 1991). Many of these neurons contain catecholamine-biosynthetic enzymes although a larger percentage synthesize neuropeptides (Millhorn et al., 1987) and other chemically unidentified neurotransmitters. Collateral projections provide neuroanatomical substrates for physiologically identified circuits which simultaneously modulate cardiopulmonary reflex excitability within the NTS and the discharge of cardiorespiratory lower motor neurons in the spinal cord

(Mtui et al., 1994).

The catecholaminergic innervation of the NTS is thought to contribute to mechanisms through which α-adrenergic receptors modulate synaptic excitability by altering their responsiveness to primary sensory afferent input (Snyder et al., 1978; Feldman & Felder, 1989). Selective destruction of one of these sources, the A2 noradrenergic nucleus in the caudal NTS, produces chronic lability of blood pressure and heart rate. The A2 area is, therefore, thought to gate baroreceptor or chemosensory inputs to sensory relay neurons in NTS and parasympathetic motor neurons in the DMX. Local circuits within the NTS are also formed by adrenergic neurons in the dorsal subnucleus of NTS (C2 area) and the neighbouring periventricular fibre system (C3 area) (Hokfelt et al., 1985; Ruggiero et al., 1985b) (Fig. 15). The precise functions of these neurons, while unclear, can be implicitly inferred by their bidirectional projections along the periventricular grey and extensive interconnections with members of the 'endogenous pain control system (EPCS)' (Pieribone et al., 1988). The EPCS includes noradrenergic neurons in the pons (locus ceruleus, subceruleal and A7 areas) and serotonergic nuclei in the pontomesencephalic central grey (n. raphe dorsalis) and medullary raphe (Fig. 16). All structures of the EPCS receive visceral afferent inputs from the NTS and convey multimodal sensory signals to widespread areas of the CNS including a direct and diffuse innervation of the cerebral hemispheres. Their descending projections are thought to modulate the responsiveness of spinal sensory neurons to noxious and non-

noxious stimulation (Clark & Proudfit, 1991). The precise descending patterns of projection within the spinal cord appear to be specific to different species or strains of the same species. Perhaps one of the functions of the descending pontine projection to the NTS and spinal grey is the modification of somatic (e.g. pain) and autonomic (e.g. baro-receptor) reflex responses associated with behavioural arousal or migraine. According to Goadsby & Lance (1988), stimulation of the nucleus locus ceruleus and the nucleus raphe dorsalis have direct and opposing effects on the cerebral circulation and may function as an intrinsic vasomotor reflex centre in a way comparable to the enteric nervous system. Cell groups of the EPCS also project to members of the visceral forebrain system (the limbic system) which encode panmodal sensory stimuli into appropriate patterns of behavioural response, while simultaneously monitoring these signals as they arrive from the periphery.

Forebrain afferents to the NTS also engage other components of visceral reflex arcs (Fig. 15). (Fig. 12 demonstrates that many of these centres also directly innervate the spinal sympathetic cell columns.) Forebrain projections to the dorsal vagal complex are topographically arranged as follows (van der Kooy et al., 1984; Gray & Magnuson, 1987). Subcortical afferents have relatively dense projections to ventral parts of the NTS surrounding the DMX. Cortical afferents to the NTS, in contrast, project to dorsal and lateral subnuclei and along a midline region largely avoided by the subcortical afferents. Afferents to the anterior NTS are exquisitely well organized: 'Lateral forebrain nuclei' (lateral prefrontal cortex and central amygdaloid nucleus) terminate predominantly laterally within the anterior (taste) nucleus whereas 'medial forebrain nuclei' (medial prefrontal cortex, bed nucleus of the stria terminalis and paraventricular hypothalamic nucleus) end on medial and ventral divisions of the anterior NTS. The latter cell groups provide a source of intramedullary projection to propriobulbaroromotor reflex nuclei in the LTF (Travers, 1988; Beckman & Whitehead, 1991).

Forebrain afferents also project to intermediate and caudal levels of the NTS which relay general visceral afferents from the thorax and abdomen (Ross *et al.*, 1981a; Ruggiero *et al.*, 1987). These structures, when stimulated, activate emotional or appetitive behaviours and elicit changes in cardiopulmonary dynamics and other autonomic reflex reactions (Smith *et al.*, 1980). Forebrain afferents to the NTS and underlying tegmentum may account for the behavioural alterations in autonomic (e.g. baroreceptor) reflex excitability which accompany genetically preprogrammed behaviours, or (due to the high degree of plasticity of the cortical receptive fields) those acquired through experience.

### Somato-visceral reflexes

It has long been recognized that somatic afferents play a crucial role in central autonomic control. Autonomic neurons are excited by stimuli from multiple sources: the organs of special sense (e.g. the pupillary light reflex) and a variety of somatic afferents, including myelinated and unmyelinated axons originating from receptors in skin and muscle and the membranous labyrinth (e.g. the vestibular-autonomic reflex). Spinal-autonomic reflexes are integrated along multiple levels of the neuraxis and, when recorded from sympathetic nerves, are characterized by early spinal and late supraspinal components (Sato & Schmidt, 1973; Mitchell *et al.*, 1983). Spinal-autonomic reflexes are mediated by short intraspinal pathways or longer supraspinal reflex arcs involving an ascending projection to brainstem autonomic nuclei. Figures 4 & 16 demonstrate potential polysynaptic pathways of a typical (hypothetical) intraspinal reflex arc.

*Intraspinal reflexes* can be provoked by stimulating myelinated (types II–III) and unmyelinated (type IV) afferents terminating in the dorsal horn. Upon entering the cord, dorsal root fibres branch into a medial myelinated bundle and a lateral bundle of finely myelinated or unmyelinated fibres. It has long been assumed that local reflexes are polysynaptic and that interneurons are intercalated between the primary afferent and visceromotor (or respiratory motor) cell group. These reflexes are organized segmentally and, unlike the supraspinal reflex, unaffected by spinal cord transections. Group IV noxious afferents will provoke autonomic responses (sudomotor or vasomotor signs) that are regional and limited to the segments proximal to the afferent inputs (Saper & De Marchena, 1986).

*Supraspinal autonomic reflex arcs* are mediated by ascending projections from the spinal dorsal horn to two brainstem systems classically subdivided by their opposing actions on sympathetic nerve discharge: a medial (spino-bulbar) system corresponds to vasodepressor sites in the NTS, raphe and medial reticular formation, and a lateral spinobulbar vasopressor system includes the C1 and A5 areas, the LTF, lateral parabrachial and lateral hypothalamic-preoptic nuclei (Mraovitch *et al.*, 1982; Ross *et al.*, 1984b; Sun & Guyenet, 1986; Barman & Gebber, 1987). Cranial nerve stimulation also elicits autonomic responses. For example, trigeminal-autonomic reflexes (Kumada *et al.*, 1977) may be mediated by a short trigemino-solitary reflex pathway to the NTS (Fig. 10) and a longer projection to the parabrachial complex (Cechetto *et al.*, 1985). It should be emphasized that neurons in the dorsal horn convey integrated signals from somatic and visceral (e.g. renal) receptors to the NTS and reticular formation (Ammons, 1988; Person, 1989). Cervero's data (1986) suggest that visceral afferents terminate in laminae I and V (and the ventral horn). Visceral and nociceptive afferents converge on somatosensory spinal neurons in the thoracic and sacral dorsal horn and may explain why visceral pain is vaguely localized to the body parts supplied by the innervated segments.

### Cerebellar-autonomic reflexes

The cerebellum or head ganglion of the somatosensory system may be conceptualized as a servocontrol mechanism subserving somato-autonomic integration (synergy). The cerebellum, of alar plate origin, is formed of a thin shell, the cerebellar cortex, and the bilaterally symmetric deep cerebellar output nuclei embedded in a large mass of white matter, the cerebellar medulla. It is generally assumed that 'the cerebellum integrates and organizes information flowing into it ...' and 'functions as a type of computer that is concerned with the smooth and effective control' and modification of ongoing movements (Eccles *et al.*, 1967). By comparing signals arriving initially from the command centre of the cerebral cortex with those issuing from differentially activated receptors (somatic and visceral), the cerebellum serves to maintain muscle tone and equilibrium while simultaneously synchronizing skeletomotor movements coordinated with a distinct pattern of visceromotor outflow to glands and smooth muscles.

Cerebellar-autonomic pathways can be conceived by imagining the environmental events leading to behavioural arousal and the assumption of an upright posture, and the hypothetical sequence of antigravity reflexes leading to postural stability and a redistribution of cardiac output (Doba & Reis, 1972). Signals to the cerebellum are directly conveyed by spinal and cranial nerves or, indirectly, by means of cerebellar relay nuclei in the brainstem reticular formation. Cerebellar afferents are delivered by three cerebellar peduncles attached to the medulla (inferior cerebellar peduncle), pons (middle cerebellar peduncle) and midbrain (superior cerebellar peduncle). The area of cerebellar-autonomic responsiveness is tightly localized to the median vermal-fastigial zone (VFZ) which consists of the vermal cortex (integrator) and the fastigial deep cerebellar nuclei (Martner, 1975). The VFZ is flanked bilaterally by sagitally organized paravermal zones and the large cerebellar hemispheres. The majority of cerebellar influences on the ANS are mediated by Purkinje cells of the vermal cortex and their principal outlet to the brainstem, the fastigial nucleus.

The rostral pole of the FN is associated with two patterns of autonomic response mediated by anatomically distinct pathways: (1) Electrical stimulation of its rostral pole elicits stereotyped patterns of behaviour, sympathetic activation (the fastigial pressor-response: FPR), baroreceptor reflex inhibition and global increases in cerebral blood flow independent of metabolism (Miura & Reis, 1970; Reis *et al.*, 1973; Nakai *et al.*, 1983). The FPR is mediated either by antidromic activation of Purkinje cell axons traversing the FN or, perhaps, by collaterals or fibres en passant of brainstem autonomic or forebrain neurons (Miura & Takayama, 1988). The full expression of the FPR is dependent on the structural integrity of the vestibular complex and nuclei which issue reticulospinal projections to the spinal IML: C1 area of the ventrolateral medulla, pontine A5 area and the paramedian reticular formation (Miura & Reis, 1970; Dormer *et al.*, 1989; Elisevich & Ciriello, 1988); (2) the vasodepressor response to excitatory amino acid activation of the FN, in contrast, appears to be directly mediated by efferent projections of the deep cerebellar neurons (Chida *et al.*, 1986).

The anatomical substrates mediating differential patterns of cerebellar response are unresolved but may be related to connections of the cerebellum with components of the somatic-autonomic reflex circuits. Somatosensory, exteroceptive and proprioceptive afferents are the principal sources of input to the VFZ and are carried by nerves which mediate somato-autonomic responses. Moreover, variable effects on arterial blood pressure and heart rate are mediated

by cortical lobules of the anterior and posterior cerebellar vermis (Henry *et al.*, 1989) which convey most processed signals to the FN and vestibular complex (Ruggiero *et al.*, 1977). Indeed it is well recognized that natural or experimental stimulation of spinal and vestibular nerves lead to potent changes in sympathetic nerve discharge and arterial blood pressure mediated by units in the nucleus RVL – the principal outlet of the medullary reticular formation to the spinal sympathetic cell columns (Miyazawa & Ishikawa, 1985; Stornetta *et al.*, 1989; Yates *et al.*, 1991). The roles of the intermediary brainstem neurons (integrators) intercalated between the peripheral receptor and the autonomic reticulospinal relay nucleus RVL have not been explored. Nonetheless, suggestive evidence that the cerebellum is incorporated within the somato-autonomic reflex pathway is consistent with the idea that afferents to the cerebellum may contribute to orthostatic control of the circulation (Miura & Reis, 1970; Mitchell *et al.*, 1983).

Figure 15 demonstrates a direct projection from the FN to the nucleus parvocellularis compactus (i.e. the nucleus parasolitarius (NPS)). This terminal field lies lateral to the area in NTS (Batten *et al.*, 1977) where most general visceral afferents from the thorax and abdomen terminate (Kalia & Mesulam, 1980). The NPS is a small and slender GABAergic nucleus (Meeley *et al.*, 1985), and as our preliminary axonal transport studies suggest, may be a major inhibitory outlet of the cerebellum to the accessory inferior olive and ventromedial thalamic nucleus (VMT). These circuits are paralleled by direct projections from the FN to the inferior olive and

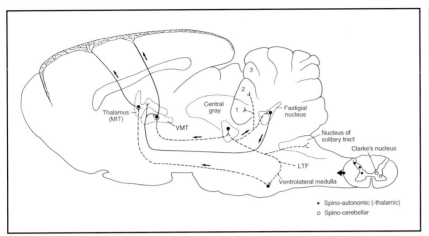

*Fig. 17. Theoretical circuits contributing to cerebellar-autonomic (behavioural) integration. Brainstem afferents to the fastigial nucleus appear to be shared by nuclei involved in central autonomic regulation (e.g., nucleus tractus solitarii and ventrolateral medulla) and diffuse thalamocortical integrative processing (e.g., midline-intralaminar complex). The fastigial nucleus also issues directs projections to loci involved in somato-autonomic reflexes (e.g. vestibular nuclei) and a major ascending projection to the thalamus (see text for details).*

VMT. The VMT in rodents shares with the midline-intralaminar complex widespread projections to the cerebral cortex (Herkenham, 1979). Electrical stimulation of FN provokes surface-negative potentials in the lower bank of the anterior ectosylvian sulcus (granular insular cortex), mediated through the suprageniculate nucleus and/or the VMT (Noda & Oka, 1985). The global cerebral blood flow response to FN stimulation might be mediated, in part, by direct and indirect cerebellar pathways to the VMT. An important component of the circulatory response, however, appears to be mediated by a system intercalated between the centromedian-parafascicular thalamic complex and the basal forebrain (Iadecola *et al.*, 1983; Mraovitch *et al.*, 1986).

The systems mediating the repertoire of responses associated with cerebellar stimulation are unknown. Figure 17 provides a theoretical schema. One possibility is that functionally distinct segments of the CNS may be integrated by shared collaterals. Connections between the vestibular complex, a major source of cerebellar afferents (Ruggiero *et al.*, 1977), and the ANS are virtually unknown. Sources of

neurons in the brainstem associated with autonomic control and which project in common to the forebrain, cerebellum and medulla oblongata are located in the dorsal pontomesencephalic tegmentum and central grey (Swanson & Hartman, 1975; Satoh & Fibiger, 1986). These include the nucleus raphe dorsalis, locus ceruleus and a continuum formed by the parabrachial and pedunculopontine cell groups surrounding the superior cerebellar peduncle. The pedunculopontine nucleus has long been recognized as a member of the extrapyramidal motor system. However, recently described projections to the cerebellum, hypothalamus and ventrolateral medulla suggest an integrative role in central autonomic control (Rye *et al.*, 1987; Yasui *et al.*, 1990). Interestingly, neurons in a similar region of the dorsal pons appear to exert intrinsic control over the cerebral circulation (Goadsby & Lance, 1988). Consistent with these observations is immunocytochemical evidence that terminals in the VFZ contain the same neurotransmitters (e.g. acetylcholine; noradrenaline) synthesized by cell bodies in this part of the brainstem. Theoretically, pontocerebellar neurons might form a unified, neurochemically heterogeneous system providing a link between the skeletomotor and autonomic nervous systems.

## Abbreviations

| | |
|---|---|
| A1 | noradrenergic-reticular cell group in the caudal ventrolateral medulla |
| A2 | noradrenergic cell group of nucleus tractus solitarii |
| A5 | noradrenergic ventrolateral pontine cell group |
| A6 | noradrenergic nucleus locus coeruleus |
| A7 | noradrenergic pontine cell group |
| ACh | acetylcholine |
| ADN | aortic depressor nerve |
| ANS | autonomic nervous system |
| C1 | adrenergic reticular cell group in rostral ventrolateral medulla |
| C2 | adrenergic cell group in the nucleus tractus solitarii |
| C3 | adrenergic reticular cell group in the rostral dorsomedial medulla |
| CNS | central nervous system |
| CVL | caudal ventrolateral reticular nucleus (= retroambigual and A1 reticular cell groups) |
| DMX | dorsal motor nucleus of the vagus |
| EPCS | endogenous pain control system |
| FN | fastigial nucleus |
| FPR | fastigial pressor response |

| | | | | | |
|---|---|---|---|---|---|
| GI | gastrointestinal tract | | ing n. of Koelliker-Fuse) | SGS | subependymal glial stratum |
| ILC | intermediolateral cell column | PNS | parasympathetic nervous | SNS | sympathetic nervous system |
| IMC | intermediomedial cell column | | system | SON | supraoptic nucleus |
| KF | nucleus of Kölliker–Fuse | PVN | paraventricular hypothalamic | STN | spinal trigeminal nucleus |
| LC | locus coeruleus | | nucleus | VFZ | vermal fastigial zone |
| LH | lateral hypothalamic nucleus | RAR | rapidly adapting pulmonary | VMT | ventromedial thalamic nucleus |
| LTF | lateral tegmental field | | receptor | VPLpc | ventral posterior lateral |
| MIT | midline-intralaminar thalamic | RVL | rostral ventrolateral reticular | | parvicellular thalamic nucleus |
| | complex | | nucleus | VPMpc | ventral posterior medial |
| NA | nucleus ambiguus | SAR | slowly adapting lung stretch | | parvicellular thalamic nucleus |
| NPS | nucleus of the tractus para- | | receptor | VRG | ventral respiratory groups |
| | solitarius | SCP | superior cerebellar peduncle | ZI | zona incerta |
| NTS | nucleus tractus solitarii | SDX | subdiaphragmatic portion of | | |
| PBC | parabrachial complex (includ- | | vagus nerve | | |

# References

Allen, G.V., Saper, C.B., Hurley, K.M. & Cechetto, D.F. (1991): Organization of visceral and limbic connections in the insular cortex of the rat. *J. Comp. Neurol.* **311**, 1–16.

Altschuler, S.M., Bao, X., Bieger, D., Hopkins, D.A. & Miselis, R.R. (1989): Viscerotopic representation of the upper alimentary tract in the rat: sensory ganglia and nuclei of the solitary and spinal trigeminal tracts. *J. Comp. Neurol.* **283**, 248–268.

Ammons, W.S. (1988): Renal and somatic input to spinal neurons antidromically activated from the ventrolateral medulla. *J. Neurophysiol.* **60**, 1967–1981.

Anders, K. & Richter, D.W. (1987): Morphology of medullary pump neurones of cats. *Pflugers Arch.* **408**, (Suppl. 1) R54.

Arango, V., Ruggiero, D.A., Callaway, J.L., Anwar, M., Mann, J.J. & Reis, D.J. (1988): Catecholaminergic neurons in the ventrolateral medulla and nucleus of the solitary tract in the human. *J. Comp. Neurol.* **273**, 224–240.

Ariëns-Kappers, C.U., Huber, C.G. & Crosby, E.C. (1936): In: *Comparative anatomy of the nervous system of vertebrates, including man.* New York: Macmillan (reprinted by Hafner, New York, 1960).

Aston-Jones, G., Ennis, M., Pieribone, V.A., Nickell, W.T. & Shipley, M.T. (1986): The brain nucleus coeruleus: restricted afferent control of a broad efferent network. *Science* **234**, 734–737.

Barman, S.M. & Gebber, G.L. (1987): Lateral tegmental field neurons of cat medulla: a source of basal activity of ventrolateral medullospinal sympathoexcitatory neurons. *J. Neurophysiol.* **57**, 1410–1424.

Batten, R.R., Jayaraman, A., Ruggiero, D.A. & Carpenter, M.B. (1977): Fastigial efferent projections in the monkey: an autoradiographic study. *J. Comp. Neurol.* **174**, 281–306.

Beckman, M.E. & Whitehead, M.C. (1991): Intramedullary connections of the rostral nucleus of the solitary tract in the hamster. *Brain Res.* **557**, 265–279.

Beckstead, R.M., Morse, J.R. & Norgen, R. (1980): The nucleus of the solitary tract in the monkey: projections to the thalamus and brain stem nuclei. *J. Comp. Neurol.* **190**, 259–282.

Benarroch, E.E., Granata, A.R., Ruggiero, D.A., Park, D.H. & Reis, D.J. (1986): Neurons of the C1 area mediate cardiovascular responses initiated from the ventral medullary surface. *Am. J. Physiol.* **250**, R932–R945.

Berger, A.J. (1979): Distribution of carotid sinus nerve afferent fibers to solitary tract nuclei of the cat using transganglionic transport of horseradish peroxidase. *Neurosci. Lett.* **14**, 153–158.

Berthoud, H.R., Niijima, A., Sauter, J. & Jeanrenaud, B. (1983): Evidence for a role of the gastric, coeliac and hepatic branches in vagally stimulated insulin secretion in the rat. *J. Auton. Nerv. Syst.* **7**, 97–110.

Bieger, D. & Hopkins, D.A. (1987): Viscerotopic representation of the upper alimentary tract in the medulla oblongata of the rat: the nucleus ambiguus, *J. Comp. Neurol.* **262**, 546–562.

Blanchard, D.C. & Blanchard, R.J. (1972): Innate and conditioned reactions to threat in rats with amygdaloid lesions. *J. Comp. Physiol. Psychol.* **81**, 281–290.

Blessing, W.W., Jaeger, C.B., Ruggiero, D.A. & Reis, D.J. (1982): Hypothalamic projections of medullary catecholamine neurons in the rabbit: a combined catecholamine fluorescence and HRP transport study. *Brain Res. Bull.* **9,** 279–286.

Block, C.H. & Estes, M.L. (1990) The cytoarchitectural organization of human parabrachial nuclear complex. *Brain Res. Bull.* **24,** 617–626.

Cechetto, D.F. (1987): Central representation of visceral function. *Fed. Proc.* **46,** 17–23.

Cechetto, D.F. & Calaresu, F.R. (1983): Response of single units in the amygdala to stimulation of buffer nerves in the cat. *Am. J. Physiol.* **244,** R646–R651.

Cechetto, D.F. & Chen, S.J. (1990): Subcortical sites mediating sympathetic responses from the insular cortex in the rat. *Am. J. Physiol.* **258,** R245–R255.

Cechetto, D.F. & Saper, C.B. (1987): Evidence for a viscerotopic sensory representation in the cortex in the rat. *J. Comp. Neurol.* **262,** 27–45.

Cechetto, D.F., Standaert, D.G. & Saper, C.B. (1985): Spinal and trigeminal dorsal horn projections to the parabrachial nucleus in the rat. *J. Comp. Neurol.* **240,** 153–160.

Cechetto, D.F., Wilson, J.X., Smith , K.E., Wolski, D., Silver, M.D. & Hachinski, V.C. (1989): Autonomic and myocardial changes in middle cerebral artery occlusion: stroke models in the rat. *Brain Res.* **502,** 296–305.

Cervero, F. (1986): Dorsal horn neurons and their sensory inputs. In: *Spinal afferent processing,* ed. T.L. Yaksh, pp. 197–216. New York, London: Plenum Press.

Chida, K., Iadecola, C., Underwood, M.D. & Reis, D.J. (1986): A novel vasodepressor response elicited from the rat cerebellar fastigial nucleus: the fastigial depressor response. *Brain Res.* **370,** 378–382.

Ciriello, J. (1983): Brainstem projections of aortic baroreceptor afferent fibers in the rat. *Neurosci. Lett.* **36,** 37–42.

Ciriello, J. & Caverson, M.M. (1986): Bidirectional cardiovascular connections between ventrolateral medulla and nucleus of the solitary tract. *Brain Res.* **367,** 273–281.

Ciriello, J. & Calaresu, F.R. (1980): Monosynaptic pathway from cardiovascular neurons in the nucleus tractus solitarii to the paraventricular nucleus in the cat. *Brain Res.* **193,** 529–533.

Claps, A. & Torrealba, F. (1988): The carotid body connections: a WGA-HRP study in the cat. *Brain Res.* **455,** 123–133.

Clark, F.M. & Proudfit, H.K. (1991): The projection of noradrenergic neurons in the A7 catecholamine cell group to the spinal cord in the rat demonstrated by anterograde tracing combined with immunocytochemistry. *Brain Res.* **547,** 279–288.

Clerc, N. & Condamin, M. (1987): Selective labeling of vagal sensory nerve fibers in the lower esophageal sphincter with anterogradely transported WGA-HRP. *Brain Res.* **424,** 216–224.

Cohen, M.I. & Feldman, J.L. (1984): Discharge properties of dorsal medullary inspiratory neurons: relation to pulmonary aferent and phrenic efferent discharge. *J. Neurophysiol.* **51,** 753–776.

Cunningham, E.T., Jr & Sawchenko, P.E. (1990): Central neural control of esophageal motility: a review. *Dysphagia* **5,** 35–51.

Dampney, R.A.L., Goodchild, A.K., Robertson, L.G. & Montgomery, W. (1982): Role of ventrolateral medulla in vasomotor regulation: a correlative anatomical and physiological study. *Brain Res.* **249,** 223–235.

De Groat, W.C. & Steers, W.D. (1990): Autonomic regulation of the urinary bladder and sexual organs. In: *Central regulation of autonomic functions,* eds. A.D. Loewy & K.M. Spyer, pp. 310–333. New York, Oxford: Oxford University Press.

Doba, N. & Reis, D.J. (1972): Changes in regional blood flow and cardiodynamics evoked by electrical stimulation of the fastigial nucleus in the cat and their similarity to orthostatic reflexes. *J. Physiol. (Lond.)* **227,** 729–747.

Donoghue, S., Felder, R.B., Jordan, D. & Spyer, K.M. (1984): The central projections of carotid baroreceptors and chemoreceptors in the cat: a neurophysiological study. *J. Physiol. (Lond.)* **347,** 397–409.

Dormer, K.J., Person, R.J., Andrezik, J.A., Foreman, R.D. & Braggio, J.P. (1989): Ventrolateral medullary lesions and fastigial cardiovascular response in beagles. *Am. J. Physiol.* **256,** H1200–H1208.

Eccles, J.C., Ito, M. & Szentagothai, J. (1967): *The cerebellum as a neuronal machine.* Berlin, Heidelberg, New York: Springer Verlag.

Edvinsson, L. (1975): Neurogenic mechanisms in the cerebrovascular bed. Autonomic nerves, amine receptors and their effect on cerebral blood flow. *Acta Physiol. Scand. (Suppl.)* **427,** 1–36.

Elisevich, K. & Ciriello, J. (1988): Cardiovascular afferent and fastigal nucleus inputs to paramedian reticulospinal neurons. *Brain Res.* **452,** 141–148.

Ellenberger, H.H. & Feldman, J.L. (1990a): Subnuclear organization of the lateral tegmental field of the rat. I: Nucleus ambiguus and ventral respiratory group. *J. Comp. Neurol.* **294,** 202–211.

Ellenberger, H.H. & Feldman, J.L. (1990b): Brainstem connections of the rostral ventral respiratory group of the rat. *Brain Res.* **513,** 35–42.

Ellenberger, H.H., Feldman, J.L. & Zhan, W.-Z. (1990): Subnuclear organization of the lateral tegmental field of the rat. II: Catecholamine neurons and ventral respiratory group. *J. Comp. Neurol.* **294,** 212–222.

Ennis, M. & Aston-Jones, G. (1989): GABA-mediated inhibition of locus coeruleus from the dorsomedial rostral medulla. *J. Neurosci.* **9,** 2973–2981.

Euler, C. von, Hayward, J.N., Marttila, I. & Wyman, R.J. (1973): Respiratory neurones of the ventrolateral nucleus of the solitary tract of cat, vagal input, spinal connections and morphological identification. *Brain Res.* **61,** 1–22.

Eyzaguirre, C., Fitzgerald, R.S., Lahiri, S. & Zapata, P. (1983): Arterial chemoreceptors. In: *Handbook of physiology, the cardiovascular system: peripheral and circulation and organ blood flow,* Vol. 3., eds. J.T. Shepherd & F.M. Abboud, pp. 557–621. Baltimore: Williams and Wilkins.

Feldman, P.D. & Felder, R.B. (1989):$\alpha_2$-Adrenergic modulation of synaptic excitability in the rat nucleus tractus solitarius. *Brain Res.* **480,** 190–197.

Fernandez de Molina, A. & Hunsperger, R.W. (1962): Organization of the subcortical system governing defence and flight reactions in the cat. *J. Physiol. (Lond.)* **160,** 200–213.

Ferssiwi, A., Cardo, B. & Velley, L. (1987a): Gustatory preference-aversion thresholds are increased by ibotenic acid lesion of the lateral hypothalamus in the rat. *Brain Res.* **437,** 142–150.

Ferssiwi, A., Cardo, B. & Velley, L. (1987b): Electrical self-stimulation in the parabrachial area is depressed after ibotenic acid lesion of the lateral hypothalamus. *Behav. Brain Res.* **25,** 109–116.

Fulwiler, C.E. & Saper, C.B. (1984): Subnuclear organization of the efferent connections of the parabrachial nucleus in the rat. *Brain Res. Rev.* **7,** 229–259.

Goadsby, P.J. & Lance, J.W. (1988): Brain stem effects on intra– and extracerebral circulations. Relation to migraine and cluster headache. In: *Basic mechanisms of headache,* eds. J. Olesen & L. Edvinsson, pp. 413–427. Amsterdam: Elsevier.

Granata, A.R. & Kitai, S.T. (1989): Intracellular study of nucleus parabrachialis and nucleus tractus solitarii interconnections. *Brain Res.* **492,** 281–292.

Gray, T.S. & Magnuson, D.J. (1987): Neuropeptide neuronal efferents from the bed nucleus of the stria terminalis and central amygdaloid nucleus to the dorsal vagal complex in the rat. *J. Comp. Neurol.* **262,** 365–374.

Hamilton, R.B. & Norgren, R. (1984): Central projections of gustatory nerves in the rat. *J. Comp. Neurol.* **222,** 560–577.

Hellner, K. & von Baumgarten, R. (1961): Über ein Endigungsgegiet afferenter, kardiovascularer Fasern des Nervus Vagus im Rautenhirn der Katze. *Pflugers Arch. ges. Physiol.* **273,** 223–234.

Henry, R.T., Connor, J.D. & Balaban, C.D. (1989): Nodulus-uvula depressor response: central GABA-mediated inhibition of –adrenergic outflow. *Am. J. Physiol.* **256,** H1601–H1608.

Herkenham, M. (1979): The afferent and efferent connections of the ventromedial thalamic nucleus in the rat. *J. Comp. Neurol.* **183,** 487–518.

Hilton, S.M., Marshall, J.M. & Timms, R.J. (1983): Ventral medullary relay neurones in the pathway from the defence areas of the cat and their effect on blood pressure. *J. Physiol. (Lond.).* **345,** 149–166.

Hilton, S.M. & Zbrozyna, A.W. (1963): Amygdaloid region for defence reactions and its efferent pathway to the brain stem. *J. Physiol. (Lond.)* **165,** 160–173.

Hökfelt, T., Johanson, O. & Goldstein, M. (1985): Central catecholamine neurons as revealed by immunohistochemistry with special reference to adrenaline neurons. In: *Handbooks of chemical neuroanatomy,* eds. A. Bjorklund, T. Hökfelt & K.J. Kuhar, pp. 157–276. Amsterdam: Elsevier/North Holland.

Hökfelt, T., Lundberg, J.M., Tatemoto, K., Mutt, V., Terenius, L., Polak, J., Bloom, S., Sasek, C., Elde, R. & Goldstein, M. (1983): Neuropeptide Y (NPY) and FMRFamide neuropeptide-like immunoreactivities in catecholamine neurons of the rat medulla oblongata. *Acta Physiol. Scand.* **117,** 315–318.

Hollenberg, M., Carriere, S. & Barger, A.C. (1965): Biphasic action of acetylcholine on ventricular myocardium. *Circ. Res.* **16,** 527–536.

Holstege, G. & Kuypers, H.G.J.M. (1977): The organizatio of the bulbar fiber connections to trigeminal, facial and hypoglossal motor nuclei. I. An anterograde degeneration study in the cat. *Brain* **100,** 239–264.

Holstege, G., Kuypers, H.G.J.M. & Dekker, J.J. (1977): The organization of the bulbar fiber connections to the trigeminal, facial hypoglossal motor nuclei. II. An autoradiographic tracing study in cat. *Brain* **100,** 265–286.

Hosoya, Y. (1980): The distribution of spinal projection neurons in the hypothalamus of the rat, studied with the HRP method. *Exp. Brain Res.* **40,** 79–87.

Housley, G.D., Martin-Body, R.L., Dawson, N.J. & Sinclair, J.D. (1987): Brain stem projections of the glossopharyngeal nerve and its carotid sinus branch in the rat. *Neuroscience* **22,** 237–250.

Iadecola, C., Mraovitch, S., Meeley, M.P. & Reis, D.J. (1983): Lesions of the basal forebrain in rat selectively impair the cortical vasodilatation elicited from cerebellar fastigial nucleus. *Brain Res.* **279,** 41–52.

Igic, R., Stern, P. & Basagio, E. (1970): Changes in emotional behaviour after application of cholinesterase inhibition in the septal and amygdala region. *Neuropharmacology* **9,** 73–75.

Jones, B.E. (1989): Basic mechanisms of sleep-wake states. In: *Principles and practice of sleep medicine*, eds. M.H. Kryger, T. Roth & W.C. Dement, pp. 121–138. Philadelphia: W.B. Saunders Company.

Jones, B.E. & Friedman, L. (1983): Atlas of catecholamine perikarya, varicosities and pathways in the brainstem of the cat. *J. Comp. Neurol.* **215,** 382–396.

Jones, B.E. & Yang, T.Z. (1985): The efferent projections from the reticular formation and the locus coeruleus studied by anterograde and retrograde axonal transport in the rat. *J. Comp. Neurol.* **242,** 56–92.

Kalia, M. & Mesulam, M.M. (1980): Brain stem projections of sensory and motor components of the vagus complex in the cat. II. Laryngeal, tracheobronchial, pulmonary, cardiac and gastrointestinal branches. *J. Comp. Neurol.* **193,** 467–508.

Kalia, M. & Richter, D. (1985): Morphology of physiologically identified slowly adapting lung stretch receptor afferents stained with intra-axonal horseradish peroxidase in the nucleus of the tractus solitarius of the cat. I. A light microscopic analysis. *J. Comp. Neurol.* **241,** 503–520.

Kalia, M. & Richter, D. (1988): Rapidly adapting pulmonary receptor afferents. I. Arborization in the nucleus of the tractus solitarius. *J. Comp. Neurol.* **274,** 560–573.

Kaneko, T., Itoh, K., Shigemoto, R. & Mizuno, N. (1989): Glutaminase-like immunoreactivity in the lower brainstem and cerebellum of the adult rat. *Neuroscience* **32,** 79–98.

Kapp, B.S., Pascoe, J.P. & Bixler, M.A. (1984): The amygdala: a neuro-anatomical systems approach to its contributions to averse conditioning. In: *The neurophysiology of memory*, eds. N. Buttlers & L.R. Squares, pp. 473–488. New York: Guilford.

King, G.W. (1980): Topology of ascending brainstem projections to nucleus parabrachialis in the cat. *J. Comp. Neurol.* **191,** 615–638.

Koizumi, K. & Brooks, C. McC. (1972): The integration of autonomic system reactions: a discussion of autonomic reflexes, their control and their association with somatic reactions. *Rev. Physiol. Biochem. Pharmacol.* **67,** 1–68.

Kooy, D. van der, Koda, L.Y., McGinty, J.F., Gerfen, C.R. & Bloom, F.E. (1984): The organization of projections from the cortex, amygdala, and hypothalamus to the nucleus of the solitary tract in rat. *J. Comp. Neurol.* **224,** 1–24.

Krettek, J.E. & Price, J.L. (1978): Amygdaloid projections to subcortical structures within the basal forebrain and brainstem in the rat and cat. *J. Comp. Neurol.* **178,** 225–253.

Krushel, A.A. & Kooy, D. van der (1988): Visceral cortex: integration of the mucosal senses with limbic information in the rat agranular insular cortex. *J. Comp. Neurol.* **270,** 39–54.

Kumada, M., Dampney, R.A.L. & Reis, D.J. (1977): The trigeminal depressor response: a novel vasodepressor response originating from the trigeminal system. *Brain Res.* **119,** 305–326.

Langley, J.N. (1921): *The autonomic nervous system*, vol. 1. Cambridge: W. Heffer & Sons.

LeDoux, J.E. (1987): Emotion. In: *Handbook of physiology – the nervous system V, higher functions of the brain*, Pt. 1., eds. V.B. Mountcastle (Section Ed.), F. Plum (Volume Ed.) & S.R. Geiger (Executive Ed.), pp. 419–459. Bethesda: American Physiological Society.

LeDoux, J.E., Del Bo, A., Tucker, L.W., Harshfield, G., Talman, W.T. & Reis, D.J. (1982): Hierarchic organization of blood pressure responses during the expression of natural behaviours in rat: mediation by sympathetic nerves. *Exp. Neurol.* **78,** 121–133.

LeDoux, J.E., Farb, C. & Ruggiero, D.A. (1990): Topographic organization of neurons in the acoustic thalamus that project to the amygdala. *J. Neurosci.* **10,** 1043–1054.

Leslie, R.A., Gwyn, D.G. & Hopkins, D.A. (1982): The central distribution of the cervical vagus nerve and gastric afferent and efferent projections in the rat. *Brain Res. Bull.* **8**, 37–43.

Levy, M.N. (1990): Autonomic interactions in cardiac control. In: *Electrocardiography. Past and future*, eds. P. Coumel & O.B. Garfein, pp. 209–221. New York: The New York Academy of Sciences.

Lioy, F. & Trzebski, A. (1984): Pressor, effect of $CO_2$ in the rat: different thresholds of the central cardiovascular and respiratory responses to $CO_2$, *J. Auton. Nerv. Sys.* **10**, 43–54.

Lipski, J., McAllen, R.M. & Spyer, K.M. (1977): The carotid chemoreceptor input to the respiratory neurones of the nucleus of tractus solitarius. *J. Physiol.* **269**, 797–810.

Lipski, J. & Trzebski, A. (1975): Bulbospinal neurones activated by baroreceptor afferents and their possible role in inhibition of preganglionic sympathetic neurones. *Pflugers Arch.* **356**, 181–192.

Lipski, J., Trzebski, A., Chodobska, J. & Kruk, P. (1984): Effects of carotid chemoreceptor excitation on medullary expiratory neurons in cats. *Respir. Physiol.* **57**, 279–291.

Loeschcke, H.H., Lattre, J. Schläfke, M.E. & Trouth, C.O. (1970): Effects on respiration and circulation of electrically stimulating the ventral surface of the medulla oblongata. *Respir. Physiol.* **10**, 184–197.

Loewy, A.D. & Burton, H. (1978): Nuclei of the solitary tract: efferent projections to the lower brain stem and spinal cord of the cat. *J. Comp. Neurol.* **181**, 421–450.

Magoun, H.W., Atlas, D., Ingersoll, E.H. & Ranson, S.W. (1937): Associated facial, vocal and respiratory components of emotional expression: An experimental study. *J. Neurol. Psychopath.* **17**, 241–255.

Martner, J. (1975): Cerebellar influences on autonomic mechanisms. An experimental study in the cat with special reference to the fastigial nucleus. *Acta Physiol. Scand.* (Supp. 425) 1–42.

McBride, R.L. & Sutin, J. (1976): Projections of the locus coeruleus and adjacent pontine tegmentum in the cat. *J. Comp. Neurol.* **165**, 265–284.

McDonald, D.M. (1981):Peripheral chemoreceptors: structure–function relationships of the carotid body. In: *Lung biology in health and disease: the regulation of breathing*, Vol. 17., ed. T.F. Hornbein, pp. 105–319. New York: Dekker.

McDonald, D.M. (1983): Morphology of the rat carotid sinus nerve. I. Course, connections, dimensions, and ultrastructure. *J. Neurocytol.* **12**, 345–372.

McDonald, D.M. & Mitchell, R.A. (1975): The innervation of glomus cells, ganglion cells and blood vessels in the rat carotid body: a quantitative ultrastructural analysis. *J. Neurocytol.* **4**, 177–230.

Meeley, M.P., Ruggiero, D.A., Ishitsuka, T. & Reis, D.J. (1985): Intrinsic GABA neurons in the nucleus of the tractus solitarius and the rostral ventrolateral medulla in the rat: an immunocytochemical and biochemical study. *Neurosci. Lett* **58**, 83–89.

Mehler, W.R. (1983): Observations on the connectivity of the parvicellular reticular formation with respect to a vomiting center. *Brain Behav. Evol.* **23**, 63–80.

Mifflin, S.W. & Felder, R.B. (1990): Synaptic mechanisms regulating cardiovascular afferent inputs to solitary tract nucleus. *Am. J. Physiol.* **259**, H653–H661.

Miller, A.D. & Kucharczyk, J. (1991): Mechanisms of nausea and emesis: Introduction and retrospective. In: *Nausea and vomiting: recent research and clinical advances.* eds. J. Kucharczyk, D.J. Stewart & A.D. Miller, pp. 1–12. Boca Raton: CRC Press.

Millhorn, D.E., Seroogy, K., Hokfelt, T., Schmued, L.C., Terenius, L., Buchan, A. & Brown, J.C. (1987): Neurons of the ventral medulla oblongata that contain both somatostatin and enkephalin immunoreactivities project to nucleus tractus solitarii and spinal cord. *Brain Res.* **424**, 99–108.

Miselis, R.R., Shapiro, R.E. & Hand, P.J. (1979): Subfornical organ efferents to neural systems for control of body water. *Science* **205**, 1022–1025.

Miselis, R.R., Weiss, M.L. & Shapiro, R.E. (1986): Modulation of the visceral neuraxis. In: *Circumventricular organs and body fluids*, Vol. III. ed. P.M. Gross, pp. 143–162. Boca Raton: CRC Press.

Mitchell, J.H., Kaufman, M.P. & Iwamoto, G.A. (1983): The exercise pressor reflex: its cardiovascular effects, afferent mechanisms and central pathways. *Annu. Rev. Physiol.* **45**, 229–242.

Miura, M. & Reis, D.J. (1970): A blood pressure response from fastigial nucleus and its relay pathway in brainstem. *Am. J. Physiol.* **219**, 1330–1336.

Miura, M. & Takayama, K. (1988): The site of the origin of the so-called fastigial pressor response. *Brain Res.* **473**, 352–358.

Miyazawa, T. & Ishikawa, T. (1985): Separation of the medullo-spinal descending pathway for somatic and autonomic outflow in the cat. *Brain Res.* **334**, 297–302.

Mondlock, J.M. & Davis, M. (1985): The role of various amygdala projection areas (bed nucleus of stria terminalis, rostral lateral hypothalamus, substantia nigra) in fear-enhanced acoustic startle. *Soc. Neurosci. Abstr.* **11**, 331.

Mraovitch, S., Calando, Y., Pinard, E., Pearce, W.J. & Seylaz, J. (1992): Differential cerebrovascular and metabolic responses in specific neural systems elicited from the centromedian-parafascicular complex. *Neuroscience* **49**, 451–466.

Mraovitch, S., Iadecola, C., Ruggiero, D.A. & Reis, D.J. (1985): Widespread reduction in cerebral blood flow and metabolism elicited by electrical stimulation of the parabrachial nucleus in rat. *Brain Res.* **341**, 283–296.

Mraovitch, S., Kumada, M. & Reis, D.J. (1982): Role of the nucleus parabrachialis in cardiovascular regulation in cat. *Brain Res.* **232**, 57–75.

Mraovitch, S., Lasbennes, F., Calando, Y. & Seylaz, J. (1986): Cerebrovascular changes elicited by electrical stimulation of the centromedian-parafascicular complex in rat. *Brain Res.* **380**, 42–53.

Mtui, E.P., Anwar, M., Gomez, R., Reis, D.A. & Ruggiero, D.A. (1993): Projections from the nucleus tractus solitarii to the spinal cord. *J. Comp. Neurol.*, **337**, 231–252.

Mtui, E.P., Anwar, M., Reis, D.J. & Ruggiero, D.A. (1995): Medullary visceral reflex circuits: local afferents to nucleus tractus solitarie synthesize catecholamines and project to thoracic spinal cord. *J. Comp. Neurol.* **351**, 5–26.

Nakai, M., Iadecola, C., Ruggiero, D.A., Tucker, L. & Reis, D.J. (1983): Electrical stimulation of the cerebellar fastigial nucleus increases cerebral cortical blood flow without changes in local metabolism: evidence for an intrinsic system in brain for primary vasodilation. *Brain Res.* **260**, 35–49.

Neil, E., Redwood, C.R.M. & Schweitzer, A. (1949): Effects of electrical stimulation of the aortic nerve on blood pressure and respiration in cats and rabbits under choralose and nembutal anesthesia. *J. Physiol. (Lond.)* **109**, 392–401.

Noda, T. & Oka, H. (1985): Fastigial inputs to the insular cortex in the cat: field potential analysis. *Neurosci. Lett.* **53**, 331–336.

Nomura, S., Mizuno, N., Itoh, K., Matsuda, K., Sugimoto, T. & Nakamura, Y. (1979): Localization of parabrachial nucleus neurons projecting to the thalamus or the amygdala in the cat using horseradish peroxidase. *Exp. Neurol.* **64**, 375–385.

Nonidez, J.F. (1937): Distribution of the aortic nerve fibers and the epithelioid bodies (supracardial 'paraganglia') in the dog. *Anat. Rec.* **69**, 299–317.

Norgen, R. & Smith, G.P. (1988): Central distribution of subdiaphragmatic vagal branches in the rat. *J. Comp. Neurol.* **273**, 207–223.

Nosaka, S., Yamamoto, T. & Yasunaga, K. (1979): Localization of vagal cardioinhibitory preganglionic neurons within rat brain stem. *J. Comp. Neurol.* **186**, 79–92.

Oppenheimer, S.M. & Cechetto, D.F. (1990): Cardiac chronotropic organization of the rat insular cortex. *Brain Res.* **533**, 66–72.

Oppenheimer, S.M., Cechetto, D.F. & Hachinski, V.C. (1990): Insular stimulation, cardia arrhythmogenesis and stroke. *Stroke* **21**, 174.

Otake, K. & Ruggiero, D.A. (1995): Monoamines and nitric oxide are employed by afferents engaged in midline thalamic regulation. *J. Neurosci.* in press.

Person, R.J. (1989): Somatic and vagal afferent convergence on solitary tract neurons in cat: electrophysiological characteristics. *Neuroscience* **30**, 283–295.

Petras, J.M. & Faden, A.I. (1977): The origin of sympathetic preganglionic neurons in the dog. *Brain Res.* **144**, 353–357.

Pieribone, V.A., Aston-Jones, G. & Bohn, M.C. (1988): Adrenergic and non-adrenergic neurons in the C1 and C3 areas project to locus coeruleus: a fluorescent double labeling study. *Neurosci. Lett.* **85**, 297–303.

Potter, E.K. (1985): Prolonged non-adrenergic inhibition of cardiac vagal action following sympathetic stimulation: neuromodulation by neuropeptide Y? *Neurosci. Lett.* **54**, 117–121.

Reis, D.J., Doba, N. & Nathan, M.A. (1973): Predatory attack, grooming and consummatory behaviours evoked by electrical stimulation of cat cerebellar nuclei. *Science* **23**, 845–847.

Reis, D.J. & Ruggiero, D.A. (1990): The bed nucleus of the transtegmental tract: a major autonomic integration center of the medulla oblongata. *Soc. Neurosci. Abstr.* **16**, 11.

Ricardo, J.A. & Koh, E.T. (1978): Anatomical evidence of direct projections from the nucleus of the solitary tract to the hypothalamus, amygdala and other forebrain structures in the rat. *Brain Res.* **153**, 1–26.

Ritcher, D.W. (1982): Generation and maintenance of the respiratory rhythm. *J. Exp. Biol.* **100**, 93–107.

Rinaman, L., Card, J.P., Schwaber, J.S. & Miselis, R.R. (1989): Ultrastructural demonstration of a gastric monosynaptic vagal circuit in the nucleus of the solitary tract in rat. *J. Neurosci.* **9**, 1985–1986.

Ross, C.A., Ruggiero, D.A., Joh, T.H., Park, D.H. & Reis, D.J. (1984a): Rostral ventrolateral medulla: selective projections to the thoracic autonomic cell column from the region containing C1 adrenaline neurons. *J. Comp. Neurol.* **228**, 168–184.

Ross, C.A., Ruggiero, D.A., Park, D.H., Joh, T.H., Sved, A.D., Fernandez-Pardal, J., Saavedra, J.M. & Reis, D.J. (1984b): Tonic vasomotor control by the rostral ventrolateral medulla: effect of electrical or chemical stimulation of the area containing C1 adrenaline neurons on arterial pressure, heart rate and plasma catecholamines and vasopressin. *J. Neurosci.* **4**, 479–494.

Ross, C.A., Ruggiero, D.A. & Reis, D.J. (1981a): Afferent projections to cardiovascular portions of the nucleus of the tractus solitarius in the rat. *Brain Res.* **223**, 402–408.

Ross, C.A., Ruggiero, D.A. & Reis, D.J. (1981b): Projections to the spinal cord from neurons close to the ventral surface of the hindbrain of rat. *Neurosci. Lett.* **21**, 143–148.

Ross, C.A., Ruggiero, D.A. & Reis, D.J. (1985): Project ions from the nucleus tractus solitarii to the rostral ventrolateral medulla. *J. Comp. Neurol.* **242**, 511–534.

Rubin, E. & Purves, D. (1980): Segmental organization of sympathetic preganglionic neurons in the mammalian spinal cord. *J. Comp. Neurol.* **192**, 163–174.

Ruggiero, D.A., Baker, H., Joh, T.H. & Reis, D.J. (1984): Distribution of catecholamine neurons in the hypothalamus and preoptic region of mouse. *J. Comp. Neurol.* **223**, 556–582.

Ruggiero, D.A., Batton, R.R., Jayaraman, A. & Carpenter, M.B. (1977): Brainstem afferents to the fastigial nucleus in the cat demonstrated by transport of horseradish peroxidase. *J. Comp. Neurol.* **172**, 189–210.

Ruggiero, D.A., Cravo, S., Arango, V. & Reis, D.J. (1989): General control of the circulation by the rostral ventrolateral reticular nucleus: anatomical substrates. In: *Prog. Brain Res., 81* ed. J. Ciriello, C. Polosa & M. Caverson, pp. 49–79. Amsterdam: Elsevier.

Ruggiero, D.A., Giuliano, R., Anwar, M., Stornetta, R. & Reis, D.J. (1990): Anatomical substrates of cholinergic-autonomic regulation in the rat. *J. Comp. Neurol.* **292**, 1–53.

Ruggiero, D.A., Gomez, R.E., Cravo, S.L., Mtui, E., Anwar, M. & Reis, D.J. (1991): The rostral ventrolateral medulla: anatomical substrates of cardiopulmonary integration. In: *Cardiorespiratory and motor coordination*, eds. H.-P. Koepchen & T. Huopaniemi, pp. 89–102. Berlin, Heidelberg, New York: Springer-Verlag.

Ruggiero, D.A., Meeley, M.P., Anwar, M. & Reis, D.J. (1985a): Newly identified GABAergic neurons in regions of the ventrolateral medulla that regulate blood pressure. *Brain Res.* **339**, 171–177.

Ruggiero, D.A. & Reis, D.J. (1988): Neurons containing phenylethanolamine *N*-methyltransferase: a component of the baroreceptor reflex? In: *Epinephrine in the central nervous system*, eds. J.M. Stolk, D.C. U'Prichard & K. Fuxe, pp. 291–307. New York: Oxford University Press.

Ruggiero, D.A., Mraovitch, S., Granata, A.R., Anwar, M. & Reis, D.J. (1987): A role of insular cortex in cardiovascular function. *J. Comp. Neurol.* **257**, 189–207.

Ruggiero, D.A., Pickel, V.M., Milner, T.A., Anwar, M., Otake, K., Mtui, E.P. & Park, D. (1995): Viscerosensory processing in nucleus tractus solitarii: Structural and neurochemical substrates. In: *Nucleus of the solitary tract*, ed. R.A. Barraco, pp. 3–34. Boca Raton: CRC Press.

Ruggiero, D.A., Ross, C.A., Anwar, M., Park, D.H., Joh, T.H. & Reis, D.J. (1985b): Distribution of neurons containing phenylethanolamine *N*-methyltransferase in medulla and hypothalamus of rat. *J. Comp. Neurol.* **239**, 127–154.

Ruggiero, D.A., Ross, C.A., Kumada, M. & Reis, D.J. (1982): Reevaluation of projections from the mesencephalic trigeminal nucleus to the medulla and spinal cord: new projections. A combined retrograde and anterograde horseradish peroxidase study. *J. Comp. Neurol.* **206**, 278–292.

Rye, D.B., Saper, C.B., Lee, H.J. & Wainer, B.H. (1987): Pedunculopontine tegmental nucleus of the rat: Cytoarchiteture, cytochemistry and some extrapyramidal connections of the mesopontine tegmentum. *J. Comp. Neurol.* **259**, 483–528.

Saper, C.B. (1982): Reciprocal parabrachial-cortical connections in the rat. *Brain Res.* **242**, 33–40.

Saper, C.B. & DeMarchena, O. (1986): Somatosympathetic reflex unilateral sweating and pupillary dilatation in a paraplegic man.

*Ann. Neurol.* **19,** 389–390.

Saper, C.B. & Loewy, A.D. (1980): Efferent connections of the parabrachial nucleus in the rat. *Brain Res.* **197,** 291–312.

Sarnat, H.B. & Netsky, M.G. (1974): *Evolution of the nervous system*, 318 p. New York: Oxford University Press.

Sato, A. & Schmidt, R.F. (1973): Somatosympathetic reflexes: afferent fibers, central pathways and discharge characteristics. *Physiol. Rev.* **53,** 916–947.

Satoh, K. & Fibiger, K. (1986): Cholinergic neurons of the laterodorsal tegmental nucleus: efferent and afferent connections. *J. Comp. Neurol.* **253,** 277–302.

Sawchenko, P.E. & Swanson, L.W. (1982): The organization of noradrenergic pathways from the brainstem to the paraventricular and supraoptic nuclei in the rat. *Brain Res.* **4,** 275–325.

Scheibel, M.E. & Scheibel, A.B. (1958): Structural substrates for integrative patterns in the brain stem reticular core. In: *Reticular formation of the brain*, ed. H.H. Jasper, pp. 31–55. Boston: Little Brown.

Schramm, L.P., Adair, H.R., Stribling, J.M. & Gray, L.P. (1975): Preganglionic innervation of the adrenal gland of the rat: a study using horseradish peroxidase. *Exp. Neurol.* **49,** 540–553.

Schwaber, J.S. & Cohen, D.H. (1978): Field potential and single unit analyses of avian dorsal motor nucleus of the vagus and criteria for identifying vagal cardiac cells of origin. *Brain Res.* **147,** 79–90.

Shapiro, R.E. & Miselis, R.R. (1985): The central organization of the vagus nerve innervating the stomach in the rat. *J. Comp. Neurol.* **238,** 473–488.

Smith, O.A., Astley, C.A., DeVito, J.L., Stein, J.M. & Walsh, R.E. (1980): Functional analysis of hypothalamic control of the cardiovascular responses accompanying emotional behaviour. *Fed. Proc.* **29,** 2487–2494.

Smith, G.P., Jerome, C. & Norgen, R. (1985): Afferent axons in the abdominal vagus mediate the satiety effect of cholecystokinin in the rat. *Am. J. Physiol.* **249,** R638–R641.

Snyder, D.W., Nathan, M.A. & Reis, D.J. (1978): Chronic lability of arterial pressure produced by selective denervation of the catecholamine innervation of the nucleus tractus solitarii in rat. *Circ. Res.* **43,** 662–671.

Stornetta, R., Morrison, S.L., Ruggiero, D.A. & Reis, D.J. (1989): Neurons in the rostral ventrolateral medulla mediate the somatic pressor reflex. *Am. J. Physiol.* **256,** R448–R462.

Strack, A.M., Sawyer, W.B., Marubio, L.M. & Loewy, A.D. (1988): Spinal origin of sympathetic preganglionic neurons in the rat. *Brain Res.* **455,** 187–191.

Sullivan, C.E. (1980): Breathing in sleep. In: *Physiology in sleep*, eds. J. Orem & C.D. Barnes, pp. 213–272. New York: Academic Press.

Sun, M.-K. (1992): Medullospinal vasomotor neurones mediate hypotension from stimulation of prefrontal cortex. *J. Auton. Nerv. Syst.* **38,** 209–218.

Sun, M.-K. & Guyenet, P.G. (1986): Hypothalamic glutamatergic input to medullary sympathexcitatory neurons in rats. *Am. J. Physiol.* **251,** R798–R810.

Swanson, L.W. & Hartman, B.K. (1975): The central adrenergic system. An immunofluorescence study of the location of cell bodies and their efferent connections in the rat utilizing dopamine-β-hydroxylase as a marker. *J. Comp. Neurol.* **163,** 467–506.

Swanson, L.W. & Mogenson, G.J. (1981): Neural mechanisms for the functional coupling of autonomic, endocrine and somatomotor responses and adaptive behaviour. *Brain Res. Rev.* **3,** 1–34.

Swanson, L.W., Sawchenko, P.E., Wiegand, S.J. & Price, J.L. (1980): Separate neurons in the paraventricular nucleus project to the median eminence and to the medulla or spinal cord. *Brain Res.* **197,** 207–212.

Sweazy, R.D. & Bradley, R.M. (1986): Central connections of the lingual-tonsillar branch of the glossopharyngeal nerve and the superior laryngeal nerve in lamb. *J. Comp. Neurol.* **245,** 471–482.

Ter Horst, G.J., Toes, G.J. & Van Willigen, J.D. (1991): Locus coeruleus projections to the dorsal motor vagus nucleus in the rat. *Neuroscience* **45,** 153–160.

Torrealba, F. & Claps, A. (1988): The carotid sinus connections: a WGA-HRP study in the cat. *Brain Res.* **455,** 134–143.

Travers, J.B. (1988): Efferent projections from the anterior nucleus of the solitary tract of the hamster. *Brain Res.* **457,** 1–11.

Tucker, D.C., Saper, C.B., Ruggiero, D.A. & Reis, D.J. (1987): Organization of central adrenergic pathways: I. Relationships of ventrolateral medullary projections to the hypothalamus and spinal cord. *J. Comp. Neurol.* **259,** 591–603.

Valverde, F. (1962): Reticular formation of the albino rat's brain stem. Cytoarchitecture and corticofugal connections. *J. Comp. Neurol.* **119**, 25–54.

Van Bockstaele, E.J., Pieribone, V.A. & Aston-Jones, G. (1989): Diverse afferents converge on the nucleus paragigantocellularis in the rat ventrolateral medulla: retrograde and anterograde tracing studies. *J. Comp. Neurol.* **290**, 561–584.

Vierck, C.J., Jr., Greenspan, J.D., Ritz, L.A. & Yeomans, D.C. (1986): The spinal pathways contributing to the the ascending conduction and the descending modulation of pain sensations and reactions. In: *Afferent processing*, ed. T.L. Yaksh, pp. 275–329. New York: Plenum Press.

Vigorito, M., Sclafani, A. & Jacquin, M.F. (1987): Effects of gustatory deafferentation on polycose and sucrose appetite in the rat. *Neurosci. Biobehav. Rev.* **11**, 201–209.

Ward, D.G. (1988): Stimulation of the parabrachial nuclei with monosodium glutamate increases arterial pressure. *Brain Res.* **462**, 383–390.

Yasui, Y., Breder, C.D., Saper, C.B. & Cechetto, D.F. (1991): Autonomic responses and efferent pathways from the insular cortex in the rat. *J. Comp. Neurol.* **303**, 355–374.

Yasui, Y., Cechetto, D.F. & Saper, C.B. (1990): Evidence for a cholinergic projection from the pedunculopontine tegmental nucleus to the rostral ventrolateral medulla in the rat. *Brain Res.* **517**, 19–24.

Yates, B.J., Yamagata, Y. & Bolton, P.S. (1991): The ventrolateral medulla of the cat mediates vestibulosympathetic reflexes. *Brain Res.* **552**, 265–272.

# Chapter 2
# Elements of cellular neurophysiology

**Michel Lamarche**

*INSERM U97, 2ter rue d'Alésia, F-75014 Paris, France*

## Resting potential

### Definition

It can readily be shown that there is a potential difference (pd) between the interior and exterior of the cell by introducing a microelectrode into a cell. The interior is negative with respect to the exterior and the pd is generally about 60–80 mV. This transmembrane potential is not the exclusive property of excitable cells; it is found in all living cells, and has been particularly well studied in red blood cells. However, this membrane potential may undergo large variations, and even transient inversions, in excitable cells. The term 'resting-potential' is used to refer to the stable pd in the absence of any external or internal perturbation. We will see that it does not indicate that all membrane activities are in a resting state.

### Origin of the resting potential

#### Concentration differences and selective permeability

Analysis of the intra- and extracellular milieux shows that they contain very different concentrations of certain constituents, particularly certain ions (see Table 1). The pd has been shown to be directly linked to this difference between the composition of the intra- and extracellular fluid

This situation can be illustrated by a simple hydrostatic analogy. This analogy depends on osmotic phenomena, which, although they are important in this context, will not be covered here. In this model, two arms of a U-tube are separated by a membrane that is permeable to water but not to a solute S; the left-hand arm of the U-tube contains a much higher concentration of S than the right-hand arm, and both are filled to the same level. Water

**Table 1. Examples of the distribution of the principal ions across the plasma membrane of excitable cells (concentrations in mM)**

| Ion | Squid axon | | Frog muscle | |
|-----|---------|--------|---------|--------|
| | Outside | Inside | Outside | Inside |
| $Na^+$ | 460 | 50 | 120 | 10 |
| $K^+$ | 10 | 400 | 2.5 | 140 |
| $Cl^-$ | 540 | 40–150 | 120 | 3 – 5 |
| $Ca^{2+}$ | 10 | $10^{-4}$ | 1 | $10^{-4}$ |

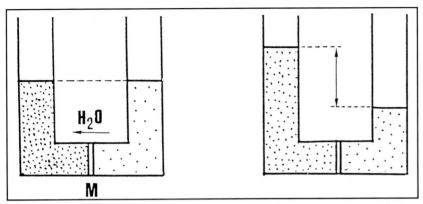

*Fig. 1. Demonstration of potential energy due to a concentration difference across a selectively permeable membrane. In the initial state (left) two arms of a U-tube are filled to the same level with aqueous solutions of different concentrations. As the membrane M is selectively permeable to water but not to solute, there is a net flux of water from the dilute to the concentrated solution until the hydrostatic pressure created balances the flux driven by the osmotic pressure difference.*

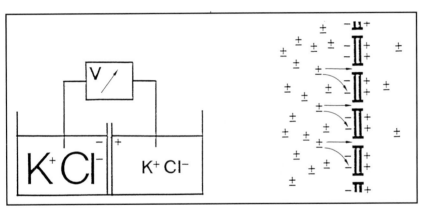

*Fig. 2. Two compartments containing solutions of potassium chloride at different concentrations (concentrated solution on the left) are separated by a membrane permeable only to potassium. Simple diffusion tends to drive solute from left to right. As chloride cannot accompany the potassium across the membrane pores (see detail right), a potential difference develops, which itself tends to keep the cations in the left-hand compartment. As a result an equilibrium situation is created in which the flux linked to a diffusion gradient is exactly balanced by the flux due to the electrical gradient – giving a net flux of zero.*

passes from the right-hand arm to the left one until the concentrations of S on both sides of the membrane are almost the same (Fig. 1). But the liquid level in the left-hand tube is higher, creating a hydrostatic pressure. Finally, an equilibrium is reached in which the flux of water due to the concentration difference is exactly counterbalanced by the flux in the op-posite direction due to the difference in hydrostatic pressure. Hence the net flux is zero.

## Electrochemical gradient in a single diffusible-ion model

What will happen if solute S is re-placed by potassium chloride and if the membrane remains permeable to water but is selectively permeable to the K⁺ ion? Figure 2 shows the ex-perimental situation in which the KCl solution is more concentrated on the left than on the right. The potassium tends to pass from the left hand com-partment to the right, according to the law of simple diffusion. But the membrane is only permeable to potas-sium, and the chloride cannot follow it. As electroneutrality must, above all, be maintained, potassium cannot dif-fuse. But electroneutrality is not strictly respected at the microscopic level, and potassium ions can cross the membrane and remain attached to its left-hand surface, attracted by the chloride ions that, in attempting to follow them, are attached to the right-hand surface of the membrane. Hence the two surfaces of the mem-brane are equivalent to the two plates of a condenser, and the build-up of excess charge results in a potential dif-ference (Fig. 2, right). This situation is produced almost instantly, unlike the osmotic model, and also represents an equilibrium: the left-right flux of potassium linked to a concentration difference, is effectively compensated for by the right-left flux due to the potential difference. Hence, the net flux is zero.

## Equilibrium potential of an ion: the Nernst equation

The equilibrium potential of a given diffusible ion is the value which the membrane potential must reach to cancel the net ion flux. This value, which is a function of the concentra-tion difference between the two com-partments, is determined by the Nernst equation:

$$E_K = RT/ZF.\log_n ([K]right/[K]left)$$

where R is the gas, T is the absolute temperature, Z the valence of the ion,

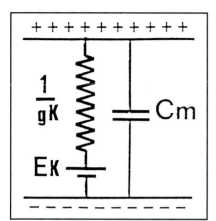

*Fig. 3. Diagram showing the electrical properties of the membrane in Fig. 2, as components of an electrical circuit. The concentration difference driving the movement of potassium, and hence the transmembrane current, is shown as a battery whose value is the equilibrium potential of this single diffusible ion. The membrane, although theoretically permeable to potassium, restricts passage of the ion with a certain resistance, which is the inverse of the conductance gK. Chloride, which cannot accompany the potassium (in this theoretical model which differs from biological membranes in this), accumulates on one side of the membrane, giving it the properties of a condenser (Cm). The system becomes somewhat more complicated when we consider a biological membrane rather than this model, as there are several diffusible ions.*

F is Faraday constant or the number of coulombs carried by one mole of a monovalent ion, [K]right is the potassium concentration in the right-hand compartment and [K]left the concentration in the left. For a monovalent cation at 20 °C, the relationship becomes:

$$E_{ion} \text{ (mV)} = 58.\log_{10} [ion]right/[ion]left$$

In a theoretical model in which the membrane is permeable to a single ion, the membrane potential is thus directly proportional to the concentration difference of the particular ion on either side of the membrane.

## Conductance and membrane current linked to an electrochemical gradient

A membrane creates by its very presence a certain 'resistance' to the passage of an ion even if it is permeable to this ion. Then, in speaking of pd, membrane resistance and capacitative properties, it is convenient to indicate this membrane as an 'equivalent electrical circuit' in which the concentration difference of the diffusible ion is the generator, and the membrane is like resistances and capacitances in parallel (Fig. 3). Similarly, we can define the membrane conductance as a measure of the ease with which charges move across the membrane. In other words, it is the inverse of the resistance of the membrane to the passage of ions; it is expressed in Siemens. It is clear that the conductance of a membrane directly reflects its permeability to ions. However it can be shown that the relationship between these two values is not necessarily linear. For example, for a given permeability, the conductance may vary with the ionic concentrations and with the electrical field. Nevertheless any increase in permeability produces an increase in conductance.

It can readily be seen, in the above model, that any change that is made in the membrane potential (by injecting current) creates a situation of disequilibrium. Thus, provided the potassium conductance remains high, there is a potassium current which tends to return the membrane potential to the potassium equilibrium potential. According to Ohm's law the potassium current intensity is proportional to both the inverse of the resistance (i.e. in our model the potassium conductance, gK) and also the difference between the membrane potential and the potassium ion equilibrium potential (Em–EK), which difference is called the electrochemical gradient or driving force. Thus:

$$I = 1/R \times V \text{ or } IK = gK.(Em-EK)$$

where IK is the potassium current intensity.

For a given constant value of the conductance gK, the current increases as the difference between Em and EK increases. Inversely, the net flux is cancelled once the displacement of K ions returns Em to the value of EK.

## The practical consequences in excitable cells

### Active transport

Let us now leave the theoretical model and examine biological membranes themselves. This situation is much more complex as the system includes several types of diffusible ions which may interfere. However, the situation for potassium seems to be affected very little by the presence of these other ions. Calculation of its equilibrium potential using its internal and external concentrations and the Nernst equation gives a value of about –80 mV, not far from the measured resting potential. This suggests that potassium is a major determinant of the Em. But the situation is much more complex for sodium. Its extracellular concentration is much higher, so that it tends to enter the cell down its concentration gradient, and this tendency is aggravated by the internal negative charge. In contrast to potassium, the two fluxes are acting in the same direction (Fig. 4). But we can measure no significant change in the

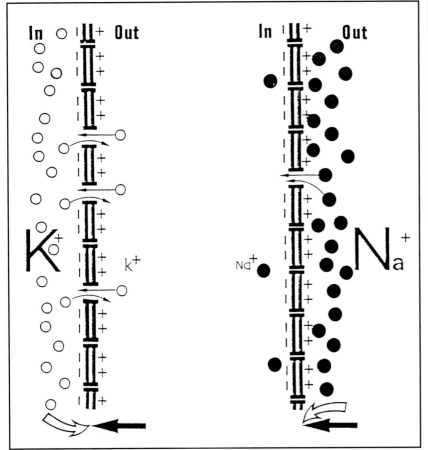

*Fig. 4. Simplified diagrams showing a single ionic species and a single type of specific channel for this ion. In a resting neuronal membrane the potassium (left) is close to equilibrium, and tends to leave the cell (curved arrows) down its concentration gradient, but it also tends to enter the cell (straight arrows) along the electrical gradient; the overall flux (black and white arrows at the bottom) is balanced. Note that in this resting situation, not all the potassium channels (which do not all have the same properties) are open. Thus the potassium conductance can increase, particularly during an action potential. The sodium ions are far from equilibrium. The two gradients, chemical and electrical, act in the same direction, and if the resting sodium conductance were not very small, there would be a large influx. As this conductance is not zero, there is a small but continuous influx of sodium, which is offset by active extrusion of the ion.*

membrane potential or in the concentrations of sodium over time. The simplest interpretation of this is to assume that the membrane is impermeable to sodium. In fact, although the resting sodium conductance is very small, it is not zero. It can be shown using radioactive sodium that transmembrane exchange does take place in the resting state. The important point is that the fluxes are in both directions. While sodium entry can be readily explained by diffusion, its exit against a concentration gradient cannot be by simple passive mechanisms. This is why it was necessary to postulate an active mechanism for exporting sodium from the cell. The mechanism can be thought of as a pump using metabolic energy (ATP) to compensate for the passive entry of sodium. As it is metabolically dependent, it can be blocked by metabolic inhibitors such as dinitrophenol (DNP) (Fig. 5).

While potassium is close to equilibrium in the resting state, it is also involved in this active pumping. When sodium is pumped out of the cell, potassium is simultaneously pumped in. When there is no extracellular potassium, the enzyme which hydrolyses ATP to provide the energy for active transport are no longer active (Fig. 5, right). They are similarly inactive when sodium ions are absent. This enzyme, which is consequently called $Na^+$-$K^+$-ATPase, is thought to form part of the pump itself. However, one sodium is not always exchanged for one potassium. Three $Na^+$ ions are generally exported for every two $K^+$ ions that enter the cell, so that the pump contributes to the production of the membrane potential: it is thus an electrogenic pump.

The situation for calcium ions is similar to that of sodium. It is much more abundant in the extracellular fluid than inside the cell, insofar as we must only consider free ions, bound ions being evidently unable to participate in electrical phenomena. We should even consider not the simple ionic concentration, but a parameter which, although linked to concentration, may differ from it significantly – ionic activity. Calcium tends to enter the cell passively, despite having a very low resting conductance, and it is actively pumped out by a $Ca^{2+}$-ATPase.

## Anions

The ratio between the internal and ex-

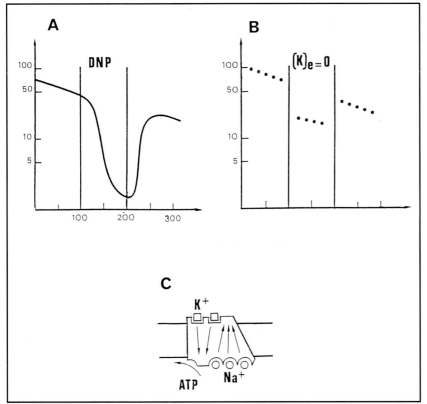

*Fig. 5. A. Demonstration of active extrusion of sodium, using metabolic energy. A squid axon is loaded with radioactive sodium by placing it in sea water containing the tracer. The axon is replaced in sea water with no radioactive tracer and the efflux of radioactive sodium followed in counts per minute (abscissa, log scale) with time. When the metabolic inhibitor dinitrophenol is placed in the medium, at between 100 and 200 min, the outflow of radioactive sodium drops, showing that it is an active process.*

*B. Demonstration that the activity of the sodium pump is partly dependent on extracellular potassium. Sodium efflux, under the same conditions, and with the same axes as in A, is markedly reduced when placed in potassium-depleted medium. The enzyme responsible for this transport, $Na^+$-$K^+$-ATPase, catalyses coupled transport, generally two potassium ions for three sodium ions (shown in C). Hence the pump is electrogenic.*

ternal concentrations of chloride ions, which are by far the most abundant, is opposite to that of potassium ions, and its movement is generally not controlled by active transport mechanisms. However, there are certain analogies with potassium ions. In particular, the resting chloride conductance, although generally smaller than that of $K^+$, is much greater than that of $Na^+$, and the chloride equilibrium potential is very close to the resting potential, suggesting a causal relationship between these two values. This may seem paradoxical if we recall that, in the theoretical model with only one cation and one anion discussed above, the mobility of the anion was zero, or much smaller than that of the cation: otherwise the concentration would rapidly reach equilibrium and no pd could be maintained. This is not the case here for chloride as its resting conductance is

far from zero. But the transmembrane pd is maintained by the presence of other ionic species with a low resting conductance (e.g. $Na^+$) and negative charges carried by non-diffusible intracellular proteins, in addition to the active transport systems.

## Ion fluxes and the origin of the resting potential

If we analyse the ionic movements within a given multi-ion system, restricting ourselves to the most important ions ($K^+$, $Na^+$ and $Cl^-$), we can say that the maintenance of a stable state reflects the fact that the net flux of positive and negative charges is zero at this potential. But this does not correspond to a thermodynamic equilibrium, because this stable resting state may be transiently disturbed towards a very different state, indicating a latent state of disequilibrium. We can therefore write:

$$IK + INa + ICl = 0$$

Each of these currents may be expressed as a function of the specific conductance and the divergence from the equilibrium potential of each ion:

$$IK = gK (Em - EK);$$

$$INa = gNa (Em - ENa);$$

$$ICl = gCl (Em - ECl)$$

As their sum is zero, we can write:

$$Em(gK + gNA + gCl) - gK.EK -$$
$$gNa.ENa - gCl.ECl = 0)$$

And hence the expression for resting membrane potential:

$$Em = \frac{(gK.EK + gNa.ENa + gCl.ECl)}{(gK + gNa + gCl)}$$

If the resting gCl and gNa are zero,

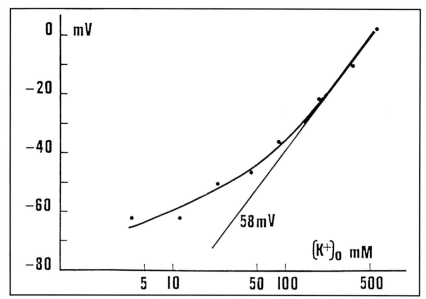

*Fig. 6. Relationship between the membrane potential (abscissa) and the extracellular potassium concentration (log scale, ordinate) in the squid axon. Calculations made using the Nernst equation considering only potassium give the theoretical straight line with a slope of 58 mV/log unit. In fact, because the resting membrane is not completely impermeable to sodium ions, the measured curve departs a little from the theoretical one, especially at low extracellular potassium concentrations.*

then we have the situation in the theoretical single mobile ion model, and Em is exactly equal to EK. But we know that this is not so, and we also know that gK is very much greater than gNa and gCl (gNa is only about 3 per cent of gK at rest). Consequently, if we consider gNa and gCl to be negligible (this is not quite true for gCl in some cells), the above equation shows that Em is almost the same as EK.

As the EK varies with the potassium concentrations on either side of the membrane (Nernst equation), we might expect a change in extracellular potassium to cause a change in the value of Em. This has been shown in the squid axon. The membrane potential increases (i.e. becomes less negative) with the log of the extracellular potassium concentration (Fig. 6). However, as the membrane is not completely impermeable to sodium

and chloride, the curve deviates from the calculated one, especially at low extracellular potassium concentrations. The low, but not zero, sodium conductance is also in part the reason why the measured resting potential is slightly offset towards a depolarization, with respect to EK. Although gNa is very small, the powerful driving force Em–ENa allows more sodium ions to enter the cell than can immediately be actively pumped out. This leads to a reduction in the internal negative charge of about 10 mV.

The above calculation takes no account of non-diffusible intracellular anions. A theoretical model can be used to predict the consequences of their presence, especially on chloride distribution. One such model is the equilibrium state described by the Donnan relationship, in which the ratio [cations]internal/[cations]external is equal to the ratio [Cl]external/[Cl]internal. This equilibrium is theoretically achieved at the price of ion movements, and these could lead to an increase in intracellular osmotic pressure. In other words, the cell should swell and finally burst. In fact, cells have regulatory systems, especially the Na+-K+-pump, which allow it to achieve a slightly different equilibrium without endangering cell integrity.

## Membrane rectifying properties

Starting from this equilibrium situation, with a certain potential and a specific membrane conductance, we can artificially impose a different membrane potential and verify that the currents resulting from the change vary as might be expected from the equivalent electrical model. Current-voltage curves can be constructed in this way, which are assumed to define the ohmic characteristics of the membrane. If the current-voltage relationship is linear, then the conductance is independent of the potential. In reality, the relationship is curved, and the membrane is said to be rectifying, i.e. at certain potentials it can be more conducting for ions passing in one direction than the other. This property involves membrane channels which are open at the different voltages tested (i.e. permeability is constant but the conductance varies). But if, in addition, certain channels are 'voltage-gated', the probability of a channel opening also depends on the membrane potential. This property is most important in the generation of the transient changes in potential discussed in the next section.

## The action potential

### The ionic basis

#### Transient inversion of membrane polarization

Excitable cells are characterized by the fact that certain stimuli can lead to sudden transient changes in membrane potential. Our knowledge of the origin of the membrane potential suggests that these changes are linked to ion movements, and hence to changes in conductance. It was first thought that the membrane became instantaneously non-selectively permeable to a range of small ions, leading to a state of partial equilibrium giving a membrane potential close to zero. But more accurate measurements have shown that the action potential is not a simple membrane depolarization. The internal surface of the membrane becomes briefly positive. We must therefore assume that ion movements (confirmed by the overall increase in membrane conductance during the action potential) are specific.

#### Role of sodium in polarization inversion

Early experiments showed that replacing the sodium in the external medium with a larger ion, such as choline, which is assumed to be unable to cross the membrane, considerably alters the action potential. Later studies with radiolabelled tracers confirmed the fundamental role of sodium in the reversal of polarization. However, it should be noted that all action potentials are not sodium-driven – there are also calcium potentials (see later). Nevertheless, the fol-

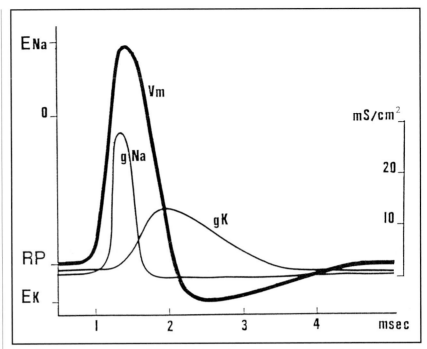

*Fig. 7. Changes in sodium (gNa) and potassium (gK) conductances (right hand scale, millisiemens/cm$^2$) in a squid axon during an action potential. The sudden increase in gNa causes the membrane potential (Vm) to change from its resting value (rp = 60 mV) to a value close to the sodium equilibrium potential (ENa +45 mV). This is called the ascending phase of the action potential. Then, when the gNa is rapidly inactivated, gK begins to increase, facilitating the return to the resting polarization. This increase in gK, although triggered by depolarization like that of gNa, occurs later and persists longer (there is no inactivation of potassium channels implicated here at this time scale). It is much longer than the repolarization phase itself, so that Vm is transiently close to EK. This is the after hyperpolarization phase, or AHP.*

lowing description is concerned mainly with mechanisms involving sodium.

We shall assume that at least the initial phase of the action potential is due to a sudden selective increase in sodium conductance. As this cation is far from equilibrium, the sodium entry flux will initially be large since the two driving forces of this flux, the concentration and potential difference, work in the same direction. If the sodium conductance gNa is durably increased, the net flux INa, determined by the relationship INa = gNa(Em – ENa), should not reach zero until the moment when the charge movement

brings the membrane potential to the sodium equilibrium potential, i.e. when Em – ENa = 0. The value of the membrane potential reached during the overshoot (passing zero) is close to the sodium equilibrium potential (Fig. 7). This implies that for this short period, the sodium current is much greater than the potassium and chloride currents.

#### Potassium and repolarization

But this situation is unstable, because the membrane returns to its resting potential in less than a millisecond in nerve fibres. The return of gNa to a low value, even if it is necessary for repolarization, is not enough to explain its rapidity. Although the phenomenon remains juxta-membranary and the entry of sodium is not 'massive' as was previously thought (only repetitive stimulation can produce detectable alterations in concentration on both sides of the membrane), such a rapid return to the resting polarization cannot be explained by active pumping out of sodium. It is therefore believed to be a second ion movement (which is not absolutely necessary since certain nodes of Ranvier function without it). The second ion is potassium, which is far from equilibrium during the overshoot. If there is an increase in gK soon after that of gNa, then there would be an efflux of potassium because of both the concentration gradient and the transient negative external charge. If the increase in gK continues, this flux continues until Em – EK is balanced, resulting in a return to the resting potential (Fig. 7).

## Voltage clamp measurements: current identification

Measurement of membrane resistance clearly shows that the action potential is accompanied by a marked increase in overall conductance, but it cannot identify the ionic species involved, or the sequence of events during an action potential. Voltage clamp measurements do, however, provide such information. Voltage clamping involves maintaining the membrane potential at a constant value despite its natural tendency to fluctuate. Precautions are taken to ensure that longitudinal currents do not disturb measurements by clamping the potential on the whole membrane surface under examination. It is then relatively simple, in view of the par-

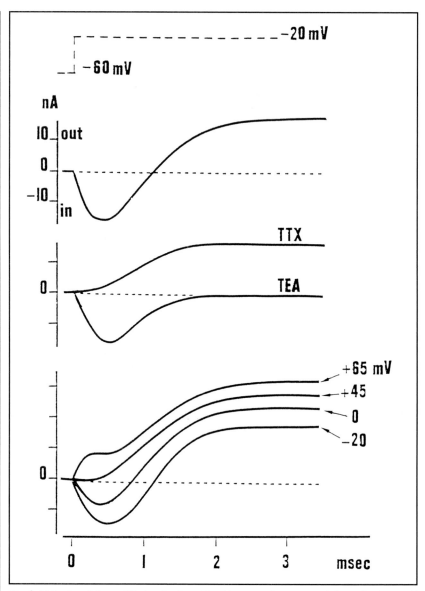

*Fig. 8. Upper panel: Inward (in) and outward (out) transmembrane currents (nanoAmperes, nA) in a voltage-clamped squid axon following step increase in potential from –60 to –20 mV. There is a transient inward current followed by an outward current which lasts as long as the voltage increase is maintained.*

*Middle panel: Selective ion channel blockers show that the transient inward current is carried by sodium ions (it is lost when tetrodotoxin (TTX) is in the medium) and the outward current is due to potassium ions (it is lost in the presence of tetra-ethyl-ammonium (TEA)).*

*Lower panel: This family of curves confirms the origin of these currents. When the clamp voltage is 0 mV the inward current is smaller than that found when the clamp is –20 mV because the factor Em-ENa in the relationship INa gNa(Em-ENa) is smaller. The outward current is higher because Em-EK is greater. When the potential is clamped at +45 mV, close to the equilibrium potential of the sodium ENa, the inward current disappears as the net flux of sodium is zero. Lastly, when the membrane is clamped at +65 mV, above the ENa, the large potassium outward current is preceded by a short sodium outward current.*

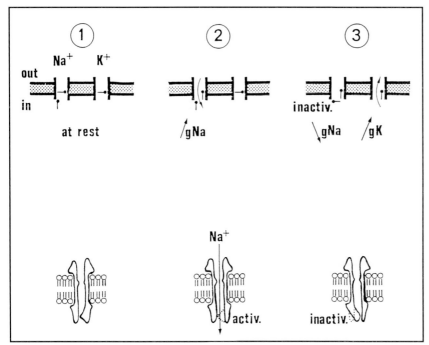

*Fig. 9. Schematic representation of the opening and closing of voltage-sensitive membrane channels during an action potential.*
*1: At the resting potential, most of the Na and K channels are closed.*
*2: Depolarization first causes opening of Na channels.*
*3: The increase in gNa is only transient as a second slower-acting gate, the inactivation gate, is soon closed by the same depolarization. At the same time, the K channel opens, allowing repolarization and return to the resting situation.*
*The relative slowness of the reopening of the inactivation gates is responsible for the absolute refractory period, i.e. the inefficiency of any new stimulus during a few ms following an initial action potential.*
*The bottom series of drawings of a section through the sodium channel protein are based on electron micrographs and show the conformational changes of activation and inactivation following depolarization.*

ameters of external currents employed to maintain the membrane potential, to evaluate endogenous transmembrane currents. In this way it can be shown that a clamped depolarization of the membrane of about 20–30 mV, which is the sort of stimulus which should trigger an action potential, triggers an initial inward current which disappears after a few milliseconds, although the membrane potential remains unchanged, and is followed by a long-lasting outward current (Fig. 8).

While these results support the hypothesis proposed above, better precision can be obtained by replacing the sodium in the extracellular fluid by an ion such as choline. When this is done, an inward current is no longer seen. Similarly, if tetrodotoxin, which blocks voltage-sensitive sodium channels, is placed in the extracellular fluid, the inward current is suppressed and only the outward current is seen. Conversely, the presence of tetra-ethyl-ammonium in the bath, which selectively blocks one type of

potassium channel, results in a loss of outward current while the inward current remains unaffected (Fig. 8). We can therefore conclude that the initial phase of the action potential is linked to a sodium inward current, while repolarization is linked to a potassium outward current. Selective blockade also shows that the sodium and potassium do not cross the membrane via the same channels. One last piece of important information is the transient nature of the entry current even when the voltage is kept constant. We must assume that this current, which seems to be triggered almost instantly by membrane depolarization, is very rapidly abolished. It is said to be inactivated; the conductance gNa, whose increase appears to be linked to depolarization, drops quickly despite the continued depolarization. This is not the case for the potassium current, which can last as long as the clamping potential is continued.

## Inactivation of sodium channels

The patch-clamp technique (see Chapter 4) can be used to study the behaviour of a limited number of membrane channels. Such studies have confirmed that depolarization is the factor which increases the probability of sodium and potassium channels opening. The kinetics of potassium channels are slightly slower than those of Na⁺ channels. Thus we can say that depolarization first opens the sodium channels, then the potassium channels. We can explain the inactivation of the sodium current, despite the continued presence of the cause of channel opening, by postulating the presence of a second gate, the inactivation gate. This gate is itself driven by depolarization, but operates more slowly (Fig. 9). This inactivation gate is statistically more likely to be open than closed at the resting potential, which has no influence on gNa because the other gate of the sodium channel is very likely to be closed. During depolarization, the sodium inward current is only seen during the brief instant separating the opening of one gate and the closure of the other. The presence of this inactivation gate, which probably lies on the internal side of the membrane, is confirmed by the fact that intracellular injection of proteolytic enzymes, such as pronase, markedly reduces this inactivation.

It is noteworthy that as long as there is no sodium current inactivation, repolarization is strongly counteracted, and the action potential greatly prolonged. Thus, if the sodium channels are activated by depolarization, the sodium current is likely to be autoamplified. Even a small depolarization opens sodium channels. If there are enough of them, this allows passage of a sodium current, leading to increased depolarization and the recruitment of more sodium channels. This is the principle of threshold depolarization: below this threshold, the process diminishes and remains local. This threshold explains the all-or-nothing nature of the action potential. Once depolarization passes a certain threshold, it continues to open sodium channels until Em reaches the value of ENa. Only inactivation can cancel this mechanism which will otherwise be an obstacle to the repolarizing action of the potassium current. In contrast, the potassium current is self-limiting: the outflow of potassium tends to return Em towards EK, so reducing IK and gradually removing its cause. However, before re-turning to the resting potential, some fibres go through a phase of slight hyperpolarization because the increase in gK lasts considerably longer that the simple repolarization phase (Fig. 7). As a result, the small difference between Em and EK under resting conditions is abolished by the potassium flux, and Em is transiently very close to EK.

Inactivation, which is triggered by membrane depolarization, affects a certain number of sodium channels even when the membrane is at its resting potential. It has been shown that under voltage clamping conditions, if a depolarization is preceded by a short pulse of hyperpolarizing voltage, the inward current is significantly greater, indicating that some sodium channels were de-inactivated by the hyperpolarization. The maximal current obtained for a conditioning hyperpolarization of about 40 mV is about 1.7 times greater than that obtained under control conditions.

## Refractory period

In contrast, a short conditioning depolarization reduces the inward current triggered by the clamping potential, and the sodium current linked to the clamping potential is completely lost when the conditioning depolarization reaches 40 mV. This explains the refractory period which follows an action potential. A second stimulus on a fibre within the milliseconds following a first action potential finds most of the sodium channels inactivated. Hence the fibre is unable to produce a second action potential. This is the absolute refractory period. In the following milliseconds, enough channels become de-inactivated for a second action potential to occur. But because many channels remain unavailable and because the gK is still high from the initial depolarization, the trigger threshold is higher and the size of the response reduced because the sodium current is too small to reach the ENa. This relative refractory period can last several milliseconds.

## Action potential propagation

Another major characteristic of action potentials is their capacity for propagation. If the potential difference across one region of the membrane is different from that of neighbouring regions, local lateral currents are established between these regions. This will only occur if the medium is a conductor – nerve fibres placed in oil no longer conduct the influx. One region is the origin of the charge, the 'source', and the other the 'sink' into which it flows, on both the internal and the external surfaces of the membrane. The overall result of these local currents is to reduce the transmembrane electrical gradient. Thus the areas around the initial site transiently affected by the action potential are depolarized, so creating the conditions for generation of an action potential. The action potential is therefore a regenerating process which is propagated along the whole length of the axon (Fig. 10). Because of the refractory period which follows the action potential, propagation normally occurs only in one direction.

The mechanism is slightly different for myelinated fibres because the electrically resistant external medium no longer allows the development of local currents in adjacent areas. This difficulty is overcome by the presence of nodes of Ranvier, which regularly interrupt the myelinated sheath. The current passes from one node to the next, so conduction occurs in a series of jumps, hence its name – saltatory. This has the double advantage of being much faster (up to 120 m/s) and requiring less energy as most of the membrane is unaffected by the ionic changes.

## Calcium potentials

There are also action potentials whose general characteristics differ little from those described above, except that the inward current is carried by calcium and the kinetics are generally slower. Although these potentials were demonstrated in invertebrates almost as long ago as sodium potentials, the properties of the calcium channels involved were only described once patch-clamp experiments became available. We now know that certain calcium channels can coexist with sodium channels, producing plateau-type action potentials, such as those of cardiac muscle. We also know that a single neuron, such as the Purkinje cell of the mammalian cerebellum, may have calcium action potentials on its huge dendritic arborization which are substituted, at the initial segment of the axon, by shorter, high-frequency, sodium action potentials. Lastly, one of the characteristics which separate voltage-dependent calcium channels from their sodium homologues is the fact that inactivation is generally linked to the entry of calcium itself. It may also be recalled here that sodium action potentials may be accompanied by calcium movements, with the consequences discussed below.

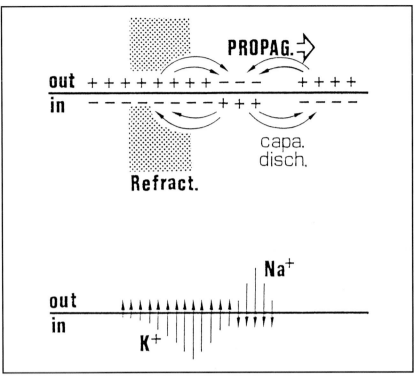

Fig. 10. Lateral currents (upper panel) and transmembrane currents (lower panel) implicated in action potential propagation. The zone affected by the action potential, i.e. by reversal of membrane potential, behaves as a charge sink (in terms of positive charge) on the external surface, or a source on the internal surface with respect to the adjacent regions up- or downstream, resulting in the appearance of local currents which cause a local discharge of membrane capacitance (Capa. disch), and hence a local depolarization. This would be without effect in the upstream area because it is in the refractory period, but it causes an increase in gNa downstream and thus creates the conditions for development of an action potential. The nerve influx thus moves from one area to the next by continually recreating the conditions of its origin. It is therefore propagated without alteration along the whole length of the axon. The lower panel shows the sequence of transmembrane currents in the corresponding regions. The arrows represent the direction and intensity of current change (cf. Fig. 7).

## Synaptic transmission

### The basic mechanisms

#### The chemical step

In most cases (see later for exceptions) the action potential cannot pass simply from the axon terminal (presynaptic element) to the post-synaptic element in the same way as it jumps from one node of Ranvier to another. There are two reasons for this. One is that the low-resistance extracellular space in the synaptic region is relatively large and thus likely to shunt the local currents, making them ineffective. The other more fundamental reason is that there are no voltage-sensitive channels in the junction area of the post-synaptic element that could be activated by any electrical stimulus. It was suspected for over a century, and has now been know for several decades, that most synapses require a chemical mediator. The pioneering studies on the role of acetylcholine (ACh), both at the neuromuscular junction and in the function of the peripheral autonomic nervous system, plus studies on adrenaline, were later confirmed in other structures and it was shown that with very few exceptions all communication between excitable tissues is chemically mediated. Neurochemical studies have identified a wide range of substances that act as neuromediators or neuromodulators, but we shall pursue this no further here.

## The neuromuscular junction

The early histological studies were carried out on the neuromuscular junction because it is relatively accessible. They showed that spherical structures – the synaptic vesicles – accumulated in the pre-synaptic terminal. These terminal vesicles are surrounded by a membrane much like the cytoplasmic membrane, and cytochemical studies have shown that they contain ACh. The sub-synaptic membrane bears regular folds that are rich in receptors and align with the synaptic vesicles. The folds result in a considerable increase in junctional area. While some workers believe that the mediator normally mobilized is the one found free in the cytoplasm, with the vesicles being only a storage form, it is generally agreed that a nervous influx results in vesicle exocytosis following fusion of the plasmic membrane with membranes of the closest vesicles. If a neuromuscular membrane preparation is instantly fixed, it can be shown that exocytotic figures are more numerous in stimulated junctions. The transmitter is thus released into the synaptic cleft and becomes transiently bound to the post-synaptic receptors causing

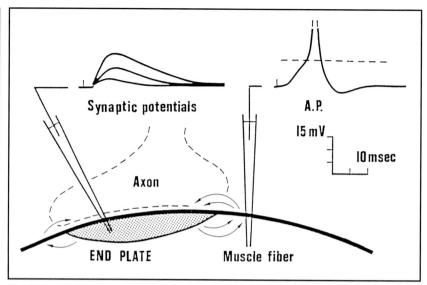

*Fig. 11. Intracellular recording of synaptic potentials at the motor end-plate of the neuromuscular junction. These synaptic potentials are graduable, summable and result from the opening of channels permeable to both sodium and potassium. They bear receptor sites activated by neurotransmitters (in this case acetylcholine). These channels are not voltage-dependent and are not selectively permeable to a single cation. The conditions for the development of an action potential are thus not all satisfied in this area. The extra-junctional area, stimulated by synaptic currents when the end-plate is depolarized, can produce a propagated muscle action potential once they rise above the threshold. Post-junctional potentials in smooth muscle can activate contractile mechanisms without triggering action potentials.*

changes in conductance as these receptors are coupled to ion channels. Finally, the transmitter is hydrolysed by the choline esterase in the synaptic cleft, and the metabolites may be recovered for synthesis of new ACh in the presynaptic terminal.

## Motor end-plate potential

Electrical recordings in the sub-synaptic region after nerve stimulation show that there is a depolarization – the end-plate potential – which is fundamentally different from the action potential. The motor end-plate potential is strictly local and extra-junctional recordings indicate only a trace of electrotonic activity which decreases with distance from the end-plate. The potential is relatively small (about 20 to 40 mV maximum), but its size

varies. The response is thus said to be graduable, depending on the amount of transmitter released. It is not followed by a refractory period, so that there can be summation of several successive responses. It can last 20–30 ms and be very asymmetrical; the rising phase is slower than the action potential, but is much shorter than the repolarization phase. The local zone of depolarization is clearly the origin of local currents analogous to those underlying action potential propagation, and it can be seen that above a certain threshold the end-plate potential sufficiently stimulates the voltage-sensitive extra-junctional channels to trigger a post-synaptic action potential (Fig. 11). Thus, an all-or-none phenomenon from the pre-synaptic fibre does not automatically pass to the post-synaptic element. It must pass via the local depolarization or post-synaptic potential, and reach a threshold value before it is transmitted. The situation of transmission in smooth muscle is different, however (Chapter 6), since excitatory junction potentials (equivalent of motor end-plate potentials) can induce smooth muscle contraction without triggering action potentials.

## Quantal transmitter release and the role of calcium

The existence of local spontaneous depolarizations having the same general characteristics as the end-plate potentials but much smaller amplitudes, hence called miniature end-plate potentials, have helped us understand the chemical step in synaptic transmission. These miniature end-plate potentials have amplitudes of about 0.4 mV in the frog. They are considered to be due to the random release of the contents of a synaptic vesicle; hence the end-plate potential is simply the synchronization of several hundred miniature end-plate potentials. Thus by artificially reducing the exocytosis produced by a pre-synaptic action potential, it can be shown that the amplitude of the end-plate potentials so reduced are made up of multiples of 0.4 mV, indicating that the end-plate potential is the sum of many miniature potentials. Such a reduction in neurotransmitter release may be obtained by using specific toxins (botulinus toxin) or more simply by reducing the extracellular calcium concentration. The arrival of the action potential at the presynaptic terminal is accompanied by a calcium influx. This influx is a necessary general precondition for the fusion of vesicle and plasma membranes.

## Ionic mechanisms involved in the end-plate potential

The ionic movements underlying this depolarization have been identified. Thus, binding of the neurotransmitter to post-synaptic receptors alters their properties so that the ion channels associated with them allow the temporary passage of ions. Voltage clamp studies show that the reversal potential (level beyond which the net influx gives way to a net efflux) of the end-plate potential is about –10 mV. None of the commonly implicated ions has an equilibrium potential of this magnitude. Tracer ion studies show that the exchange of sodium and potassium increases during synaptic activation and that gNa and gK both increase simultaneously, as the Na influx current and the potassium efflux current use the same channel. This simultaneous movement explains both the relatively slow time course of the depolarizing and repolarizing current, since the two currents flow in opposite directions, and the value of the reversal potential which lies between ENa and EK.

## Types of transmission

### Excitatory synapses

We can thus summarize the neuromuscular junction function as the arrival of a pre-synaptic action potential causing a calcium influx which allows vesicle exocytosis. The ACh released is temporarily bound to receptors which are themselves voltage-independent cation channels allowing the counterflow of sodium and potassium. The resulting local, gradual depolarization may stimulate voltage-dependent channels in the extra-junctional regions. With minor variations,

this is also the model for excitatory interneuronal synapses. There may be different neurotransmitters (glutamate, monoamines, peptides, etc.) and variations in the receptor (NMDA vs AMPA glutamate receptors, etc.) and particularly in the sometimes indirect relationship between the receptors and the ion channels. The receptor-channel is relatively rare, and in most cases there is a second messenger within the post-synaptic element which triggers a cascade reaction that may involve G proteins, cyclic AMP, GTP, arachidonic acid, a protein kinase, etc., between the receptor and the channel. The final response of the channel may be the opening of a closed channel or the closing of an open one, resulting in a decrease in conductance: excitatory post-synaptic potentials (EPSP) linked to a drop in potassium conductance have been described. The passive entry of sodium tends to depolarize the resting membrane, and at resting conditions this is opposed by the simultaneous efflux of $K^+$. If this is blocked by a drop in the resting gK, depolarization is facilitated, leading to a true EPSP. The time scale of these phenomena may be, but need not be, slower, and the polarization changes may last several seconds or even minutes. We therefore speak of these long-term changes in polarization that generally have no effect themselves on the propagated post-synaptic activity as neuromodulation. They can alter the response to other transmitters by shifting the resting potential.

### Inhibitor mechanisms

Inhibitor mechanisms are of two types. In post-synaptic inhibition, the transmitter (the best known example is GABA) causes the appearance of an

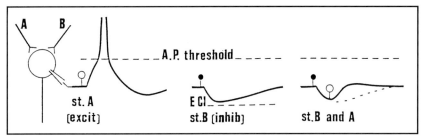

*Fig. 12. Post-synaptic inhibition. Stimulation of an excitatory pathway (st A) produces an excitatory post-synaptic potential (EPSP) to which an action potential can be added, if the threshold is reached. Stimulation of an inhibitory pathway (st B) leads, in this case, to opening of chloride channels, producing a hyperpolarization, as ECl is here lower than the resting potential. When the excitatory stimulus is given a few ms after the inhibitory stimulus, the EPSP no longer reaches the threshold, not just because of the basic hyperpolarization, but also because of the shunt effect resulting from the lower membrane resistance in the area affected by the post-synaptic inhibitory potential.*

inhibitory post-synaptic potential (IPSP) in the sub-synaptic membrane. The IPSP can be thought of as a mirror image of the EPSP. It is generally a hyperpolarization, and it may be thought to oppose the EPSP by moving the membrane potential away from its action potential triggering threshold (Fig. 12). However, the interpretation of the effects of IPSP is not limited to this simple algebraeic addition of two opposing local events. The IPSP is also the result of an increase in ionic conductance, so that the membrane resistance is locally reduced during the IPSP, acting as a shunt for the local currents triggered by the EPSP. These currents crossing a smaller resistance produce a smaller change in potential. Thus even if the IPSP causes no hyperpolarization, as when the equilibrium potential of the particular moving ion is equal to the resting potential, the inhibitory effect is nevertheless present because of the local reduction in membrane resistance.

The ion implicated in a hyperpolarization must have an equilibrium potential more negative that the resting potential. This is generally chloride, but could also be potassium; in some cases, it can be both sequentially.

However, inhibition may also be not associated with any hyperpolarization or reduction in post-synaptic resistance. Thus we must assume that a presynaptic mechanism can produce analogous effects. It has been shown that pre-synaptic inhibition acts across an axon–axon synapse leading to a small-amplitude, long-lasting depolarization (of about 100 ms) at the pre-synaptic terminal. This depolarization leads to the inactivation of some of the sodium channels as described earlier. The terminal action potential is thus reduced and the subsequent calcium entry reduced, resulting in the release of less excitatory transmitter. It should be noted that facilitation effects may be obtained by prolonging the pre-synaptic action potential by another type of axon–axon contact.

## Gap junctions and electrical synapses

Lastly, there are very special contact areas called gap-junctions, composed of ionic channels of low selectivity arranged opposite each other on two apposed plasma membranes. They directly connect the cytoplasms of two adjacent cells. These junctional channels are generally open at rest and allow direct electrical coupling between neurons or smooth muscle cells. They are particularly numerous at electrical synapses and allow the passage of an action potential from one cell to another with no synaptic delay. Electrical synapses were first described in invertebrates, but they have also been found in mammals. They are not very common except in certain nuclei, such as the inferior olive. These channels also allow intercellular exchange of various substances, particularly second messengers, providing for the transfer of supplementary information.

## Summary

In this chapter it is shown how the resting membrane potential (negative intracellularly) is created by the existence of concentration differences and selective permeability to certain ions (higher permeability to $K^+$, lower permeability to $Na^+$ and $Cl^-$), associated with an active ionic pump to extrude $Na^+$ and the presence of negatively charged intracellular proteins. The equilibrium potential of each ionic species (the potential at which the net flux of the ion in question is zero) can be calculated by the Nernst equation. The electrochemical gradients present can be momentarily reduced by the opening of specific channels: thus, if the membrane is moderately depolarized, sodium channels open briefly allowing rapid sodium influx so that the membrane potential tends towards the sodium equilibrium potential (positive intracellularly). The opening of potassium channels, while sodium channels are inactivated, then causes an efflux of this ion which returns the membrane potential on its resting level. This sequential opening and closing of sodium and potassium channels explains the action potential, its all-or-nothing nature, and the current spread resulting from the potential difference between active and inactive zones which creates the conditions of its propagation in the adjacent region.

At the neuromuscular (or interneuronal) junction, the prejunctional action potential is associated with calcium entry which triggers vesicular/plasmic membrane fusion (exocytosis) and hence quantal release of transmitter. This molecule causes ionic conductance changes in the specialized post-junctional membrane by activating specific receptors gating ionic channels, either directly or indirectly through second messengers. In the case of excitatory synapses the membrane is depolarized by simultaneously increased sodium and potassium conductance, causing local depolarization which, if sufficient, triggers an action potential. Inhibitory post-synaptic conductance changes involve especially chloride and sometimes potassium ions. Gap junctions, and the related electrical synapses, are also described.

## References for further reading

Alberts, B., Bray, D., Lewis, J., Raff, M., Roberts, K. & Watson, J.D. (1989): *Molecular biology of the cell*, 2nd edn, New York: Garland.

Eccles, J.C. (1964): *The physiology of synapses*. Berlin: Springer.

Fawcett, D.W. (1981): *The cell*, 2nd edn. Philadelphia: Saunders.

Hammond, C. & Tritsch, D. (1990): *Neurobiologie cellulaire*. Paris: Doin.

Hille, B. (1984): *Ionic channels of excitable membranes*. Sunderland, MA: Sinauer.

Hodgkin, A.L. (1964): *The conduction of the nervous impulse*. Springfield, IL.: Thomas.

Hubbard, J.I., Llinas, R. & Quastel, D.M.J. (1969): *Electrophysiological analysis of synaptic transmission*. Baltimore: William and Wilkins.

Hulme, E.C. (1990): *Receptor biochemistry: a practical approach*. Oxford: IRL Press.

Jack, J.J.B., Noble, D. & Tsien, R.W. (1975): *Electric current flow in excitable cells*. Oxford: Clarendon Press.

Kandel, E.R. (1977): *Handbook of physiology; Section 1: The nervous system*, Vol. 1: Cellular biology of neurons. Bethesda, MD: American Physiological Society.

Kandel, E.R., Schwartz, J.H. & Jessell, T.M. (1991): *Principles of neural sciences* 3rd edn. New York: Elsevier.

Katz, B. (1966): *Nerve, muscle and synapse*. New York: McGraw Hill.

Katz, B. (1969): *The release of neural transmitter substances*. Springfield, IL.: Thomas.

Kuffler, S.W., Nicholls, J.G. & Martin, A.R. (1984): *From neuron to brain: a cellular approach to the function of the nervous system* (Second edn. Sunderland, MA.: Sinauer.

Sakmann, B. & Neher, E. (1983): *Single channel recording*. New York: Plenum Press.

Shepherd, G.M. (1974): *The synaptic organization of the brain*. New York: Oxford University Press.

Shepherd, G.M. (1988): *Neurobiology*. New York: Oxford University Press.

# Chapter 3

# Neurotransmitter systems

Thérèse M. Jay

*CNRS URA 1491, Université Paris-Sud, 91405 Orsay, France*

## Introduction

Research early in the century showed that chemical agents could transmit the activity of peripheral nerves to their target organs. Acetylcholine was the first chemical neurotransmitter to be demonstrated mediating the effect of the inhibitory vagus nerve on the frog's heart (Loewi, 1921). Later, it was shown that acetylcholine present in nerve terminals in contact with muscle was also involved as a neurotransmitter in the initiation of the muscle contraction (Dale, 1935). Chemical transmission was then accepted as being the communication between a neuron and its target organ or between neurons. After acetylcholine, the catecholamines (dopamine, norepinephrine, epinephrine) were identified as neurotransmitters in the peripheral and central nervous system. Another amine, serotonin (5-hydroxytryptamine) was found to be present in many organs, especially in the brain and it was also shown that serotonin was a neurotransmitter in central and peripheral nervous systems.

In the 1950s, five aminoacids (glutamate, aspartate, γ–aminobutyrate, glycine and taurine) were suggested as transmitter candidates due to their presence in high concentrations in cells and organs. The physiological evaluation was not facilitated because they were also involved in metabolic pathways and biosynthesis in both neurons and glial cells.

In the 1970s, a number of other substances from a completely different chemical category were considered as neurotransmitters. These substances were the neural small peptides. Even though substance P had been described early in 1931 by Von Euler and Gaddum, it was only in the 1970s that this small peptide was recognized as a neurotransmitter (Leeman & Mroz, 1975). Other peptides of different sizes ranging from carnosine (2 amino acids) to somatostatin (14 amino acids), were shown to be present like the amines in small quantities in localized regions of the central nervous system. Enkephalins, endorphins, insulin, angiotensin I and II, vasoactive intestinal polypeptide, cholecystokinin, prolactin, vasopressin and ocytocin are among the active putative peptide neurotransmitters which have been described.

Recently, other small molecules like the calcium ion, adenosine, adenosine triphosphate (ATP), cyclic adenosine monophosphate (cAMP), guanosine triphosphate (GTP), cyclic guanosine monophosphate (cGMP), prostaglandins and histamine have been suggested as neurotransmitters. These compounds control the communication between neurons through the chemical synapses, and at their junctions with their various target organs.

The number of substances suggested as neurotransmitters is presently at least 50 and is still increasing. Nevertheless, a number of criteria must be met to consider a substance as a proven neurotransmitter and these criteria have not been clearly established for each substance.

## Identification criteria for neurotransmission and general scheme

## Identification criteria for a neurotransmitter

Over the years, different approaches have formalized a set of requirements which must be satisfied before a substance can be considered as a neurotransmitter. These criteria are the following:

– The neurotransmitter is present in the neurons from which it is released.

– The transmitter must be stored within the neuron from which it is released.

– The enzymes which synthesize the neurotransmitter and its precursors are present in the same neurons.

– The transmitter is released upon presynaptic nerve stimulation.

– The same postsynaptic response as with nerve stimulation should be elicited when the transmitter is applied locally to the region of the postsynaptic membrane.

– The postsynaptic response to the exogenously applied transmitter should terminate rapidly.

– Agents which block the response to presynaptic stimulation should also block the response to exogenously applied neurotransmitter.

These requirements can be summarized as followed:

(1) presence,
(2) synthesis,
(3) release,
(4) inactivation,
(5) pharmacological action,
(6) reception.

## General scheme in synaptic transmission

Like every other cell, neurons are characterized by an inside relative deficiency of positively charged ions which result in a net negativity measured across the cell membrane (Fig. 1). Some ions, such as potassium ($K^+$) can normally enter through the membrane much more easily than sodium ($Na^+$) or calcium ($Ca^{2+}$). The total positive ions inside neurons do

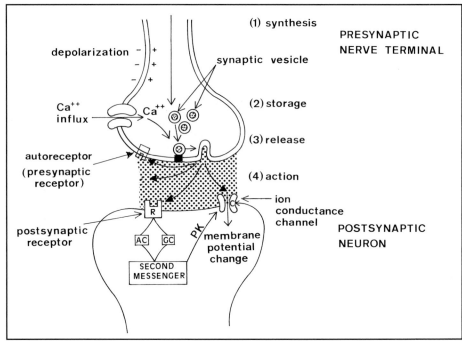

*Fig. 1. General scheme for neurotransmission. Different steps in the synaptic transmission: (1) synthesis (2) storage (3) release of the transmitter and (4) action. Neurotransmitters once released into the synaptic cleft bind to their specific receptors on the postsynaptic cell; the receptor either activates a selective ion channel (C) and the signal is transferred as an electrical message (change in the membrane potential) or the receptor may activate a selective enzyme system (R): adenylate or guanylate cyclase (AC or GC), initiating the formation of second messengers (cyclic nucleotides) which will phosphorylate a membrane protein via a proteine kinase (PK). This protein is linked to an ion channel and the final message will also be a change in the membrane potential.*

not quite equal the negative charges contributed by the chloride, phosphate and organic acids inside the cytoplasm. Neurons are electrically excitable cells, i.e. they have the ability to regulate their internal ionic polarity. In a typical neuron that has an axon projecting to some distant point, there is an integration of synaptic inputs from dendrites and soma at the axon hillock. At this site, the action potential or discharge is initiated at a certain level of depolarization. The wave of depolarization temporarily reverses the internal negativity of the resting potential and as it moves down the axon, progressive segments of the axon also undergo this transient reversal. It is the action potential initiated at the axon hillock which will trigger transmitter release at the axon terminals (Eccles, 1957). Arrival of the action potential at the synaptic bouton results in a rapid rise in the internal $Ca^{2+}$, the membrane opens, and the transmitter stored within synaptic vesicles is released by exocytosis.

Once neurotransmitters are released into the synapse, they bind to their specific receptors on the postsynaptic cell and their action is communicated either as an excitatory or an inhibitory message. It is a membrane potential change caused by an activation of ion gates or channels due to the close association of the transmitter with the receptor which is the signal. Sometimes, a chemical signal like the activation of an enzyme system (adenylate cyclase, guanylate cyclase) will produce cyclic 3′, 5′-AMP and cyclic 3′, 5′-GMP as the second messengers. These cyclic nucleotides activate specific protein kinases which will phosphorylate other proteins located on the inner face of the postsynaptic cell. Modifications of

these proteins probably alter the permeability of the membranes and the final message will still be an electrical signal. Another second-messenger system which involves the inositol phospholipids (phosphatidylinositol and polyphosphoinositides) also plays an active role in the signal transduction. The inositol lipid breakdown initiated by a phosphodiesterase (phospholipase C) is closely associated with a rise in intracellular $Ca^{2+}$ and triggers enzyme activity. When a stimulus activates a receptor that is coupled to phospholipase C, the enzyme degrades the phosphoinositides to produce two intracellular signals, inositol triphosphate that will raise the intracellular $Ca^{2+}$ and diacylglycerol that will activate an enzyme, protein kinase C, to phosphorylate certain proteins. Degradation of diacylglycerol yields arachidonic acid, another intracellular signal. All these messengers that alter ion channels all play active roles in determining the responses to stimuli.

## Catecholaminergic pathways

The group of catecholamines includes norepinephrine, epinephrine and dopamine. Dopamine is the precursor for norepinephrine and N-methylated norepinephrine forms epinephrine. The three catecholamines are localized in three distinct neuronal pathways.

## Synthesis

Catecholamines are formed (Fig. 2) from their amino-acid precursor tyrosine by a sequence of enzymatic steps. Tyrosine is normally present in the cir-

culation, taken up from the bloodstream by an active transport mechanism and concentrated within the brain. Within catecholaminergic neurons, tyrosine is converted to 3,4-dihydroxyphenylalanine (DOPA) by tyrosine hydroxylase. Tyrosine hydroxylase is the initial and rate-limiting enzyme in the catecholamine synthesis that is controlled by multiple processes. This key enzyme is controlled by dopa and dopamine, a feedback inhibition that involves the pteridine cofactor. Reduced tyrosine hydroxylase is the active form of the enzyme required for the enzymatic reaction. The feedback effect of catecholamines is possible because these substances oxidize the pteridine cofactor, preventing it from generating the active form of tyrosine hydroxylase. Tyrosine hydroxylase is also controlled by an allosteric mechanism through phosphorylation of the enzyme by a $Ca^{2+}$-stimulated or cyclic AMP-mediated protein kinase. There are drugs that specifically inhibit tyrosine hydroxylase like N-methyl-p-tyrosine, a competitive inhibitor.

L-DOPA, the product of tyrosine hydroxylase activity, is decarboxylated to dopamine in both dopamine and norepinephrine neurons by the enzyme dopa decarboxylase. This enzyme requires pyridoxal phosphate for its activity. It acts on all naturally occurring aromatic L-amino acids, like histidine, tyrosine, tryptophan and phenylalanine as well as both dopa and 5-hydroxytryptophan. The antibodies raised to dopa decarboxylase have been used for immunohistochemical localization of the enzyme in the three types of neurons: noradrenergic, dopaminergic and serotoninergic. A drug that specifically in-

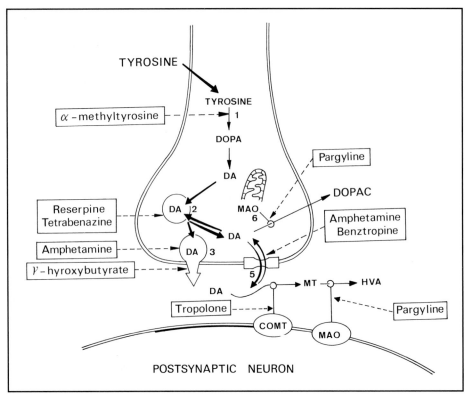

*Fig. 2. Schematic drawing of a dopaminergic neuron indicating the different steps of the synthesis and the sites of drug action. In the first step of the synthesis, tyrosine is converted to 3,4-dihydroxyphenylalanine (DOPA) by tyrosine hydroxylase. L-DOPA, the product of tyrosine hydroxylase activity is decarboxylated to dopamine (DA) by the enzyme DOPA decarboxylase. Within the presynaptic terminal, DA can be degraded in dihydroxyphenylacetic (DOPAC) by monoamine oxidase (MAO), located in the mitochondria. Outside the presynaptic neuron, DA can be taken up into the presynaptic terminal or inactivated by the enzyme catechol-O-methyl transferase (COMT) which converts DA in 3-methoxytyramine (MT) and homovanillic acid (HVA). There are drugs that inhibit specifically tyrosine hydroxylase like α-methyl-p-tyrosine, a competitive inhibitor. Reserpine causes a profound and long-lasting depletion of endogenous DA in neurons and the storage vesicles appeared to be irreversibly damaged. Tetrabenazine also affects the uptake-storage mechanisms but these effects are not irreversible. γ-Hydroxybutyrate blocks the release of DA in DA neurons and amphetamine releases DA by blocking the DA re-uptake. Amphetamine as well as benztropine are potent inhibitors of the re-uptake mechanism. Pargyline is an inhibitor of MAO. Tropolone is an inhibitor of COMT.*

ionine to norepinephrine. The level of PNMT in adrenergic tissue is regulated by the corticosteroids. Adrenal medullary cells use epinephrine as a neurohormone which is released directly into the blood stream. In the central nervous system neurons that utilize epinephrine as their transmitter have been mapped with an antiserum raised against purified PNMT from bovine adrenal glands.

## Storage and release

Catecholamines are found within storage vesicles, called dense-cored or granular vesicles that are present in high density within axons and nerve terminals. The mechanism that concentrates the catecholamines inside the vesicles is an ATP-dependent process linked to a proton pump (Holz, 1978). Reserpine, a specific potent inhibitor of the vesicular amine pump causes a profound and long lasting depletion of endogenous catecholamines in neurons and the storage vesicles appeared to be irreversibly damaged. Tetrabenzamine also affects the uptake-storage mechanisms but these effects are not irreversible.

The vesicles not only have a role of storage but also they mediate the process of their release. When an action potential arrives at the axon terminal, the $Ca^{2+}$ channels open allowing a $Ca^{2+}$ influx. A fusion of the vesicles with the neuronal membrane will

hibits this enzyme is α–methyldopa.

Dopamine is transformed into norepinephrine by hydroxylation of the ethylamine chain and the enzyme responsible for this conversion is called dopamine β-hydroxylase. This enzyme requires molecular oxygen and utilizes copper and ascorbic acid as cofactors to effect the hydroxyla-

tion. Dopamine β-hydroxylase is found only in noradrenergic neurons within the vesicles that store catecholamines.

The final enzyme that transforms norepinephrine into epinephrine is phenylethanolamine N-methyltransferase (PNMT), by the transfer of a methyl group from S-adenosyl-meth-

occur and the vesicles will discharge their content into the extracellular space. This $Ca^{2+}$ dependent release has been demonstrated in different preparations, in peripheral and sympathetically innervated tissues, in corpus striatum and other brain regions. By means of superfusion through a push-pull cannula, release of dopamine can be followed from caudate nucleus or substantia nigra (Glowinski, 1982) and it is possible to study the neural influences controlling dopamine release. This technique has been improved recently by fixing dialysis membranes over the tips of the push-pull cannulae. Catecholamine release may also be detected *in vivo* using electrochemical methods (*in vivo* voltametry). By means of these different techniques, regulation of catecholamine release has been studied. Two drugs – hydroxybutyrate and amphetamine – affect the release of dopamine. γ-Hydroxybutyrate blocks the release of dopamine in dopamine neurons and amphetamine releases dopamine by blocking the dopamine re-uptake.

## Distribution of catecholaminergic pathways

In the last two decades, the development of neuroanatomical methods aided the visualization of neurons containing a given type of transmitter substance. With the formaldehyde fluorescence histochemical method of Falk and Hillarp, it has been possible to map the distribution of the catecholamine-containing cell bodies and axonal pathways in the brain. Many sensitive techniques including the glyoxylic acid fluorescence techniques, immunohistochemical

methods, retrograde transport using horseradish peroxidase, and axoplasmic transport of labelled proteins were used in complementary fashion to map these different systems.

## Dopaminergic pathways

Three major dopamine-containing nuclei have been described (A9, A8–10 and A12) that are located primarily in the midbrain (Fig. 3) (Ungerstedt, 1971; Björklund & Lindvall, 1984). One of the major dopaminergic tracts, the nigrostriatal pathway, originates in the zona compacta of the substantia nigra (A9) and projects to the caudate nucleus and putamen of the corpus striatum. There is a dense dopaminergic innervation in the corpus striatum, nearly 80 per cent of all brain dopamine is found in this structure. The substantia nigra also sends some axons to the frontal cortex and some cells in the lateral part of the substantia nigra project to the cingulum (Emson & Koob, 1978). Medial and superior to the substantia nigra, dopaminergic cell bodies are located in the ventral tegmental area of Tsaï (region A8 and A10). These neurons provide an innervation to the limbic

cortex (medial prefrontal, cingulate and entorhinal cortex) and other limbic structures (septum, nucleus accumbens, olfactory tubercle, amygdaloid complex, thalamic paraventricular nucleus, piriform cortex). These two systems are often called mesocortical and mesolimbic dopamine pathways. It is worth mentioning that this pathway (A10) exhibits some unique pharmacological properties. Indeed the mesocortical system appears to be less sensitive to the action of dopamine agonists, requiring larger doses to get the maximal biochemical effect.

Dopaminergic cell bodies located in the arcuate nucleus of the median eminence and the periventricular nuclei of the hypothalamus (A12) send axons that innervate the intermediate lobe of the pituitary and the median eminence (referred as the tuberoinfundibular or tuberohypophysial system). In addition to these major pathways, there is a short dopamine system in the retina which links inner and outer plexiform layers and in the olfactory bulb which links together mitral cell dendrites in separated adjacent glomerula.

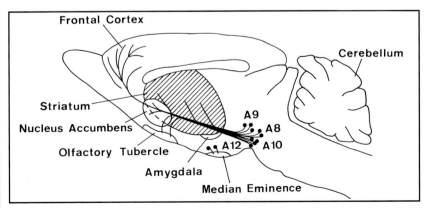

*Fig. 3. Dopaminergic neuronal pathways in the rat brain shown in sagittal section. (From Ungerstedt, 1971).*

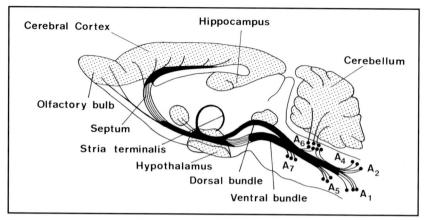

*Fig. 4. Noradrenergic neuronal pathways in the rat brain shown in sagittal section. Shaded regions represent the major terminal areas. The descending pathways are not included. (From Ungerstedt, 1971.)*

## Noradrenergic pathways

There are two major central noradrenergic pathways and one minor (Fig. 4). These systems are also termed bundles, dorsal and ventral bundles. The dorsal bundle arises in the locus ceruleus (A4, A6), a compact cell group located in the caudal pontine central grey (Dahlström & Fuxe, 1964). The majority of axons from the locus ceruleus project rostrally and have extensive collateral branches that innervate many brain regions including the brainstem, cerebellum, amygdala, hippocampus, septum, fornix, cingulum, external capsule, olfactory nuclei, and the entire cerebral cortex. The second noradrenergic pathway arises from ventrally located cell bodies diffusely distributed in the subceruleus region (A1, A2, A5 and A7) (Dahlström & Fuxe, 1964). These neurons innervate the brainstem and hypothalamus and the fibres intermingle with those arising from the locus ceruleus to form the central tegmental bundle (CTB). Then, they separate as a ventral bundle and again join each other to form the medial forebrain bundle (MFB). Because of the complex intermingling of fibres originating in various noradrenergic cell bodies, it is extremely difficult to get a specific analysis of the physiology of these projections. Other noradrenergic cell bodies located in the dorsal raphe nucleus project to the pretectal area, habenula, thalamus and hypothalamus (Lindvall & Björklund, 1974).

## Adrenergic pathways

With the development of sensitive immunoassays for phenylethanolamine-*N*-methyl transferase (PNMT) and their application in immunohistochemistry, the existence of epinephrine-containing neurons in the central nervous system could be confirmed (Fig. 5). There are two groups called C1 and C2 (Hökfelt *et al.*, 1974), whose cell bodies are located in the rostral part of the lateral reticular nucleus, adjacent to the inferior olive (C1) and in the dorsomedial reticular formation just below the fourth ventricle (C2). The axons of these epinephrine-utilizing neurons project to the thalamus and hypothalamus, to the dorsal motor nucleus of the vagus nerve. They also innervate the locus ceruleus, the nucleus of the solitary tract and caudally the intermediolateral cell columns of the spinal cord. Based on the anatomical projection, and looking at the control that epinephrine could exert on neurons in the locus coeruleus, it was shown that epinephrine iontophoresed into the locus coeruleus inhibits the firing of these cells (Fuller, 1982).

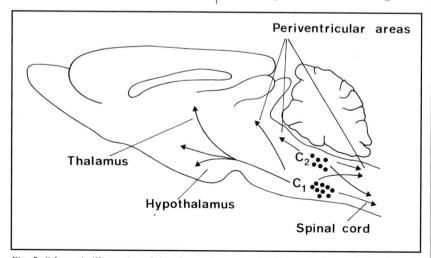

*Fig. 5. Schematic illustration of the adrenergic neuronal pathways in the rat brain shown in sagittal section. (From Hökfelt et al., 1974.)*

## Cholinergic pathways

As was mentioned at the beginning of this chapter, acetylcholine was the first chemical neurotransmitter isolated and characterized in both structure and function. Most of the earlier studies on acetylcholine were investigated in the autonomic system and on the neuromuscular junction. In the central nervous system, it is only after the detection of biogenic amines that the presence of acetylcholine could be determined in neuronal pathways.

## Synthesis

Acetylcholine is synthesized in one step from two components, acetate and choline, catalysed by the enzyme, choline acetyl transferase (Fig. 6). The choline is acetylated by acetyl Coenzyme A (acetyl CoA) mostly in nerve terminals. Choline is transported to the brain both free and in a phospholipid form (phosphatidylcholine) by the blood. The acetate moiety derives from acetyl CoA generated from glucose via pyruvate. A deficiency in glucose or other cofactor essential for acetyl CoA formation will affect the acetylcholine synthesis with deficiency in cholinergic transmission (Gibson & Blass, 1976). Choline appears to be the rate-limiting step in the synthesis and its uptake can specifically be inhibited by hemicholinium-3, troxypyrrolium (competitive) or AF64A (noncompetitive). Acetylcholine synthesis can be blocked by styryl pyridine derivatives. Choline acetyl transferase, the enzyme that catalyses acetylcholine synthesis, has been purified from many mammalian brain regions and the antibody raised against this purified enzyme was used to map the different cholinergic pathways by histochemistry.

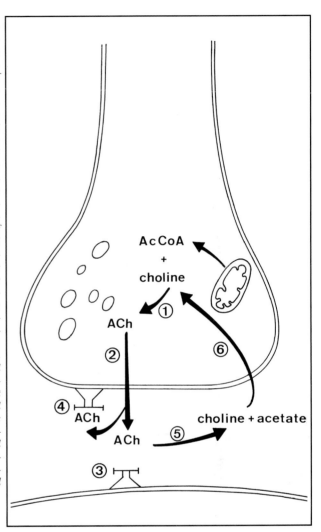

Fig. 6. Schematic drawing of a cholinergic neuron indicating the different steps of the synthesis and the sites of drug action. Acetylcholine (ACh) is synthesized in one step from acetate and choline, catalysed by the enzyme choline acetyl transferase. The choline is acetylated by acetylCoA mostly in nerve terminals. ACh is hydrolysed by cholinesterases and about half of the choline produced by cholinesterase activity is reutilized to make new ACh. The uptake of choline can specifically be inhibited by hemicholinium-3, troxypyrrolium (competitive) or AF64A (noncompetitive). ACh synthesis can be blocked by styryl pyridine derivatives. ACh transport into vesicles is blocked by vesamicol. Release is enhanced by β-bungarotoxin. Release is blocked by botulinium toxin, cytochalasin B, collagenase pretreatment and $Mg^{2+}$.

## Storage and release

In the nerve terminals, acetylcholine is present both in the cytoplasm and in synaptic coated vesicles. The mechanism of acetylcholine release requires the presence of $Ca^{2+}$ (De Belleroche & Bradford, 1972). Arrival of the action potential at the synaptic bouton results in a very rapid rise in internal $Ca^{2+}$, coated vesicles filled with transmitter bind to the carrier protein, the vesicle coat fuses with the membrane and opens to release its content. Acetylcholine release is usually followed by a new synthesis to replenish stores and newly synthesized acetylcholine is always the first to be released during nerve activity. Acetylcholine is hydrolysed by cholinesterases and about half of the choline produced by cholinesterase activity is reutilized to make new acetylcholine.

Electrophysiological experiments have shown that acetylcholine is released in multimolecular packets or quanta at some and probably all cholinergic synapses (Katz, 1966). Within each vesicle are 5000–10 000 molecules of acetylcholine and one-fifth as many molecules of ATP. For example in the neuromuscular junction, a few hundred quanta may be released at the axon terminal. In the central nervous system an impulse probably releases no more than one or two quanta from any one varicosity. The probability that an acetylcholine quantum will release depends on the synaptic potential and on the concentration of $Ca^{2+}$.

Acetylcholine transport into vesicles is blocked by vesamicol. Release is enhanced by β-bungarotoxin. Release is blocked by botulinium toxin, cytochalasin B, collagenase pretreatment and $Mg^{2+}$.

## Acetylcholine pathways

Various approaches using different components of the cholinergic system have been used to map the central cholinergic neurons but it was difficult to find one wholly reliable technique (Fig. 7). Recently with the development of specific antibodies raised to the purified choline acetyl transferase, the synthesizing enzyme of acetylcholine, the topography of these neurons have been described (Mesulam *et al.*, 1983; Cuello & Sofroniew, 1984). Six major pathways have been identified. The Ch1 and Ch2 groups are located in the medial septal nucleus and the vertical limb nucleus of the diagonal band. These two groups give the major cholinergic projections of the hippocampus. The Ch3 group in the lateral portion of the horizontal limb nucleus of the diagonal band

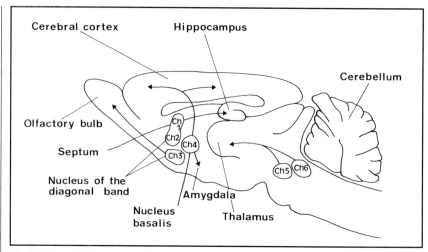

*Fig. 7. Schematic illustration of the major cholinergic pathways in the rat brain shown in sagittal section. (From Mesulam et al., 1983; Cuello & Sofroniew, 1984.)*

sends projections to the olfactory bulb. The Ch4 group including neurons in the nucleus basalis and in the diagonal band nuclei provides most of the cortical innervation. The Ch5 and Ch6 groups within the pedunculopontine nucleus (Ch5) and the laterodorsal tegmental grey (Ch6) are at the origin of the major cholinergic innervation of the thalamus. Apart from these major cholinergic groups, a small percentage of the neostriatal neuronal population uses acetylcholine (Phelps *et al.*, 1985). In addition, it appears that cholinergic elements in the spinal cord (dorsal horn, central grey matter, intermediate spinal grey region) are considerably more numerous and widespread than was previously thought (Barber *et al.*, 1984).

The intense immunoreaction to choline acetyltransferase present in the interpeduncular nucleus presumably corresponds to the cholinergic fibres from the fasciculus retroflexus, originating in the diagonal band nucleus or the medial nucleus of the habenula.

## Serotonergic pathways

Serotonin (5HT), a vasoactive substance was first discovered in serum. After characterization of this substance, it was found that 5HT was present in high concentration in chromaffin cells of the intestinal mucosa and that it was a vasoconstrictor. 5HT acts on many organs of the body (cardiovascular, respiratory, and gastrointestinal systems) and this substance is not unique to mammals, it is also present among plants and animals and found in fruits as bananas, pineapples, plums and nuts. The central nervous system contains a small part of the 5HT in the body (about 1 per cent). As 5HT cannot cross the blood–brain barrier, brain cells must synthesize their own.

The distribution of 5HT pathways can not be mapped with the same histofluorescence method used for the cate-

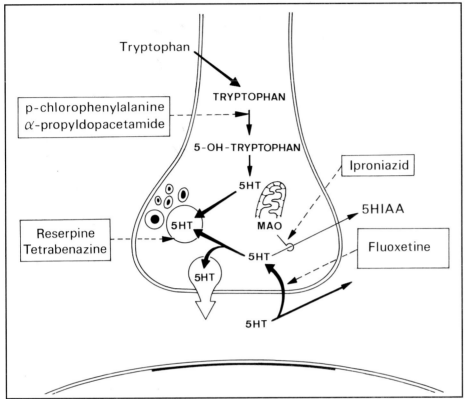

*Fig. 8. Schematic drawing of a serotonergic neuron indicating the different steps of the synthesis and the sites of drug action. The first step in the synthesis is the hydroxylation of tryptophan by tryptophan hydroxylase to form 5-OH-tryptophan. This enzyme can be inhibited by p-chlorophenylalanine and α-propyldopacetamide. The next step is the decarboxylation of 5-OH-tryptophan which forms serotonin (5HT). Within the presynaptic terminal, 5HT can be degraded in 5-hydroxy-indole-acetic acid (5HIAA) by monoamine oxidase (MAO), located in the mitochondria. Reserpine and tetrabenazine affect the uptake-storage mechanism, causing a marked depletion of 5HT. Outside the presynaptic neuron, 5HT can be taken up into the presynaptic terminal. Fluoxetine and tricyclic drugs with a tertiary nitrogen such as imipramine and amitryptyline appear to be potent inhibitors of this uptake mechanism.*

tion $Ca^{2+}$-dependent or cyclic-AMP-dependent kinases of the enzyme. The next and last step in the 5HT synthesis is the 5-hydroxytryptophan decarboxylation, catalysed by dopa decarboxylase, presumed to be the same enzyme that decarboxylates L-DOPA to form dopamine. There is some evidence that these two enzymes are the same, for example, immune sera raised against the decarboxylase purified from kidney crossreact with both decarboxylases from brain.

## Storage and release

Serotonin is found in nerve terminals, stored in dense-cored or granular vesicles like the other biogenic amines. It appears to be synthesized outside the vesicles, trapped inside (Joh *et al.*, 1975), and released by exocytosis. Reserpine and tetrabenazine affect the uptake-storage mechanism, causing a marked depletion of 5HT.

Serotonin can be released from brain slices and synaptosomes in response to depolarizing stimuli and it can be released in the cerebrospinal fluid when raphe neurons (a serotonergic nucleus) are electrically excited. A specific serotonin-binding protein has been isolated from regions of the brain enriched in serotonergic vesicles (Tamir & Liu, 1982). This protein is released with 5HT during synaptic activity. There are no drugs which selectively block the release of 5HT. However lysergic acid diethylamide by blocking

cholamine neurons. A modified method has to be applied to localize these pathways.

## Synthesis

The primary substrate for the synthesis of 5HT is an essential dietary amino acid, tryptophan (Fig. 8). As it cannot be synthesized *de novo*, elimination of tryptophan can lower the levels of brain 5HT. The first step in the synthesis is a hydroxylation of trypto-

phan by tryptophan hydroxylase to form 5-hydroxytryptophan which uses the same cofactor (pteridine) as tyrosine hydroxylase. This enzyme can be inhibited by *p*-chlorophenylalanine and α–propyldopacetamide. Tryptophan availability and its rate of entry into the brain control the synthesis; the reduced cofactor pteridine is also a rate-limiting step. As for tyrosine hydroxylase, an allosteric control of tryptophan hydroxylase appears to be mediated through a phosphoryla-

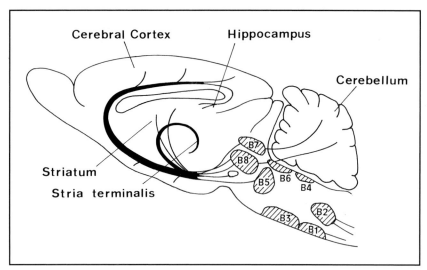

*Fig. 9. Serotonergic neuronal pathways in the rat brain shown in sagital section. (From Fuxe & Johnsson, 1974.)*

the firing rate of 5HT causes a reduction in 5HT release.

## Serotonergic pathways

As previously mentioned, the fluorescence histochemical method used to reveal the catecholamine neurons was not valid for serotonin neurons. This technique had to be modified and combined with some pharmacology. Newer techniques like localization of tryptophan hydroxylase, orthograde transport of labelled amino acids, retrograde transport of horseradish peroxidase and autoradiographic localization of tritiated 5HT have also contributed to the mapping of central serotonin pathways (Fig. 9). By means of specific antibodies raised against 5HT, the localization and morphology of the 5HT neurons in the brainstem of the rat has also been described (Steinbush, 1981). Serotonin-containing neurons are restricted to clusters of cells, the raphe nucleus, situated in the midline of the brain stem and me-

dulla oblongata. This raphe nucleus consists of nine cell groups ($B_1$–$B_9$), originally described by Dahlström & Fuxe (1964). Groups $B_1$ to $B_6$ are located in the pons, the floor of the fourth ventricle and the medulla. The most caudal groups $B_1$ to $B_3$ project backwards to the ventral and intermediate horns of the spinal cord. Groups $B_7$ to $B_9$ located in the midbrain (raphe dorsalis, medianus, and centralis superior) provide the extensive serotonin innervation to the telencepalon and the diencephalon. These cells produce a medial ascending pathway projecting to the hypothalamus and preoptic area and a lateral ascending pathway that innervates the cerebral cortex and the corpus striatum. There is an overlapping between fibres and terminals from different raphe neurons except in group $B_8$ (raphe medianus) whose neurons give preferentially the serotonin innervation of the limbic structures and in group $B_7$ (dorsal raphe) whose neurons project to the striatum, cerebral and cerebellar cortices

and thalamus. In addition immunocytochemical localization of 5HT has also detected serotonin-producing cells outside the raphe nucleus (area postrema, caudal locus ceruleus, region of the interpeduncular nucleus). In the hypothalamus, 5HT is present in neurons, in mast cells, and in tanycytes which are specialized ependymal cells located around the third ventricle (Smith & Ariëns Kapper, 1975). In the diencephalon, the pineal gland is also very rich in 5HT.

## Peptide-containing cell groups and pathways

Neuropeptides are the most recent group of compounds identified as substances mediating neural activity, i.e. neurotransmitters. Peptides have been recognized in several ways: as the releasing hormones, specialized in endocrinological functions through the hypothalamus and the pituitary gland, others were first recognized as hormones produced in the intestinal tract, some were identified in search of a mechanism to explain pain sensation or sleep induction and others were only isolated as chemical compounds present in brain.

Like the catecholamines, neuropeptides serve as synaptic transmitters and as hormones that communicate information via the circulation from their cells of origin to target cells elsewhere. On the other hand, these agents differ in several ways from the other transmitters so far described. They are present in small quantities, their mode of synthesis appears quite different and they usually have a potent biological function which is often employed as a bioassay to screen dif-

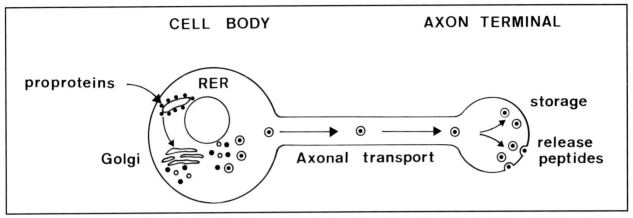

*Fig. 10. Illustration of the biosynthesis and release of peptides in neurons. Synthesis is located in rough endoplasmic reticulum (RER) where precursors (proproteins) are produced. The proproteins are transported to the Golgi apparatus, and packaged into neurosecretory granules which are transported to the nerve terminals for storage and release.*

ferential extraction or purification. In addition, they may coexist in neurons with other peptides or neurotransmitters, this subject will be discussed under 'Cotransmission'.

About 25 peptides have been found in the central nervous system in recent years, either localizing these peptides inside the neurons by immunocytochemistry or pharmacologically looking at their action following iontophoretic application (Snyder, 1980).

Neuropeptide research is an area in neurobiology which has been expanding very fast, so it is difficult to give an overview of this field. We will give some background and describe the distribution of some of these peptides in brain. References are given to the reader for more information.

## Synthesis of neuropeptides

As we mentioned previously, neuropeptides differ from other transmitters in their mode of synthesis. Synthesis is directed by messenger RNA in rough endoplasmic reticulum,

where large precursors (proproteins) are produced (Fig. 10). These proproteins are transported to the Golgi apparatus where they may be cleaved, then they are packaged into neurosecretary granules and transported to the nerve terminals for storage and release. The proteolytic cleavage by special processing peptidases to produce shorter peptides occurs only after packaging, during transport and before release. A special feature of neuropeptides is that they can be formed only in the cell body.

## Secretion of neuropeptides

Synaptic release of peptides which has been demonstrated does not differ from those of classical neurotransmitters. The peptides appear to be released in a calcium-dependent manner. The propagated action potential depolarizes the nerve terminal and induces an influx of $Ca^{2+}$, which produces exocytosis and the content of the secretory granules is extruded into the extracellular space (Berelowitz *et al.*, 1978).

## Are peptides neurotransmitters, neuromodulators or hormones?

At the beginning of this chapter, we considered the different criteria of identification for a substance to be a transmitter. Whether all neuropeptides have met these criteria and are neurotransmitters is an interesting question. Some of them have met most of these criteria but no peptide has met all of them. Some peptides which are released at a considerable distance from the site of action, either directly into blood or in the extracellular fluid, and are carried to the target cells where they act are identified as hormones. Some peptides can also act as neurotransmitters, i.e. substances released directly onto the site of action, in a synapse where they bind to presynaptic or postsynaptic receptors. Other peptides are identified as neuromodulators, i.e. released from a synapse and acting at a local or distant postsynaptic site to modify the action of a neurotransmitter on the same postsynaptic site, amplifying or attenuating the effect. The neuromodulator may also influence the release of the neurotransmitter by acting on presynaptic receptors. Another possibility is that the peptide may coexist with the neurotransmitter and be coreleased with it.

## Peptides' distribution in the central nervous system

### Substance P

Substance P was the first peptide to be proposed as a transmitter. Present in most areas of the central nervous system, substance P is also active in the gut, stimulating vascular and extravascular smooth muscle, and in salivary glands, enhancing salivation.

Radioimmunoassay and immunohistochemical techniques have shown several central substance P pathways (Emson, 1979). This peptide is present at very high concentration in the substantia nigra and the hypothalamus. Substance P has been proposed as an excitatory neurotransmitter in the striatonigral pathway and in the dorsal horn. Its release in the substantia nigra is regulated by another transmitter, GABA (Hong *et al.*, 1977). Substance P functions as a primary afferent (sensory) transmitter in the spinal cord and its presence in the so-called C fibres, suggests substance P as one of the neurotransmitters involved in pain (Henry, 1980). In regulating pain perception, perhaps substance P may be associated with serotonin. Another action of substance P which may be relevant in this book, is that this peptide may regulate axon reflexes in the skin which are responsible for the local vasodilation mediated by sensory axons around an injured area. This proposal (Dale, 1935) is consistent with the potent vasodilatory action of substance P.

## Cholecystokinin

Cholecystokinin (CCK) was first isolated in the duodenum as a substance that contracted the gall blader. CCK has been localized in both brain and periphery. This compound has a COOH-terminal octapeptide, CCK-8 which appears to be the predominant form in gut and brain. Radioimmunoassay and immunohistochemistry methods have been used to demonstrate that CCK is widely distributed (for reviews, Beinfeld, 1983) in rat brain. The cerebral cortex contains the highest content of CCK and the periaqueductal grey matter the greatest density of cells containing CCK. Because of the extent of the cerebral cortex, the total brain content is far greater than that of any of the other peptide. CCK staining cells, fibres and terminals are present in cerebral cortex, olfactory bulb, hypothalamus, thalamus, hippocampus, amygdala, periaqueductal grey matter, substantia nigra, ventral tegmental area, dorsal raphe, nucleus linearis rostralis, nucleus parabrachialis dorsalis and spinal cord. Some areas contain only CCK-staining fibres and terminals: olfactory tubercle, nucleus accumbens, caudate/putamen, globus pallidus, thalamus, septum and most of the spinal cord. Among the specific CCK neuronal pathways which have been observed are the hypothalamo-hypophysial system whose cell bodies are located in the supraoptic and paraventricular nuclei of the hypothalamus, and the fibres and terminals in both the median eminence and the posterior lobe of the pituitary. In this pathway, CCK was found to be co-localized with oxytocin and enkephalin. It was shown recently that the terminals' fields in the nucleus accumbens, olfactory tubercle, and the septal and prefrontal cortical projections arose primarily from CCK perikarya in the rat ventral tegmental area whereas the projections to the caudate-putamen and anterior cingulate cortex arose predominantly from neurons in the substantia nigra compacta (Seroogy & Fallon, 1989). CCK appears to coexist with dopamine in the ventral tegmental area neurons which project to the limbic structures. The amygdala receives innervation mainly from cell bodies located in the substantia nigra lateralis (Seroogy & Fallon, 1989). In the thalamo-cortical pathway, CCK appears to coexist with vasoactive intestinal peptide (Ogawa *et al.*, 1989). There is some suggestion that hippocampal neurons containing CCK project via the fornix to the septum, nucleus accumbens (Handelmann *et al.*, 1983). So far, CCK has been shown to have an excitatory action on neurons but it remains uncertain whether CCK has a neurotransmitter or neuromodulator capacity.

## Enkephalins and endorphins

Enkephalins and endorphins are opiate-like peptides which were identified after the discovery of the opiate receptors, for which they have a high affinity. These opiate-like peptides include a family of six structurally related peptides (dynorphin, Leu- and Met-enkephalin, α, β and γ-endorphin) and two pituitary peptides (morphine-like factors, anodynin). The following description will be restricted to the enkephalins and endorphins. Studies of the distribution of these peptides by immunohistochemistry have shown that enkephalins are widely distributed throughout the brain and spinal cord while endorphins are limited to the hypothalamus.

β-Endorphin-containing neurons are restricted to the basomedial and basolateral nucleus of the hypothalamus with projections to the ventral septum, nucleus accumbens, and paraventricular nucleus of the thalamus. Descending fibres from the same group of neurons are present in the periaqueductal grey matter, the locus ceruleus, and the reticular formation (Hökfelt *et al.*, 1977a; Bloom *et al.*, 1978). β-endorphin is probably released from the pituitary gland and acts as an hormone but its target cells have not yet been identified.

Brain regions rich in enkephalin include the limbic system and the regions involved in the transmission and modulation of pain. These areas are the lateral septum, amygdala, substantia nigra, striatum, nucleus reticularis gigantocellularis, nucleus raphe magnus and pallidus, nucleus caudalis of the spinal tract of the trigeminal nerve, and the substantia gelatinosa of the spinal cord (Barchas *et al.*, 1978). Brain regions rich in enkephalins are also rich in opiate receptors except the globus pallidus which contains a high level of enkephalin and is relatively low in opiate receptors. There is a long enkephalin-containing pathway from the caudate-putamen to the globus pallidus. Enkephalins unlike endorphins do not function as circulating substances, but like a neurotransmitter.

## Neurotensin

Neurotensin was isolated from the hy-pothalamus as a biological active substance lowering blood pressure (Carraway & Leeman, 1975). It is also a potent endogenous hypothermic substance, properties related to the C-terminal of the molecule (Nemeroff *et al.*, 1980). Again, radioimmunoassays and immunocytochemistry have indicated large amounts of neurotensin in the central nervous system in structures such as the nucleus accumbens, septum, hypothalamus, substantia nigra, amygdala, central grey matter of the mesencephalon, locus ceruleus, motor trigeminal nucleus and substantia gelatinosa (Kobayashi *et al.*, 1977). The rich content of neurotensin in the substantia nigra and ventral tegmentum have suggested a possible interaction with dopaminergic systems. Very little neurotensin is found in the cerebral cortex or the hippocampus. When applied locally neurotensin either inhibits or excites neurons (locus ceruleus and spinal cord). These pharmacological effects together with other biochemical data have shown that neurotensin indeed acts as a neurotransmitter or a modulator.

## Oxytocin and vasopressin

Oxytocin and vasopressin are two major hormones secreted by the posterior pituitary gland. These hormones are synthesized in the hypothalamus (paraventricular and supraoptic nuclei) and have excitatory and inhibitory effects on neurons in these nuclei. Vasopressin and oxytocin are present in many different brain regions (Sofroniew, 1983; Swanson & Sawchenko, 1983). Neurons are located in the hypothalamus and fibres in the neocortex, hippocampus, septum, amygdala, locus ceruleus, substantia nigra and spinal cord. It is clear that these two peptides have endocrine functions but their presence in brain appear to be unrelated to these functions. Different pharmacological and electrophysiological studies have shown that these two peptides behave as transmitters or modulators, whether they serve any paracrine function remains to be established. Vasopressin and oxytocin are involved in autonomic regulation and nociception and influence some behavioural performance.

## Angiotensin

Angiotensin is a prohormone present in blood which derives from angiotensinogen and the action of renin. The two major active products are angiotensin I and angiotensin II. Angiotensin I is known to stimulate thirst and drinking and also to raise the blood pressure. Angiotensin II is a potent vasoconstrictor and causes renal sodium retention by stimulating aldosterone secretion. These peripheral actions are complementary to its central action.

Immunohistochemical studies have indicated angiotensin systems in brain (Fuxe *et al.*, 1976; Ramsey, 1979). It is interesting to note that some of the angiotensin-containing brain regions are the circumventricular organs like the area postrema, the subfornical organ, and the organ vasculosum of the lamina terminalis which are known to lack the blood–brain barrier allowing them access to angiotensin.

## Vasoactive intestinal peptide

Vasoactive intestinal peptide (VIP) was isolated from gut as a 28 amino acid peptide that causes vasodilation. VIP also stimulates the conversion of glycogen to glucose, the secretion of insulin, glucagon and somatostatin and inhibits the secretion of gastric enzymes. After its isolation in intestine, VIP was demonstrated in brain, with the highest levels in the cerebral cortex. VIP is present mostly in nerve terminals (central amygdaloid nucleus, suprachiasmatic nucleus, medial preoptic and anterior hypothalamic nuclei) while it appears to be contained in small neurons in the neocortex and the limbic cortex (Larsson *et al.*, 1976; Fahrenkrug, 1980). VIP nerve terminals in the cerebral cortex are in close association with blood vessels, suggesting a VIP control of the cerebral blood flow by vasodilation. In this same brain region, a colocalization of VIP and acetylcholine has been reported (Eckenstein & Baughman, 1984). Like CCK, VIP is a neuronal excitant of hippocampal and cortical neurons.

## Neuropeptide Y

Neuropeptide Y (NPY) is also an intestinal peptide which is present in brain. Radioimmunoassays and immunohistochemistry have revealed its presence in the hypothalamus, olfactory tubercle, nucleus accumbens, amygdala, spinal cord and cerebral cortex (Emson & De Quidt, 1984). Co-localization of NPY with catecholamines and with other peptides like somatostatin have been suggested. NPY is known for its vasoconstrictor activity on cerebral vessels (Lundberg & Tatemoto, 1982). NPY serves an important neuroregulatory role, whether it is a neurotransmitter or a neuromodulator still remains to be determined.

## Bombesin

Isolated from the skin of the frog,

bombesin was later found in brain tissues. The peripheral actions of bombesin are similar to those of CCK or VIP and with its vasoconstrictor properties, bombesin resembles angiotensin. By increasing plasma glucose, bombesin appears to control carbohydrate metabolism. Hypothalamus is the richest brain region in bombesin. When bombesin is applied by iontophoresis on cortical neurons, it causes an augmentation of firing rates.

## Somatostatin

Somatostatin is one of these peptides, originally detected in the hypothalamus, controlling the release of hormone through the pituitary gland. Somatostatin was known for its property to inhibit the secretion of growth hormone. Later, it has been shown that somatostatin was also controlling the release of other peptides by a similar inhibitory action. This peptide is widely distributed in the intestine and the pancreas. Radioimmunoassays, immunohistochemistry and bioassays have revealed the presence of somatostatin in brain outside the hypothalamus. Cell bodies are present in the neocortex, cerebral cortex, hippocampus, amygdala, interpeduncular area, and the dorsal root ganglia (Brownstein *et al.*, 1975; Kobayashi *et al.*, 1977; Eppelbaum, 1977, for review, Eppelbaum, 1986). Short interneurons are mostly found in the neocortex and the hippocampus. Locally projecting neuronal tracts are represented in the hypothalamus. There are also long projecting neuronal pathways, from the periventricular hypothalamic system to most of the limbic regions, i.e. the olfactory tubercle, the lateral septum, the habenula and the hippocampus. Caudally, through

the medial forebrain bundle, the hypothalamic projections reach the locus coeruleus and the substantia nigra. A long descending projection has also been described from the amygdala to the nucleus reticularis of the medulla oblongata and some periaqueductal neurons containing somatostatin project to the nucleus raphe magnus. The central action of somatostatin is complex, this peptide can elicit stimulatory and inhibitory electrophysiological effects.

## Other peptides

Other hypothalamic-releasing factors contained within the hypothalamus that regulate the synthesis and release of pituitary hormones have been shown to be distributed outside the hypothalamus.

Thyrotropin-releasing hormone (TRH) releases thyrotropin from the pituitary. TRH nerve terminals have been detected in the cerebral cortex, the nuclei of the facial, trigeminal, and hypoglossal nerves, the ventral horn of the spinal cord, the nucleus accumbens, the lateral septum, the bed nucleus of the stria terminalis and the retina (Ogawa *et al.*, 1981). Evidence suggests that the TRH peptide is a neurotransmitter or a neuromodulator.

Luteinizing hormone-releasing hormone (LHRH) releases luteinizing hormone and follicle-stimulating hormone. LHRH neurons are present in the arcuate and preoptic area, in the periventricular and suprachiasmatic nuclei, in the lateral septum, the mammillary bodies, the diagonal band nuclei plus the interstitial nucleus of the stria terminalis. Axonal processes can be observed in the median eminence, supraoptic crest, the posterior pituitary, mammillary bodies, septum, thalamus, ventral tegmental area habenula and amygdala (Barry *et al.*, 1974). Some LHRH neurons visualized in the medial septal area, diagonal band, and olfactory tubercle seem to contact blood vessels, an observation which suggest a possible involvement of these neurons in the regulation of blood supply.

Other peptides are under study for their pharmacological properties and distribution in brain: bradykinin, corticotropin, prolactin, sleep-inducing peptides; the list is far from complete (for review: Krieger *et al.*, 1983). Carnosine is the smallest peptide identified as transmitter in the brain and concentrated in the olfactory pathway.

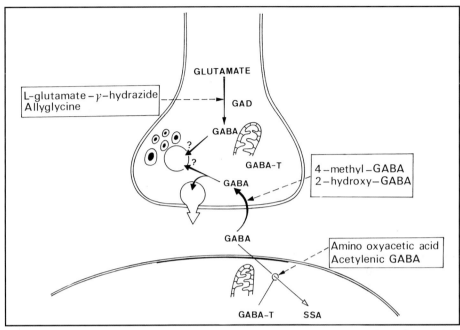

*Fig. 11. Schematic drawing of a GABAergic neuron indicating the different steps of the synthesis and the sites of drug action. The first step is the synthesis, i.e. the decarboxylation of glutamate by the enzyme glutamic acid decarboxylase (GAD). GAD is inhibited by a number of various hydrazines which are nonspecific inhibitors. L-Glutamate-γ-hydrazide and allylglycine are more selective inhibitors. No selective inhibitors of GABA release have been found. The uptake mechanism can be inhibited by agents such as 4-methyl-GABA and 2-hydroxy-GABA which are not completely specific in their inhibitory effects. GABA is metabolized in succinic semialdehyde (SSA) by transamination by GABA-transaminase. There are inhibitors of GABA-transaminase such as gabaculline and acetylenic GABA.*

## Amino acid neurotransmitters

Several amino acids are recognized as neurotransmitters in the central nervous system. Since these substances are also involved in the intermediary metabolism, it has been difficult to fulfil all the criteria required for a substance to be a transmitter. On the basis of physiological studies, amino acids have been separated in two groups: inhibitory amino acids (GABA, glycine, taurine, β-alanine) which hyperpolarize neurons, and excitatory amino acids (glutamic acid, aspartic acid, cysteic acid, homocysteic acid) which depolarize neurons.

## Inhibitory amino acids

### γ-aminobutyric acid (GABA)

GABA is quantitatively the principal inhibitory transmitter of the nervous system. Its mode of action was discovered in studies of the inhibitory neuron to a lobster muscle. In brain, GABA has been discovered in the lateral vestibular nucleus of Deiters which receives its input from the Purkinje fibres. Stimulation of the inhibitory GABA neuron increases the conductance for $Cl^-$ of the postsynaptic membrane.

### *Synthesis and release (Fig. 11)*

One of the precursors of GABA is glu-cose, the main energy source of the brain. GABA is biosynthesized by the enzyme glutamic acid decarboxylase (GAD) which acts on glutamate (produced via the citric acid cycle) by removing a γ-carboxyl group as $CO_2$ to produce a γ-amino acid. GAD is present in nerve terminals and is inhibited by a number of various hydrazines which are nonspecific inhibitors. L-glutamate-γ-hydrazide and allylglycine are more selective inhibitors.

Studies have demonstrated that GABA release is $Ca^{2+}$ dependent (Nicholls, 1989). The amount of GABA present in synaptic vesicles represents a fractionable percentage of what is present in the terminal but GABA may also be released from the cytosol. The transmitter is synthesized, released constantly, and removed from the synapse by a $Na^+$-dependent transport system of either the postsynaptic cell, the original axon terminal or the surrounding glial cell. A specific transaminase then catalyses the reaction to form succinic semialdehyde which is oxidized to succinic acid (the citric acid cycle). At present, there are no selective inhibitors of GABA release. The uptake mechanism can be inhibited by agents such as 4-methyl-GABA and 2-hydroxy-GABA which are not completely specific in their inhibitory effects. GABA is metabolized in succinic semialdehyde (SSA) by transamination by GABA-transaminase which is localized in mitochondria. There are inhibitors of GABA-transaminase such as gabaculline and acetylenic GABA.

### Distribution of GABA pathways

As GABA is supposed to be present in the inhibitory interneurons found in most areas of the brain, it is widely distributed (Fagg & Foster, 1983). The most convincing evidence for specific GABA neurons comes from immunohistochemical studies of GAD and GABA. GABA-neurons are present in the spinal cord, brainstem and in defined regions of the cerebellum (Purkinje cells, cerebellar golgi, stellate and basket cells) of the hippocampus (basket cells), thalamus, hypothalamus, olfactory bulb, and basal ganglia and in the retina (McGeer et al., 1987). The cerebral cortex contains GABA-terminals in layer IV. Substantia nigra is also very rich in GABA terminals which arise from the globus pallidus and control the activity of dopaminergic cell bodies.

## Glycine

Glycine was not considered as a possible neurotransmitter since a study of its distribution in the spinal cord showed that it could be a postsynaptic inhibitory transmitter. Then, subcellular fractionation studies indicated that it was preferentially localized to synaptosomes, an in vitro model of synapse. Glycine inhibition is also achieved by increasing conductance for $Cl^-$.

### Synthesis and release

Glycine is a nonessential amino acid that crosses the blood–brain barrier and is incorporated in peptides, proteins, nucleotides and nucleic acids. It must be synthesized de novo in brain tissue since the uptake of glycine from blood is too slow. The immediate precursor of glycine is serine and some studies have shown that glycine in brain is synthesized from glucose through serine and not from transport. The enzyme serine hydroxymethyltransferase (SHMT) converts serine to glycine. A $Ca^{2+}$-dependent and a $K^+$-stimulated release has been demonstrated in retina and in spinal cord, providing some evidence that glycine is a transmitter.

### Distribution of glycine

Unlike GABA, the activity of SHMT cannot be used to map glycine-containing neurons in brain. Regional distribution data suggests the presence of glycine in nerve endings in the pons, medulla and midbrain and possibly the diencephalon. Renshaw interneurons in the spinal cord are glycinergic. Recently, glycine-like immunoreactivity has been used to map glycinergic neurons and evidence for a transmitter role of glycine

in the cerebellum was shown (Ottersen et al., 1988; Ottersen, 1989). The authors mentioned that even if glycine serve metabolic functions in addition to a transmitter role, it is easy to interpret the immunolabelling pattern for glycine due to a great difference between the transmitter and metabolic pools.

## Excitatory amino acids

Numerous observations suggest that the major excitatory transmitters in the brain may be glutamate and aspartate, two amino acids which are present in high concentration. Glutamate and aspartate have largely been shown to fulfil the requirements for a substance to be a transmitter, i.e. $Ca^{2+}$-dependent release upon stimulation, high affinity uptake into nerve terminals, presence of the amino acids and synthetic enzyme in nerve terminals, blockade of synaptic transmission by excitatory amino acid antagonists.

## Aspartate and glutamate

Only a few investigations discriminate between the action of these two amino acids. They are metabolically closely related and they are taken up by the same high-affinity uptake mechanism. They are released by the same mechanisms and can act as false transmitter for each other.

### Synthesis, storage and release

The influx of glutamate from plasma across the blood–brain barrier is much lower than the efflux of glutamate from brain, i.e. glutamate must be synthesized in brain tissue. Both glutamate and aspartate can be biosynthesized by transamination of their corresponding ketoacids (2 oxoglutarate and oxaloacetate) produced in the Krebs cycle, located in the mitochondria. Nerve terminals contain mitochondria, i.e. they can generate aspartate and glutamate from glucose and pyruvate. Another source of glutamate is glutamine which is converted to glutamate by a mitochondrial enzyme, glutaminase. Such an additional pathway does not exist for aspartate; asparagine is absent or almost undetectable in brain. The relative contribution of glutamine or glucose to the transmitter glutamate remains an open question. For glutamate there is also a small compartment confined to astroglial cells where glutamine is synthesized from glutamate via glutamine synthetase and glutamine can then be transported from glial cells into nerve terminals and acts as the major precursor for newly synthesized glutamate via glutaminase.

The release of both aspartate and glutamate has been demonstrated to be $Ca^{2+}$-dependent in different brain and spinal cord regions and also in glial preparations (Fonnum, 1984). Whether amino acid neurotransmitters are synaptically released from the cytosol or from the synaptic vesicles by exocytosis remains controversial (Nicholls, 1989; Burger *et al.*, 1989). Glutamate-like immunoreactivity has been indicated in some synaptic vesicles and an ATP-dependent uptake of glutamate into protein-associated synaptic vesicles has been demonstrated. Similar levels of glutamate and aspartate are present in synatic vesicles and they represent a fractionable percentage of what is present in the terminal. Some data suggest that part of the amino acids released come from the cytosol.

### *Distribution of aspartate and glutamate*

The ubiquitous presence and high concentration of glutamate and aspartate in brain makes difficult the localization of the specific neurons using these amino acids. Different techniques have been used to identify glutamate or aspartate pathways. As mentioned above, glutamate and aspartate are taken up by a high-affinity uptake mechanism. One of the techniques used to identify glutamate and/or aspartate pathways is the combination of lesions and high-affinity uptake studies. Also lesions combined with $Ca^{2+}$-dependent release or with the level of the transmitter itself have been of considerable value (Fonnum *et al.*, 1979).

Sensitivity to iontophoresed glutamate and aspartate on the postsynaptic site may also indicate a pathway whose transmitter is aspartate or glutamate.

Retrograde transport of tritiated D-aspartate, after microinjection into the nerve terminals is another method based on the high affinity transport of aspartate and glutamate (Streit, 1980). D-Aspartate, which is not metabolized acts as a false transmitter and is taken up by the same high-affinity mechanisms as the L-forms of aspartate and glutamate.

Immunohistochemical methods with aspartate aminotransferase and glutaminase have also been used to map these amino acids' pathways but the most promising technique today is the histochemical localization of glutamate-like immunoreactivity (Storm-Mathisen *et al.*, 1983). Recent methodological developments have made possible the utilization of specific anti-

sera to explore the cellular and subcellular distribution of neuroactive amino acids in a quantitative manner (Ottersen, 1989). This quantitative immunocytochemistry looks promising in distinguishing between the different pools of amino acids: transmitter and metabolic pools.

Using these approaches, a number of pathways have been proposed to utilize excitatory amino acids: all the sensory as well as most of the corticofugal pathways (Ottersen & Storm-Mathisen, 1984), allocortical and subcortical pathways (Taxt & Storm-Mathisen, 1979), various populations of cells in the retina (Miller & Slaughter, 1986), and ganglion cells projecting to the lateral geniculate and superior colliculus.

Among the corticofugal pathways from the neocortex, distribution of glutamate fibres has been proposed in the nucleus accumbens, the neostriatum, the thalamus, the lateral geniculate body, the colliculus superior, the substantia nigra, the amygdala and the olfactory tubercle. The projection to the accumbens, the olfactory tubercle and the neostriatum comes from the prefrontal and frontal cortex, the projection to the substantia nigra comes from the frontal cortex, the projection to the thalamus from the entire cortex (particularly from the pyriform cortex) and the projection to the colliculus superior and the lateral geniculate from the visual cortex.

In the hippocampus, glutamate is the transmitter of the perforant path which originates in the entorhinal cortex and projects to the fascia dentata (granular cell dendrites and mossy fibre endings) and to the pyramidal cell dendrites of CA3. Another hippocampal system that utilizes glutamate is the Schaffer collaterals (axons from the pyramidal neurons of CA3 projecting to the stratum radiatum of CA1). The fornix/fimbria fibres which originate in the hippocampus-subiculum are part of another important glutamate pathway (Jay & Witter, 1991; Jay et al., 1992); these fibres reach the lateral septum, the nucleus accumbens, the prefrontal cortex, the nucleus of the diagonal band, the mediobasal hypothalamus, and the mammillary body.

Within the cerebellum, climbing and parallel fibres are among the proposed systems using excitatory amino acids, in this case aspartate seems to have a preferential role (Wiklund et al., 1982; Sandoval & Cotman, 1978).

## Cotransmission

In the last ten years the discovery that individual neuron could contain two active compounds or transmitters has raised some doubt about the Dale's principle widely accepted by a generation of neurobiologists. The concept that each nerve cell makes and releases only one neurotransmitter had to be reexamined. The term cotransmission can be defined as the action of two transmitters simultaneously released from the same neuron on a single target cell.

One evidence of cotransmission has been shown in individual sympathetic neurons maintained in tissue culture that may contain and release either acetylcholine, norepinephrine or both (Furshpan et al., 1976). There is also biochemical evidence that large single invertebrate neurons contain and release more than one transmitter (Brownstein et al., 1974; Cottrell, 1976). In the mammalian central nervous system, there is morphological evidence that neurons contain monoamine and peptide in the same terminal (Chan-Palay et al., 1978; Hökfelt et al., 1977b; Pickel et al., 1977). Mesencephalic dopamine neurons contain a cholecystokinin-like peptide (Hökfelt et al., 1980). Substance P has been detected in serotonin-containing neurons in the brainstem. Vasoactive intestinal polypeptide (VIP) appears to be colocalized with acetylcholine in the cerebral cortex (Eckenstein & Baughman, 1984) and neuropeptide Y with epinephrine neurons in the locus ceruleus (Hökfelt et al., 1983). Galanin immunoreactivity was demonstrated in cholinergic neurons of the septum-basal forebrain complex projecting to the hippocampus in the rat (Melander et al., 1985).

These studies raise a semantic problem. The presence of two transmitters does not mean that the two substances are released. Coexistence and corelease are the necessary conditions for cotransmission. Whether the two transmitters are always released together or only in certain cases remains to be established. Release of classical neurotransmitters (acetylcholine, norepinephrine, serotonin) occurs at lower frequencies (1–10 Hz) than release of the coexisting peptide neurotransmitters (5–40 Hz). In the example showing evidence of corelease (VIP and acetylcholine) from sympathetic nerves in the salivary glands, acetylcholine is released at low frequencies, producing salivation and vasodilation and VIP is released at higher frequencies, causing a marked vasodilation (Lundberg, 1981). In this case, VIP is enhancing the effect of acetylcholine. This differential release of classical neurotransmitters and peptides may be the result of tissue stores of classical neurotransmitters which are 50–1000 times greater than those of the peptides neurotransmitters.

The release of classical neurotransmitters is subject to regulation by presynaptic receptors. Can a presynaptic receptor known to regulate the release of the coexisting neurotransmitter also regulate the release of the other neurotransmitter from the same terminal? This question has been studied (for review see Bartfai et al., 1988). Figure 12 summarizes the results which in different systems indicate that the inhibition or enhancement of the release of one of the coexisting neurotransmitters through a presynaptic receptor regulates the release of the other coexisting neurotransmitter. In addition to the autoreceptors for classical neurotransmitter there are peptidergic autoreceptors which may regulate the release of the classical neurotransmitter. There is the possibility of as many types of autoreceptors in a nerve terminal as the terminal contains neurotransmitters.

Cotransmission does occur in experimental conditions but numerous questions about its function are still unanswered. It appears that the terminals do use more than one transmitter but it remains to be shown that this occurs during normal behaviour. Not all transmissions seem to be mediated by several transmitters. The cotransmission may affect the size and the nature of the signal. One transmitter may facilitate the actions of the other

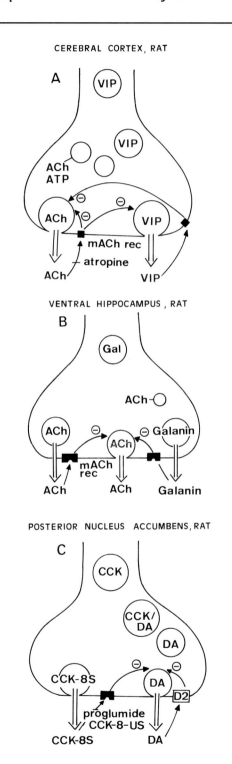

CEREBRAL CORTEX, RAT

A

VENTRAL HIPPOCAMPUS, RAT

B

POSTERIOR NUCLEUS ACCUMBENS, RAT

C

*Fig. 12. Schematic drawing indicating the regulation of coexisting neurotransmitters via presynaptic receptors. (From Bartfai et al., 1988.)*

*(A) In rat cerebral cortex slices, the muscarinic autoreceptor inhibits the release of acetylcholine (ACh) and of the coexisting vasoactive intestinal peptide (VIP). VIP inhibits the release of ACh.*

*(B) In the ventral hippocampus of the rat where high concentration of galanin specific binding sites was demonstrated on cholinergic terminal, galanin inhibits the ACh release. ACh release is also modulated by presynaptic muscarinic receptor.*

*(C) In the posterior nucleus accumbens and in the rat nucleus caudatus-putamen, where dopamine (DA)/cholecystokinin (CCK)-8S coexistence was demonstrated, CCK-8S inhibits the release of DA. DA also regulates its own release via D2 presynaptic receptors.*

by modulation of its receptors. It may have quite different postsynaptic effects and act on different targets.

## Parasynaptic transmission

A common hypothesis in neurobiology is that neurotransmitters bind to postsynaptic receptors located at the terminals of pathways containing the relevant transmitter. In the recent literature, it appears that the distribution of receptors mapped by autoradiography and the location of transmitter release studied by immunohistochemistry of nerve terminals do not regularly display the expected perfect correspondences (for review, Herkenham, 1987). This disparity of localization may reflect another mode of communication used by transmitters and receptors, namely parasynaptic. The parasynaptic mode involves the diffusion of substances from release sites through the extracellular fluid to target cells that possess specific receptors for these substances.

The first morphological evidence that a transmitter could act at a distance was examined with monoamines (Descarries *et al.*, 1975; Beaudet & Descarries, 1978; Molliver *et al.*, 1982). These authors found that a large proportion of nerve terminals did not make synaptic contacts in rat cortex. In addition, pharmacological studies have shown for the noradrenergic system a nonsynaptic/presynaptic action where transmitters are released into the extracellular fluid from local synaptic release sites and reach the receptor after diffusion (Mobley & Greengard, 1985). The nonsynaptic release of monoamines in the central nervous system is analogous to the fact that in the peripheral autonomous nervous system, the axon terminals that release norepinephrine or acetylcholine rarely make synaptic contacts with effector tissues. Transmitters diffuse for distances of several micrometers to reach their targets such as smooth muscle (Iversen, 1979).

Evidence that transmitters are nonsynaptically released from dendrites and axons was also demonstrated from ultrastructural studies for transmitters like substance P, serotonin, somatostatin, neuropeptide Y and vasoactive intestinal peptide. On the other hand, ultrastructural studies of receptors have shown a nonsynaptic localization for neurotensin and opiate receptors but more work is needed to show the extent of this localization.

Support for the hypothesis of parasynaptic transmission also comes from studies in the cerebrospinal fluid which contain every transmitter yet identified. The hormonal role of peptides was shown in studies of cerebrospinal fluid peptide changes associated with electrical stimulation. Receptors localized on epithelial cells of the choroid plexus which secretes cerebrospinal fluid are at a distance from nerve terminals which are localized on blood vessels.

The parasynaptic transmission which occurs in parallel with the neuronal circuitry and has the same degree of selectivity is a new concept of interneural communication.

## Summary

When a neuron fires, it releases a chemical substance from special storage vesicles into the synapse. This chemical mediator is called a neurotransmitter. It crosses the synapse to the membrane of the next neuron (or muscle cell) where it interacts with specific receptors. A number of criteria have to be defined for a substance to be considered as a neurotransmitter. To summarize, it has to be synthesized in a neuron, released by nerve action and specifically bound by the postsynaptic membrane where it causes stimulation or inhibition of the postsynaptic cell by depolarization and hyperpolarization respectively. In addition, transmitters are localized in specific neural pathways and most of these different systems have been mapped.

The classical transmitters are the amines: acetylcholine, dopamine, adrenaline, noradrenaline, serotonin and histamine and the amino acids: γ-amino-butyric acid (GABA), glycine, glutamate and aspartate. Other potent neurotransmitters, the neural peptides, have been added to the list of compounds. These released substances are present in small quantities in localized regions of the nervous system. It seems they act locally as transmitters at specific synapses and at a distance as modulators or regulators.

Later, the concept that a neuron releases only one transmitter had to be reexamined and the term cotransmission was defined as the action of two transmitters simultaneously released from the same neuron. When the two transmitters are stored in separate vesicles they can be released differentially at different stimulation frequencies. When cotransmitters exist in a common vesicle they cannot be differentially released and one facilitates the action of the other. Cotransmission occurs in experimental conditions, but numerous questions about its function are unanswered.

Finally, the heterogeneous distribution of the sites of release of neurotransmitters and the receptors to which they bind have raised the question as to whether there is only a conventional synaptic junction. Transmitters can diffuse over some distance to bind to receptors in a parasynaptic mode, i.e. in parallel with the synapse-linked circuitry. Two modes of communication could be used by transmitters and receptors, and transmitters could operate in both modes.

## References

Barber, R.P., Phelps, P.E., Houser, C.R., Crawford, G.D., Salvaterra, P.M. & Vaughn, J.E. (1984): The morphology and distribution of neurons containing choline acetyltransferase in the adult rat spinal cord: an immunocytochemical study. *J. Comp. Neurol.* **229**, 329–346.

Barchas, J.D., Akil, H., Elliott, G.R., Holman, R.B. & Watson, S.J. (1978): Behavioral neurochemistry: neuroregulators and behavioural states. *Science* **200**, 964–973.

Barry, J., Dubois, M.P. & Carette, B. (1974): Immunofluorescence study of the preoptic-infundibular LRF neurosecretory pathway in the normal gastrated or testosterone-treated male guinea pig. *Endocrinology* **95**, 1416–1423.

Bartfai, T., Iverfeldt, K., Fisone G. & Serfözö, P. (1988): Regulation of the release of coexisting neurotransmitters. *Ann. Rev. Pharmacol. Toxicol.* **28**, 285–310.

Beaudet, A. & Descarries, L. (1978): The monoamine innervation of rat cerebral cortex: synaptic and nonsynaptic axon terminals. *Neuroscience* **3**, 851–860.

Beinfeld, M.C. (1983): Cholecystokinin in the central nervous system: a minireview. *Neuropeptides* **3**, 411–427.

Berelowitz, M., Kronheim, S., Pimstone, B. & Sheppard, M. (1978): Potassium stimulated calcium dependent release of immunoreactive somatostatin from incubated rat hypothalamus. *J. Neurochem.* **31**, 1537–1539.

Björklund, A. & Lindvall, O. (1984): Dopamine-containing systems in the CNS. In: *Handbook of chemical neuroanatomy*, Vol. 2, eds. A. Björklund & O. Lindvall, pp. 55–122. Amsterdam: Elsevier.

Bloom, F., Battenberg, E., Rossier, J., Ling, N. & Guillemin, R. (1978): Neurons containing β–endorphin in rat brain exist separately from those containing enkephalin: immunocytochemical studies. *Proc. Natl. Acad. Sci. USA* **75**, 1591–1595.

Brownstein, M., Arimura, A., Sabo, H., Schally, A.V. & Kizer, J.S. (1975): The regional distribution of somatostatin in the rat brain. *Endocrinology* **96**, 1456–1461.

Brownstein, M.J., Saavedra, J.M., Axelrod, J., Zeman, G.H. & Carpenter, D.O. (1974): Coexistence of several putative neurotransmitters in single identified neurons of aplysia. *Proc. Natl. Acad. Sci. USA* **71**, 4662–4665.

Burger, P.M., Mehl, E., Cameron, P.L., Maycox, P.R., Baumert, M., Lottspeich, F., De Camilli, P. & Jahn, R. (1989): Synaptic vesicles immunoisolated from rat cerebral cortex contain high levels of glutamate. *Neuron* **3**, 715–720.

Carraway, R. & Leeman, S.E. (1975): The amino acid sequence of a hypothalamic peptide, neurotensin. *J. Biol. Chem.* **250**, 1907–1911.

Chan-Palay, V., Jonsson, G. & Palay, S.L. (1978): Serotonin and substance P coexist in neurons of the rat's central nervous system. *Proc. Natl. Acad. Sci. USA* **75**, 1582–1586.

Cottrell, G.A. (1976): Does the giant cerebral neurone of *Helix* release two transmitters: ACh and serotonin? *J. Physiol.* (Lond.) **259**, 44–45P.

Cuello, A.C. & Sofroniew, M.V. (1984): The anatomy of the CNS cholinergic neurons. *Trends Neurosci.* **7**, 74–78.

Dahlström, A. & Fuxe, K. (1964): Evidence for the existence of monoamine-containing neurons in the central nervous system. I Demonstration of monoamines in the cell bodies of brainstem neurons. *Acta Physiol. Scand.* **232**, (Suppl.) 1–55.

Dale, H.H. (1935): Pharmacology and nerve endings. *Proc. R. Soc. Med.* **28**, 319–332.

De Belleroche, J.S. & Bradford, H.F. (1972): The stimulus induced release of acetylcholine from synaptosome beds and its calcium dependence. *J. Neurochem.* **19**, 1817–1819.

Descarries, L., Beaudet, A. & Watkins, K. (1975): Serotonin nerve terminals in adult rat neocortex. *Brain Res.* **100**, 563–588.

Eccles, J.C. (1957): *The physiology of nerve cells.* Baltimore: Johns Hopkins Press.

Eckenstein, F. & Baughman, R.W. (1984): Two types of cholinergic innervation in cortex, one co-localized with vasoactive intestinal peptide. *Nature* (Lond.) **309**, 153–155.

Emson, P.C. & Koob, G.F. (1978): The origin and distribution of dopamine-containing afferents to the rat frontal cortex. *Brain Res.* **142**, 249–267.

Emson, P.C. (1979): Peptides as neurotransmitter candidates in the CNS. *Prog. Neurobiol.* **13**, 61–116.

Emson, P.C. & De Quidt, M.E. (1984): NPY – a new member of the pancreatic polypeptide family. *Trends Neurosci.* **7**, 31–35.

Eppelbaum, J., Willoughby, J.O., Brazeau, P. & Martin, J.B. (1977): Effects of brain lesions and hypothalamic deafferentation on somatostatin distribution in rat brain. *Endocrinology* **101**, 1495–1502.

Eppelbaum, J. (1986): Somatostatin in the central nervous system: physiology and pathological modifications. *Prog. Neurobiol.* **27**,: 63–100.

Fagg, G.E. & Foster, A.C. (1983): Amino acid neurotransmitters and their pathways in the mammalian central nervous system. *Neuroscience* **9**, 701–719.

Fahrenkrug, J. (1980): Vasoactive intestinal polypeptide. *Trends Neurosci.* **3**, 1–2.

Fonnum, F. (1984): Glutamate: a neurotransmitter in mammalian brain. *J. Neurochem.* **42**, 1–11.

Fonnum, F., Lund-Karlsen, R., Malthe-Sorenssen, D., Sterri, S. & Walaas, I. (1979): Localization of neurotransmitters, particularly glutamate, in hippocampus, septum, nucleus accumbens, and superior colliculus. *Prog. Brain Res.* **51**, 167–191.

Fuller, R.W. (1982): Pharmacology of brain epinephrine neurons. *Ann. Rev. Pharmacol. Toxicol.* **22**, 31–55.

Furshpan, E.J., Macleish, P.R., O'Lague, P.H. & Potter, D.D. (1976): Chemical transmission between rat sympathetic neurons and cardiac myocytes developing in microcultures: evidence for cholinergic, adrenergic, and dual-function neurons. *Proc. Natl. Acad. Sci.*

*USA* **73**, 4225–4229.

Fuxe, K., Ganten, D., Hökfelt, T. & Bolme, P. (1976): Immunohistochemical evidence for the existence of angiotensin II-containing nerve terminals in the brain and spinal cord in the rat. *Neurosci. Lett.* **5**, 241–246.

Gibson, G.E. & Blass, V.P. (1976): Impaired synthesis of acetylcholine accompanying mild hypoxia and hypoglycemia. *J. Neurochem.* **27**, 37–42.

Glowinski, J. (1982): *In vivo* regulation of nigro-striatal and mesocortico-prefrontal dopaminergic neurons. In: *Neurotransmitters interaction and compartmentation*, ed. H.F. Bradford, pp. 219–234. New York: Plenum Press.

Handelmann, G.E., Beinfeld, M.C., O'Donohue, T.L., Nelson, J.B. & Brenneman, D.E. (1983): Extra-hippocampal projections of CCK neurons of the hippocampus and subiculum. *Peptides* **4**, 331–334.

Henry, J.L. (1980): Substance P and pain: an updating. *Trends Neurosci.* **3**, 95–97.

Herkenham, M. (1987): Mismatches between neurotransmitter and receptor localizations in brain: observations and implications. *Neurosci.* **23**, 1–38.

Hökfelt, T., Fuxe, K., Goldstein, M. & Johansson, O. (1974): Immunohistochemical evidence for the existence of adrenaline neurons in the rat brain. *Brain Res.* **66**, 235–251.

Hökfelt, T. Elde, R., Johanson, O., Terenius, L. & Stein, L. (1977a): The distribution of enkephalin immunoreactive cell bodies in the rat central nervous system. *Neurosci. Lett.* **5**, 25–31.

Hökfelt, T., Elfvin, L.G., Elde, R., Schultzberg, M., Goldstein, M. & Luft, R. (1977b): Occurence of somatostatin-like immunoreactivity in some peripheral sympathetic noradrenergic neurons. *Proc. Natl. Acad. Sci. USA* **74**, 3587–3591.

Hökfelt, T., Skirboll, L., Rehfeld, J.F., Goldstein, M., Markey, K. & Dann, O. (1980): A subpopulation of mesencephalic dopamine neurons projecting to limbic areas contains a cholecystokininlike peptide. Evidence from immunohistochemistry combined with retrograde tracing. *Neuroscience* **5**, 2093–2124.

Hökfelt, T., Lundberg, J.M., Langercrantz, H., Tatemoto, K., Mutt, V., Terenius, L., Polak, J., Bloom, S., Sasek, C., Elde, R. & Goldstein, M. (1983): Neuropeptide Y (NPY)-and FMRF amide neuropeptide-like immunoreactivities in catecholaminergic neurons of the rat medulla oblongata. *Acta Physiol. Scand.* **117**, 315–318.

Holz, R.W. (1978): Evidence that catecholamines transport in chromaffin vesicles is coupled to vesicle membrane potential. *Proc. Natl. Acad. Sci. USA* **75**, 5190–5194.

Hong, J.S., Yang, H.Y.T., Racagni, G. & Costa, E. (1977): Projections of substance P containing neurons from neostriatum to substantia nigra. *Brain Res.* **122**, 541–544.

Iversen, L.L. (1979): Co-transmitters-modulation, and the peripheral nervous system. *Behav. Brain Sci.* **2**, 430.

Jay, T.M. & Witter, M.P. (1991): Distribution of hippocampal CA1 and subicular afferents in the prefrontal cortex of the rat studied by means of anterograde transport of *Phaseolus vulgaris-Leucoagglutinin*. *J. Comp. Neurol.* **313**, 574–586.

Jay, T.M., Thierry, A.M., Wiklund, L. & Glowinski, J. (1992):Excitatory amino acid pathway from the hippocampus to the prefrontal cortex. Contribution of AMPA receptors in hippocampo-prefrontal cortex transmission. *Eur. J. Neurosci.* **4**, 1285–1295.

Joh, T.H., Shikimi, T., Pickel, V.M. & Reis, D.J. (1975): Brain tryptophan hydroxylase: purification of production of antibodies to, and cellular and ultrastructural localization in serotonergic neurons of rat midbrain. *Proc. Natl. Acad. Sci. USA* **72**, 3575–3580.

Katz, B. (1966): *Nerve, muscle and synapse*. New York: McGraw-Hill.

Kobayashi, R.M., Brown, M. & Vale, W. (1977): Regional distribution of neurotensin and somatostatin in rat brain. *Brain Res.* **126**, 584–589.

Krieger, D.T., Brownstein, M.J. & Martin, J.B. (1983): *Brain peptides*. New York: Wiley Press.

Larsson, L.-I., Fahrenkrug, J., Schaffalitzky De Muckadell, O., Sundler, F., Hakanson, R. & Rehfeld, J.F. (1976): Localization of vasoactive intestinal polypeptide (VIP) to central and peripheral neurons *Proc. Natl. Acad. Sci. USA* **73**, 3197–3200.

Leeman, S.E. & Mroz, E.A. (1975): Substance P. *Life Science* **15**, 2033–2044.

Lindvall, O. & Björklund, A. (1974): The organization of the ascending catecholamine neuron systems in the rat brain as revealed by the glyoxylic acid fluorescence method. *Acta Physiol. Scand.* Suppl. **412**, 1–48.

Loewi, O. (1921): Über humorole Übertragborkeit der herznervenwirkung. I. Mitteilung. *Pflügers Arch.* **189**, 239–242.

Lundberg, J.M. (1981): Evidence for coexistence of vasoactive (VIP) and acetylcholine in neurons of cat exocrine glands.

Morohological, biochemical and functional studies. *Acta Physiol. Scand.* Suppl. **496,** 1-57.

Lundberg, J.M. & Tatemoto, K. (1982) Pancreatic polypeptide family (APP, Bpp, NPY and PYY) in relation to sympathetic vasoconstriction resistant to alpha-adrenoceptor blockade. *Acta Physiol. Scand.* **116,** 393-402.

McGeer, P.L., Eccles, J.C. & McGeer, E.G. (1987): Inhibitory amino acids. In: *Molecular neurobiology of the mammalian brain*, eds. P.L. McGeer, J.C. Eccles and E.G. McGeer, pp. 197-234. New York: Plenum Press.

Melander, T., Staines, W.A., Hökfelt, T., Rökaeus, A. & Eckenstein, F. (1985): Galanin-like immunoreactivityin cholinergic neurons of the septum-basal forebrain complex projecting to the hippocampus of the rat. *Brain Res.* **360,** 130-138.

Mesulam, M.M., Mufson, E.J., Wainer, B.H. & Levey, A.I. (1983): Central cholinergic pathways in the rat: an overview based on an alternative nomenclature (Ch1-Ch6). *Neuroscience* **10,** 1185-1201.

Miller, R.F. & Slaughter, M.M. (1986): Excitatory amino acid receptors of the retina: diversity of subtypes and conductance mechanisms. *Trends Neurosci.* **9,** 211-218.

Mobley, P. & Greengard, P. (1985): Evidence for widespread effects of noradrenaline on axon terminals in the rat frontal cortex. *Proc. Natl Acad. Sci. USA* **82,** 945-947.

Molliver, M. E., Grzanna, R., Lidov, H.G.W., Morrison, J.H. & Olschowka, J.A. (1982): Monoamine systems in the cerebral cortex. In: *Cytochemical methods in neuroanatomy*, eds V. Chan-Palay & S.L. Palay, pp. 255-277. New York: Alan R. Liss.

Nemeroff, C., Luttinger, D. & Pranje, A.J. (1980): Neurotensin: central nervous system effects of a neuropeptide. *Trends Neurosci.* **3,** 212-215.

Nicholls, D.G. (1989): Release of glutamate, aspartate, and γ-aminobutyric acid from isolated nerve terminals. *J. Neurochem.* **52,** 331-341.

Ogawa, N., Yamakawi, Y., Kuroda, H., Ofuji, T., Itoga, E. & Kito, S. (1981): Discrete regional distributions of thyrotropin releasing hormone (TRH) receptor binding in monkey central nervous system. *Brain Res.* **205,** 169-174.

Ogawa, R., Itoh, K., Kaneko, T. & Mizuno. N. (1989): Co-existence of vasoactive intestinal polypeptide (VIP)- and cholecystokinin (CCK)-like immunoreactivities in thalamocortical neuron in the ventrolateral nucleus of the rat. *Brain Res.* **490,** 152-156.

Ottersen, O.P. & Storm-Mathisen, J. (1984): Neurons containing or accumulating transmitter amino acids. In: *Handbook of chemical neuroanatomy*, Vol. 3, eds. A. Björklund, T. Hökfelt & M.J. Kuhar, pp. 141-246. Amsterdam: Elsevier.

Ottersen, O.P., Storm-Mathisen, J. & Somogyi, P. (1988): Colocalization of glycine-like and GABA-like immunoreactivities in Golgi cell terminals in the rat cerebellum: a postembedding light and electron microscopic study. *Brain Res.* **450,** 342-353.

Ottersen, O.P. (1989): Quantitative electron microscopic immunocytochemistry of neuroactive amino acids. *Anat. Embryol.* **180,** 1-15.

Phelps, P.E., Houser, C.R. & Vaughn, J.E. (1985): Immunocytochemical localization of choline acetyltransferase within the rat neostriatum: a correlated light and electron microscopic study of cholinergic neurons and synapses. *J. Comp. Neurol.* **238,** 286-307.

Pickel, V.M., Reis, D.J. & Leeman, S.E. (1977): Ultrastructural localization of substance P in neurons of spinal cord. *Brain Res.* **122,** 534-540.

Ramsey, D.J. (1979): The brain renin angiotensin system: a re-evaluation. *Neuroscience* **4,** 313-321.

Sandoval, M.E. & Cotman, C.W. (1978): Evaluation of glutamate as a neurotransmitter of cerebellar fibres. *Neuroscience* **3,** 199-206.

Seroogy, K.B. & Fallon, J.H. (1989): Forebrain projections from cholecystokinin-like immunoreactive neurons in the rat brain. *J. Comp. Neurol.* **279,** 415-435.

Sofroniew, M.V. (1983): Vasopressin and oxytocin in the mammalian brain and spinal cord. *Trends Neurosci.* **6,** 467-472.

Storm-Mathisen, J., Leknes, A.K., Bore, A.T., Vaaland, J.L., Edminson, P., Haug, F.M.S. & Ottersen, O.P. (1983): First visualization of glutamate and GABA in neurones by immunocytochemistry. *Nature (Lond.)* **301,** 517-520.

Streit, P. (1980): Selective retrograde labeling indicating the transmitter of neuronal pathways. *J. Comp. Neurol.* **191,** 429-463.

Swanson, W.L. & Sawchenko, P.E. (1983): Hypothalamic integration: organization of the paraventricular and supraoptic nuclei. *Ann. Rev. Neurosci.* **6,** 269-324.

Smith, A.R. & Ariëns Kapper, J. (1975): Effect of pinealectomy, pCPA, and pineal extracts on the rat parvocellular neurosecretory hypothalamic system: a fluorescence histochemical investigation. *Brain Res.* **86,** 353-371.

Snyder, S. (1980): Brain peptides as neurotransmitters. *Science* **209,** 976-983.

Steinbush, H.W.M. (1981): Distribution of serotonin immunoreactivity in the central nervous system of rat of the rat cell bodies and terminals. *Neuroscience* **6,** 557–618.

Tamir, H. & Liu K.P. (1982): On the nature of the interaction between serotonin and serotonin binding protein. Effects of nucleotides, ions and sulfhydryl reagents. *J. Neurochem.* **38,** 135–141.

Taxt, T. & Storm-Mathisen, J. (1979): Tentative localization of glutamergic and aspartergic nerve endings in brain. *J. Physiol. (Paris)* **75,** 677–684.

Ungerstedt, U. (1971): Striatal dopamine release after amphetamine or nerve degeneration revealed by rotational behaviour. *Acta Physiol. Scand.* Suppl. **367,** 49–67.

Von Euler, U.S. & Gaddum, J.H. (1931): An unidentified depressor substance in certain tissue extracts. *J. Physiol. (Lond.)* **72,** 74–87.

Wiklund, L., Toggenburger, G. & Cuenod, M. (1982): Aspartate: possible neurotransmitter in cerebellar climbing fibers. *Science* **216,** 78–80.

# Chapter 4

# Receptors, ionic channels and ion pumps

Robert C. Miller, Martin Galvan and C. Robin Hiley*

*Marion Merrell Dow Research Institute, 16 rue d'Ankara, 67084 Strasbourg, France; and*
*\*Department of Pharmacology, University of Cambridge, Tennis Court Road, Cambridge CB2 1QJ, England*

## Introduction

With recent advances in understanding of pharmacology and molecular biology, the word 'receptor' has come to have several distinct meanings but, in this chapter, only those receptors that are found on cell membranes will be considered. In the simplest case of this type, a receptor is a macromolecule that spans the membrane and contains within its structure a specific recognition site (or binding site), for a group of chemical structures called ligands. Such recognition sites are often referred to as receptors by pharmacologists, and ligand in this case refers to a molecule that can induce a pharmacological response (an agonist), or inhibit the induction of a pharmacological response (an antagonist). A comprehensive list of recep-

tors and their specific agonists, antagonists and second messenger systems has been published (Watson & Abbot, 1991), and a simplified table is given in an annex at the end of the book. Receptor macromolecules whose physiological regulation depends primarily on the presence of a ligand and the activation of intracellular enzymes are referred to as ligand-operated receptors. Receptor macromolecules that incorporate an ion channel may be ligand-operated but they are often regulated physiologically by changes in membrane potential, that is they may be voltage-operated, and regulation by ligands is of secondary importance. Also, some ion channels may regulated by intracellular components such as ATP.

## Receptors

### Role

To induce a pharmacological response, the receptor macromolecules function as transducers that, when activated, transfer information from the outside to the inside of cells. This activation can take the form of induction of enzyme activity or of ion movements across the membrane which will usually change the potential difference across the cell membrane.

### Summary of receptor structures

Receptors structures can be divided into several classes of which the major types are described in detail below. Among these groups there are some receptors typified by the nicotinic re-

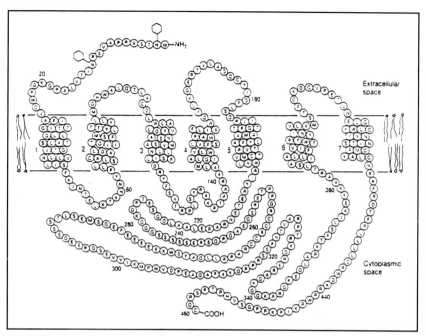

single protein molecule. These proteins are characterized by the presence of seven largely helical membrane-spanning regions and have an extracellular amino terminus with an intracellular carboxy terminus (Fig. 1). Extracellular regions may be glycosylated. These receptors are also characterized by their association with membrane-bound protein complexes that bind guanosine nucleotides, the G-proteins. The G-proteins are enzymes and transmit information from the receptors to other membrane-associated enzymes or ion channels (see also Chapter 6).

Receptor macromolecules may also have a simpler primary structure. For example the receptors for nerve growth factor, epidermal growth factor and platelet-derived growth factor, have only one membrane spanning domain. Receptors of this type are not linked to G-proteins, but some exhibit kinase activity, either threonine or serine/threonine specific, as is the case for the receptors for platelet-derived growth factor, epider-

*Fig. 1. The amino acid sequence of the muscarinic $m_1$ receptor for acetylcholine (represented by the single letter code for amino acids) showing the seven putative transmembrane helices. The transmembrane helical sequences are joined by relatively short loops (except for the loop joining Helices 5 and 6 in the intracellular space) in the intra- and extra-cellular spaces. The N-terminal region projects into the extracellular space and carries oligosaccharide chains attached to asparagine residues (represented by N). The C-terminus is in the intracellular space. Although the general arrangement is similar in different members of the rhodopsin superfamily of receptors, the lengths of the interhelical loops can vary. The $\beta_2$-adrenoceptor has a shorter loop between Helices 5 and 6 than the muscarinic receptor shown here but the intracellular C-terminal region is much longer.*

ceptor for acetylcholine, which incorporate an ion channel composed of several macromolecular subunits, all of which might span the cell membrane. In other cases, such as calcium channels (see below and Bertolino & Llinás, 1992), some of the subunits might not span the membrane. One or more ligand binding sites may be contained in each subunit.

A large group of receptors, which because of their resemblance to the photoreceptor rhodopsin is termed the rhodosphin superfamily, when stimulated elicit rapid mechanical responses in cells and these consist of a

**Table 1. Receptors present in the cerebral vasculature and identified as members of the rhodopsin superfamily**

| Receptor | Identified subtypes | Cloned receptors |
|---|---|---|
| 5-HT | 1A, 1B, 1C, 1D, 2 | 1a, 1b, 1c, 1d, 2 |
| Adrenoceptor | $\alpha_{1A}$, $\alpha_{1B}$, $\alpha_{2A}$, $\alpha_{2B}$, $\beta_1$, $\beta_2$ | $\alpha_{1A}$, $\alpha_{1B}$, $\alpha_{2A}$, $\alpha_{2B}$, $\beta_1$, $\beta_2$ |
| Angiotensin | | 1 structure |
| Dopamine | $D_1$, $D_2$ | $d_1$, $d_2$, $d_3$, $d_4$, $d_5$ |
| Endothelin | $ET_A$, $ET_B$ | $ET_A$, $ET_B$ |
| Muscarinic | $M_1$, $M_2$, $M_3$ | $m_1$, $m_2$, $m_3$, $m_4$, $m_5$ |
| Tachykinin | $NK_1$, $NK_2$, $NK_3$ | $NK_1$, $NK_2$, $NK_3$ |

All the cloned receptors have been characterized as being single proteins, containing seven membrane spanning regions, and are linked to G-proteins.

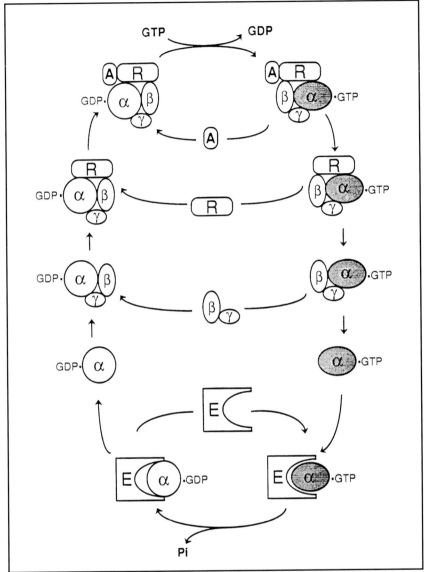

*Fig. 2. Stimulation of adenylyl cyclase by a receptor (R) belonging to the rhodopsin superfamily (e.g. the β-adrenoceptor). The G-protein consists of an α subunit together with a βγ subunit complex which is highly hydrophobic and might serve, amongst other functions (such as having an enzyme regulatory action of its own) to bind the G-protein in the membrane.. In the unstimulated state, the G-protein binds GDP and is associated with the receptor which is in a high affinity state for the agonist (A). Binding of the agonist to the receptor causes a change in the conformation of the receptor and of the associated G-protein (and to a decrease in affinity of the agonist for the receptor promoting the dissociation of the agonist from the receptor). The G-protein/receptor complex dissociates and, either just before this happens or just afterwards, the α subunit releases the GDP and takes up a molecule of GTP. In turn, the activated α subunit then dissociates from the βγ subunit and binds to membrane- bound adenylyl cyclase (E) which begins to generate cyclic AMP from ATP. Hydrolysis of the bound GTP by the GTPase activity of the α subunit ends the stimulation of the cyclase and the α subunit then recombines with the βγ subunits. The cycling of the G-protein can continue as long as the agonist is present.*

mal growth factor, IGF-1 (insulin-like growth factor-1), CSF-1 (colony-stimulating factor-1) and FGF (fibroblast growth factor), or tyrosine specific, for example insulin receptors. The intracellular messenger associated with others is unkown.

Receptor macromolecules are named either by the ion that they preferentially pass or the transmitter to which they normally respond *in vivo*. This transmitter is thus the *primary ligand* for the receptor.

## G-protein coupled (rhodopsin-type) receptors

The seven membrane spanning regions of the rhodopsin-type receptor macromolecules have been modelled on the published X-ray structure of bacteriorhodopsin, a non-G-protein linked proton pump that specifically binds the ligand, retinal, in a membrane (Henderson *et al.*, 1990). In this model, the transmembrane spanning domains of receptors linked to a G-protein are though to be arranged, in a plan view, more or less in an ellipse in the membrane and they interact with each other such that there exists in the structure a cleft open to the extracellular space. From the results of experiments to label irreversibly the ligand-binding sites of muscarinic receptors and of site-directed mutagenesis in receptor macromolecules, the transmitter binding site is thought to be located in this cleft, about one-third of the way through the membrane. In general, like ion channels, the rhodopsin superfamily type of receptors may also carry one or more ligand binding sites other than that for the transmitter. Binding of a ligand to these 'accessory' sites might influence the binding of ligands to the transmitter binding site.

## Transmitter–receptor interactions

The transducer function of receptors occurs by virtue of their coupling to intracellular, or 'second', messenger systems within cells through a cascade of enzymes. It is thought that, in the normal unactivated state of the rhodopsin-type receptor macromolecule, a G-protein complex is bound to its intracelluar surface (Fig. 2). This G-protein is composed of three subunits termed α, β and γ and the α-subunit has a molecule GDP (guanosine diphosphate) bound to it. Activation of the receptor by an agonist results in a conformational change in the receptor macromolecule leading to the release of the G-protein and dissociation of the α-subunit from the βγ complex.

Either just before, or after, dissociation, the GDP is replaced by a GTP (guanosine triphosphate) molecule and this combination can activate other enzymes such as phospholipase C or adenylyl cyclase. Its ability to activate enzymes is terminated when its own enzymatic activity metabolizes the GTP to GDP. The liberated α subunit of the G-protein thus functions as a cofactor. The βγ subunit of G proteins may also activate or modulate cellular systems such as the form of adenylyl cyclase currently designated Type II. So far about 20 α-, four β- and three γ-subunits of G-proteins have been identified. There are therefore many different possible G-proteins and a number have been identified that bind selectively to certain receptors. Since there are very many α-sub-

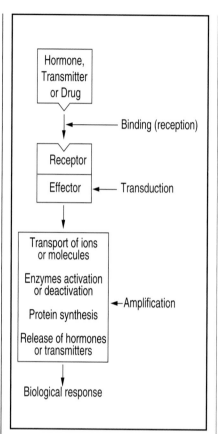

*Fig. 3. Scheme showing the stages in the production of a cellular response by the activation of a receptor. The receptor acts as a specific recognition site for the agonist and the message is conveyed to the cell by means of a transducer mechanism involving an effector (e.g. a G-protein or an ion channel). Most systems produce amplification since a single agonist/receptor complex can produce many molecules of an intracellular messenger, such as cyclic AMP or inositol trisphosphate. A single effector may also be activated or inhibited by other receptors which produces convergence of extracellular signals on a single intracellular mechanism.*

units, several different types of receptors may activate the same enzymes; this is a type of convergence which results in the cell only giving a single type of response as a result of activation by several different agonists. It also means that the cell will give the

same maximal response whether it is stimulated by one or more neurotransmitters or hormones. At the same time, the use of an enzyme as the end point of transduction of an extracellular signal provides amplification of that signal; i.e. combination of a single molecule of an agonist with a single receptor macromolecule generates several intracellular messenger molecules to interact with their targets within the cell (Fig. 3).

## Agonist–receptor interactions

This has been the subject of constant theoretical analysis over the last 60 years and has been reviewed many times (for example Ariëns, 1964; Van Rossum, 1977; Bowman & Rand, 1980; Kenakin, 1987; Black & Leff, 1983). A brief outline of current concepts is given here. The basis of these analyses is an assumption that the law of mass action applies, that the interaction of the agonist with the receptor is reversible, that agonist, receptors and agonist–receptor complexes are in equilibrium and that the response produced is proportional to the number of these agonist–receptor interactions. It is further often assumed that there is an interaction of one molecule of agonist with each receptor molecule.

Agonist + Receptor ⇔ Agonist/ receptor complex ⇒ Response

This can be written:

$$A + R \underset{k_{-1}}{\overset{k_{+1}}{\Longleftrightarrow}} AR \Rightarrow Response$$

($k_{+1}$ and $k_{-1}$ are rate constants for the association and dissociation of the ligand–receptor complex). The rate of formation of agonist-receptor complex is:

$$k_{+1}[A][R]$$

and the rate of breakdown is:

$$k_{-1}[A][R]$$

where the brackets denote respectively the concentration of agonist, receptor and agonist–receptor complex. If the total amount of the receptor in the system is $[R_T]$ then the number of unoccupied $[R]$ available to interact with the agonist is

$$[R] = [R_T] - [AR]$$

and the rate of association can be written as

$$= k_{+1}[A]([R_T] - [AR])$$

With any given concentration of agonist, the rates of association and dissociation of the agonist–receptor complex are equal at equilibrium, so that

$$= k_{+1}[A]([R_T] - [AR]) = k_{-1}[AR]$$

and

$$k_{+1}/k_{-1} = [AR]/\{[A]([R_T]-[AR])\}$$

This ratio has been called the affinity constant ($K_{aff}$, or the equilibrium association constant) of the agonist in a particular system and has the units $M^{-1}$ (i.e. $1 \, mol^{-1}$).

The dissociation constant, $K_d$, the reciprocal of the affinity constant therefore has units M (i.e. $mol \, l^{-1}$) and represents what is usually called the affinity of the agonist for its receptor.

$$K_d = k_{-1}/k_{+1} = \{[A]([R_T]-[AR])\}/[AR]$$

and the proportion of receptors, r, occupied at any agonist concentration

[A] is:

$$r = [AR]/[R_T] = [A]/([A] + K_d)$$
$$\text{Equation A}$$

and this can be expressed as:

$$[A] = \{r/(1-r)\}K_d$$

If however there are n ligand sites on each receptor and all must be occupied to produce the relevant conformational change then:

$$[A]^n = \{r/(1 - r)\}K_d$$

These latter two equations describe a hyperbola when n is > 1. If n = 1 then

the response is proportional to [AR], the proportion of receptors interacting with the agonist, and half the maximal response will be produced when half the receptors are occupied, i.e. when $[AR] = 0.5[R_T]$ or r = 0.5, thus the concentration eliciting the half maximal response (the $EC_{50}$) will be equal to $K_d$.

## 'Spare' receptors; efficacy of agonists; partial agonists

Most observations of agonist–receptor interactions cannot be explained

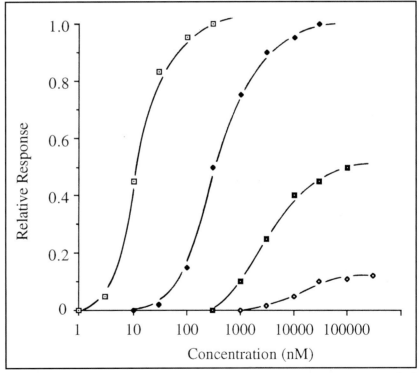

Fig. 4. *Effect of increasing alkylation of receptors on agonist concentration-effect curves. Such curves might be seen in the guinea-pig ileum in response to acetylcholine before and after incubation of the tissue with benzilylcholine mustard, an irreversible antagonist at muscarinic receptors. The control curve (shown by the open squares) and the first curve obtained after exposure to the alkylating agent show the same maximal response; thus it is not necessary to activate all the receptors to obtain the maximum response that can be generated by a tissue. In these circumstances the tissue is said is said to contain 'spare receptors'. Subsequent exposures to the irreversible antagonist produce depression of the maximal response.*

by a linear relationship between receptor occupancy and response (see Clarke, 1937). Progressive irreversible alkylation of a fraction of a receptor population (i.e. their effective removal from the system) sometimes results in a progressive reduction in the apparent potency (increases in the $EC_{50}$ value) of the agonist before any reduction in the maximal response is seen. It would be predicted that further alkylation would then progressively reduce the maximal response without further increasing the $EC_{50}$ (Fig. 4), although according to the operational analysis of Black & Leff (1983), and at least in some experimental circumstances, the $EC_{50}$ can increase to some extent. These observations led to the idea that 'spare' receptors exist, i.e. activation of only a proportion of receptors is necessary to evoke a maximal response. However, the number of receptors present will determine the potency of any given agonist. Thus, changes in the receptor number, 'up-' or 'down-regulation', in a tissue allows respectively an increase or a reduction in sensitivity to an agonist, without necessarily a change in the ability of the tissue to produce an absolute maximal response.

Another early observation was that some agonists in a series, acting on a single type of receptor, elicited smaller maximal responses than other agonosts acting on the same receptors. If the law of mass action does apply, then any agonist eliciting a smaller maximal response than another must be occupying all receptors when the maximum response is obtained and such an agonist is referred to as a partial agonist. The $EC_{50}$ value of a partial agonist should therefore be numeri-

cally equal to its $K_d$ for the receptors. These observations imply that all agonist–receptor interactions are not equal and led to the introduction of the concept of the efficacy of an agonist. A partial agonist has a lower efficacy than an agonist eliciting a greater maximal response (a full agonist) by activation of the same receptor population. It should be noted that efficacy and affinity are independent and that a partial agonist might have a high affinity (higher even than that of a full agonist) for its receptors. The exact nature of efficacy is not known. It might be expected that activation of a receptor by any agonist would result in exactly the same effect inside a cell and that activation is essentially an all-or-nothing event if activation of a G-protein complex is involved. One possible explanation might be that a single ligand-receptor interaction is capable of activating more than one G-protein, the number depending on the ligand. It is also possible that all agonists for a particular type of receptor do not interact with it at exactly the same site. In fact it might be expected that agonists of markedly different physical size would at least have differing secondary interactions with the receptor macromolecule (i.e. interactions around the primary binding site) and that these could favour transduction to differing extents. Such considerations mean that concentration–response relationships rarely fit the equations given above and that quantitative information on agonist–receptor interactions obtained in functional studies must usually be confined to determination of the maximum response and $EC_{50}$. For agonists, determination of binding constants such $K_d$ requires the use of ligand-binding studies (see below).

## Agonist and antagonist interactions

Agonists and antagonists can interact with the receptor macromolecule at the same site, in which case the antagonism can be of either a competitive or irreversible type. In these cases the interaction of the antagonist with the receptor macromolecule probably does not itself induce any conformational change of significance to the response being studied.

## Competitive antagonism

By definition a competitive antagonist interacts reversibly with the agonist recognition site without causing a response. At any given instant therefore there are fewer recognition sites available to interact with an agonist A in the presence of an antagonist D.

$$A + D + R \Leftrightarrow AR + DR \Rightarrow Response$$

Since

$$[R] = [R_T] - [DR] - [AR]$$

then

$$[R]/[AR] = [R_T]/[AR] - 1 - [DR]/[AR]$$

Since, from the law of mass action,

$$[R]/[AR] = K_A/[A]$$

and

$$[DR] = [D][R]/K_D$$

where $K_A$ and $K_D$ are the respective dissociation constants ($K_d$ values) of the agonist and antagonist, then

$$K_A/[A] = [RT]/[AR] - 1 - [D]K_A/K_D[A]$$

or

$$[AR]/[R_T] = K_D[A]/(K_A K_D + K_A[D] + K_D[A])$$

Equation B

If $[AR]/[R_T]$ determines the fractional response, and the relationship between this ratio and the fractional response is not changed by the presence of the antagonist, Equations A and B then give:

$$[A]_0/([A]_0 + K_A) = K_D[A]_D/(K_AK_D + K_A[D] + K_D[A]_D)$$

where $[A]_0$ and $[A]_D$ are the concentrations of agonist producing the same fractional response in the absence and presence of D, respectively. This can be expressed as:

$$([A]_D/[A]_0) - 1 = [D]/K_D$$

where:

$$[A]_D/[A]_0 = \text{dose ratio.}$$

If the concentration of antagonist is such that $[A]_D = 2[A]_0$ then $K_D = D$. That is, the dissociation constant of the antagonist is equal to the concentration which produces a dose ratio of 2. It should be noted that the dose ratio given by an antagonist is dependent only on its concentration and is independent of the agonist used.

The last equation describes a straight line and data obtained at several antagonist concentrations can be plotted in the form of (dose ratio–1) versus [D]; if simple competitive antagonism is taking place the plot should be a straight line passing through the origin with a slope equal to $1/K_D$. However, even if such a graphical relationship can be plotted it *does not prove* that the interaction is competitive.

Schild developed this approach further by considering the log transformation of this relationship.

$$\log (\text{dose ratio} - 1) = \log [D] - \log K_D$$

Thus, when log (dose ratio –1) is plotted against log [D] (the Schild plot; see Arunlakshana & Schild, 1959), the result should be a straight line with a slope of unity and an intercept on the abscissa of log $K_D$. In some cases the Schild plot does not have a unitary slope and, since this implies that the antagonism is not competitive, the intercept cannot be taken to be log $K_D$. To allow quantitation in these circumstances, use may be made of the concept of $pA_x$ introduced by Schild. $pA_x$ is the negative logarithm (to base 10) of the concentration of antagonist causing a dose ratio of x and the most commonly used value is that of the $pA_2$. Since the dose ratio is 2 when the Schild plot meets the abscissa, then this intercept is equal to $-pA_2$. When simple competition is taking place, $pA_2 = -\log K_D$ (= $\log K_{aff}$ for the antagonist), but $pA_2$ may be quoted when results not compatible with competition (e.g. a Schild plot slope $\neq 1$) have been obtained.

## Non-competitive antagonism

The concept of non-competitive antagonism is one that has been borrowed from enzyme kinetics where it describes an inhibitor whose effects cannot be overcome by increasing the substrate concentration. The inhibitor might bind to the active site of the enzyme or to another allosteric site. Pharmacologists often use the term non-competitive when irreversible antagonism or a type of allosteric effect is actually meant.

## Ligand interactions via distinct sites on receptors – 'allosteric' interactions

There may exist on receptor macromolecules secondary specific sites for the interaction of compounds distinct, and perhaps remote, from the transmitter recognition site. If a ligand, on binding to such a secondary site, induces a conformational change in the transmitter binding site, then the affinity of the transmitter binding site for agonists or antagonists may be increased or decreased. If affinity is decreased then ligands interacting with this secondary site behave as antagonists and are usually referred to as 'allosteric' antagonists, in analogy with enzyme terminology, but are more properly called metaffinoid or metacoid (for example, Van Rossum, 1977) antagonists. If affinity is increased, the term 'inverse agonist' is often used.

Metacoid refers to the situation where two ligands interact separately with two distinct recognition sites and these recognition or receptor sites are independent with regard to their interactions with their respective ligands but the receptor–effector systems are interrelated beyond the level of the recognition sites.

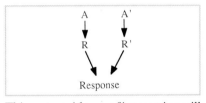

This metactoid type of interaction will not be detected by ligand interactions in binding experiments.

Metaffinoid refers to the case where the interaction of a ligand with its specific recognition site changes the affinity of a second site for its specific ligand.

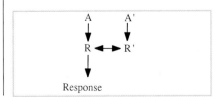

This type of interaction will be detected by binding experiments, but in the usual inhibition or displacement experiments may well be indistinguishable from competitive interactions.

There may also exist ligands specific for the secondary (R′) binding site that induce no conformational change in the receptor macromolecule. Such compounds will have no effect alone and will not be detectable by an interaction with the binding of a ligand for a specific primary (R) site, but will antagonize the effects of compounds that exert their effects on the primary site via a metaffinoid site. They will of course antagonize the binding of ligands specific for the secondary site.

## Equilibrium and kinetic binding studies

It can be seen from the above discussion that quantification of ligand–receptor interactions using functional studies in isolated tissues or the whole animal, is complicated by the lack of knowledge of the exact relationship between receptor occupancy by an agonist and the response generated. It is only with antagonists that reliable estimation of binding constants can be made, although an ingenious method for determining agonist affinity for receptors sensitive to alkylation by, for example, phenoxybenzamine has been devised (Furchgott & Bursztyn, 1967).

Thus the development of radioligand binding techniques, in which the binding site is studied directly with a radiolabelled probe, provided great impetus to receptor pharmacology on

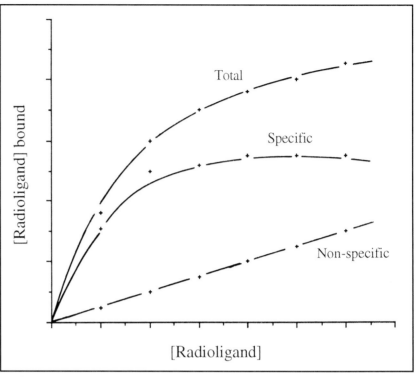

*Fig. 5. Total and non-specific binding of a radiolabelled ligand to a tissue and the derived specific binding as a function of ligand concentration. Note that the specific binding has the form of a rectangular hyperbola and that the non-specific binding increases linearly with ligand concentration.*

their introduction in the late 1960s and early 1970s. In view of the relatively low abundance of receptors in most tissues studied (of the order of pmol/g wet weight), the successful ligand must exhibit a high affinity for the receptor in question. This allows use of a very low concentration of the labelled ligand and consequently a reduction of the 'non-specific' binding to the tissue. Non-specific binding is a poorly studied and largely ignored phenomenon, in which the labelled ligands bind to membrane sites unrelated to the receptor under study. In addition to the required high affinity, a high specific activity of at least 20 Ci mmol$^{-1}$ is required to enable low quantities of specific binding to be ac-

curately measured at low concentrations of ligand. Although tritiated ligands are commonly employed, iodinated compounds might be preferred due to their inherent higher specific activity.

There are two basic types of experiment carried out with receptor radioligands – saturation studies and 'competition' studies. In saturation studies the tissue, or other preparation containing the binding site of interest, is incubated with increasing concentrations of radioligand. Since the ligand will 'bind' to the tissue at sites other than that being investigated (for example, hydrophobic ligands will partition into the lipid of the mem-

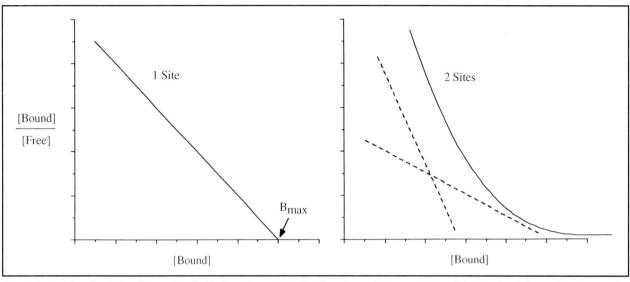

*Fig. 6. Typical Scatchard plots illustrating the binding of a radiolabelled ligand to a single population of binding sites (left panel) and to two sites (right panel). The plot is made by plotting the ratio of bound ligand to free ligand on the ordinate against the concentration of bound ligand (usually expressed as the amount of binding per unit weight of protein in the preparation) on the abscissa. The $K_{aff}$ ($1/K_d$) of the ligand is given by the gradient of the plot and the maximal binding capacity ($B_{max}$) by the intercept on the abscissa. The total binding to two sites (solid line) is the sum of binding to a high-affinity site (steeper dashed line) and to a low-affinity site (less steep dashed line); in the case of a plot like this, the slopes of the two components shown by the dashed lines are equal to the $K_{aff}$ values for each of the two sites and the intercepts on the abscissa show the amounts of the high affinity site (left intercept) and the $B_{max}$ (right intercept); the amount of the low affinity site is therefore the difference between these two quantities.*

branes present), then the specific binding has to be distinguished from this non-specific binding by an appropriate pharmacological method. Usually this is achieved by incubating a parallel series of samples with an excess concentration ($\approx 100 \times K_d$) of an unlabelled ligand for the receptor under investigation. Ideally a different ligand to the radioligand is used to minimize the chances of interaction at sites other than the receptor. The second series of samples will usually yield a linear increase in non-specific binding with increasing radioligand concentration whilst the series with no addition save the radioligand will show a curvilinear relationship (Fig. 5). Subtraction of the non-specific binding from the total binding at each concentration of radioligand gives the specific binding to the receptor site and this shows a curvilinear relationship with ligand concentration. Indeed, it is usually a rectangular hyperbola described by Equation A above. The total number of specific binding sites ($B_{max}$) is obtained by reading it directly from the graph, or by use of either computer-aided curve fitting or by Scatchard analysis (Fig. 6). All these analyses allow determination of $K_d$ for the radioligand, which at its simplest is equal to the concentration at which the specific binding = $B_{max}/2$. In principle, this method allows detection of the presence of multiple binding sites, together with their affinities, or of cooperativity. However, unless the sites are clearly separated in affinity, and are present in quantities not too different from each other, it may prove difficult in practice to obtain accurate results.

In addition to equilibrium studies like these, in which the radioligand and binding site are incubated for sufficient time for the quantity of binding to be stabilized, it is possible to determine the kinetic rate constants $k_{+1}$ and $k_{-1}$. The dissociation rate constant, $k_{-1}$, can be measured by allowing the radioligand and receptor to come to equilibrium and then, at time 0, either by dilution or by addition of a large excess of unlabelled ligand, the diminution of binding with time can be followed. A semi-logarithmic plot of binding against time yields a straight line with a slope numerically equal to $k_{-1}$. The association rate constant cannot be determined quite as easily since the rate of formation of the ligand–receptor complex, $k_{obs}$, is dependent on the ligand concentration and both $k_{+1}$ and $k_{-1}$. A series of

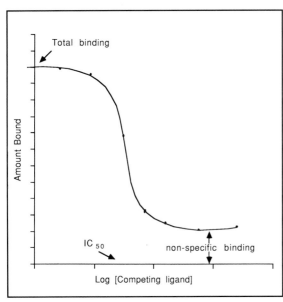

*Fig. 7. Inhibition of the binding of a radioligand by an unlabelled competitor. The amount of radioligand bound decreases as the concentration of the competing ligand increases until the residual binding remains constant. This residue is the non-specific binding and is assumed to be to sites other than the receptor being investigated. The $IC_{50}$ is the concentration of unlabelled ligand which inhibits 50 per cent of the specific binding of the radioligand. Note that the initial, total, binding is not necessarily equal to the maximal binding capacity of the receptor, and is most often chosen to be some 5–10 per cent of the $B_{max}$.*

experiments is done at different radio-ligand concentrations and each is allowed to continue until equilibrium is reached, with specific binding being determined at several intermediate time points. A plot of $\ln[1-(B_t/B_{max})]$, where $B_t$ is the specific binding at time t, against time yields a line with a slope of $k_{obs}$. Plotting the values of $k_{obs}$ against the ligand concentrations at which they were obtained gives a straight line with a slope equal to $k_{+1}$. Of course, if $k_{-1}$ is known then a value of $k_{+1}$ can be calculated from $k_{obs}$ at any single radioligand concentration.

A major problem with saturation experiments is that they use large quantities of radioligand.

Competition experiments use a fixed concentration of radioligand and the binding to the preparation is determined in the absence, and in the presence of increasing concentrations, of unlabelled ligand. If the competing ligand only binds to a single site on the receptor then the concentration at

which half-displacement of the specific binding of the ligand occurs, the $IC_{50}$ (Fig. 7), is simply related to its equilibrium dissociation constant $(K_D)$ by:

$$K_D = IC_{50}/[(A/K_A) + 1]$$

where A and $K_A$ are respectively the concentration and equilibrium dissociation constant $(K_d)$ of the radioliand. If the binding of the competing, unlabelled compound, is complex, for example if it is to two binding sites, then the competition curve shown in Fig. 5 will not be a simple sigmoid shape and more detailed analysis is required. At this point it is usually best to use a computer-aided fitting procedure to obtain estimates of various $K_d$ values and relative proportions of the different binding sites. Agonists for G-protein linked receptors commonly show multiple binding sites. A useful tool in this respect is the Hill plot (Fig. 8). When the slope of this double log plot is unity, then the binding of the competing ligand most probably obeys the law of mass action

for a single binding site. A slope > 1 implies that binding is positively co-operative but a Hill slope < 1 might indicate either negative co-operativity or the presence of more than one binding site (Fig. 8). Thus, in the case of there being more than one binding site, the Hill plot will usually show one or more points of inflection.

## Ionic channels

### Definition

In the course of evolution, nature has evolved advanced systems for regulating the passage of ions across cell membranes. In particular, the movement of ions present in high concentrations in intra- and extracellular fluids (sodium, potassium, calcium and chloride) takes place via the opening and closing of specific channels found in all types of excitable tissue (nerve and muscle) and also many other cell types (e.g. neuroglia, adipose tissue, epithelia).

For many years the existence of ionic channels was a hypothesis firmly based on measurements of ionic currents across the membrane of single cells. Such recordings give information regarding the behaviour of many thousands of channels in an ensemble. However, in recent years, the introduction of the so-called 'patch clamp' technique (Sakmann & Neher, 1984) has made it possible to record the current flowing through a single membrane channel and thus our knowledge and understanding of the function of ionic channel proteins has taken a great leap forward (Hille, 1984; Narahashi, 1988, 1990).

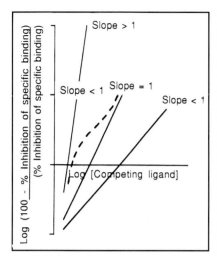

*Fig. 8. Hill plots for compounds competing for binding with radioligands when the binding is competitve (slope = 1), positively co-operative (slope > 1) or perhaps either negatively co-operative (slope < 1 for a straight line) or to more than one binding site (slope < 1 but for a line showing one or more points of inflection).*

When a single ion channel opens, a small current flows across the membrane with a magnitude in the picoampere range and corresponding to the passage of around $10^8$ ions s$^{-1}$. In order to measure these small quantities, it is necessary to isolate a tiny (about 1 $\mu$m$^2$) patch of membrane. This can be achieved by sealing the heat-polished end of a micropipette onto the cell surface by the application of negative pressure through the pipette. A smooth surface on the pipette is essential to form a tight seal on the membrane with a resistance of the order of tens of gigaohms. This technique, as illustrated in Fig. 9, can be employed for recording membrane currents not only from single ionic channels, but also from whole cells including small cells which were largely inaccessible to intracellular recording techniques. By appropriate manipulation of the pipette and the cell membrane it is possible to produce a patch

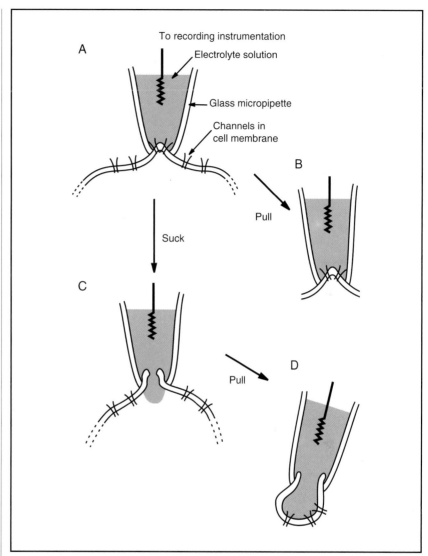

*Fig. 9. In the patch clamp technique, the membrane is sealed to the end of the glass recording micropipette by application of negative pressure (A) to give the 'cell-attached' configuration. The attached patch can then be pulled off the membrane (B) to give an 'inside-out' patch in which the intracellular face of the membrane is exposed to the bath solution and the extracellular face to the solution in the microelectrode. If, instead, the cell-attached preparation is subjected to stronger suction than was used to attach the cell to the micropipette, the membrane ruptures to give a preparation from which 'whole cell recording' may be made (C); the internal environment of the cell will equilibrate with the contents of the micropipette. Finally, an 'outside-out' patch can be made by pulling sharply at a whole cell preparation (D); the act of pulling the electrode breaks the cell membrane which then spontaneously reseals with the outside face facing the bathing solution and the inside face towards the interior of the micropipette.*

of membrane adhering to the end of the pipette which can be removed from the rest of the cell. This can be done such that either the intracellular surface of the membrane faces the bath solution (inside-out configuration; suitable for studying the effects of intracellular messengers such as cyclic nucleotides or calcium) or the extracellular surface of the membrane faces the bath solution (outside-out configuration; for the investigation of the effects of agonists on membrane permeability).

## Mechanisms

The opening and closing of channels may be controlled by a number of distinct systems (Fig. 10): (1) the voltage across the cell membrane; (2) the presence of neurotransmitter substances in the extracellular space at the cell membrane; (3) the concentration of 'messenger' molecules inside the cell. Some channels are highly specialized and respond only to changes in one of these systems, whereas others respond to changes in all three. These different possibilities will be outlined and this will be followed by a detailed description of individual types of channel.

## Voltage-dependent channels

Voltage-dependent ionic channels can exist in at least three states (Fig. 11): open, closed and inactivated (also a closed state). It now seems clear that in response to a change in potential across the cell membrane, a voltage-sensing region initiates a change in the three-dimensional structure of the channel protein such that an ion-conducting pore forms. The pore might remain open for as long as the membrane potential deviates from its initial value or the channel might close in a time- and voltage-dependent manner despite the continuation of the voltage stimulus. This might lead to the establishment of the third,

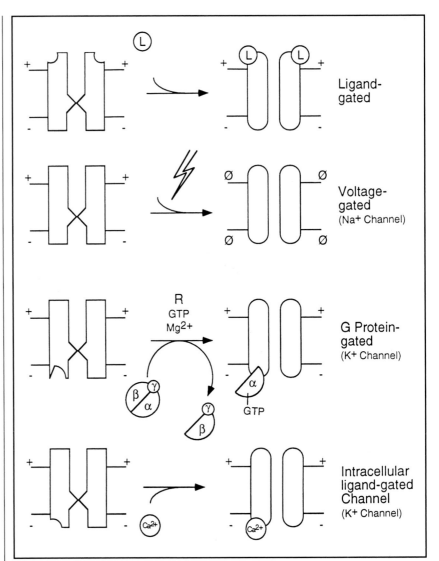

Fig. 10. Illustration of four means of controlling the permeability of ion channels. The channel (e.g. the nicotinic acetylcholine receptor, the GABA$_A$ receptor) may be controlled directly by the binding of a neurotransmitter (L); control may be through binding of an intracellular ligand (e.g. the calcium-activated potassium channel); the channel might be directly sensitive to the transmembrane voltage (e.g. the voltage-operated sodium channel); or the channel might be controlled through the action of a G-protein (e.g. the acetylcholine-dependent potassium channel). Some channels are controlled by more than one process (see text).

'inactivated', state in which the channel is closed but cannot easily reopen. A return to potentials at which the incidence of the inactivated state is low allows the channel protein to revert to the normal closed state.

## Receptor-coupled channels

Receptor-coupled ionic channels are macromolecules or aggregations of macromolecules that form a conducting pore following the binding of a neurotransmitter or hormone to the extracellular surface of the channel protein. There are probably two basic forms, one in which the binding of agonist molecule directly signals the opening of the channel and a second in which binding of the agonist leads to production of an intracellular messenger molecule which itself acts on the pore-forming protein to elicit opening. Closure can follow dissociation of the agonist from the receptor in a time- and concentration-dependent manner, or both, which may correspond to the phenomenon known as desensitization.

## Dependency on intracellular constituents

Changes in the concentration of intracellular constituents can also influence channel status. For example,

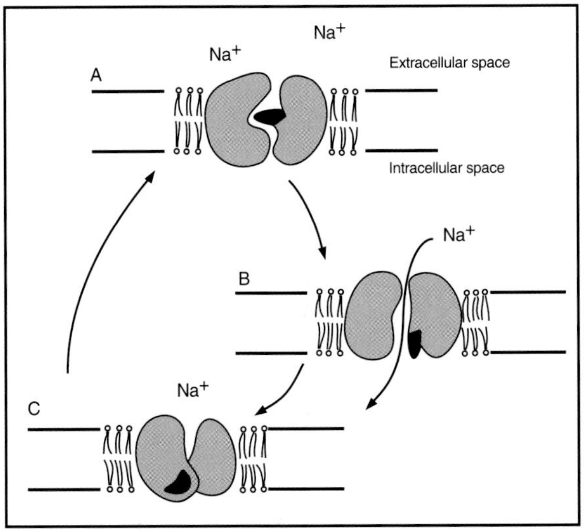

*Fig. 11. The voltage-operated sodium channel has three conformational states. In (A), the channel is closed, but in the 'resting' conformation, such that when the membrane depolarizes it swiftly moves to the 'open' state (B). About 1 ms later, the channel moves to a third, 'inactivated', state in which it is closed (C) but cannot reopen until the membrane has repolarized. At rest, the proportion of channels in the resting and the inactivated states is determined by membrane potential. If the resting membrane is at a depolarized potential, then the number of channels in the resting state is reduced so that a rapid depolarization only produces a small amount of current.*

the metabolic state of the cell (high or low intracellular ATP levels) can significantly influence the opening of one type of potassium channel (see below). Of the numerous substances known to influence ionic channels, the most important are: $Ca^{2+}$, cyclic AMP (which activates a specific kinase), inositol trisphosphate ($IP_3$; probably through changes in intracellular calcium concentration) and ATP.

## Voltage-dependent channels

### Sodium channels

In all excitable cells, the concentration of sodium ions is substantially lower (1–10 mM) than in the surrounding extracellular space (ca. 100–140 mM). Thus when sodium channels open, sodium ions flow down the concentration gradient leading to a depolarization of the cell membrane potential. The first formal description of sodium channels presented by Hodgkin & Huxley in 1952 was based on voltage clamp recordings from squid giant axons; however, the basic principles are applicable to almost all excitatory cells. Sodium channels exhibit a marked voltage-dependence, being almost all closed at potentials negative to −60 mV. Depolarization increases the probability of channel opening, but the time

for which sodium channels remain open is extremely short (< 1 ms), an inactivated state being established within this time.

The primary consequence of the opening of voltage-dependent sodium channels is the generation of the upstroke of the action potential in most types of nerve, cardiac muscle and skeletal muscle. The opening of these channels results in a rapid depolarization (from ca. −70 to +30 mV) which lasts approximately 1 ms. The rapid onset of inactivation (closure) terminates the permeability change to sodium ions as does the later opening of repolarizing potassium channels (see below).

Sodium channels can be blocked by tetrodotoxin, found in puffer fish (the *Tetraodontidae*) and certain types of American newt of group *Salamandridae*. The toxin binds to a site near, or in, the mouth of the channel and so blocks the pore. A large number of other, therapeutically used, compounds also block sodium channels. In particular the local anaesthetics procaine and lignocaine exert their pain-relieving action via the block of sodium channels in axonal membranes normally transmitting action potentials from peripheral pain receptors to the central nervous system. Finally, some drugs (e.g. veratrum alkaloids and the insecticide DDT) reduce specifically the channel inactivation process and in this way greatly prolong the open time of the channels. Some fractions of scorpion toxin act in this manner as do several synthetic agents developed for the treatment of cardiac failure.

The channels consists of up to three subunits of which the α subunit is thought to form the ion pore since it has four putative transmembrane domains, each composed of six transmembrane sequences (Fig. 12); this subunit carries the binding site for tetrodotoxin. In mammalian brain there are also $\beta_1$ and $\beta_2$ subunits but mammalian skeletal muscle comprises only the $\alpha$ and $\beta_1$ subunit; all three types of subunit are glycosylated with up to 30 per cent of their mass being carbohydrate.

## Potassium channels

Potassium channels are a large family of membrane ionic channels intimately involved in the control of cell membrane potential and action potential discharge frequency (Cook, 1990; Weston & Hamilton, 1991). Since the intracellular potassium concentration (ca. 140 mM) greatly exceeds the extracellular concentration (3–6 mM), potassium ions leave the cell during channel opening, despite the electrical gradient, and thus a membrane hyperpolarization ensues. The potassium permeability of excitable cell membranes is usually high in the resting state and this can account almost entirely for the observed resting membrane potential (–50 to –85 mV). During the depolarized phase of the action potential, potassium permeability rises and thus drives the membrane potential back towards more negative values. This process of repolarization usually leads to a so-called action potential after-hyperpolarization, a period during which the membrane potential transiently overshoots its resting value. Such after-hyperpolarizations last between 2 ms and several seconds depending on the cell studied and the particular subpopulation of potassium channels in-

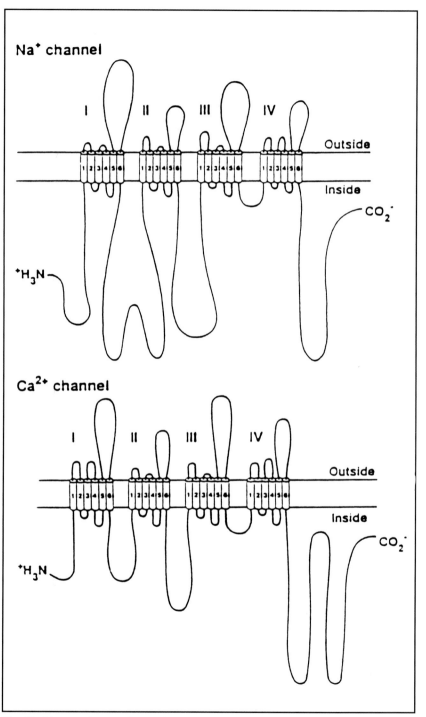

*Fig. 12. Diagram showing the α subunit of the voltage-sensitive sodium channel and the $\alpha_1$ subunit of the L-type calcium channel in the membrane. Note that there are thought to be four transmembrane domains, each consisting of six transmembrane helices.*

volved in the repolarization process.

In addition to their role in membrane repolarization after an action potential, potassium channels are involved in the regulation of action potential discharge frequency, increases in permeability being inhibitory and decreases being excitatory. Note that a decrease in outward potassium current can result in cell depolarization and the initiation of action potentials. The most important discrete potassium currents found in mammalian neurons and muscle cells are as follows.

### Delayed potassium channels

The delayed rectifier is the current responsible for action potential repolarization in many types of axon, neuronal cell somata, skeletal and cardiac muscle. It is termed delayed since it activates shortly after the onset of depolarization and is a rectifier because the opening of the channels allows current to flow more easily in an outward direction across the cell membrane than in an inward direction. This behaviour is similar to that of a common diode. The neuronal delayed rectifier is blocked by high (mM) concentrations of tetraethylammonium (TEA) and 4-aminopyridine (4-AP). The corresponding current in cardiac muscle cells can be blocked by a number of novel synthetic agents, which may be of value in the treatment of life-threatening arrhythmias. Novel blockers of the cardiac delayed rectifier currently under development include UK-68,798, tedisamil and sematilide.

### Fast potassium channels

These channels carry the transient outward (A-type) current which is a rapidly activating, rapidly inactivating, species of potassium current involved in action potential repolarization and the determination of interspike intervals. The current is activated by depolarization and inactivates (channels close) within 10–1000 ms. Several different types of transient outward current have been distinguished on the basis of different kinetic behaviour. Most forms are blocked by 4-AP and some by dendrotoxin (isolated from green mamba venom). Modulation by activation of neurotransmitter receptors, such as β-adrenoceptors and muscarinic receptors, is also possible.

## Calcium channels

The regulation of intracellular calcium concentration is an important homoeostatic mechanism in both excitable and non-excitable cells. Under normal circumstances, intracellular calcium concentration is maintained at very low levels ($< 1$ μM). Small, transient rises (lasting several seconds) most usually serve to alter some aspect of cell function whereas large, more permanent, increases in intracellular calcium concentration can have fatal effects on the cell in question. In the nerve terminals at synaptic junctions, transient increases in intracellular calcium initiate neurotransmitter release into the synaptic cleft. Although the exact mechanism by which calcium ions promote the exocytosis of synaptic vesicles is as yet unknown, it is certain that the calcium ions necessary for this process pass from the extra- to the intracellular space via specific voltage-dependent ionic channels. Thus removal of calcium ions from the extracellular space blocks synaptic transmission.

In muscle cells, increases in intracellular calcium initiate the coupling of the actin and myosin filaments and so bring about a contraction of the muscle fibre (see Chapter 6). The origin of the calcium ions involved in this process differs, depending on the type of muscle (and also the species) in question. In skeletal muscle, a depolarization of the muscle membrane leads to a release of calcium from intracellular stores in the sarcoplasmic reticulum. This release takes place via channels located on the membranes of intracellular organelles. In contrast, during the action potential of cardiac muscle, most of the calcium necessary for contraction flows into the cell from the extracellular space, through specific plasma membrane channels. Smooth muscle exhibits a mixture of these two systems: intracellularly stored calcium can be released by depolarization and transmembrane calcium movement takes place via membrane channels.

Since in terrestrial animals, extracellular calcium is present at a concentration of between 1 and 2.5 mM, there is a transmembrane concentration gradient of about 2000 to 1. Calcium influx depolarizes cells and so can contribute to electrical activity such as pacemaker depolarizations, seizure-type discharges and calcium-mediated action potentials. An elaborate system of membrane calcium channels has evolved to control the movement of calcium from the extracellular to the intracellular space. Four types of voltage-dependent calcium channels can be distinguished: the low-threshold or T(transient) type, the high-threshold, N(neuronal) type

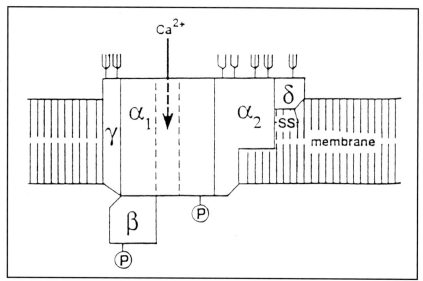

*Fig. 13. The transmembrane subunit organization of the L-type calcium channel. It can be seen that the β and δ subunits do not cross the membrane. The α subunit forms the pore through which the calcium ions cross the membrane and carries the binding site for dihydropyridines (e.g. nifedipine) and phenylalkylamines (e.g. diltiazem). Both the $\alpha_1$ subunit and the β subunit can be phosphorylated by cyclic AMP-dependent protein kinase.*

and the L(long-lasting) channels (Tsien *et al.*, 1988) with the P(Purkinje)-type as a recent addition.

### L-type channels

L-type calcium channels provide the link between electrical excitation and contraction in the heart and smooth muscle. The antihypertensive dihydropyridine drugs (e.g. nifedipine and nitrendipine) block these channels and so reduce arterial muscle tension and lower blood pressure. Chemically related dihydropyridines such as Bay K 8644 can increase the opening probability of L-type calcium channels and are termed calcium channel agonists or openers. The cardiac stimulatory action of adrenaline and noradrenaline is also mediated by an increase in L-type calcium current following activation of $\beta_1$-adrenoceptors on the heart muscle membrane.

The channel consists of five different subunits (Fig. 13); the $\alpha_1$ subunit has four transmembrane domains (each with six transmembrane segments, a structure also found in the α subunit of the voltage-sensitive sodium channel) and carries the binding site for the dihydropyridines whilst the β subunit is probably only found on the intracellular face of the membrane and has a site which can be phosphorylated by cyclic AMP-dependent protein kinase. The $\alpha_2$ and γ subunits cross the membrane whilst the δ subunit is only found on the extracellular face of the membrane.

### N-type channels

The N-type channels are involved in controlling neurotransmitter release. They are resistant to block by dihydropyridines but are sensitive to cadmium, ω-conotoxin and a high concentration of some antibiotics (e.g. neomycin and kanamycin), but not to

nickel. N-type channels are also subject to regulation by a variety of neurotransmitters and peptides which act on receptors located on the presynaptic nerve terminals (e.g. $\alpha_2$-adrenoceptors) to cause a reduction of calcium channel opening and consequently a reduction in neurotransmitter release (prejunctional inhibition).

### T-type channels

T-channels have been shown to be involved in pacemaker activity in heart muscle and CNS neurons and in the contraction of certain veins. These channels are relatively resistant to cadmium but may be blocked by the related divalent cation nickel and by the ω-Aga-IIIA fraction of funnel web spider venom.

### P-type channels

The channels carry a calcium current that is blocked by fractions of funnel web spider venom (FTX, ω-Aga-IVA), cadmium and cobalt but is not affected by other organic channel blockers. Although first found in the Purkinje cells of the cerebellum, they are now known to be distributed widely in the brain. These channels might correspond to the functional channels produced by expression of the protein coded by a single clone isolated from guinea-pig cerebellum.

## Ligand-gated channels

This is a group of receptors that has grown considerably over the last 20 years and which have a common motif in that their subunits include four transmembrane domains. The nicotinic acetylcholine receptor was the first of these to be identified and characterized but it has been joined by excitatory receptors for glutamate and by inhibitory receptors activated by glycine and GABA. In all cases the receptor combines one or more ligand recognition sites with an intrinsic ion channel and so antagonists can act either at the transmitter binding site (competitive antagonists) or within the ion channel (channel blockers which could also be termed non-competitive antagonists).

## Nicotinic acetylcholine receptor

This was the first ligand-gated channel to be isolated and fully characterized, a process that was aided by its great abundance in the electric organ of *Torpedo* and *Electrophorus* species (see review by Galzi *et al.*, 1991). It consists of a pentamer (Fig. 14) composed of two $\alpha$-subunits (each of which possesses an acetylcholine binding site) and one each of the $\beta$-, $\gamma$- and $\delta$-subunits (electric organ) or $\beta$-, $\varepsilon$- and $\delta$-subunits (mammalian skeletal muscle). Each subunit has four membrane-spanning regions and the functional receptor allows passage of sodium, potassium and calcium ions, though the first of these predominates in physiological conditions. The nicotinic receptor found in neurons, especially the autonomic ganglion, probably consists of $\alpha$- and $\delta$ subunits. Both types of receptor can be activated by acetylcholine or carbachol and blocked by d-tubocurarine. The neuronal receptors can be differentiated pharmacologically by their selective activation by dimethylpiperazinium and by being subject to channel block with hexamethonium. Gallamine and decamethonium produce channel block of the muscle receptor which has a higher conductance than the neuronal type.

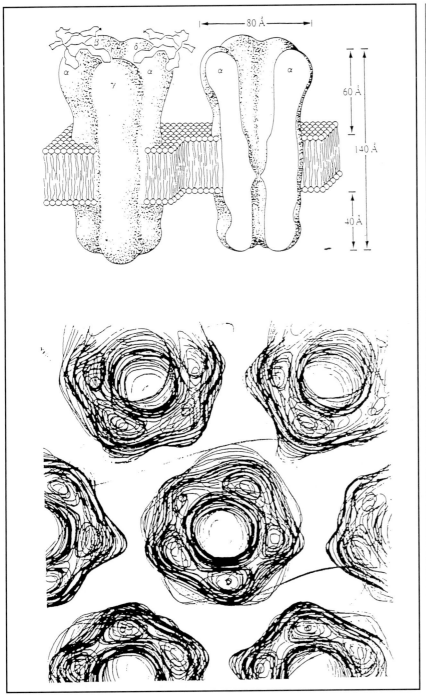

*Fig. 14. Proposed structure for the nicotinic receptor for acetylcholine, from electric organ, consisting of two $\alpha$ subunits, and one each of the $\beta$, $\gamma$ and $\delta$ subunits. The pentamer forms a channel which is permeable to sodium and other monovalent cations and it is opened by the binding of two molecules of acetylcholine, one to each of the $\alpha$ subunits. The receptor may be inactivated by the binding of $\alpha$-bungarotoxin to the $\alpha$ subunits, as shown in the diagram.*

## Excitatory amino acid (glutamate) receptors

These receptors are widely distributed throughout the central nervous system and have been categorized into five main types. On the basis of iontophoretic application of drugs and radioligand binding studies studies three types have been identified as being selective for NMDA (*N*-methyl-D-aspartate), kainate and AMPA (DL-α-amino-3-hyroxy-5-methyl-4-isoxalone propionic acid). The actions of agonists at all these sites can be antagonized by kynurenate. There is also a fourth type of glutamate receptor associated with an ion channel that is selectively activated by L-AP4 (L-amino-4-phosphonobutanoate) and this appears to be a prejunctional receptor mediating inhibition of transmitter release. These receptors are sometimes termed 'ionotropic', that is activation causes a change in ionic permeability of the cell and this is to distinguish them from a different type of receptor, the so-called 'metabotropic' glutamate receptor (for which a selective ligand is 1-aminocyclopentane-1S, 3R-dicarboxylic acid), which is a G-protein-coupled receptor that, when activated, causes a change in intracellular inositol phosphate and diacylglycerol concentration. More than 10 subunits of ionotropic receptors for excitatory amino acids have been cloned and the relationship between these and the pharmacologically characterized receptors has not been fully resolved. The following discussion describes the most important features of three of the naturally occurring receptors and relates this, where possible, to the subunits so far identified.

### NMDA receptor

These receptors contain an intrinsic ion channel with a large conductance when open and which allows passage of a current that is carried by calcium as well as sodium and potassium which pass through the other glutamate-gated ion channels (recent studies, however suggest that all ionotropic glutamate receptors may be able to carry calcium). This calcium current also shows voltage dependency and the receptor is thought to be involved in the process of use-dependent synaptic plasticity within the central nervous system. Cloning studies suggest that the ability to allow a high permeability for calcium could be related to a single substitution (asparagine for glutamine/arginine) in the second transmembrane region of the receptor subunit forming the ion channel. These same studies show that functional receptors, of low conductance, can be formed from a single type of subunit, although formation of receptors with more than one type of subunit produces channels with greater conductance.

The NMDA receptor can be blocked by 2-amino-5-phosphonovaleric acid and it has auxiliary binding sites that allow modulation by glycine and by phencyclidine. Glycine facilitates depolarization induced by glutamate acting at NMDA receptors, indeed its presence is essential for an agonist to open the ion channel, but has no effect on its own; it is therefore a positive allosteric modulator of this receptor. Phencyclidine and MK-801, on the other hand, reduce excitation of cells activated by NMDA and also act at a site distinct from the NMDA recognition site, probably inside the ion channel itself. They can be regarded as non-competitive inhibitors of the NMDA receptor and their action shows use-dependence (a greater block from a given concentration of phencyclidine in the presence of a greater frequency of excitation of the cells). The NMDA receptor also exhibits voltage-dependent inhibition by magnesium (being more marked at membrane potentials more negative than $-20$ or $-30$ mV) at physiologically relevant magnesium concentrations and there is evidence for allosteric interaction between the magnesium and phencyclidine sites within the ion channel.

### AMPA receptor

Formerly this was called the quisqualate receptor and is perhaps the best understood in terms of the subunits of which it is composed (Sommer & Seeberg, 1992). On activation, this receptor forms a channel with a lower conductance (5–15 pS) than the NMDA receptor (40–50 pS). Four closely related subunits of this receptor have been cloned and they have four presumed transmembrane domains, but the specific subunit composition of physiological AMPA receptors in different tissues has not been finally determined although functional receptor/ion channel systems can be made with four different or four identical subunits. The subunit composition is important for determining the electrophysiological properties of the channel; three of the four subunits cloned have a glutamine residue in the second transmembrane domain which in the fourth type is replaced by the positively charged arginine. The result of this change is that channels assembled from the fourth type of subunit do not show the rectification properties associated with channels constructed from the other three types.

Antagonists at the AMPA receptor include several quinoxaline derivatives of which the most potent are DNQX and NBQX.

### Kainate receptor

Pharmacologically this receptor shares selective antagonists (relative to the NMDA receptor) with the AMPA receptor, although it shows differential activation by kainate rather than AMPA and quisqualate. Two separate groups of subunits have been cloned for this receptor, one of which can form functional receptors, coupled to ion channels when a single type of subunit is expressed and other of which only forms ligand recognition sites without an associated ion channel. Although these various subunits are widely expressed in the central nervous system, the subunit composition of physiological kainate receptors within the nervous system is not yet known.

## Amino acid-activated chloride channels

The distribution of chloride ions across the cell membrane is largely dependent on the presence or absence of chloride co-transport mechanisms (see below). In neurons, a net outward movement of chloride establishes a concentration gradient from extra- to intracellular space such that when the membrane-bound chloride channels open, chloride ions flow into the cell and hyperpolarize the membrane. This hyperpolarization is the basis of neuronal inhibition and is an important mechanism controlling the excitability of single neurons and neuronal

ensembles (Hille, 1984).

Although some types of chloride channel have been shown to exhibit voltage-dependent opening and closing, by far the most important stimulus for the opening of such channels is the synaptic release of an inhibitory neurotransmitter, such as $\gamma$-aminobutyric acid (GABA) or glycine, which then acts on a ligand-gated ion channel.

### GABA$_A$ receptor

This is the best studied of the receptor/chloride ion channel complexes. Whilst the chloride channel protein complex has only one binding site for the natural neurotransmitter, there are several other binding sites present, which can influence the function of the receptor/channel complex. In addition these accessory sites are of considerable clinical significance.

Historically, it was first recognized that the convulsant agent picrotoxin antagonized the action of GABA but could not displace GABA from its receptors. It was therefore postulated that picrotoxin directly bound to, and blocked, the GABA-coupled chloride channel without affecting the GABA binding site. Subsequently it was found that the benzodiazepine drugs, such as diazepam and fluorazepam, facilitated GABA-mediated neuronal inhibition and this is now known to be due to an additional benzodiazepine binding site present on the GABA-receptor chloride ion channel complex. The presence of benzodiazepine molecules facilitates the binding of GABA molecules and thus increases neuronal inhibition.

Nowadays, a complex pharmacology of this multi-site structure has

evolved, with the characterization of agonists and antagonists for the three available binding sites. At the receptor level, GABA and the natural product muscimol act as full agonists, whereas the convulsant alkaloid bicuculline is the most useful competitive antagonist. As previously mentioned, the convulsant agent picrotoxin is a channel blocker; direct channel openers have not yet been identified. There appear to be two subtypes of benzodiazepine binding site, the characteristics of which are determined by the subunit composition of the receptor. Diazepam is an agonist at both these sites and its effects can be blocked by an antagonist such as flumazenil. Of particular interest is the observation that there exists a third type of ligand, called an inverse agonist (e.g. $\beta$-carbolines such as DMCM), which has the opposite physiological effect to diazepam and the other agonists at this site; the exact pharmacological mechanism underlying this phenomenon remains the object of much investigation.

### Glycine receptor

The chloride channel which forms part of this receptor can be distinguished from that of the GABA$_A$ receptor by its sensitivity to the convulsant strychnine. Several subunits of this receptor have been cloned, falling into $\alpha$ and $\beta$ types, and all have four transmembrane domains; the receptor/ion channel complex consists of a heterogenous pentamer and the glycine binding site is found on the $\alpha$ subunits.

## Ligand-gated potassium channels

Potassium channels that are ligand-gated have similar functions in the cell to the voltage-gated channels for this ion. One type of ligand-gated channel is sensitive to acetylcholine in the extracellular space and so conforms to well-established ideas of a receptor mediating the actions of a neurotransmitter. However, there are two important types of channel which are affected by the intracellular concentrations of ligand, in one case calcium and in the other ATP. Thus the types of ligand that can modify channel gating are not restricted either to substances outside the cell or to neurotransmitters and hormones.

### Acetylcholine-dependent potassium channels

These carry a potassium current characterized by a lack of time- or voltage-dependent inactivation which is blocked by the action of acetylcholine on muscarinic receptors (hence M-current; Brown, 1990). The M-current contributes to the resting membrane potential in many neurons and blockade of these channels leads to depolarization. As a result of the coupling to muscarinic receptors, the M-current is subject to regulation via the release of acetylcholine. Some forms of slow excitatory postsynaptic potentials result from blockade of these channels. A number of other neurotransmitters can lead to closure of M-channels, for example susbstance P, luteinizing hormone releasing factor and 5-HT, implying that the receptor coupling to the ion channel indirect and it is thought that it involves a G-protein as an intermediary.

### Calcium-activated potassium channels

These are so termed since the opening of the channels is initiated by an increase in intracellular calcium concentration. (Marty, 1989). Conver-

sely, a fall in intracellular calcium concentration leads to channel closing. Influx of calcium into excitable cells usually has important functional consequences, such as the release of neurotransmitter, the secretion of hormone or the initiation of muscle contraction. Calcium-activated potassium channels can contribute to the repolarizing phase of the action potential and also play an important role in regulating repetitive action potential discharges. There are a number of subgroups, which can be distinguished on physiological as well as pharamacological grounds. Notable blockers of calcium-activated potassium channels are: TEA, charybdotoxin (from scorpion venom) and apamin (from bee venom). In addition all substances which block entry of calcium into cells (see below) will indirectly prevent activation of calcium-dependent potassium channels. Finally, many receptors for neurotransmitters are coupled to calcium-activated $K^+$-channels such that activation of the receptor leads to channel blockade. The receptors that have been shown to be involved in this system are the muscarinic receptor for acetylcholine, $\beta_1$-adrenoceptor, 5-$HT_{1A}$, adenosine $A_1$ receptors and that for corticotrophin releasing factor (Nicoll, 1990).

### ATP-coupled potassium channels

The coupling of potassium channels to the metabolic state of the cell is mediated via the so-called ATP-dependent potassium channels (Ashcroft, 1988). These channels are open when intracellular ATP levels are low and close as the concentration of this nucleotide rises. This system has important physiological consequences

as illustrated by the following example. Under conditions of low plasma glucose, the ATP-coupled $K^+$-channels of the insulin-secreting, pancreatic $\beta$-cells are open and hold the membrane at a subthreshold potential for insulin secretion. When the plasma glucose concentration rises, the cells accumulate the sugar and its subsequent metabolism generates intracellular ATP which closes the channels. The resulting depolarization leads to trains of action potentials which trigger insulin secretion.

ATP-coupled potassium channels have been found in many different types of smooth muscle cells, heart muscle and some types of neurons. Apart from the above-mentioned example, their function is as yet unclear but they are blocked by the oral antidiabetic drugs such as tolbutamide and glibenclamide and opened by vasodilator drugs such as cromakalim and diazoxide.

## The 5-$HT_3$ receptor

The 5-$HT_1$ group of receptors can affect the ion permeability of a membrane (e.g. the 5-$HT_{1A}$ receptor causes hyperpolarization by increasing potassium conductance) but this is indirect and mediated through G-protein dependent mechanisms. On the other hand, the 5-$HT_3$ receptor is a ligand-gated ion channel which directly controls membrane permeability. Activation of this receptor causes rapid depolarization of the cell (less than 30 ms latency) which is short-lived (100–300 ms) and carried largely by sodium ions, although the channel itself is non-selective between sodium, potassium and similar cations.

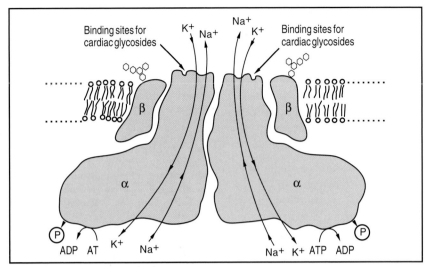

*Fig. 15. Diagram of the $Na^+/K^+$ ATPase which acts as the sodium pump. The protein is a heterotetramer consisting of two α subunits and two β subunits (which are glycoproteins). Cardiac glycosides bind at the extracellular surface of the α subunit whilst the ATPase activity resides at the intracellular surface. The β subunit might also be sensitive to cardiac glycosides.*

## Ionic pumps and exchange mechanisms

The passage of ions across the cell membrane through ion channels requires the restoration of the transmembrane ionic gradient. Since the original disturbance of the ionic gradient was the result of ions passing down their electrochemical gradients, the restoration process usully requires the input of energy. When this energy is supplied directly by ATP, the process is said to comprise a pump and the transport protein is thus an ion-dependent ATPase; the best known example is the sodium/potassium ATPase which forms the 'sodium pump'. The sodium gradient thus established can then be used to drive exchange mechanisms in which the energy liberated by the sodium passing down its gradient is used to move other ions against their gradients(s); such processes include a sodium/calcium exchange mechanism

found widely in the body and a sodium/potassium/chloride co-transport. The body possesses a veritable menagerie of ion exchange and co-transport mechanisms and it is not appropriate to describe here all those so far discovered. Thus this discussion will comprise only the most important systems with a widespread distribution in the body.

## Sodium pump

The pump, which is more correctly called sodium/potassium ATPase, is a heteromeric trimer having two subunits of relative molecular mass of 100 000 and another of about 50,000 (Fig. 15). It is found in all active cells and conveys two potassium ions into the cell in exchange for three sodium ions with the hydrolysis of one molecule of ATP; that is, its action is electrogenic and causes the cell to hyperpolarize. If the ionic gradients are reversed in an artificial system, the pump can be

driven 'backwards' with the generation of ATP from ADP. The mechanism of action of the pump, and its relationship to the hydrolysis of ATP, is well characterized and the pump is inhibited reversibly by the binding of cardiac glycosides such as ouabain to the potassium binding site when this is facing the extracellular space.

## Calcium pump and sodium/calcium exchange

Both the sarcoplasmic reticulum and the plasma membranes of many cells contain calcium-dependent ATPases which extrude calcium from the cytoplasm. Although not identical, these two proteins have much in common. Both consist of homodimers of a single type of polypeptide chain, are magnesium-dependent and have to operate against gradients of the order of 1 to 10 000. The activity of both pumps can be stimulated by the presence of sodium or potassium on the cytoplasmic side of the membrane that they traverse. However, the plasma membrane pump is calmodulin-sensitive; if three out of its four calcium binding sites are occupied, then calmodulin binding increases the rate of pumping.

The sodium/calcium exchange mechanism is found in most active cells and brings about the exchange of one calcium ion for three sodium ions, thus it is electrogenic causing depolarization of the cell. The process is facilitated by potassium but only in the retinal rod (where the stoichiometry is 4 sodium: 1 calcium + 1 potassium) is potassium obligatory. The exchanger cloned from mammalian heart plasma membranes has been modelled as having 11 membrane-spanning domains and a single molecule may be able to act as a functional transporter. It is not yet known whether the transport mechanism is sequential, with the transport of one ion following the movement of another, or is a true antiport with both ions moving at the same time.

## Chloride pump

There are several different transport processes that can regulate transmembrane gradients of chloride and, although an ATP-dependent chloride pump exists in bacteria, these usually use the inwardly directed sodium concentration gradient to provide the energy for chloride movement. In secretory cells and widely throughout the body (including vertebrate skeletal muscle) there is found a sodium/potassium/chloride cotransport mechanism, with the stoichiometry 1:1:2, which is sensitive to inhibition by the loop diuretics frusemide and bumetanide. All three ions move into the cell and transepithelial transport of chloride in the kidney and elsewhere occurs as a result of the chloride leaving the cell through cyclic AMP and calcium dependent chloride channels. Another chloride entry pathway is provided by a sodium/chloride cotransport which is sensitive to inhibition by the thiazide diuretics such as hydrochlorothiazide. This too is encountered in other cells apart from the renal epithelium.

Net inward movement of chloride in neurons is brought about by a chloride/bicarbonate exchange mechanism which is sensitive to inhibition by SITS (4-acetamido-4′-iso-thiocyano-stilbene-2,2′-disulphonic acid). Both bicarbonate and chloride are carried against their concentration gradients and the energy needed appears to be provided by parallel sodium movement.

## Summary

Neurotransmitters and circulating hormones (except steroids) interact with specific cell surface receptors to change the electrical, contractile and/or metabolic activity of neurons and smooth muscle cells. The manner in which transmitters and exogenously applied drugs bind to receptors can be described by a series of linear kinetic equations. The molecular structure and pharmacological selectivity of receptors has been elucidated and classified using advanced molecular biological and classical pharmacological techniques. In most cases, the result of transmitter binding to a specific receptor is a change in membrane ionic permeability brought about via the opening or closing of membrane bound ionic channels. Such channels are ion-selective pores with a unique molecular structure and pharmacology. Thus, cerebral blood flow can be increased or decreased by substances acting directly on the vessel wall smooth muscle or indirectly either on the endothelium or on the membrane of the neurons innervating the smooth muscle. At both sites cellular activity can be modulated by drugs acting on receptors or on ionic channels. Restoration of transmembrane ionic gradients, after changing the activity of a cell by the transfer of ions, requires the activity of translocation processes which, when they require energy supplied by ATP, are commonly termed pumps. The most important pump is that responsible for the transfer of sodium out of, and potassium into, the cell and the sodium gradient thus built up is used to drive other translocation processes.

## References

Ariëns, E.J. (1964): *Molecular pharmacology*, Vol. 1: *The mode of action of biologically active compounds*. New York: Academic Press.

Arunlakshana, O. & Schild, H.O. (1959): Some quantitative uses of drug antagonists. *Br. J. Pharmacol.* **14,** 48–58.

Ashcroft, F.M. (1988): Adenosine, 5′-triphosphate-sensitive potassium channels. *Ann. Rev. Neurosci.* **11,** 97–118.

Bertolino, M. & Llinás, R.R. (1992): The central role of voltage-activated and receptor-operated calcium channels in neuronal cells. *Ann. Rev. Pharmacol.* **32,** 399–421.

Black, J.W. & Leff, P. (1983): Operational models of pharmacological agonism. *Proc. R. Soc. Lond.* **B220,** 141–162.

Bowman, W.C. & Rand, M.J. (1980): *Textbook of pharmacology*. 2nd edition, Oxford: Blackwell Scientific Publications.

Brown, D.A. (1990): G-proteins and potassium currents in neurons. *Ann. Rev. Physiol.* **52,** 215–242.

Clarke, A.J. (1937): *General pharmacology. Handbook of experimental pharmakology.* (reprinted 1973). Berlin: Springer-Verlag: Berlin.

Cook, N.S. (1990): *Potassium channels.* Chichester: Ellis Horwood.

Furchgott, R.F., & Bursztyn, P. (1967): Comparison of dissociation constants and of relative efficacies of selected agonists acting on parasympathetic receptors. *Ann. N.Y. Acad. Sci.* **144,** 882–899.

Galzi, J.-L., Revah, F., Bessis, A. & Changeux, J.-P. (1991): Functional architecture of the nicotinic acetylcholine receptor: from electric organ to brain. *Ann. Rev. Pharmacol. Toxicol.* **31**, 37–72.

Henderson, R., Baldwin, J.M., Ceska, T.A., Zemlin, F., Beckmann, E., Downing, K.H. (1990): Model for the structure of bacteriorhodopsin based on high-resolution electron cryomicroscopy. *J. Mol. Biol.* **213**, 899–929.

Hille, B. (1984): *Ionic channels of excitable membrane*s. Sunderland, MA: Sinauer Associates Inc.

Hodgkin, A.L. & Huxley, A.F. (1952): Currents carried by sodium and potassium through the membrane of the giant axon of *Loligo*. *J. Physiol.* **116**, 449–472.

Kenakin, T.P. (1987): *Pharmacologic analysis of drug–receptor interaction*. New York: Raven Press.

Marty, A. (1989): The physiological role of calcium-dependent channels. *Trends Neurosci.* **12**, 420–424.

Nicoll, R.A. (1990): Functional comparison of neurotransmitter receptor subtypes in mammalian central nervous system. *Physiol. Rev.* **70**, 513–565.

Narahashi, T. (1988): *Ion channels*, Vol. 1. New York: Plenum.

Narahashi, T. (1990): *Ion channels*, Vol. 2. New York: Plenum.

Sakmann, B. & Neher, E. (1984): Patch clamp techniques for studying ionic channels in excitable membranes. *Ann. Rev. Physiol.* **46**, 455–472.

Sommer, B. & Seeberg, P.H. (1992): Glutamate receptor channels: novel properties and new clones. *Tr. Pharmacol. Sci.* **13**, 291–296.

Tsien, R.W., Lipscombe, D., Madison, D.V., Bley, K.R. & Fox, A.P. (1988): Multiple types of neuronal calcium channels and their selective modulation. *Trends Neurosci.* **11**, 431–438.

Van Rossum, J.M. (1977): *Kinetics of drug action. Handbook of experimental pharmacology*, Vol. 47. Berlin: Springer-Verlag.

Watson, S. & Abbott, A. (1991): Receptor nomenclature supplement. *Tr. Pharmacol. Sci.* **11**, Suppl. 1.

Weston, A.H. & Hamilton, T.C. (1991): *Potassium channel modulators – pharmacological, molecular and clinical aspects*. Oxford: Blackwell Scientific Publications Ltd.

# Chapter 5

# The anatomy of the brain vasculature

## K.C. Hodde[*] and R. Sercombe[†]

[*]*Academic Medical Centre, Department of Experimental Surgery, Meibergdreef 9, 1105 AZ Amsterdam, The Netherlands;*
[†]*Laboratoire de Recherches Cérébrovasculaires, CNRS UA 641, Université Paris VII, 10 Avenue de Verdun, 75010 Paris, France*

## Introduction

This chapter describes the basic anatomy of brain vascularization, in particular in the rat, with comparative remarks to that in the rabbit, dog, cat and man.

In various atlases of the rat the vascular system of the head region is dealt with in differing degrees of detail. For a general atlas, Greene's (1935) goes into considerable detail in meticulously executed drawings. *Craigie's Neuroanatomy* (Zeman & Innes, 1963) has a comparatively brief section on the vascularity of the brain and refers to the work of Craigie who published extensively over a period of more than two decades on the comparative vascularity of various brain regions and compared those with other species. His work is still used for reference material.

Wuenscher *et al.* (1965) deal with the intrinsic vascularity of the rat brain stem in a thorough manner as does Duvernoy (1978) for man. Hebel & Stromberg (1986) give a complete survey of the literature.

## Comparative anatomy

The works of Tandler (1899) and Hofmann (1900, 1901) are rated as classics in the field and are still used as sources for reference. Van Gelderen (1933 and earlier work), although less well known, has published thorough investigations on the comparative aspects of the cephalic veins, including those of the rat. Butler (1967) gives a detailed account of the development of the dural venous sinuses and emissary veins of various mammals with emphasis on the transverse sinus and related vessels in the rat. Stephan (1975) in his work on the allocortex,

has a comprehensive section on the comparative aspect of the vascularization of the forebrain in all vertebrate classes, both the arteries and the veins. Bugge (1974) studied the cephalic arterial system in various mammals and used the anastomotic relations, especially between the internal and external carotids, as a criterion for systematic classification. Simoens *et al.* (1979, 1984) provide a very complete and lucidly illustrated source of comparative information on the cephalic arteries and veins in domestic mammals.

## Methods used

A multitude of approaches in various combinations with a wide array of instruments ranging from the naked eye to sophisticated optical, X-ray and electronic equipment has been used for *in vivo*, *in vitro* and post-mortem studies of blood vessels. A short overview of commonly used methods is given below.

With intra-vital microscopy blood vessels can be observed *in vivo*. The pial vessels have been studied extensively with this method by creating a cranial window, replacing a part of the dura by translucent material. Flow and vessel dynamics then can be recorded. Similarly, the hypophyseal portal vessels have been studied in the living animal.

In fixed tissue, the vessels can be observed macroscopically in dissection preparations after injecting liquid latex or plastic which solidifies *in situ*. Light microscopically, vessels are visualized in unstained material by phase contrast; in stained sections, 5–25 µm, or semithin sections, 0.5–1 µm thick, by colour contrast in the tissue or by injecting the vessels with a contrasting medium (gelatin ink or rubber compounds such as Microfil™).

Transmission electron microscopy (TEM) of ultra-thin (50 nm thick) sections use metal-impregnation for selective and yield ultrastructural details in the nanometerscale.

Scanning electron microscopy (SEM) gives a pseudo-three-dimensional image of a surface by reflected electrons, the number of which is determined by the surface geometry. Specimens are not embedded, have to be dry and electron conductive and have the surface of interest exposed to the electron beam. Its range of magnification typically is from macroscopic to the nanometer scale with a large depth of field.

X-ray contact or projection images (microradioangiograms) are obtained by filling the vessels with a radio-opaque medium (e.g. Micropaque) and cutting slices in millimeter thickness for X-ray exposure.

Corrosion cast preparations are made by injecting a liquid compound (rubber or pre-polymerized resin) which solidifies *in situ*. The surrounding tissue is dissolved and the vascular cast can be studied macroscopically and microscopically. The wide range of magnification and great depth of field of the SEM are especially useful in these three-dimensional specimens. The compounds available reproduce surface features below the µm order and give a good mirror image of endothelial cells.

## Blood supply to the brain

Generally stated, the mammalian brain is supplied with blood from two paired sets of arteries: the vertebrals and the carotids which enter the skull at the base. The vertebral arteries are branches of the right and left subclavian arteries and fuse to become the basilar artery, mainly supplying the brainstem and the cerebellum. The right and left carotid arteries have an internal and an external branch either from a common stem or directly from the aortic arch varying with the species and with intra-specific variations as well.

The basic branching pattern of the cephalic arterial supply (of the vertebral–basilar and internal-external carotids) has been described by Bugge (1978 and earlier work). He demonstrated seven possible anastomotic channels in the basic arterial pattern and their phylogenetic and ontogenetic development through persistence, reduction, obliteration and secondary anastomoses into species-specific patterns. This he used as a criterion for systematic classification of the species studied in his extensive and meticulous work.

The primary area supplied by the vertebral-internal carotid artery system is the brain and parts of the inner ear, the eyeball and the nose. In the Ursid family the brain indeed is exclusively supplied by those arteries. In the Felid family, on the other hand, the external carotid artery system has annexed almost the whole intracranial supply of the internal carotid. The rat, rabbit, dog and man take a position somewhere between those two extremes.

## Extracranial supply of the intracranial circulation in dog and cat

In the dog, anastomotic branches connect the (extracranial) maxillary artery with the internal carotid, sometimes forming a beginning of an intracranial rete. Its contribution to the total brain circulation has been measured in conscious dogs to be about equal to the proximal part of the internal carotid.

In the cat, the proximal part of the internal carotid is obliterated and the distal part is supplied exclusively by branches of the maxillary artery (Figs. 1, 2). These form an extracranial rete mirabile and intracranially two sets of inter-anastomosing branches which come together in a single vessel, joining to the distal internal carotid or directly the circle of Willis (Simoens *et al.*, 1979).

In the rat, rabbit and man there is no rete mirabile between the external

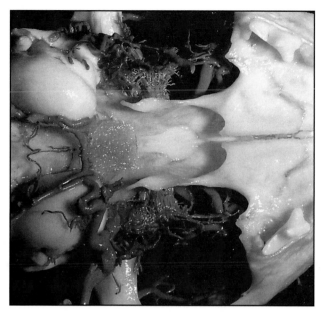 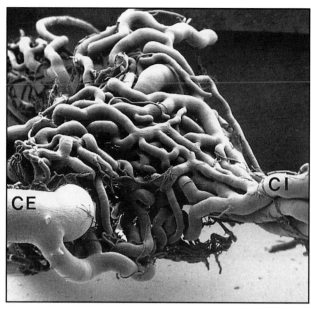

*Fig. 1 (left). Ventral view of cat skull with resin-injected rete mirabile (extracranial part) visible in relation to the skull. Picture width 5 cm.*
*Fig. 2 (right). Resin corrosion cast of cat rete mirabile, arterial part, the cavernous vein through which it passes not being injected. Left: the external carotid (CE) develops into numerous branches, joining together in the non-obliterated part of the internal (intracranial) carotid artery (CI) on the right. Magnification 15 ×.*

and internal carotid systems, and the internal carotid and vertebral arteries virtually supply all the blood which goes to the brain.

## The circle of Willis

This closed arterial circuit at the base of the brain comes about through the interconnection of the vertebral-basilar and the carotid systems. The basilar artery in which the left and right vertebrals come together develops *in situ* from plexiform networks as does the common anterior trunk when present which is the midline fusion of the anterior cerebral arteries, end-branches of the internal carotids. The basilar artery rostrally branches into the left and right posterior cerebral arteries which connect with the left and right internal carotids, respectively, through secondary anastomoses – the posterior communicating arteries – thus completing the circle of Willis.

Apart from the interspecific pattern variations, the circle of Willis is a highly variable structure. This has been extensively studied particularly in man, rat and mouse. In man, abnormalities have been reported in about half of the specimens studied. Anomalies of the antero-posterior links were more common than of the left–right anastomoses. Findings in the rat and mouse correspond with this. In the cat, incidentally, the consistent absence of the fusion of the anterior cerebral arteries as well as of an anterior communicating artery has been reported by Klein (1980).

There is a genetic component involved in the intra-species variation. Comparing the differences in shape of the circle of Willis between different strains of inbred mice, Ward *et al.*

(1990) concluded that these were due to additive genetic variation between these strains. Genetically defined heterogeneous mice consistently show anomalies similar to those previously described in genetically undefined rodents, whereas inbred mice do not.

The variability in vessel diameters in the circle of Willis and its main afferent and efferent branches in man was analysed by Hillen (1987) with multivariate statistical techniques. Relations of the vessel sizes found were inferred to be consequential to flow patterns. Van der Zwan (1991), in continuation of this, demonstrated that the peripheral resistances are represented by the sizes of the efferent vessels of the circle. This suggests that the variability of the circle of Willis is predominantly determined by haemodynamic factors.

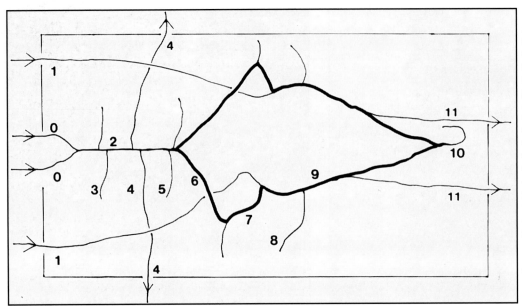

*Fig. 3. Diagram of the circle of Willis of the rat. 0: vertebral A.; 1: internal carotid A.; 2: basilar A.; 3: A. cerebelli post.; 4: labyrinthine A.; 5: A. cerebelli ant.; 6: A. cerebri ant.; 7: A. communicans ant.; 8: A. cerebri media; 9: A. cerebri ant.; 10: truncus communicans ant.; 11: A. olfactoria. The rectangular outline represents the skull. Note the intra-extracranial passage of the labyrinthine and olfactory arteries, supplying the inner ear and the nasal area respectively.*

## The arterial supply to the brain and the circle of Willis in the rat, and the descriptive terminology used

In the rat, the arterial supply to the brain is through the internal carotid and the vertebral arteries (Fig. 3). The latter merge together into the basilar artery which runs rostrad on the ventral side of the brainstem, and divides into the two AA posteriores. Those, in turn, join the posterior communicating arteries, which are branches of the internal carotid arteries and thus form the caudal part of the circle of Willis. Figure 4 shows the ventral aspect of a latex-injected and fixed brain. The internal carotid arteries have been cut away here at the point of division into the medial and anterior cerebral arteries. Figure 5 shows its dorsal counterpart.

Each vertebral artery, before fusing into the basilar, gives off a medially directed branch which unites with the one from the opposite side to form the anterior spinal artery. Approxi-

mately at midpoint of the basilar artery the prominent posterior cerebellar arteries can be seen; at a variable distance, mostly anterior from these, the labyrinthine arteries branch off from the basilar. The terminal arteries of the basilar are the anterior cerebellar and the posterior cerebral arteries.

From the anterior cerebral artery a large olfactory artery branches off which supplies most of the nasal mucosa. The anterior cerebral arteries fuse at the midline into the common anterior trunk. This trunk continues in the perpendicular plane and passes dorsad and then caudad over the corpus callosum where it divides again in two. Each terminal branch runs to the medial side of its own hemisphere and anastomoses with branches of the middle and posterior cerebral arteries. It then sends out a branch which curves over the posterior aspect of the corpus callosum. It terminates by supplying the anterior extremity of the choroid plexuses of the lateral and third ventricles while

anastomosing with one of the choroidal branches of the posterior cerebral artery.

Upon cursory examination, the literature appears congruent, but when reviewed carefully several inconsistencies become apparent. These are due to differences in terminology and, more importantly, to differences in concepts, primarily those referring to the posterior cerebral and communicating arteries. Greene (1935) considers the anterior cerebellar arteries to be the terminal branches of the basilar artery: 'The posterior communicating artery arises from the internal carotid on the basal surface of the brain ... to reach the median basilar artery formed by the union of the vertebral arteries of either side', and, 'The posterior cerebral artery is a branch of the posterior communicating artery' (p. 185). This she concludes by comparing the rat specimens with a description by Adachi (1928, as quoted by Greene): '... by the A. communicans posterior is under-

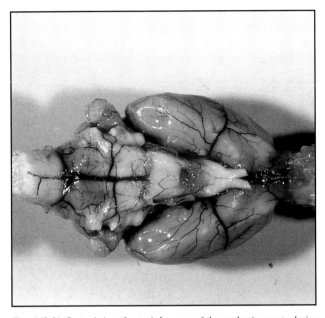 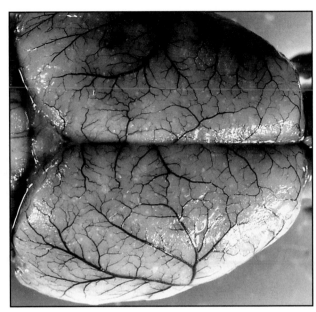

*Fig. 4 (left). Latex-injected arterial system of the rat brain, ventral view.*
*Fig. 5 (right). Latex-injected arterial system of the rat, dorsal view of the cerebral hemispheres.*

stood the connection between the internal carotid and the basilar arteries, so that the posterior cerebral does not, in the ordinary sense, spring from the basilar, but from the posterior communicating artery'. Adachi concluded this from his observations of human material, where the calibres of the carotid and of the basilar sections of the thus defined posterior communicating artery vary in size. The posterior cerebral artery seemingly takes its origin from either the basilar, or from the carotid artery, whichever has the largest diameter.

Moffat considers the posterior cerebral artery as a branch of the basilar from the embryological point of view. It joins the posterior communicating artery which is a branch of the internal carotid. But Moffat also states that 'Often, the posterior communicating artery is larger than the first part of the posterior cerebral artery, so that the latter then may be regarded as a branch of the internal carotid' (Moffat, 1961, p. 486). This last point of view he shares with Greene, who agrees with Adachi, and it can be found in nearly the same words again in the description of the circle of Willis in the current edition of *Gray's Anatomy* for the human.

Hofmann (1900) described the place of origin of the posterior cerebral artery as varying among the cranial ramus of the cerebral carotid, the caudal ramus of the cerebral carotid (which Tandler had called the A. communis posterior or the r. posterior A. carotis internae), and the cranial end of the A. basilaris. He demonstrated in the comparative section of his work that several branches representing all different places of origin can co-exist in one specimen in which the various calibres add up to about the same cross-sectional aggregate.

He also mentioned an r. anastomoticus R. caudalis of the cerebral carotid, which does not figure prominently in his descriptions. For his terminology he apparently looked at the total pattern from the viewpoint of the comparative anatomist.

From the foregoing it appears that the morphological descriptions are very much alike, but that the descriptive terminology differs because of varying points of departure taken by the authors. It should be mentioned that the asymmetry of the origin of the posterior cerebral arteries has been well documented in many species.

In the recent literature, Zeman & Innes (1963) mention the disparity in nomenclature and describe the two situations in which the terminal branches of the basilar are called either the posterior cerebral or the posterior communicating arteries. While refraining from taking a stand themselves they mention that 'the latter view seems the more in keeping with embryonic development' (p. 32). Al-

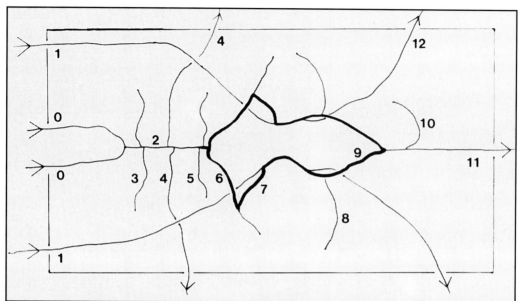

*Fig. 6. Diagram of the circle of Willis of the rabbit. Numbers as in Fig. 3, except for the unpaired 11: A. ethmoidalis interna, and the paired 12: AA. ophthalmica internae. Note the intra-extra-cranial passage of labyrinthine, internal ophthalmic and internal ethmoidal arteries which supply parts of the inner ear, eye and nose.*

though they rightly point out the disparity they also add to the confusion by misinterpreting the view of Moffat who had arrived at the opposite conclusion. Wuenscher *et al.* (1965) mention without comment the ultimate branching of the basilar into the posterior communicating arteries, and quote Tandler, Hofmann (incorrectly), and Greene. Hebel & Stromberg (1986) on the other hand, reintroduce Hofmann's terms to some extent (although using them sometimes for different entities) and add some new ones. They describe the internal carotid artery as dividing into a rostral and a caudal cerebral artery. The caudal one (in their schematic rendering at least) joins the caudal communicating artery, while the rostral cerebral artery after branching off the medial cerebral artery divides into the olfactory artery and the rostral communicating artery which fuses with the corresponding vessel from the other side. There is something to be said for their naming that part of the anterior cerebral artery between the branch-

ing of the olfactory artery and the common anterior trunk the 'rostral communicating artery'. It expresses the analogy between the embryonic development of this part of the circle of Willis and the posterior communicating artery as described by Moffat. Both the basilar artery and the common anterior trunk develop *in situ* from plexiform networks, and the later connecting vessels develop from several anastomosing constituents. Unfortunately, these authors do not provide any kind of explanation as to why they chose these particular terms.

The midline fusion of the two anterior cerebral arteries has been variously called the A. cerebri anterior communis (Hofmann); azygos vessel (Greene); the midline anterior trunk (Moffat); truncus cerebri anterius (Zeman & Innes); azygos anterior cerebral artery (Brown); a. communicans rostralis (Firbas *et al.*); truncus cerebri rostralis (Hebel & Stromberg).

Firbas *et al.* are the only authors who

consistently describe a communicating *artery* as opposed to vessels fusing into a common *trunk*. Most authors view the anterior communication of the circle of Willis as a fusion in the midline of the two anterior cerebral arteries. Occasionally additional anastomosing vessels have been described, and only rarely anterior communicating arteries as such. This view was supported by Hodde (1981) from a large series of specimens.

The occurrence of so-called 'arterial islets' in the circle of Willis and the basilar artery has been noted by many workers in various vertebrate species, including man. These are synonymously called 'Arterieninsel'; biradicular origin; duplicated loops; perforations; islands; cleft-like formations and button-hole formations. The morphology of these commonly occurring phenomena suggests incomplete fusion of double or plexiform channels during embryological development.

In conclusion, it can be stated that

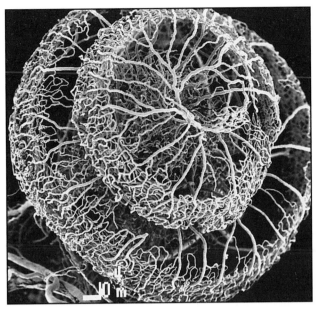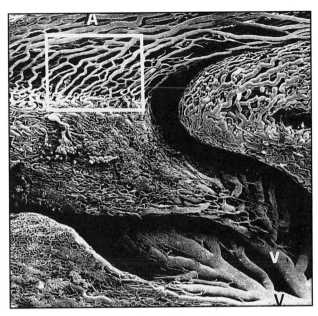

*Fig. 7 (left). Resin cast of the rat cochlea vascular bed, supplied by a branch of the basilar artery. Magnification 15×.*
*Fig. 8 (right). Part of the rat nasal turbinate vascular bed, exclusively supplied by the olfactory artery, a branch from the circle of Willis. Top: arterial branches of the A. olfactoria (A), bottom: venous outflow (V) into V. sphenopalatina, leading to the internal maxillary vein. Rectangle: numerous arterio-venous shunts (A-V-A), 20–80 µm lumenal diameter. There are a few hundred of these shunts in the whole nasal circulation, all provided with specialized muscular cuffs.*

there is no uniform nomenclature used to describe the circle of Willis in the rat. Moreover, identical names are used to describe different entities, and vice versa, which makes it difficult to compare data in the literature.

## Extracerebral supply by intracranial sources

Some extracerebral tissues are supplied by branches of intracranial arteries in a recognizable pattern which is species-specific. These tissues are: the inner ear, by a branch of the basilar artery (the labyrinthine A.); the eye, by a branch of the internal carotid artery (the inner ophthalmic A.) and the nose, by a branch of the anterior cerebral artery (the olfactory or inner ethmoidal A.).

This is a general mammalian pattern and its existence is not surprising be-

cause the tissues supplied are in fact part of the nervous system. The pattern is ontogenetically modified by reduction, obliteration, secondary anastomoses and persistence which determine the inter-specific differences observed. This is exemplified here in the rabbit (Fig. 6) and in the rat (Figs 3, 7, 8).

In the rabbit, part of the inner ear is supplied by the labyrinthine artery, a branch of the basilar artery, similar to the situation in the rat. A large part of the eye is supplied by a branch of the internal carotid which comes off proximal to the medial cerebral artery and has more than half its diameter. This inner ophthalmic artery also forms a sizable anastomosis with the external ethmoidal artery which supplies most of the nasal region.

From the fused anterior cerebral ar-

teries, the common anterior trunk, an unpaired inner ethmoidal artery (about half the diameter of the trunk) passes through the cribroid plate to contribute to the nasal vascular plexus.

In the rat most of the inner ear is supplied by the labyrinthine artery which is a direct branch of the basilar artery. Unlike the rabbit, there is no internal ophthalmic artery so that the eye is supplied by extracranial sources only. On the other hand, there is a major intracranial vascular contribution to the nasal circulation by branches of the left and right anterior cerebral arteries. These paired olfactory arteries have about half the diameter of the internal carotids and form the exclusive supply of the nasal septum and endoturbinate regions of the nose. Moreover, in both these regions extended zones with arteriovenous anastomoses are present which may shunt (intracranial) arterial blood directly into the systemic venous circulation (Fig. 8).

The existence of these configurations has to be taken into account when considering flow measuring models, also being a potential source of significant extracerebral contamination.

## Distribution of flow to the brain

The vertebral and carotid systems have distinct territories of supply which vary with the species. Studies done since the turn of this century mainly in dogs and primates were extended by McDonald & Potter (1951) in the rabbit. What was seen was that all four arteries supply their own region. In the basilar artery (and probably similarly in the anterior trunk when present), due to the laminar flow, there is no mixing of blood from left and right. The meeting points between vertebral and carotid systems do not show mixing at the 'dead point' interfaces in the left and right posterior communicating arteries (and probably similarly so in the anterior communicating artery when present). In cases of occlusion of one or more arteries the connections showed mixing or forward/backward flow as a result of the altered haemodynamic conditions.

The distribution of the vertebral and carotid supply in dogs and cats was determined by Wellens *et al.* (1975 and earlier work). They showed consistent differences between the two species, a considerable extracranial supply by the vertebral arteries in the dog and left–right transmission of blood in both arterial systems for both

species.

The variability of the cortical and intracerebral territories supplied by the major cerebral arteries (posterior, middle and anterier cerebral AA.) and its relation to the circle of Willis was investigated in man by Van der Zwan (1991) and Van der Zwan & Hillen (1991). He demonstrated by simultaneous injection of differently coloured injection media in these arteries that there was a large variability in size and asymmetry in the supplied territories, in distinct contrast to the general concept of relative invariance as stated in the literature of the last hundred years.

The pial vessels show many inter-arterial anastomoses between the branches of the cerebral arteries. This has been seen by many investigators in various species and has been studied specifically in the rat. Arteriovenous anastomoses do not occur.

## Venous drainage

The blood in the compressible veins of the brain flows into the rigid dural sinuses. These are connected either with the jugular veins or with the (spinal epidural) vertebral veins (Fig. 9). The relative distribution of the brain drainage through the internal and external jugulars and the vertebral veins is very different among mammalian species.

The (intracranial) dural venous sinuses and the external jugular venous system are connected by the emissary veins which run through the various fontanelles and foramina of the skull.

Because of the many intra-extra-

cranial venous vascular communications there is a considerable species-dependent and variable degree of mixing of intra- and extracerebral blood in the various sinuses and veins.

In the rat, the venous drainage of the brain is presented here in simplified form as an example of the complexity of the venous interrelationships. In general terms, the brain is drained via three routes which are all interconnected: the transverse sinuses, the cavernous sinuses and the occipital sinuses. These eventually drain into the external jugular veins, the internal jugular veins and the vertebral veins respectively. These may be interconnected, as in cat and rabbit but are not in the rat. The transverse sinus is dorsally continuous with the confluens sinuum (also named torcular herophili). Here the superior sagittal, transverse (paired), post-torcular sinuses and the sinus rectus or v. cerebri magna of Galen flow together. These sinuses drain most of the dorsal superficial veins of the brain. Ventrally, the transverse sinus divides into two equal-sized sinuses: the (rostrally situated) petrosquamous sinus and the sigmoid sinus. The first leaves the skull via the post-glenoid emissary vein to join the internal maxillary vein which drains into the external jugular vein. The sigmoid sinus runs ventro-caudally and is joined by the occipital sinus. It then continues in a much reduced size after branching off a large vessel which leaves the skull as the vertebral vein via the foramen magnum.

The cavernous sinus (also paired) located ventrally is formed by the junction of the ophthalmic vein and the anterior cerebral vein. It divides into two sinuses of which the superior petrosal sinus joins the transverse sinus,

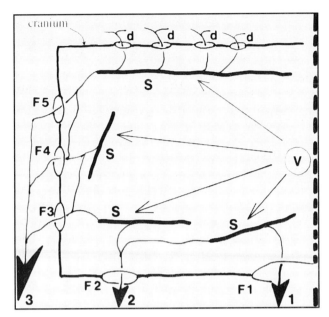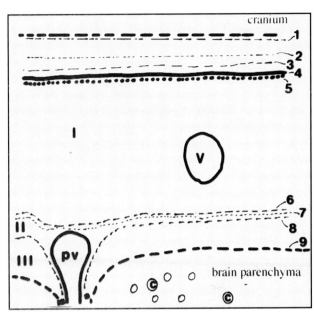

*Fig. 9 (left). Schematic diagram of the principal routes of venous drainage of the brain. V: venous drainage of brain parenchyma into the dural sinuses (S). In turn, these drain through foramina in the skull into the vertebral vein (1), (F1, Foramen magnum), into the v. jugularis interna (2) (F2, Foramen V. jug. int.), and into the V. jug. externa (3) via various foramina (F3, 4, 5, etc.), depending on the species. Veins passing through these foramina are called emissary veins. Top: diploic veins (d) connect veins of the scalp with dural sinuses through the cranium. Right vertical indicates midsagittal plane.*

*Fig. 10 (right). Diagram of meningeal layers and spaces. 1: periosteal dura; 2: meningeal dura; 3: dural border cells; 4: arachnoid barrier cells; 5: external arachnoid; 6: internal arachnoid; 7: external pia; 8: internal pia; 9: glia limitans; I: arachnoid space; II: pial space; III: subpial space; V: pial vessel (artery resp. vein); pv: penetrating vessel (arteriole resp. venule); *: perivascular space of Virchow–Robin; c: parenchymal capillary. Solid lines indicate main sites of blood–brain barrier. After Krisch et al., 1984; Krisch, 1988; Haines 1991. See text for further explanation.*

and the inferior petrosal sinus joins the smaller sigmoid sinus and continues as the internal jugular vein, leaving the skull through the jugular foramen.

Of all available sources, Greene (1935, reprinted in 1968) gives the most detailed and accurate description of the venous system as a whole. Recently, the anatomical pattern and development of the venous system of the cranial base in the rat was described by Szabo (1990) who found a transverse basal sinus system ('sinus transversus basalis') and described its venous connections.

As in the rat, the main venous outflow channels in the dog, cat and rabbit are the external jugular and the vertebral veins, the internal jugular being diminutive in size or sometimes absent (dog).

The dog shows patterns of dural sinuses and emissary veins similar to the rat. In the cat the external jugular and vertebral veins have ipsilateral communications, and these in turn anastomose across the midline (Du Boulay 1974).

In the rabbit there is no post-glenoid emissary vein but a large mastoid emissary vein eventually draining into the external jugular via the superficial temporal vein. The vertebral vein has communications both with the external as with the (much smaller) internal jugular veins.

Despite wide phylogenetic differences there are many similarities with man in the pattern of dural venous sinuses and emissary veins in the rabbit.

In man, almost all cerebral blood is drained through the internal jugular veins. The internal jugular bulbs contain blood, two-thirds of which originates from the ipsilateral side, with less than 3 per cent of extracerebral origin. The vertebral veins are poorly developed and carry blood from the medulla oblongata.

To withdraw cerebral blood with a minimal admixture of extracerebral origin, it can be taken from the internal jugular bulb in man (through percutaneous puncture) and in the

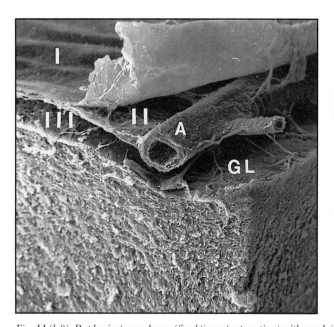
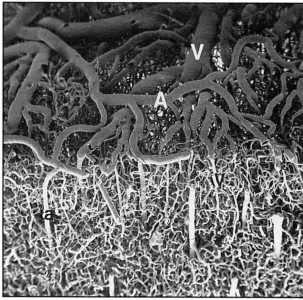

*Fig. 11 (left). Rat brain parenchyma (fixed tissue preparation) with overlying pial artery (A) illustrating the vascular–meningeal relationship. GL: glia limitans; I: arachnoid space. Pial space (II) surrounds artery. III: sub-pial space; 1000 ×.*
*Fig. 12 (right). Oblique view on corrosion cast of rat brain cortical vessels. Bottom half: perpendicular cross-section, penetrating arterioles (a) and venules (v) with intercalated capillary network. Top half: pial arteries (A) seen to cross over veins (V).*

cat. In the dog it should be withdrawn from the posterior third of the superior sagittal sinus through a boneflap on the skull-top.

## Vascular-meningeal relations

The three meningeal layers, the dura, arachnoid and pia mater envelop the brain. The dura has a periosteal interface with the cranium, and the pia overlies the brain parenchyma following its contours. The arachnoid lies between, displaying many trabeculae which gives it its spider-web aspect. Together pia and arachnoid are called the leptomeninges.

Traditionally, in or under all three meninges fluid-filled extracellular cavities have been described: the (potential) subdural space, the (sub)arachnoid and the (sub)pial spaces. At the interface of the dura

with the arachnoid, the dural border cells and the arachnoidal, tight-junctioned barrier cells have been described. It has been assumed so far that between these two cell layers a fluid-filled or otherwise 'potential' space, the subdural space, existed. It now is concluded that 'the creation of a cleft in this area of the meninges is the result of tissue damage' (Haines, 1991) disrupting the dural border cell layer by tissue handling or (in the clinical situation) bleeding. 'Subdural haematoma' should be called dural border haematoma, a haematoma at the dura–arachnoid interface but within the (dural border cell layer of the) dura. Recent studies have shown that both the arachnoid and the pial spaces are completely lined with respectively arachnoidal and pial lining cells. Thus, only in the leptomeninges are these fluid-filled spaces present together with the subpial space be-

tween the pia mater and the neuropileum lining the surface of the brain parenchyma (Figs. 10, 11). Dura mater and leptomeninges develop from an embryonic network of connective tissue-forming cells, and the formation of cerebrospinal fluid (CSF)-containing spaces accompanies the differentiation of the meningeal cellular layers.

The superficial blood vessels supplying and draining the brain traverse the arachnoid space and are covered with a pial cell layer. Around the vessels proper is a (perivascular) space, which is also the case around the (pre-capillary) arterioles and (post-capillary) venules entering the brain parenchyma and emerging from it, also called the Virchow–Robin space.

Originally it was thought that a free communication existed between the

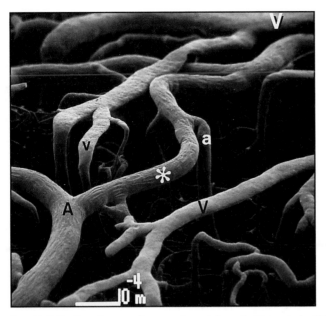 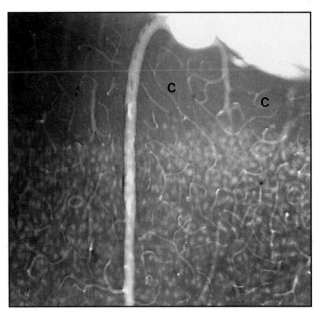

*Fig. 13 (left). Vascular corrosion cast of rat pial vessels. Note arteries (A), veins (V) and intra-arterial anastomosis (\*). Bar = 100 μm.*
*Fig. 14 (right). Penetrating cortical vessel, rat cortex, 200 μm thick section perpendicular to the surface showing layers I, II and III. Fixed tissue, DAB-stained and epon embedded. Light micrograph. Capillaries (c) and cell outlines visible.*

perivascular spaces and the arachnoid space. It now has been established from tracer experiments that these spaces are separated anatomically. This is also true for the pial space and for the intercellular compartment formed by the intercellular clefts of the neuropil, the subpial space and the leptomeninx. A thin layer of identified pial cells encloses the perivascular space of arteries and arterioles in the arachnoid, pial and subpial spaces and in the brain parenchyma. It becomes incomplete at the pre-capillary level. No similar pial sheath was observed around the intracerebral and subpial venules. This suggests that a direct continuity exists between the perivascular spaces around arachnoid arteries and intracerebral arterioles (Zhang *et al.*, 1990).

## Microcirculatory patterns

### Cerebral and cerebellar cortex

The vessels traversing the leptomeningeal space are distributed over the cerebral cortex. Frequently arteries cross over veins but no arteriovenous anastomoses have been seen. However, there are inter-arterial anastomoses between branches of the middle, anterior and posterior cerebral arteries, and between branches of these vessels themselves (Figs. 12, 13). The veins, to a lesser extent, also interconnect; moreover, there are indications that the points of junction with the superior sagittal and transverse sinuses are genetically predetermined. On the cerebellar cortex, the arteries and veins run together in the fissures and branch out from there at regular intervals over the surface.

The parenchymal arterioles and ve-

nules enter the cortex separately (Fig. 14). They can be distinguished in cast preparations (Figs. 15, 16) by the distinct difference in the imprinted shape of the endothelial cell nuclei: arterial – elongated, venous – round (Miodonski *et al.*, 1976; Hodde *et al.*, 1977). See also Scharrer (1940) for a discussion on the difficulty and ensuing confusion on the recognition of arteries and veins at the time.

The parenchymal capillaries branch from the penetrating arterioles and interconnect to form arcades, draining through emerging venules which alternate regularly with the penetrating arterioles. In the cerebellar cortex, the capillary arrangement in the molecular layer appears as hairpin loops perpendicular to the pial surface (Figs. 17, 18). It is not clear whether this is a phylogenetically primitive feature or a secondary phenomenon, brought about by factors intrinsic to

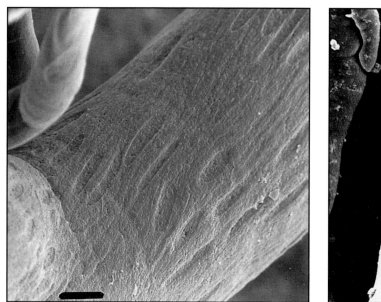 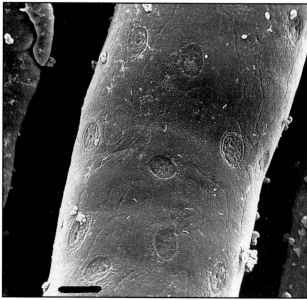

*Fig. 15 (left). Cast of an artery: ovoid nuclear imprints of endothelial cells oriented in vessel axis. Cell borders faintly visible. Bar= 10 μm.*
*Fig. 16 (right). Cast of a vein. Round nuclear imprints and randomly oriented cell borders (arrows). Bar = 10 μm.*

*Fig. 17 (left). Vascular corrosion cast of rat cerebellum (paraflocculus). Note pial arteries (A) and veins (V) running together in fissures.*
*Fig. 18 (right). Detail of Fig. 17. Note capillary loop configuration (c) which occupies the granular layer. A: artery; V: vein; pa: penetrating arteriole;*
*pv: penetrating venule. Bar = 20 μm.*

 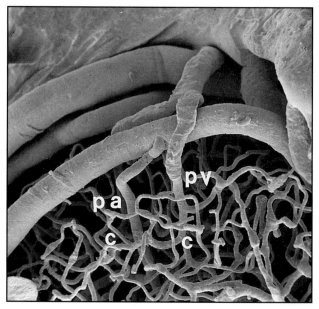

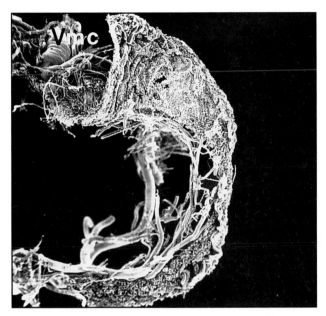 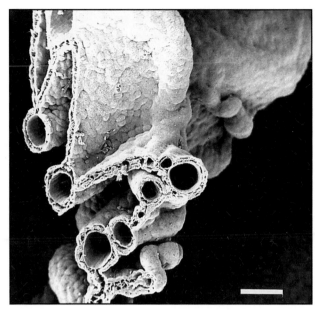

*Fig. 19 (left). Vascular cast of left lateral ventricular choroid plexus (rat), frontal view. Top left: median cerebral vein (Vmc).*
*Fig. 20 (right). Choroid plexus, lateral ventricle, rat , scanning electron micrograph of fixed tissue showing choroidal epithelial cells (a specialized ependyma) and transsected blood vessels: veins, arteries and sinusoids. Bar = 50 μm.*

the cerebellum. The diameter of capillaries is typically reported to be between 4 and 7 μm, depending on the method used. In the cortex, arteriovenous (pre-capillary) shunts have been seen to occur but rarely whereas ramified, pre-capillary 'preferred pathways' are more common (Ravens, 1974; Duvernoy *et al.*, 1981; 1983; Motti *et al.*, 1986).

The capillary arrangements in both cerebral and cerebellar cortices can be seen as a continuously interconnected meshwork. Recruitment of parenchymal capillaries is unlikely (Baer, 1980). The vascular density and preferential orientation varies with the various cortical layers.

## Choroid plexus

From studies made in the last decade with corrosion cast specimens in the rat, rabbit, cat, dog and man, several capillary patterns have been recognized. Typically, the diameter of the capillaries is generally larger (10–15 μm) than found elsewhere and should be considered to be sinusoids. The fenestrated endothelial cells show discontinuous gap junctions and are invested in a continuous thin basement membrane. The plexus is covered by cuboidal epithelial cells (a specialized form of the ependymal cells lining the ventricles) which show continuous tight junctions near the ventricular lumen. Between the ependyma and the capillaries is a layer of loose connective tissue of varying thickness.

The capillary patterns distinguished are, respectively, meshes around the larger arteries and veins, villous structures and glomerulus-like formations (Figs. 19–22). Arteriovenous connections have been found at the hilus of the glomerular formations in human specimens. The functional implication of these consistently recognized patterns in the species studied, possibly subserving different functions, needs further study.

## Portal vessels and venous drainage of the hypophysis

The arrangements of the vascular circulation in the hypophysis are the most complex in the body, according to Wislocki & King (1936, p. 444). The large body of knowledge accumulated since has not reduced this complexity but rather increased it. For a recent review, see the study of Murakami and co-workers (1987) in which the system in the rat is exemplified from corrosion cast specimens.

Simply stated, the general mammalian pattern here can be described as

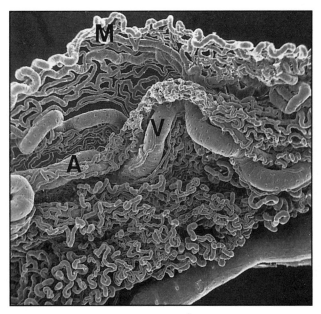 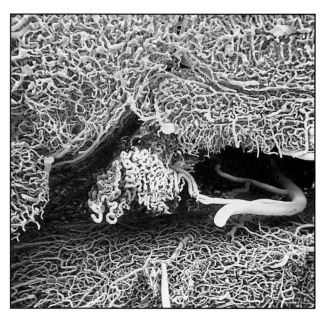

*Fig. 21 (left). Detail of Fig. 19. Note various sinusoidal arrangements: garland around artery (A), veins (V), fronds (F) and marginal bed (M). Bar = 50 μm.*
*Fig. 22 (right). Vascular cast of the 4th ventricular choroid plexus with supplying arteries between cerebellar (top) and brain stem (bottom) vessel cast.*

the different anatomical substructures possessing separate capillary beds: the primary capillary plexus (of the median eminence, the infundibular stalk and the infundibular process or neurohypophysis), the secondary capillary plexus (of the adenohypophysis) and the subependymal plexus, all with their supply and drainage.

The supply of the primary and the subependymal plexus is via hypophyseal arteries, whereas the sole supply of the secondary plexus in most mammals studied is through the portal vessels, originating in the primary plexus, an arrangement found nowhere else in the brain.

Long and short portal vessels are distinguished, the latter arising from the distal part of the stalk and from the neurohypophyseal capillary bed. Both sets of portal vessels show a similar asymmetry in form: capillary

diameter at the primary plexus side, and the wider sinusoids at the secondary plexus side.

Drainage of the primary plexus is through the portal vessels via the secondary plexus into the systemic circulation via the adenohypophyseal veins, as well as directly through the neurohypophyseal veins (Figs 23–26).

Basically the direction of the bloodstream in the portal vessels has been described towards the secondary plexus. A possible retrograde flow has been surmised by presupposing various vasomotor states in the various parts of the vascular beds in combination with a seemingly restricted outflow area of the adenohypophysis. Thus the portal vessels would act as a vascular switch, transporting pituitary hormones directly to the brain.

However, in the rat no direct vascular connection between the median eminence and the hypothalamus has been found (Murakami *et al.*, 1987) and a possible route via the cerebrospinal fluid has been excluded (Krisch *et al.*, 1983). Moreover, the seemingly small diameter of the outflow channels of the adenohypophysis might be an artefactual misinterpretation of negative results and the cross-sectional area of the hypophyseal veins has yet to be assessed quantitatively. Also, the retrograde flow observed *in vivo* could have been the flow in arterioles which are present between the long portal vessels. Therefore, the actual flow direction in the portal vessels still remains a matter of argument rather than of demonstration.

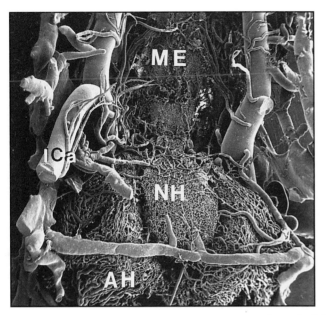

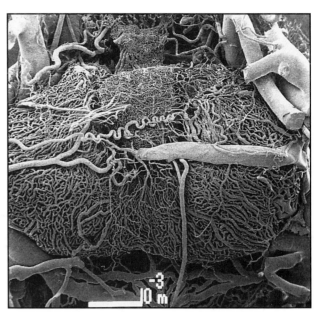

*Fig. 23 (left). Dorsal view of completely injected hypophyseal vascular bed, rat. ICa: internal carotid; CS: cavernous sinus; ME: median eminence; NH: neurohypophysis; AH: adenohypophysis. Note prominent neurohypophyseal veins draing directly into the cavernous sinus.*
*Fig. 24 (right). Same, dog, posterior view. Bar = 1 mm.*

*Fig. 25 (left). Short portal vessels (SPV) connecting neurohypophyseal capillaries (NH) and adenohypophyseal sinusoids (AH). Rat hypophysis, dorsal view. Top left: long portal vessels (LPV) visible underlying infundibular capillaries. Bar = 50 μm.*
*Fig. 26 (right). Diagram of the hypophyseal circulation. Arcuate area: median eminence (ME), stalk and neurohypophysis (NH). CS: cavernous sinuses. Between NH and CS is the outline of the adenohypophysis. nhv: neurohypophyseal veins, directly draining the primary capillary plexus into the systemic circulation (CS); ahv: adenohypophyseal veins, draining the secondary (sinusoidal) plexus into the CS. LPV and SPV: long and short portal vessels, connecting the primary capillary plexus of ME and NH capillaries (c) with the secondary sinusoid (s) plexus of the adenohypophysis. The arterial supply is exclusively to the primary capillary plexus of the ME, stalk and neurohypophysis.*

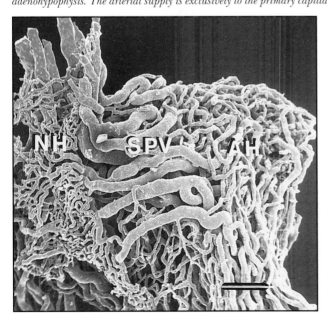

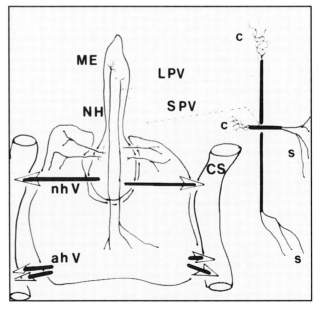

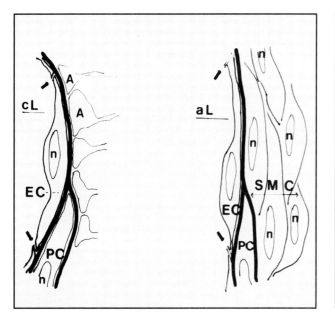
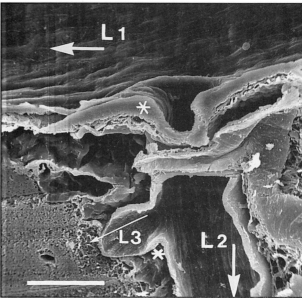

*Fig. 27 (left). Diagrams of capillary (left) and arteriolar vessel walls. cL and aL: capillary resp. arteriolar lumina; n: nucleus (closed ovals); EC: endothelial cell; PC: pericyte; A: astrocytic processes; SMC: smooth muscle cells; BM: basement membrane. Arrows indicate endothelial tight junctions.*
*Fig. 28 (right). Scanning electron micrograph of hemisected pial artery (L1), branching into penetrating arteriole (L2), with side branch (L3). Note tissue configurations forming subendothelial cell cushions (\*). Arrows indicate flow direction. Bar = 50 μm.*

## The vessel wall

### General structure

Unlike most capillary beds, the brain capillaries are not freely permeable to small solutes due to meandering networks of tight junctional bands between the endothelial cells. Only in the periventricular organs fenestrations are present. As in other capillaries, the endothelial cells secrete vasoactive compounds (see Abbott *et al.*, 1989 and Chapter 6). There is a thick basement membrane which also involves the pericytes. Both endothelial cells and pericytes may be contractile, containing myofilaments and myofibrillar elements. Parenchymal arterioles lack an internal elastic lamina and have one to three layers of smooth muscle cells (Fig. 27). Between endothelial cells and smooth muscle cells

various types of myoendothelial junctions have been seen with a gap width of 20 nm or smaller. The presence of tight junctions is considered to be doubtful and has yet not been confirmed. The larger arteries in the brain have a fenestrated internal elastic lamina and up to 20 layers of smooth muscle cells, an adventitia and leptomeningeal coverings. Vasa vasorum have not been seen, the necessary nutrients probably also being provided by the cerebrospinal fluid (Liszczak *et al.*, 1984). Extradural cerebral arteries possess vasa vasorum exclusively in the adventitia, originating directly from the vessel lumen. The carotid sinus wall has an extensive adventitial capillary plexus. In the Virchow–Robin space a special type of perivascular cells different from pericytes, astrocytes and microglia have been described in the rat: fluorescent granular perithelial cells with

many inclusion bodies which, with increasing age, can narrow the vessel lumen.

### Subendothelial arterial cushions

At the site of certain arterial collateral branching points, cushion-like structures can be seen to protrude into the lumen of the parent vessel (Figs. 28–30). These were noted in the last century and have since been described in different vascular beds under a variety of names: Arterienwuelste, arterial valves, intimal cushions or pads, arterial branch pads or polsters, pial arterial sphincters and subendothelial, arterial or intra-arterial cushions. According to the literature, the cushions may consist of various components: smooth muscle cells, fibroblasts and/or abundant metachromatic ground substance containing muco-

 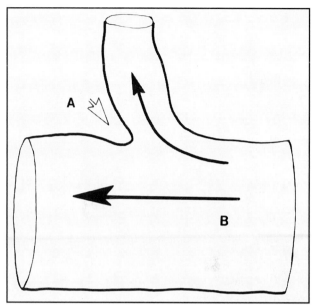

*Fig. 29 (left). Cast of an artery with two side branches. Note subendothelial cell cushion indentation deepening in the downstream direction (arrows). Flow was from left to right. Bar = 50.8 μm.*

*Fig. 30 (right). Cross-section of a sub-endothelial cell cushion profile (a), impinging downstream into the lumen of the parent artery (b). Arrow indicates direction of flow. Note the gentle slope of the upstream side of the branch.*

polysaccharides. Direct innervation has not been seen.

In regard to its function, the effect on the discharge haematocrit has been measured *in vivo* by Fourman & Moffat (1961) by comparing vascular beds with and without these cushions. They found a cell-skimming effect where cushions are present which sample blood from the cell-rich axial stream in the parent vessel and keep the haematocrit at the same level. The continually decreasing haematocrit in the absence of cushions by plasma skimming has been analysed mathematically by Yan *et al.* (1991).

## Sphincters

In addition to the subendothelial cushions, local action of vascular smooth muscle cells has been observed as strictures in corrosion casts

(Figs 31, 32) as have spastic features, ascribed to local myogenic reactions (constriction plus dilatation) to high injection pressure.

Precapillary sphincters have been defined as metarteriolar smooth muscle cells myogenically monitoring capillary flow, whereas systemic resistance is probably regulated by the proximal (innervated) larger diameter arterioles and arteries. An extended discussion and definitions are given by Faraci *et al.* (1989) and by other authors in the same supplement.

## The vascular innervation

### Generalities

Cerebral vessels are richly endowed with nerve fibres of several types, run-

ning in the adventitia in a variety of patterns. Their demonstration and detailed investigation, performed at first largely by silver staining methods (before the 1960s), has been extended enormously, first by the use of specific staining and fluorescence histochemical methods (e.g. the Falck & Hillarp method of formation of fluorophores by chemical treatment of catecholamines with formaldehyde gas) and more recently by the use of specific antibodies which recognize a continually increasing variety of putative transmitters or transmitter-related agents (frequently enzymes). Transmission electron microscopy has been the major tool in examining the intricate relationships of these nerve fibres with other elements of the vessel wall, i.e. smooth muscle cells, mast cells, and other nerve fibres, but a recently-developed technique, confocal microscopy, promises to contribute further

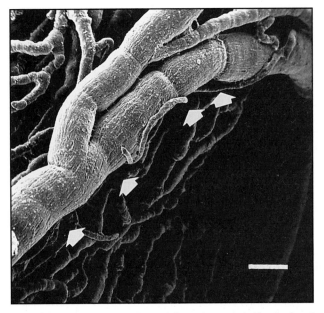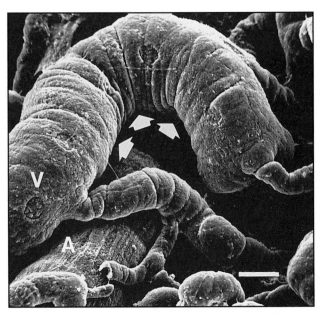

*Fig. 31 (left). Constrictions in arterial cast (arrows), indicating locations of smooth muscle cell action at the time of the setting of the proprietary methacrylate injection medium (Mercox™). Bar = 20 μm.*

*Fig. 32 (right). Constrictions in a venous (V) cast (arrows). Note round endothelial nuclear impressions (circles) and the underlying artery (A) with ovoid nuclear impressions. Bar = 20 μm.*

information. Useful reviews of investigations of cerebrovascular innervation can be found in Owman & Edvinsson (1977), MacKenzie *et al.* (1984),Owman & Hardebo (1986), and Seylaz & Sercombe (1989).

This section will deal with the 'local' aspects of innervation, to the exclusion of the pathways, which are described in Chapter 11. Extrinsic innervation (i.e., originating outside the central nervous system) of many segments of the vascular tree have been observed, to varying degrees (see below). The existence of a characteristic autonomic innervation of intraparenchymal arterioles continuous with that on pial arteries is well established for the sympathetic nerves. Whether cerebral vessels in general (including capillaries) receive an intrinsic innervation (i.e. originating inside the central nervous sys-

tem) is a controversial question relating to the hypothesis of intrinsic nervous control of the cerebral circulation. Some elements of response (mainly anatomical) are given below, and functional aspects of the question are dealt with in Chapter 12.

The general morphological features of the nerves on rat cerebral arteries have recently been described in a confocal microscopy study (Mathiau *et al.*, 1993a). This study was made with two antibodies to neuronal proteins, one against the low molecular mass neurofilament protein (NF-L) and the other against an intermediate filament protein (peripherin). All the fibres observed were labelled with both antibodies, suggesting that all fibres observed on these vessels are extrinsic in origin (see also 'Serotonergic Innervation'). It was found that the global density of the innervation decreases

in proportion to the diameter of the vessels. On the major pial arteries the nerve fibres are isolated or grouped in bundles of various size (Fig. 33), and the great majority of fibres are varicose, even in the largest bundles. At high magnification three main types of nerves can be defined morphologically by the appearance of their varicosities and intervaricose segments, but what these types correspond to on a functional basis is not yet known. The results of this study also suggest that close contacts may exist between varicosities of adjacent fibres (see later).

## The sympathetic innervation

### General features

Nerve bundles and fibres, positively identified as sympathetic by their disappearance after sympathetic gan-

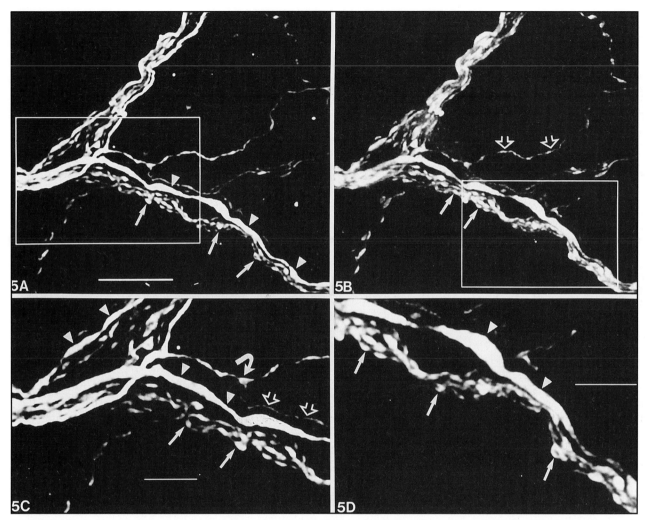

*Fig. 33. Confocal micrographs of a complex nerve bundle occurring in a rat pial artery. Immunofluorescence labelling by anti-peripherin/fluorescein isothiocyanate (A) and by anti-NF-L/rhodamine (B). C and D are enlargements of the rectangular areas in A and B respectively. The nerve bundle is composed of different types of nerve fibres, all varicose. Most of them are thin with small round varicosities (arrows). In the others, varicosities appear as large swellings occurring irregularly along the axon (arrowheads). With one exception, individual fibres appear not to divide at the collateral branching of the bundle, but rather to form two groups running in different directions. Summation of 10 tomographs, 0.5 μm interval. Scale bar 25 μm (A), 10 μm (C), 12.5 μm (D). From Mathiau et al. (1993c), with permission.*

glionectomy, have been extensively characterized in the adventitia of cerebral vessels by studies with specific histochemical methods (chemically-induced monoamine fluorescence, or immunocytochemical recognition of dopamine β-hydroxylase, tyrosine hydroxylase, or monoamine oxidase). Although other transmitters or putative transmitters (serotonin, neuropeptide Y or NPY) have been identified in cerebrovascular sympathetic fibres, noradrenaline is the only totally characteristic marker. NPY is also found in a small proportion of non-adrenergic fibres containing vasoactive intestinal polypeptide (VIP) (see Chapter 11). A large proportion of NPY-immunoreactive fibres disappear on sympathectomy and 6-hydroxydopamine treatment; correspondingly, sequential immunocytochemical labelling for dopamine-hydroxylase and NPY, or electron microscopy immunolabelling of NPY and identification of adrenergic varicosities by 5-hydroxydopamine treat-

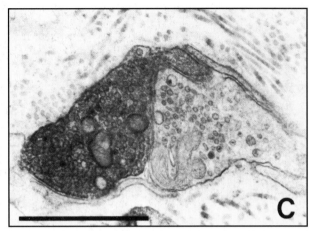

*Fig. 34. Immunoelectron photomicrograph from the wall of a rat pial artery showing two noradrenergic terminals, with both small and large granulated vesicles, enclosed within a Schwann cell sheath. One is NPY-immunoreactive (left), and shows colocalization, but the other is NPY-negative. Scale bar 1 μm. From Matsuyama et al,. 1985, with permission.*

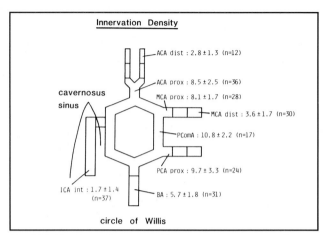

*Fig. 35A: Numbers of fluorescent fibres counted per unit area (0.1 mm² ) on the canine main cerebral arteries. Means ± SD, n = number of samples. ACA, anterior cerebral artery; MCA, middle cerebral artery; PComA, posterior communicating artery; PCA, posterior cerebral artery; BA, basilar artery; ICA, internal carotid artery. From Kawai & Ohhashi (1986), with permission.*

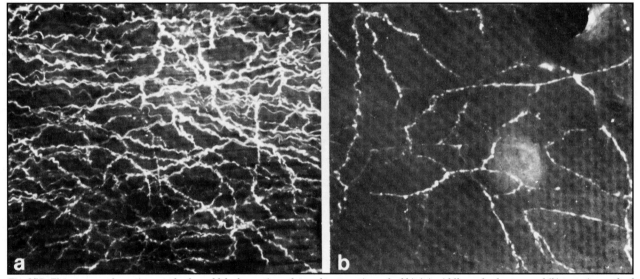

*Fig. 35B. Fluorescence photomicrographs, formaldehyde reaction, of spread preparations of rabbit (a) middle cerebral artery and (b) posterior cerebral artery. The plexus of noradrenergic nerves is denser in the wall of the middle cerebral artery. ×160. From Sercombe et al. (1975), with permission.*

ment, have both indicated colocalization of NPY and noradrenaline (Edvinsson *et al.*, 1987; Matsuyama *et al.*, 1985) (Fig. 34).

The plexus of adrenergic nerves in large cerebral arteries count among the densest in arteries of a given diameter. The appearance of the nerves varies according to the part of the vascular tree examined: many nerves in the vertebral and basilar arteries have a smooth appearance, with few varicosities, and a lower fluorescence intensity (as observed by formaldehyde histochemistry, whereas in other arteries, especially those close to the internal carotid artery, there are many varicosities with intense fluorescence. The general pattern of the nerves on the larger vessels is that

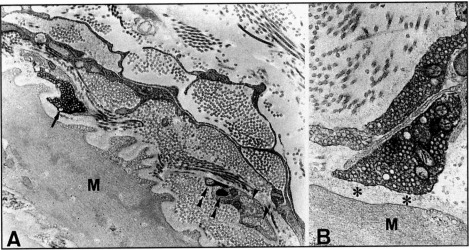

*Fig. 36. A: Electron micrograph of monoamine oxide (MAO)-positive axons at the inner margin of the adventitia of a rat cerebral artery (arrow). Several axons lacking synaptic vesicles also contain MAO-reactive substances (double arrowheads). Single arrowheads indicate MAO-negative axons. M, smooth muscle cell. × 15200.*
*B: High magnificaiton of MAO-containing unmyelinated nerve terminal. Coalescence of axonal and smooth muscle basement membranes (asterisks). MAO-reativity is present in the cytoplasm. M, smooth muscle cell. × 30400. From Shigematsu et al. (1989), with permission.*

of a complex meshwork with many circumferential fibres, in most laboratory species being of greater density in rostral than caudal arteries (Fig. 35). This meshwork decreases in density in relation to the diameter of the artery so that in small pial arteries of 100 μm or less only two to three fibres appear to run irregularly along the adventitia (e.g., Hill *et al.*, 1986), but single fibres may still occur in vessels down to 15 μm. However, if one examines branching points (not arterial dichotomies) it can be seen that some degree of concentration of the fibres occurs apparently in relation to the cushions at the entry to the small arteries (Baramidze *et al.*, 1982). Intraparenchymal arteries and arterioles are also frequently innervated, at least in larger species (from the rabbit upwards). Those investigations which have compared different regions have revealed considerable variation in the proportion of innervated arterioles, with a general tendency towards a higher proportion in structures irrigated via the anterior and middle cerebral arteries (e.g. Owman *et al.*, 1978). Up to 70–90 per cent of arterioles may be innervated. It should be noted that this represents a genuine innervation continuous with the sympathetic fibres on pial vessels, with varicose fibres in close relation to the smooth muscle. According to Lindvall *et al.* (1975) all such fibres detectable by electron microscopy disappear after sympathectomy.

Apart from pial and intraparenchymal arteries and arterioles, a dense plexus of sympathetic fibres is present in pial and collecting veins (though less dense than equivalent arteries) and in the sinuses, the choroidal arteries and veins and in the dural arteries. However, in the case of the dural arteries they are not confined to the adventitia, but may course independently across the dural connective tissue.

## Ultrastructure

Many of the nerve fibres on cerebral arteries are found in bundles of nonmyelinated nerves, enclosed in a Schwann cell plasmodium and containing microtubuli, running at the surface of the adventitia. Fibres present deeper within the adventitia are interrupted by numerous varicosities containing either small (50 nm) dense-cored or electron-lucent synaptic vesicles, and both types may also contain larger vesicles (100 nm) with a dense core. With permanganate fixation, or after administration of 5-hydroxydopamine which is taken up into adrenergic nerves, the sympathetic varicosities have been shown to contain the small, electron-dense vesicles. Estimations of the proportion of varicosities of this type in cerebral arteries vary from about 30–50 per cent.

Such sympathetic varicosities may occur in close relation to smooth muscle cells of the *outermost* layer, within a distance of only 80–100 nm (Fig. 36). However, a recent study (Luff & McLachlan, 1989; Dodge et al, 1994) suggest that some major cerebral arteries, such as the rabbit basilar and middle cerebral arteries, may possess only few, if any, such close contacts characteristic of neuromuscular junctions, so that a high proportion of these varicosities are in fact

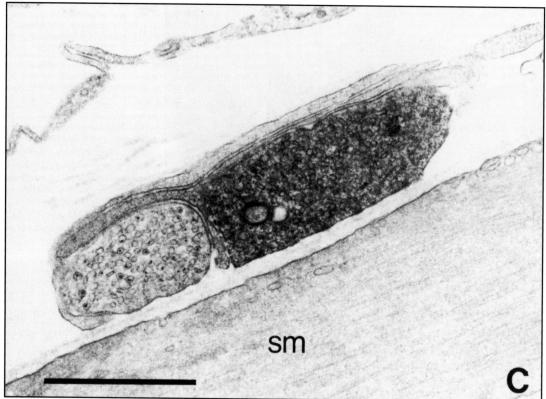

*Fig. 37. Immuno-electron photomicrograph from the wall of a rat pial artery showing a VIP-immunoreactive terminal (right) and a noradrenergic terminal (left) in close apposition, both being directly apposed to the smooth muscle cell (sm). Scale bar 1 μm. From Matsuyama et al., 1985, with permission.*

situated distant from the vascular media (1 μm or more). Variation between different cerebral arteries and different species is probably considerable in this respect.

Near the smooth muscle, the varicosities lose part or all of their Schwann cell sheath. In addition, frequent close contacts (within 25 nm) can be observed between a varicosity containing small electron-dense vesicles and one containing electron-lucent ones, or between two varicosities containing electron-dense vesicles. Such contacts also occur in the region of the smooth muscle, such that a functional relationship may exist in parallel between sympathetic and non-sympathetic axons and the smooth muscle (Fig. 37). It was originally supposed that the varicosities containing elec-

tron-lucent vesicles were cholinergic, but this may not always be the case (Fig. 37).

Further functional contacts have been found in the shape of close apposition between adventitial mast cells and not only sympathetic but also parasympathetic and trigeminal (peptidergic) fibres (Fig. 8, Chapter 11).

## Parasympathetic innervation

Parasympathetic fibres innervating cerebral vessels are believed to be of two main types. Those containing choline acetyltransferase (ChAT), i.e. capable of synthesizing acetylcholine (ACh), and those containing vasoactive intestinal polypeptide (VIP). Much exploration has also been performed by studying fibres stained for

acetylcholine esterase (AChE), although this is undoubtedly a less specific marker of cholinergic nerves than ChAT (Rossier, 1977). Butyryl cholinesterase, an almost ubiquitous enzyme, must be efficiently inhibited, or most nerves and even the endothelium may be stained.

The earlier investigations with AChE staining (e.g. Edvinsson et al., 1972) showed in rat and cat major cerebral arteries a meshwork of fibres, denser in rostral than caudal vessels, an observation generally confirmed by ChAT immunocytochemistry. In the same publication, it was shown by sequential labelling that the pattern of noradrenaline fluorescence and AChE staining was largely identical, suggesting that the different fibres ran frequently in the same bundles,

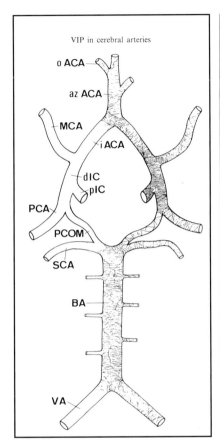

VIP in cerebral arteries

*Fig. 38. Schematic representation showing the innervation by VIP-immunoreactive fibres in the wall of cerebral arteries of the rat. ACA, anterior cerebral artery; MCA, middle cerebral artery; IC, internal carotid artery; PCA, posterior cerebral artery; PCOM, posterior communicating artery; SCA, superior cerebellar artery; BA, basilar artery; VA, vertebral artery. From Matsuyama et al., 1983, with permission.*

which is certainly compatible with the close contacts found by electron microscopy between the adrenergic and (presumed) cholinergic varicosities (see above, 'Ultrastructure'). Some confirmation has been given by later studies. Intraparenchymal vessels have not been found with AChE fibres, and it has been reported that small pial arteries and arterioles

have fewer AChE fibres than sympathetic fibres. Cerebral veins are sparsely supplied with AChE fibres, if at all.

Studies with ChAT immunocytochemistry have given general confirmation of the earlier work with AChE staining (but no double labelling experiments on ChAT/noradrenaline nerves have been done). VIP immunocytochemistry has revealed a distribution of VIP-immunoreactive fibres on cerebral arteries generally similar to that of AChE and adrenergic fibres, though perhaps not identical (Fig. 38). On the basis of the similar distribution of VIP-immunoreactive and AChE-stained fibres in pial arteries and cell bodies in the sphenopalatine ganglion, some investigators have suggested that VIP and ACh are colocalized in the same parasympathetic fibres. More recent studies with ChAT and VIP immunocytochemistry (e.g. Suzuki et al., 1990) including simultaneous double labelling (Miao & Lee, 1990), suggest that the proportion of cerebrovascular parasympathetic nerves containing both transmitters may be variable, and sometimes rather minor (even 5 per cent). VIP nerves, however, are undoubtedly present in cerebral veins, though less dense than in equivalent arteries, and a few intracerebral arterioles have been reported to possess VIP-immunoreactive fibres. Dural arteries, including human vessels, also possess VIP nerves (Jansen et al., 1992).

VIP-labelled varicosities have been found by electron microscopy to form close contacts with the vascular smooth muscle, at distances of 80–300 nm, and to be frequently enclosed with adrenergic varicosities in the same Schwann cell sheath (Fig. 37).

## Trigeminal (peptidergic, sensory) innervation

Substance P (SP) and calcitonin gene-related peptide (CGRP) have been identified in cerebrovascular nerves by immunocytochemistry. Such fibres are fine and varicose, forming a meshwork in the larger arteries and veins (denser in the former), and a few fibres may be present on pial arterioles. It is clearly established that SP and CGRP coexist in most fibres (Suzuki et al., 1989), but a minor fraction of CGRP-positive fibres is SP-negative. Projections from the trigeminal ganglion to the rat superficial temporal artery originate mostly from pure CGRP containing neurons. Careful electron microscopic study by immunolabelling suggests that the SP-containing varicosities (which must also contain CGRP) are found not only in mixed nerve bundles in the outer part of the adventitia in arteries, but also (isolated or with other nerves) in close apposition to the smooth muscle. Large vesicles (100 nm) are seen in some of the SP-labelled varicosities. The density of this innervation is considerably greater in the rostral, compared to the caudal, cerebral arteries. Neurokinin A, a tachykinin closely related to SP, is also present.

SP/CGRP-positive nerves are also present in meningeal arteries, including human vessels, where they play an important role in the manifestation of pain and inflammation (see Chapters 11 and 14). Their close relation to mast cells, allowing functional interactions, is certainly an important element in their involvement in pathological states.

## Nitric oxide synthase-containing nerves

In recent studies, many fibres containing the enzyme responsible for the synthesis of nitric oxide from arginine (nitric oxide synthase, NOS) have been identified on cerebral arteries in rats (Nozaki *et al.*, 1993; Minami *et al.*, 1994) and humans (Tomimoto *et al.*, 1994). The relative density of NOS-containing fibres appears up to 2–4 times greater in rostral than in caudal arteries. A high proportion of positively labelled fibres in the rostral circulation (75 per cent) appear to originate in the sphenopalatine and otic ganglia and are presumably parasympathetic; colocalization of NOS with VIP was found in a substantial proportion of VIP-positive cells. Electron microscopy has revealed that NOS-positive axon varicosities contain spherical agranular vesicles 40–50 mm in diameter, and a common axon–smooth muscle separation is 0.5–0.8 μm (Loesch *et al.*, 1994).

The remaining NOS-positive fibres (about 25 per cent) seem to originate in the trigeminal ganglion, but colocalization with CGRP was not found. There is some controversy about the presence of NOS-positive cells in the superior cervical ganglion.

## 'Intrinsic' innervation

### Serotonergic innervation

In 1983 it was proposed by Edvinsson *et al.*, that a serotonergic innervation, originating in certain raphe nuclei, could explain the presence of serotonin-immunoreactive (5-HT-I) fibres which had been demonstrated in cerebral arteries by several groups. This hypothesis was based not only on various histochemical studies but also on neurochemical investigation of 5-HT and its principal metabolite, 5-hydroxyindole acetic acid in the pial vessels. Since then, there has been a spate of studies both in support of and against this hypothesis. The debate is not yet closed, and it is thus necessary to summarize briefly the present state of our knowledge.

The early immunohistochemical studies on cerebral arteries demonstrated a network of 5-HT-I fibres on large pial arteries, and a few or single fibres on small arteries. These results were often obtained on immersion-fixed preparations. A large number of subsequent studies showed that such fibres could not be identified in perfusion-fixed preparations, and complementary experiments confirmed that the presence of blood elements was necessary for the observation of 5-HT-I fibres. It can be shown that 5-HT, released in the presence of the arteries, can be taken up specifically by sympathetic nerve fibres (Saito & Lee, 1987). Several other studies have confirmed this finding, and an investigation by electron microscopy indicated that the 5-HT was localized exclusively in catecholaminergic fibres (Jackowski *et al.*, 1988). However, retrograde tracing of the nerve fibres projecting to the middle cerebral artery had suggested that there was a contingent of fibres from the raphe nuclei (Tsai *et al.*, 1985), but recently an anterograde tracing study of the projections from the dorsal raphe nucleus failed to discover any labelled fibres on the pial vessels (Mathiau *et al.*, 1993b). A possible explanation for the results of Tsai *et al.* is that the horseradish peroxidase placed on the middle cerebral artery also diffused to the underlying cerebral cortex which is richly supplied with serotonergic axons from the dorsal and medial raphe nuclei.

As mentioned above, neurochemical measurements have also contributed to the controversy: the 5-HT content of pial vessels was at first found to be reduced by raphe destruction, but not by sympathectomy, and 5-HT uptake into pial vessels (*in vitro*) has been found to be reduced on the one hand by lesion of the raphe (Amenta *et al.*, 1985) or on the other hand by sympathectomy (Chang *et al.*, 1990). Recent studies in the rat have attempted to distinguish two vascular compartments, the major pial arteries and the small pial vessels (including arachnoid membranes). Indeed, the 5-HT content of both compartments was reduced by lesion of the raphe, but only in the major pial artery fraction was it reduced by sympathectomy (Bonvento *et al.*, 1991). Although this apparently suggests that, at least in the small vessel fraction, 5-HT fibres of central origin may exist, this result may be influenced by factors such as the presence of a high concentration of raphe-derived axons in the cerebral cortex releasing large quantities of 5-HT into the cerebro-spinal fluid, which would be taken up by the sympathetic fibres.

Other recent studies have looked at the possibility of 5-HT synthesis in pial vessels. Immunohistochemical identification of tryptophan hydroxylase-immunoreactive (TPH-I) nerves in pial arteries apparently demonstrated the presence of true serotonergic nerves (Chedotal & Hamel, 1990). A following study showed that, although the distribution of these fibres seemed not to be quite identical to

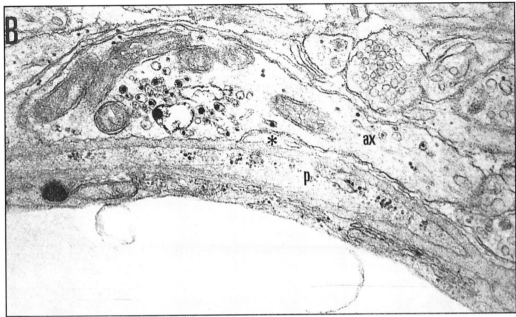

*Fig. 39. Electron micrograph of the paraventricular nucleus (hypothalamus) showing a small dense-core vesicle-containing varicosity lying directly on the basal lamina of a pericyte (p) associated with a capillary endothelial cell. The preterminal axon (ax) is surrounded by astrocytic processes (\*). Note the presence of fibrils in the pericyte cytoplasm. Ganglionectomized, 5-OHDA pretreated rat. Scale bar 0.5 μm. From Swanson et al. (1977), Brain Res. 136: 166–173, with permission.*

that of dopamine β-hydroxylase-immunoreactive fibres (i.e. sympathetic), sympathectomy led to their disappearance (Cohen *et al.*, 1992). In addition, one *in vitro* study of the pial vessels failed to demonstrate biochemically any TPH activity whereas in another such study TPH activity was found, and was reduced by dorsal raphe destruction (Mathiau *et al.*, 1993c; Moreno *et al.*, 1994).

Although highly contradictory, the balance of the evidence seems rather against the presence of an intrinsic serotonergic innervation, especially for the large pial arteries. In confirmation, another recent study failed to discover any nerve fibres on pial vessels of various size which were *not* labelled for the specific protein of peripheral neurons, peripherin (a 57 kDa intermediate filament protein) (Mathiau *et al.*, 1993a). Whether intraparenchymal arterioles can be influenced by serotonergic neurons from the raphe is a question considered in Chapter 12.

## Noradrenergic innervation

It has been reported that after sympathectomy the adrenergic fibres associated with intraparenchymal arterioles disappeared, as seen by fluorescence histochemistry and electron microscopy. However, careful studies by electron microscopy have provided evidence that in rat and cat presumed monoaminergic (adrenaline or noradrenaline-containing) varicosities can be found in close apposition to capillary endothelial cells or pericytes (Rennels & Nelson, 1975; Swanson *et al.*, 1977). They were clearly identified as non-sympathetic and monoaminergic in the hypothalamus of rats, where the vesicles were of the small dense-cored variety (25–40 nm). The varicosities lay on the basal lamina within about 60 nm of either a pericyte or an endothelial cell where there was no intervening astrocytic process (Fig. 39). Interestingly, the same study re-

ported that there were as many varicosities containing small (50 nm) electron-lucent vesicles in the same situation, i.e. one for about 40 complete capillary profiles examined. The latter vesicles were deemed to be of the cholinergic type. Electron and light microscopic evidence was reviewed by Hartman *et al.* (1980).

It seems likely that the origin of such adrenergic terminals is the locus coeruleus, and the distance seems appropriate for the transmitter, if released, to have some effect on the vascular elements (see Chapter 12). A study with horseradish peroxidase axon tracing (Rennels *et al.*, 1983) has suggested that local neurons of the hypothalamus may make some close contacts with the capillaries, but no such contacts were observed in the cerebral or cerebellar cortices. An extensive study (Jones, 1982) with catecholamine fluorescence, combined with Evans blue fluorescence to identify the vessels, concluded that, although capillary innervation might exist in specific regions, the lack of systematic correlation did not support the idea of a systematic centrally-derived innervation.

## Cholinergic innervation

An extensive literature has lent support to the hypothesis that an intrinsic cholinergic regulation of cerebral blood flow may exist (see Chapter 12), Recent anatomical evidence concords with older isolated studies which showed the presence of varicosities containing electron-lucent cholinergic-type vesicles in close apposition to hypothalamic capillaries (Swanson *et al.*, 1977) and ChAT-labelled varicosities similarly placed in the amygdala (Armstrong, 1986). Light and electron microscopic immuno-cytochemistry have shown both ChAT- and VIP-immunolabelled fibres in close contact with rat cortical microvessels (capillaries and arterioles) (Arneric *et al.*, 1988; Chedotal *et al.*, 1994a, b). A sizeable proportion of such fibres was situated at $\leq 0.25$ µm from the outer basal lamina, but they usually abutted onto perivascular glia processes. Biochemical measurements of ChAT activity have shown that isolated microvessels display specific activity as high as or higher than a nerve-terminal enriched cortical fraction or cortical homogenate. An anterograde tracing study has demonstrated a close association between fibres projecting from the basal forebrain and cortical blood vessels (Luiten *et al.*, 1987), but further studies are awaited to determine the origin of such cortical microvascular cholinergic fibres, intracortical or from the basal forebrain.

## Blood–brain barrier and brain volume regulation

### The barrier

The endothelial cells of the brain microvessels form the barrier between the blood and the interstitial cerebrospinal fluid of the brain in the brain parenchyma.

Typical features are: asymmetric luminal and abluminal cell membranes which give the endothelium its polarized properties, tight continuous quintuple-layered intercellular junctions, no fenestrations, few vesicles, no fluid-filled bulk transport channels, low wall thickness (0.2 µm), a higher mitochondrial content than in other tissues and a thick basement membrane.

The brain endothelial properties appear to be induced by brain tissue, notably by glial cells.

The barrier formed by the tight-junctioned endothelium keeps neurotransmitters inside and noxious agents outside the brain, maintaining a homoeostasis in the cerebral interstitial fluid.

Comparative studies have shown this barrier in some invertebrates to be formed by tight-junctioned perivascular glial end feet, which otherwise are not a barrier. This is also seen in vertebrates, but only in elasmobranch fish where it is considered to be a secondary evolutionary development (see Abbott *et al.*, 1989 for review).

At the onset of the barrier function in early development, when the transendothelial thickness of 0.6 µm decreases sharply in the last prenatal week in the rat to then gradually reach the adult thickness of 0.2 µm, the onset of ion transport control was not seen to coincide with a clear change in mitochondrial content. This is in contrast to the choroid epithelium (which in the adult has a greater porosity than endothelium) where it increases (Keep & Jones, 1990). It has been suggested that the increase in barrier function is correlated with the increase of completeness of fusion between adjacent membranes (Abbott *et al.*, 1989).

The main other sites where continuous tight junctions occur are choroidal epithelial cells and specialized areas of ependymal cells (e.g. of the median eminence, the area postrema and of all periventricular organs), where the capillaries are fenestrated and the endothelial barrier is completely absent. The high permeability appears to be an adaptation for the release of (neuro)secretory products into the bloodstream (neurohypophysis, median eminence, pineal gland) or for a chemoreceptor function (area postrema: vomiting and cardiovascular responses; subfornical organ: thirst and drinking response to circulating angiotensin (Bradbury, 1979)). Tight junctions are also seen between the arachnoid barrier cells where the arachnoid is bordering the dura which is well vascularized.

The leptomeningeal fluid compartments (subpial, pial and arachnoidal) are not in open communication with one another and the (non-vascularized) leptomeninges show numerous complex junctions, probably also including tight junctions.

Moreover, around the leptomeninges as well as subependymally where haemal and CSF milieu are in continuity, as for instance in the median eminence, phagocytic cells are seen. These might play a role in selective barrier functions in the circumventricular organs where a blood–brain barrier is lacking.

## Cerebrospinal fluid (CSF) production and drainage

The choroid plexuses produce most if not all CSF. Across the choroid epithelium with its asymmetric properties of the opposite-facing cell membranes, a net transport of water and solutes takes place. The cuboidal epithelial cells have many microvilli on the apical side and extensive folds on the basal side. They are apically joined by tight junctions which are somewhat porous compared to brain endothelial cells. The various plexuses (of the lateral, third and fourth ventricle) might produce CSF with different efficiency (Gomez & Potts, 1981) whereas extrachoroidal sources might also contribute. The CSF circulates through the ventricles and reaches the meningeal aspect of the brain via the foramina of Luschka and Magendie. It circulates over the outer surface of the brain and drains through the arachnoid villi or (Pacchionian) granulations into the dural venous sinuses, especially the superior sagittal sinus.

The arachnoid granulations are covered by endothelial cells, continuous with those of the adjacent sinus wall. No clearly defined tight junctions have been seen here. The surface of the villi also shows fissures which lead to the formation of endothelial-lined tubular configurations (Gomez et al., 1974). A unidirectional, hydrostatic pressure-dependent fluid transport through and between the endothelial cells takes place draining most of the CSF into the venous system. All dissolved solutes including those of high molecular weight, pass non-selectively.

The subpial compartment which comprises the intercellular clefts of the brain and of the leptomeninges does not communicate with the pial or arachnoid spaces, and its interstitial CSF drains preferentially into the upper cervical lymph nodes (Krisch et al., 1984).

## Brain volume regulation and the intracranial pressure

The brain volume is basically regulated by the low hydraulic conductivity of the brain capillaries and by the high osmotic activity of the major solutes in the blood–brain system, limiting bulk water flow across the blood–brain barrier (see Fenstermacher, 1984 for full review).

Intracranial pressure changes are met with cerebral vessel autoregulation towards an appropriate change of flow resistance. Volume changes may occur when the autoregulatory capacity (60–150 mmHg) is transgressed, blood flow gets compromised and metabolic changes then induce tissue swelling. In establishing the intracranial pressure the CSF production and uptake play a major role. The steady production (500 ml per 24 h in the human) is hardly influenced by changes in transchoroidal hydrostatic or osmotic presssure gradients. The large re-uptake capacity easily accomodates excess fluid from brain oedema which reaches the CSF compartment.

Although both volume and pressure changes of the brain are linked to the vascular and CSF systems because they are contained within the cranium, they are regulated differently and have to be considered accordingly. The intracranial pressure is not necessarily an indicator of the brain volume.

**Acknowledgement:** This chapter is dedicated to Professor J. Ariens Kappers on the occasion of his 85th birthday

## Summary

In this chapter we have shown what common patterns and variations exist in the brain vasculature in some commonly used laboratory animals and in man. A number of key references have been brought together as a help for further consultation in specific questions.

The notion has been stressed that there can be extensive extracranial contributions to the intracranial circulation depending on the species, and also an intracranial supply to various extracranial areas. The species-specific patterns of the circle of Willis and the intra-specific and individual variations of the interconnecting segments and their role in the distribution of flow to the brain are mentioned. The sometimes confusing differences in the descriptive terminology used are discussed.

Various recognizable microvascular patterns are described, the singular properties of the blood vessel wall and the vascular-meningeal relations including those in relation to the blood–brain barrier are dealt with.

An overview of the innervation is given covering the established extrinsic innervations (sympathetic, parasympathetic and trigeminal) and putative intrinsic innervations. The chapter ends with an outline of the general principles of the blood–brain barrier and the regulation of the brain volume.

# References

## *Introduction*

### Comparative anatomy

Bugge, J. (1974): The cephalic arterial system in insectivores, primates, rodents and lagomorphs, with special reference to the systemic classification. *Acta Anat.* **87**, (Suppl. 62) 1–160.

Duvernoy, H.M. (1978): *Human brainstem vessels*, p. 188. Berlin: Springer.

Greene, E.Ch. (1935): *Anatomy of the rat*, 2nd edn., reprinted in 1968, New York: Hafner.

Hebel R. & Stromberg, M.W. (1986): *Anatomy and embryology of the laboratory rat*. Worthsee: BioMed Verlag.

Hofmann, M. (1900): Zur vergleichende Anatomie der Gehirn- und Rueckenmarksarterien der Vertebraten. *Z. Morph. Anthrop.* **2**, 247–322.

Hofmann, M. (1901): Zur vergleichende Anatomie der Gehirn- und Rueckenmarksvenen der Vertebraten. *Z. Morph. Anthrop.* **3**, 239–299.

Simoens, P., de Vos, N.R. & H. Lauwers (1979): *Illustrated anatomical nomenclature of the heart and the arteries of head and neck in the domestic mammals.* Mededelingen Fac. Diergeneeskunde, State University Ghent, Belgium. **21**, 1–100.

Simoens, P., de Vos, N.R. & Lauwers, H. (1984): Illustrated anatomical nomenclature of the venous system in the domestic mammals. Mededelingen Fac. Diergeneeskunde, State University Ghent, Belgium, **26**, 1–91.

Stephan, H. (1975): Allocortex. In: *Handbuch der mikroskopischen Anatomie des Menschen*, ed. W. Von Moellendorf & W. Bargmann, 4.B and 9. Teil, pp. 96–123. Berlin: Springer.

Tandler, J. (1899): Zur vergleichende Anatomie der Kopfarterien bei den Mammalia. *Denkschr. Acad. Wiss. Wien.* **67**, 677–784.

Van Gelderen, Chr. (1933): Die vergleichende Anatomie des Venensystems. In: *Handbuch der vergleichende Anatomie* eds. L. Bolk & E. Goeppert u.A. pp. 685–744. Berlin: Urban u. Schwarzenberg.

Zeman, W. & Innes, J.R.M. (1963): *Craigie's neuroanatomy of the rat*. New York: Academic Press.

### Methods used

Auer, L.M. (1987): Measurement of pial vessel hemodynamics. In: *Physico-chemical techniques, neuromethods*, Vol. 8., eds. A.A. Boulton, G.B. Baker & D. Boisvert. Clifton, NJ: Humana Press.

Eddy, H.A. (1976): Microangiographic techniques in the study of normal and tumor tissue vascular systems. *Microvasc. Res.* **11**, 391–413.

Gabaldon, M. (1987): Methodological approaches for the study of the aortic endothelium of the rat. *Atherosclerosis* **65**, 139–149.

Hasegawa, T. & Ravens, J.R. (1968): A new metallic impregnation method for demonstration of cerebral vascular patterns. *Acta Neuropath.* **10**, 183–188.

Hodde, K.C., Steeber, D.A. & Albrecht, R.M. (1990): Advances in corrosion casting methods. *Scanning Microscopy* **4**, 693–704.

Lametschwandtner, A., Lametschwandtner, U. & Weiger, T. (1990): Scanning electron microscopy of vascular corrosion casts – technique and applications: updated review. *Scanning Microscopy* **4**, 889–941.

Saunders, R.L. de C.H. & Bell, M.A. (1971): X-ray microscopy and histochemistry of the human cerebral blood vessels. *J. Neurosurg.* **35**, 128–140.

Tompsett, D.H. (1970): *Anatomical techniques*, 2nd edn, pp. 93–212. London: E. & S. Livingstone.

## *Blood supply to the brain*

Brown, J.O. (1966): The morphology of circulus arteriosus cerebri in rats. *Anat. Rec.* **156**, 99–106.

Bugge, J. (1978): The cephalic arterial system in carnivores, with special reference to the systematic classification. *Acta Anat.* **101**, 45–61.

Firbas, W., Sinzinger, H. & Schlemmer, M. (1973): Ueber den Circulus arteriosus bei Ratte, Maus und Goldhamster. *Anat. Histol. Embryol.* **2**, 243–251.

Gillilan, L.A. (1976): Extra- and intra-cranial blood supply to brains of dog and cat. *Am. J. Anat.* **146,** 237–254.

Hillen, B. (1987): The variability of the Circulus arteriosus (Willisii): order or anarchy? *Acta Anat.* **129,** 74–80.

Hodde, K.C. & De Blecourt, C.V. (1979): Extracerebral supply by the internal carotid artery in the rabbit and the rat. *Acta Neurol. Scand.* **60,** (Suppl. 72) 636–637.

Klein, Th. (1980): Korrosionsanatomische Untersuchungen am Blutgefaesssystem de Encephalon und der Meninges bei *Felis domestica. Zentralbl. Vet. C. Anat. Histol. Embryol.* **9,** 236–279.

Moffat, D.B. (1961): The development of the posterior cerebral artery. *J. Anat.* **95,** 485–494.

Lee, M.-W., Reid, I.A. & Ramsay, D.J. (1986): Blood flows in the maxillocarotid anastomoses and internal carotid artery of conscious dogs. *Anat. Rec.* **215,** 192–197.

McDonald & Potter, J.M. (1951): The distribution of blood to the brain. *J.Physiol.* **114,** 356–371.

Van der Zwan, A. & Hillen, B. (1991): Review of the territories of the major cerebral arteries. *Stroke* **22,** 1078–1084.

Van der Zwan, A. (1991): The variability of the major vascular territories of the human brain. p. 128. Thesis, State University of Utrecht.

Ward, R., Collins, R.L., Tanguay, G. & Miceli, D. (1990): A quantitative study of cerebrovascular variation in inbred mice. *J. Anat.* **173,** 87–95.

Wellens, D.L.F., Wouters, L.J.M.R., de Reese, R.J.J., Beirnaert, P. & Reneman, R.S. (1975): The cerebral blood distribution in dogs and cats. An anatomical and functional study. *Brain Res.* **86,** 429–438.

## Venousdrainage

Auer, L.M. & Loew, F. (Eds) (1983): *The cerebral veins.* Vienna: Springer.

Butler, H. (1967): The development of mammalian dural venous sinuses with especial reference to the post-glenoid vein. *J. Anat.* **102,** 33–56.

Du Boulay, G.H. (1976): Anatomy of the collateral cerebral circulation in man and animals. In: *Effect of carotid artery surgery on cerebral blood flow (clinical and experimental studies),* eds. W.R. Jennett, J.D. Miller & A.M. Harper, pp. 21–39. Amsterdam: Exerpta Medica, Elsevier NH Biomed. Press.

Hegedus, S.A. & Shackelford, R.T. (1959): A comparative-anatomical study of the cranio-cervical venous systems in mammals, with special reference to the dog. Relationships of anatomy to measurements of cerebral blood flow. *Am. J. Anat.* **116,** 375–386.

Szabo, K. (1990): The cranial venous system in the rat: anatomical pattern and ontogenetic development. *Anat Embryol.* **182,** 225–234.

## Vascular-meningeal relations

Angelov, D.N. & Vasilev, V.A. (1989): Morphogenesis of rat cranial meninges. *Cell Tissue Res.* **257,** 207–216.

Haines, D.E. (1991): On the question of a subdural space. *Anat. Rec.* **230,** 3–21.

Krahn, V. (1982): The pia mater at the site of entry of blood vessels into the central nervous system. *Anat. Embryol.* **164,** 257–263.

Krisch, B., Leonhardt, H. & Oksche, A. (1984): Compartments and perivascular arrangement of the meninges covering the cerebral cortex of the rat. *Cell Tissue Res.* **238,** 459–474.

Krisch, B. (1988): Ultrastructure of the meninges at the site of penetration of veins through the dura mater, with particular reference to Pacchionian granulations. *Cell Tiss. Res.* **251,** 621–631.

Leiderer, R. & Hammersen, F. (1990): On the fine structure of arc-capillaries: true a.v. anastomoses or sphincter capillaries? *Anat. Embryol.* **182,** 79–84. (discusses dural circulation)

Zhang, E.T., Inman, C.B.E. & Weller, R.O. (1990): Interrelationships of the pia mater and the perivascular (Virchow–Robin) spaces in the human cerebrum. *J. Anat.* **170,** 111–123.

## Microcirculatory patterns

### Cerebral and cerebellar cortex

Baer, Th. (1980): The vascular system of the cerebral cortex. Changes during ontogenesis, aging and oxygen deprivation. *Adv. Anat. Embryol. Cell Biol.* **59,** 1–62.

Bardosi, A. & Ambach, G. (1985): Constant position of the superficial cerebral veins of the rat: a quantitative analysis. *Anat. Rec.* **211**, 338–341.

Coyle, P. & Jokelainen, P.T. (1982): Dorsal cerebral arterial collaterals of the rat. *Anat. Rec.* **203**, 397–404.

Duvernoy, H.M., Delon, S. & Vannson, J.L. (1981): Cortical blood vessels of the human brain. *Brain Res. Bull.* **7**, 519–579.

Duvernoy, H.M., Delon, S. & Vannson, J.L. (1983): The vascularization of the human cerebellar cortex. *Brain Res. Bull.* **11**, 419–480.

Hodde, K.C., Miodonski, A., Bakker, C. & Veltman, W.A.M. (1977): Scanning electron microscopy of micro-corrosion casts with special attention on arterio-venous differences and application to the rat's cochlea. *Scan. Electron. Microsc.* **II**, 477–484.

Mato, M., Ookawara, S. & Namiki, T. (1989): Studies on the vasculogenesis in rat cerebral cortex. *Anat. Rec.* **224**, 355–364.

Miodonski, A., Hodde, K.C. & Bakker, C. (1976): Rasterelektronenmikroskopie von Plastik-Korrosions-Praeparaten: morphologische Unterschiede zwischen Arterien und Venen. *Beitr. Elektronenmikroskop. Direktabb. Oberfl.* **9**, 435–442.

Motti, E.D.F., Imhof, H.-G. & Yasargil, M.G. (1986): The terminal vascular bed in the superficial cortex of the rat. *J. Neurosurg.* **65**, 834–846.

Ravens, J.R. (1974): Anastomoses in the vascular bed of the human cerebrum. In: *Pathology of cerebral microcirculation*, ed. Cervos-Navarro, J., pp. 26–38 Berlin: De Gruyter.

Scharrer, E. (1940): Arteries and veins in the mammalian brain. *Anat. Rec.* **78**, 173–196.

Wuenscher, W., Schober, W. & Werner, L. (1965): Archtektonischer Atlas vom Hirnstamm der Ratte. p. 61. Leipzig: S. Hirzel Verlag.

### Choroid plexus

Hodde, K.C. (1979): The vascularization of the choroid plexus of the lateral ventricle of the rat. *Beitr. elektronenmikrosk. Direktabb. Oberfl.* (BEDO) **12**, 395–400.

Millen, J.W. & Woollam, D.H.M. (1953): Vascular patterns in the choroid plexus. *J. Anat.* **87**, 114–123.

Miodonski, A., Poborowaska, J. & Friedhuber de Grubenthal, J. (1979): Scanning electron microscopy study of the choroid plexus of the lateral ventricle in the cat. *Anat. Embryol.* **155**, 323–331.

Motti, E.D.F., Imhof, H.-G., Janzer, R.C., Marquardt, K. & Yasargil, G.M. (1986): The capillary bed in the choroid plexus of the lateral ventricles: a study of luminal casts. *Scan. Electron. Microsc.* **IV**, 1501–1513.

Weiger, T., Lametschwandtner, A., Hodde, K.C. & Adam, H. (1986): The angioarchitecture of the choroid plexus of the lateral ventricle of the rabbit. A SEM study of vascular corrosion casts. *Brain Res.* **378**, 285–296.

### Portal vessels and venous drainage of the hypophysis

Bergland, R.M. & Page, R.B. (1979): Pituitary–brain vascular relations: a new paradigm. *Science* **204**, 18–24.

Daniel, P.M. & Prichard, M.M.L. (1975): Studies of the hypothalamus and the pituitary gland. With special reference to the effects of transsection of the pituitary stalk. *Acta Endocrinol.* (*Kbh*) **80**, (Suppl. 201), 213.

Hodde, K.C. (1981): Vascular casts of the rat pituitary gland. *Beitr. elektronenmikr. Direktabb. Oberfl.* (BEDO) **14**, 593–602.

Krisch, B., Leonhardt, H. & Oksche, A. (1983): The meningeal compartments of the median eminence and the cortex. *Cell Tiss. Res.* **228**, 597–640.

Murakami, T. Kikuta, A., Taguchi, T., Ohtsuka, A. & Ohtani, O. (1987): Blood vascular architecture of the rat cerebral hypophysis and hypothalamus. A dissection/scanning electron microscopy of vascular casts. *Arch. Histol. Jpn.* **50**, 133–176.

## The vessel wall

### General structure

Aydin, F., Rosenblum, W.I. & Povlishock, J.T. (1991): Myoendothelial junctions in human brain arterioles. *Stroke* **22**, 1592–1597.

Barker, S.G.E., Causton, B.E., Baskerville, P.A., Gent, S. & Martin J.F. (1992): The vasa vasorum of the rabbit carotid artery. *J. Anat.* **180**, 225–231.

Bennett, H.S., Luft, J.H. & Hampton, J.C. (1959): Morphological classification of vertebrate blood capillaries. *Am. J. Physiol.* **196**, 381–390.

Boswell, C.A., Mayno, G., Joris, I. & Ostrom, K.A. (1992): Acute endothelial cell contraction *in vitro*: a comparison with vascular smooth muscle cells and fibroblasts. *Microvasc. Res.* **43**, 178–191.

Hansen, J.T. (1987): Morphology of the carotid sinus wall in normotensive and spontaneously hypertensive rats. *Anat. Rec.* **218**, 426–433.

Liszczak, T.M., Black, P. McL., Varsos, V.G. & Zervas, N.T. (1984): The microcirculation of cerebral arteries: a morphologic and morphometric examination of the major canine cerebral arteries. *Am. J. Anat.* **170**, 223–232.

Mato, M., Aikawa, E., Mato, T.K. & Kurihara, K. (1986): Tridimensional observation of fluorescent granular perithelial (FGP) cells in rat cerebral blood vessels. *Anat. Rec.* **215**, 413–419.

Rhodin, J.A.G. (1980): Architecture of the vessel wall. In: *Handbook of physiology*, Section 2, The cardiovascular system, Vol. II, vascular Smooth Muscle, ed. Bohr *et al.*, pp. 1–31. American Physiology Society: Bethesda.

Roggendorf, W. & Cervos-Navarro, J. (1977): Ultrastructure of arterioles in cat brain. *Cell Tiss. Res.* **178**, 495–515.

Roggendorf, W. & Cervos-Navarro, J. (1978): Ultrastructure of venules in cat brain. *Cell Tiss. Res.* **192**, 461–474.

Walmsley, J.G., Campling M.R. & Chertkow, H.M. (1983): Interrelationships among wall structure, smooth muscle orientation and contraction in human major cerebral arteries. *Stroke* **14**, 781–790.

### Subendothelial arterial cushions

Casellas, D., Dupont, M., Jover, B.J. & Mimran, A. (1982): SEM study of arterial cushions in rats: a novel application of the corrosion-replication technique. *Anat. Rec.* **203**, 419–428.

Fourman, J. & Moffat, D.B. (1961): The effect of intra- arterial cushions on plasma skimming in small arteries. *J. Physiol.* **158**, 374–380.

Hodde, K.C. (1981): *Cephalic vascular patterns in the rat*, pp. 36z–38. PhD Thesis, University of Amsterdam, pp. 36–38.

Jokelainen, P.T., Jokelainen, G.I. & Coyle, P. (1982): Non-random distribution of rat pial arterial sphincters. In: *Cerebral blood flow: effects of nerves and transmitters*, eds. D.D. Heistad & M.L. Marcus, pp. 107–116. Amsterdam: Elsevier North Holland.

Kardon, R.H., Farley, D.B., Heidger, P.M. & Van Orden, D.E. (1982): Intraarterial cushions of the rat uterine artery: a SEM evaluation utilizing vascular cats. *Anat. Rec.* **203**, 9–29.

Yan, Z.-Y., Acrivos, A. & Weinbaum, S. (1991): A three-dimensional analysis of plasma-skimming at microvascular bifurcations. *Microvasc. Res.* **42**, 17–38.

### Sphincters

Faraci, F.M., Baumbach, G.L. & Heistad, D.D. (1989): Myogenic mechanisms in the cerebral circulation. *J. Hypertension* **7**, (Suppl. 4): S61–S64.

Motti, E.D.F., Imhof, H.-G., Garza, J.M. & Yasargil, G.M. (1987): Vasospastic phenomena on the luminal replica of rat brain vessels. *Scanning Microsc.* **1**, 207–222.

Nakai, K., Naka, K., Yokote, H., Ikatura, T., Imai, H., Komai, N. & Maeda, T. (1989): Vascular 'sphincter' and microangioarchitecture in the central nervous system: constriction of intraparenchymal blood vessels following a treatment of vasoconstrictive neurotransmitter. *Scanning Microsc.* **3**, 337–341.

Rhodin, J.A.G. (1967): The ultrastructure of mammalian arterioles and precapillary sphincters. *J. Ultrastruct. Res.* **18**, 181–223.

Wiedeman, M.P., Tuma, R.F. & Mayrovitz, H.N. (1976): Defining the precapillary sphincter. *Microvasc. Res.* **12**, 71–75.

## *Innervations*

### General reviews

MacKenzie, E.T., Seylaz, J. & Bes, A. (eds.) (1984): *Neurotransmitters and the cerebral circulation*. New York: Raven Press.

Owman, C. & Edvinsson, L. (1977): *Neurogenic control of the brain circulation*. Oxford: Pergamon Press.

Owman, C. & Hardebo, J.E. (eds.) (1986): *Neural regulation of brain circulation*. Amsterdam: Elsevier.

Seylaz, J. & Sercombe, R. (eds.) (1989): *Neurotransmission and cerebrovascular function*, Vol. II. Amsterdam: Excerpta Medica.

## Specific papers

Amenta, F., De Rossi, M., Mione, M.C. & Geppetti, P. (1985): Characterization of [$^3$H]5-hydroxytryptamine uptake within rat cerebrovascular tree. *Eur. J. Pharmacol.* **112**, 181–186.

Armstrong, D.M. (1986): Ultrastructural characterization of acetyltransferase-containing neurons in the basal forebrain of the rat: evidence for a cholinergic innervation of intracerebral blood vessels. *J. Comp. Neurol.* **250**, 81–92.

Arneric, S.P., Honig, M.A., Milner, T.A., Greco, S., Iadecola, C. & Reis, D. (1988): Neuronal and endothelial sites of acetylcholine synthesis and release associated with microvessels in rat cerebral cortex: ultrastructural and neurochemical studies. *Brain Res.* **454**, 11–30.

Baramidze, D.G., Reidler, R.M., Gadamski, R. & Mehedlishvili, G.I. (1982): Pattern and innervation of pial microvascular effectors which control blood supply to the cortex. *Blood Vessels* **19**, 284–291.

Bonvento, G., Lacombe, P., MacKenzie, E.T., Fage, D., Benavides, J., Rouquier, L. & Scatton, B. (1991): Evidence for differing origins of the serotonergic innervation of major cerebral arteries and small pial vessels in the rat. *J. Neurochem.* **56**, 681–689.

Chang, J.Y., Hardebo, J.E. & Owman, Ch. (1990): Kinetic studies on uptake of serotonin and noradrenaline into pial arteries of rats. *J. Cereb. Blood Flow Metab.* **10**, 22–31.

Chedotal, A. & Hamel, E. (1990): Serotonin-synthesizing nerve fibers in rat and cat cerebral arteries and arterioles: Immunohistochemistry of tryptophan-5-hydroxylase. *Neurosci. Lett.* **116**, 269–274.

Chedotal, A., Cozzari, C., Faure, M.P., Hartman, B.K. & Hamel, E. (1994a): Distinct choline acetyltransferase (ChAT) and vasoactive intestinal polypeptide (VIP) bipolar neurons project to local blood vessels in the rat cerebral cortex. *Brain Res.* **646**, 181–193.

Chedotal, A., Umbriaco, D., Descarries, L., Hartman, B.K. & Hamel, E. (1994b): Light and electron microscopic immunocytochemical analysis of the neurovascular relationships of choline acetyltransferase and vasoactive intestinal polypeptide nerve terminals in the rat cerebral cortex. *J. Comp. Neurol.* **343**, 57–71.

Cohen, Z.V.I., Bonvento, G., Lacombe, P., Seylaz, J., MacKenzie, E.T. & Hamel, E. (1992): Cerebrovascular nerve fibers immunoreactive for tryptophan-5-hydroxylase in the rat: distribution, putative origin and comparison with sympathetic noradrenergic nerves. *Brain Res.* **598**, 203–214.

Dodge, J.T., Bevan, R.D. & Bevan, J.A. (1994): Comparison of density of sympathetic varicosities and their closeness to smooth muscle cells in rabbit middle cerebral and ear arteries and their branches. *Circ. Res.* **75**, 916–925.

Edvinsson, L., Degueurce, A., Duverger, D., MacKenzie, E.T. & Scatton, B. (1983): Central serotonergic nerves project to the pial vessels of the brain. *Nature* **306**, 55–57.

Edvinsson, L., Nielsen, K.C., Owman, C. & Sporrong, B. (1972): Cholinergic mechanisms in pial vessels. *Z. Zellforsch.* **134**, 311–325.

Edvinsson, L., Copeland, J.R., Emson, P.C., McCulloch, J. & Uddman, R. (1987): Nerve fibres containing neuropeptide Y in the cerebrovascular bed: immunocytochemistry, radioimmuno assay, and vasomotor effects. *J. Cereb. Blood Flow Metab.* **7**, 45–57.

Hara, H., Jansen, I., Ekman, R., Hamel, E., MacKenzie, E.T., Uddman, R. & Edvinsson, L. (1989): Acetylcholine and vasoactive intestinal peptide in cerebral blood vessels: effect of extirpation of the sphenopalatine ganglion. *J. Cereb. Blood Flow Metab.* **9**, 204–211.

Hartman, B.K., Swanson, L.W., Raichle, M.E., Preskoru, S.H. Clark, H.B. (1980): Central adrenergic regulation of cerebral microvascular permeability and blood flow: anatomic and physiologic evidence. *Adv. Exp. Med. Biol.* **131**, 113–126.

Hill, C.E., Hirst, G.D.S., Silverberg, G.D. & van Helden, D.F. (1986): Sympathetic innervation and excitability of arterioles originating from the rat middle cerebral artery. *J. Physiol.* **371**, 305–316.

Itakura, T., Okuno, T., Nakakita, K., Kamei, I., Naka, Y., Imai, H., Komai, N., Kimura, H. & Maeda, T. (1984): A light and electronmicroscopic immunohistochemical study of vasoactive intestinal polypeptide- and substance P-containing nerve fibres along the cerebral blood vessels: comparison with aminergic and cholinergic fibers. *J. Cereb. Blood Flow Metab.* **4**, 407–414.

Jackowski, A., Crockard, A. & Burnstock, G. (1988): Ultrastructure of serotonin-containing nerve fibres in the middle cerebral artery of the rat and evidence for its localization within catecholamine-containing nerve fibres by immunoelectron microscopy. *Brain Res.* **443**, 159–165.

Jansen, I., Uddman, R., Ekman, R., Olesen, J., Ottosson, A. & Edvinsson, L. (1992): Distribution and effects of neuropeptide Y, vasoactive intestinal peptide, substance P, and calcitonin gene-related peptide in human middle meningeal arteries: comparison with cerebral and temporal arteries. *Peptides* **13**, 527–537.

Jones, B.E. (1982): Relationship between catecholamine neurons and cerebral blood vessels studied by their simultaneous fluorescent revelation in the rat brain stem. *Brain Res. Bull.* **9**, 33–44.

Kawai, Y. & Ohhashi, T. (1986): Histochemical studies of the autonomic nervous system. *J. Auton. Nerv. Syst.* **15**, 103–108.

Lindvall, M., Cervos-Navarro, J., Edvinsson, L., Owman, C. & Stenevi, U. (1975): Non-sympathetic perivascular nerves in the brain. Origin and mode of innervation studied by fluorescence and electron microscopy combined with stereotaxic lesions and sympathectomy. In: *Blood flow and metabolism in the brain*, eds. A.M. Harper, W.B. Jennett, J.D. Miller, J.O. Rowan, pp. 1.7–1.9. Edinburgh: Churchill Livingstone.

Loesch, A., Belai, A. & Burnstock, G. (1994): An ultra-structural study of NADPH-diaphorase and nitric oxide synthase in the perivascular nerves and vascular endothelium of the rat basilar artery. *J. Neurocytol.* **23**, 49–59.

Luff, S.E. & McLachlan, E.M. (1989): Frequency of neuromuscular junctions on arteries of different dimensions in the rabbit, guinea pig and rat. *Blood Vessels* **26**, 95–106.

Luiten, P.G.M., Gaykema, R.P.A., Traber, J. & Spencer, D.J. Jr. (1987): Cortical projection patterns of magnocellular basal nucleus subdivisions as revealed by anterogradely transported *Phaseolus vulgaris* leuro-agglutinin. *Brain Res.* **443**, 229–250.

Mathiau, P., Riche, D., Behzadi, G., Dimitriadou, V., Wiklund, L. & Aubineau, P. (1993a): Absence of serotonergic innervation from raphe nuclei in rat cerebral blood vessels. I. Histological evidence. *Neuroscience* **52**, 645–655.

Mathiau, P., Reynier-Rebuffel, A.M., Issertial, O., Callebert, J. & Aubineau, P. (1993b): Absence of serotonergic innervation from raphe nuclei in rat cerebral blood vessels.II. Lack of tryptophan hydroxylase activity *in vitro*. *Neuroscience* **52**, 657–665.

Mathiau, P., Escurat, M. & Aubineau, P. (1993c): Immunohistochemical evidence for the absence of central neuron projection to pial blood vessels and dura mater. *Neuroscience* **52**, 667–676.

Matsuyama, T., Shiosaka, S., Matsumoto, M., Youeda, S., Kimura, K., Abe, H., Hayakawa, T., Inoue, H. & Tohyama, M. (1983): Overall distribution of vasoactive intestinal polypeptide-containing nerves on the wall of cerebral arteries: an immunohistochemical study using whole-mounts. *Neuroscience* **10**, 89–96.

Matsuyama, T., Shiosaka, S., Wanaka, A., Yoneda, S., Kimura, K., Hayaleawa, T., Emson, P.C. & Tonyama, M. (1985): Fine structure of peptidergic and catecholaminergic nerve fibers in the anteriorcerebral artery and their interrelationship: an immunselection microscopic study. *J. Comp. Neurol.* **235**, 268–276.

Minami, Y., Kimwa, H., Aimi, Y. & Vincent, S.R. (1994): Projections of nitric oxide synthase-containing fibers from the sphenopalatine ganglion to cerebral arteries in the rat. *Neuroscience* **60**, 745–759.

Miao, F.J.P. & Lee, T.J.F. (1990): Cholinergic and VIPergic innervation in cerebral arteries: a sequential double-labeling immuno-histochemical study. *J. Cereb. Blood Flow Metab.* **10**, 32–37.

Moreno, M.J., López de Pablo, A.L. & Marco, E.J. (1994): Tryptophan hydroxylase activity in rat brain base arteries related to innervation originating from the dorsal raphe nucleus. *Stroke* **25**, 1046–1049.

Nozaki, K., Moskowitz, M.A., Maynard, K.I., Koketsu, N., Dawson, T.M., Bredt, D.S. & Snyder, S.H. (1993): Possible origins and distribution of immunoreactive nitric oxide synthase containing nerve fibers in cerebral arteries. *J. Cereb. Blood Flow Metab.* **13**, 70–79.

Owman, C., Edvinsson, L. & Hardebo, J.E. (1978): Pharmacological *in vitro* analysis of amine-mediated vasomotor functions in the intracranial and extracranial vascular beds. *Blood Vessels* **15**, 128–147.

Rennels, M.L. & Nelson, E. (1975): Capillary innervation in the mammalian central nervous system: an electron microscopic demonstration. *Am. J. Anat.* **144**, 233–241.

Rennels, M.L., Gregory, T.F. & Fujimoto, K. (1983): Innervation of capillaries by local neurons in the cat hypothalamus: a light microscopic study with horseradish peroxidase. *J. Cereb. Blood Flow Metab.* **3**, 535–542.

Rossier, J. (1977): Choline acetyltransferase: a review with special reference to its cellular and subcellular localization. *Int. Res. Neurobiol.* **20**, 283–337.

Saito, A. & Lee, T.J.F. (1987): Serotonin as an alternative transmitter in sympathetic nerves of large cerebral arteries of the rabbit. *Circ. Res.* **60**, 220–228.

Sercombe, R., Aubineau, P., Edvinsson, L., Mamo, H., Owman, Ch., Pinard, E. & Seylaz, J. (1975): Neurogenic influence on local cerebral blood flow. Effect of catecholamines or sympathetic stimulation as correlated with the sympathetic innervation. *Neurology* **25**, 954–963.

Shigematsu, K., Akiguchi, I., Oka, N., Kamo, H., Matsubayashi, K., Kameyama, M., Kawamura, J. & Maeda, T. (1989): Monoamineoxidase-containing nerve fibers in the major cerebral arteries of rats. *Brain Res.* **497**, 21–29.

Suzuki, N., Hardebo, J.E. & Owman, Ch. (1989): Origins and pathways of cerebrovascular nerves storing substance P and calcitonin gene-related peptide in rat. *Neuroscience* **31**, 427–438.

Suzuki, N., Hardebo, J.E. & Owman, C. (1990): Origins and pathways of choline acetyltransferase-positive parasympathetic nerve fibers to cerebral vessels in rat. *J. Cereb. Blood Flow Metab.* **10,** 399–408.

Swanson, L.W., Connelly, M.A. & Hartman, B.K. (1977): Ultrastructural evidence for central monoaminergic innervation of blood vessels in the paraventricular nucleus of the hypothalamus. *Brain Res.* **136,** 166–173.

Tomimoto, H., Nishimura, M., Suenaga, T., Nakamura, S., Akiguchi, I., Wakita, H., Kimura, J. & Mayer, B. (1994): Distribution of nitric oxide synthase in the human cerebral blood vessels and brain tissues. *J. Cereb. Blood Flow Metab.* **14,** 930–938.

Tsai, S.H., Lin, S.Z., Wang, S.D., Liu, J.C. & Shih, C.J. (1985): Retrograde localization of the innervation of the middle cerebral artery with horseradish peroxidase in cats. *Neurosurgery* **16,** 463–467.

## *The blood–brain barrier and brain volume regulation*

Abbott, N.J., Bundgaard, M. & Hughes, C.C.W. (1989): Morphology of brain microvessels: a comparative approach. In: *Cerebral microcirculation. Progress in Applied Microcirculation*, Vol. 16, eds. F. Hammersen & K. Messmer, pp. 1–19. Basel: Karger.

Bradbury, M. (1979): *The concept of a blood–brain barrier*, p. 465. Chichester: Wiley.

Davies, M.G. & Hagen, P.-O. (1993): The vascular endothelium. A new horizon. *Ann. Surg.* **218,** 593–609 (review article).

Fenstermacher, J.D. (1984): Volume regulation of the central nervous system. In: *Edema*, eds. N.C. Staub & A.E. Taylor, pp. 383–404, New York: Raven Press.

Gomez, D.G., Potts, D.G. & Deonarine, V. (1974): Arachnoid granulations of the sheep, structural and ultrastructural changes with varying pressure differences. *Arch. Neurol.* **30,** 169–175.

Gomez, D.G. & Potts, D.G. ( 1981): The lateral, third and fourth ventricle choroid plexus of the dog: a structural and ultrastructural study. *Ann. Neurol.* **10,** 333–340.

Keep, R.F. & Jones, H.C. (1990): Cortical microvessels during brain development: a morphometric study. *Microvasc. Res.* **40,** 412–426.

Krisch, B., Leonhardt, H. &. Oksche, A. (1983): The meningeal compartments of the median eminence and the cortex. *Cell Tiss. Res.* **228,** 597–640.

Krisch, B., Leonhardt, H. & Oksche, A. (1984): Compartments and perivascular arrangement of the meninges covering the cerebral cortex of the rat. *Cell Tiss. Res.* **238,** 459–474.

Krisch, B. (1986): The functional and structural borders between the CSF- and blood-dominated milieus in the choroid plexuses and the area postrema of the rat. *Cell Tiss. Res.* **245,** 101–115.

# Chapter 6

# Cerebrovascular smooth muscle and endothelium

**William J. Pearce[1] and David R. Harder[2]**

[1]*Division of Perinatal Biology, Departments of Physiology and Pediatrics, Loma Linda University School of Medicine, Loma Linda, CA, USA 92350 and* [2]*Department of Physiology, Medical College of Wisconsin, Milwaukee, WI, USA 53226*

## Introduction

Modern methods capable of measuring cerebral perfusion and metabolism with high degrees of spatial and temporal resolution have repeatedly demonstrated in both humans and experimental animals that, despite a considerable heterogeneity of flow and metabolism in different brain areas, the ratio between flow and metabolism remains remarkably uniform. Such uniformity suggests a very precise and efficient regulatory system. As with all control systems, we can describe the systems controlling cerebral perfusion each as a composite of three parts:

(1) a 'sensor' which monitors environmental conditions;
(2) an 'integrator' which sums input from the various sensors;
(3) an 'effector' component which responds to 'integrator' signals to bring about change in the environment.

In systems controlling cerebral blood flow, the 'effector' component is always vascular smooth muscle in the walls of the cerebral arteries. In some control systems, cerebrovascular smooth muscle also serves as 'integrator' and/or 'sensor'. Thus, the physiological properties of cerebrovascular smooth muscle are critical determinants of the overall behaviour of the systems regulating cerebral blood flow. The goal of this chapter is to describe the more salient of these properties, and relate them to the mechanisms responsible for the coupling of cerebral flow and metabolism.

## The biochemistry of vascular smooth muscle contraction

As with other types of muscle, contraction of vascular smooth muscle involves a precise interaction between actin and myosin. The biochemical regulation of this interaction, and the structural framework within which it occurs, however, is quite different than found in either cardiac or skeletal muscle.

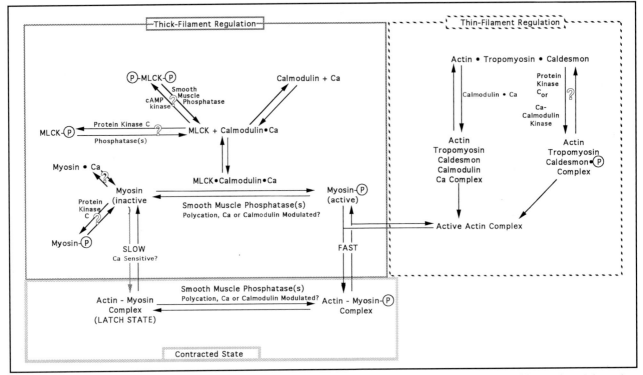

*Fig. 1. Contractile biochemistry. This diagram summarizes the pathways described in the text. On left are shown thick filament (myosin) reactions and on the right, reactions involving thin filaments. The letter P within a circle indicates an attached phosphate group. Question marks indicate uncertain or controversial reactions. MLCK = myosin light chain kinase.*

## Mechanisms of vascular contraction

All muscle contains thick filaments composed primarily of myosin, and thin filaments composed largely of actin. In skeletal and cardiac muscle, myosin and its associated ATPase are inherently active and contraction is regulated by altering the characteristics of the thin filaments which allow them to bind and interact with myosin. In contrast, the myosin AT-Pase of vascular smooth muscle is not inherently active and the initiation of contraction is effected by controlling the extent of phosphorylation, and thus the ATPase activity, of myosin (Fig. 1).

## Contractile proteins

### Myosin

The myosin of smooth muscle consists of a heavy tail and two light chains, each approximately 20 kilodaltons (kD) in mass. The light chains tonically inhibit myosin ATPase activity, and their removal by enzymatic means correspondingly increases myosin ATPase activity. Physiologically, this inhibition is removed by light chain phosphorylation, which can occur at multiple sites. The diphosphorylated form of myosin light chain yields the highest ATPase activity, but the monophosphorylated form that predominates *in vivo* is probably of greater physiological importance. The overall extent of light

chain phosphorylation controls the cross-bridge cycle rate and thus contraction velocity. As in all muscles, the other major determinant of contraction velocity is load. In general, the magnitude of the peak level of myosin phosphorylation correlates well with the magnitude of sustained contraction.

Myosin light-chain kinase (MLCK) is the main enzyme responsible for phosphorylation of myosin. It exists in multiple forms between 80 and 125 kD in mass, requires Mn-ATP or Mg-ATP as its phosphate donor, and has a broad pH optimum between 7 and 8. The activated enzyme phosphorylates the myosin light chains in random order, and can also phosphorylate it-

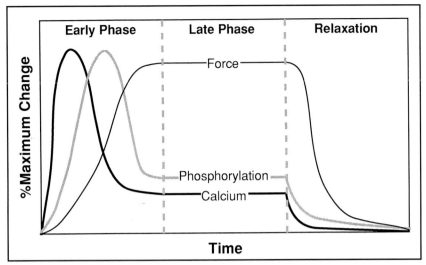

*Fig. 2. Time course of contractile events. Contraction is initiated by stimuli which trigger a rise in myoplasmic calcium. In the early phase of contraction, this rise in calcium precedes the rise in the level of phosphorylation of myosin by a delay attributable to the activation of myosin light chain kinase by calcium–calmodulin and the subsequent kinetics of myosin phosphorylation. Calcium and the extent of myosin phosphorylation both quickly reach their peaks and then fall to stable levels above baseline during the late phase of contraction. Force, however, rises steadily during the early phase of contraction and reaches an elevated plateau which is maintained during the late phase. Relaxation is triggered by withdrawal of the stimulus and a fall in myoplasmic calcium. See text for further details.*

self. All forms of MLCK are activated by the calcium–calmodulin complex, and thus it is probably the concentration of this complex, and not of calcium alone, that governs smooth muscle force generation.

Vascular smooth muscle calmodulin concentration ranges from 30 to 50 μM, but much of this is bound and the free concentration averages 5 μM. At calmodulin concentrations much below 5 μM, myosin calcium sensitivity diminishes but can be restored by addition of calmodulin. Each calmodulin molecule binds up to four calcium ions, and when bound to MLCK its affinity for calcium is increased. In many respects, calmodulin is functionally similar to troponin C, the calcium binding subunit of the skeletal muscle troponin complex.

Myosin ATPase activity is diminished

by dephosphorylation mediated by smooth muscle phosphatases. These diverse and ubiquitous enzymes vary not only in terms of their multimeric structure, but also in their requirements for magnesium, their ability to bind myosin directly, and the specificity of the site from which they cleave phosphate; some dephosphorylate both myosin light chain and MLCK. As a class, they are relatively non-specific but exhibit greater activity towards contractile than non-contractile proteins. *In vivo*, the overall level of smooth muscle phosphatase activity is relatively low and appears to be calcium insensitive. Although some studies have suggested that certain smooth muscle phosphatases can be activated by polycation peptides such as polylysine, no convincing evidence has yet suggested that regulation of phosphatase activity is of physiologi-

cal significance in vascular smooth muscle.

Although the phosphorylation of myosin light chains is of undoubted importance in the initiation of contraction, it is probably not the only factor governing contractile tone. As contraction begins, the extent of myosin phosphorylation rapidly rises to a peak and then falls to a plateau slightly greater than baseline (Fig. 2). In contrast, contractile tension rises steadily to a plateau which is maintained even though the levels of myosin phosphorylation have fallen. Under these conditions, where tension is high and the level of myosin phosphorylation is low, reduction of intracellular calcium concentration produces relaxation. Thus, the phosphorylation of myosin cannot be the only calcium-sensitive step in the system regulating contraction in vascular smooth muscle.

### Thin filament proteins

The search for the second calcium-sensitive component of the smooth muscle contractile system has produced a wide variety of hypotheses in recent years. The majority have focused on some type of thin filament mediated regulation. In smooth muscle, the ratio of thin to thick filaments is much higher than in skeletal muscle. Smooth muscle thin filaments contain no troponin as in skeletal muscle, but they do contain a unique type of tropomyosin in the form of a dimer with a helical conformation. Smooth muscle tropomyosin appears to potentiate actomyosin ATPase activity, but the potential role of this tropomyosin in regulating contraction remains controversial.

By far the largest component of

smooth muscle thin filaments is actin, mainly of the α subtype which in monomeric form is composed of approximately 375 amino acids with a total mass of near 42 kD. As in skeletal muscle, smooth muscle actin binds myosin tightly in the absence of ATP and dissociates readily when ATP is present. Smooth muscle actin may also bind MLCK, an effect inhibited by calcium–calmodulin and potentiated by tropomyosin. Phosphorylation of a 21 kD actin subunit appears to produce a number of high-affinity calcium-binding sites on actin and lowers the cytosolic calcium concentration required for half-maximal activation of myosin ATPase. The physiological importance of this phosphorylation is at present unclear.

Apart from actin and tropomyosin, smooth muscle thin filaments also contain vinculin, an actin-binding protein which anchors the thin filaments to the network of dense bodies which asymmetrically traverse the cell interior and communicate force to the ends of the cell. The dense bodies also contain α-actinin, as do the Z-disks in striated muscle, and filamin, a homodimer with a monomeric weight of about 250 kD. Filamin cross-links actin filaments and low concentrations inhibit, but high concentrations stimulate, myosin ATPase. The interactions between filamin and actin are inhibited by tropomyosin, caldesmon, and vinculin. And as for many of the known contractile proteins, the physiological significance of the chemistry of filamin is unclear but remains under investigation.

Another important component of smooth muscle thin filaments is caldesmon, a highly acidic protein which is quite similar in both primary structure and function to troponin. Caldesmon exists in two main forms: a 130–155 kD species and a 70–80 kD species. Its cytosolic concentration is highly variable, but averages about 10–12 μM which gives approximately one caldesmon molecule per 26 actin molecules, *in vivo*. At low calcium concentrations, caldesmon binds to F-actin and inhibits actin–myosin interaction. The affinity of caldesmon for actin is markedly enhanced by the presence of tropomyosin, and thus tropomyosin appears necessary for the inhibitory caldesmon–actin interaction. However, when cytosolic calcium is elevated and calcium–calmodulin complexes are present, these complexes bind to caldesmon and release its inhibition of actomyosin ATPase activity.

Elevated cytosolic calcium not only enhances the binding of caldesmon to calcium–calmodulin, but it also increases the probability that caldesmon is phosphorylated by either a calcium–calmodulin dependent protein kinase, or by protein kinase C. Phosphorylated caldesmon has a reduced affinity for actin and an increased affinity for calcium–calmodulin, and thus is less of an inhibitor of actomyosin ATPase than native caldesmon. As for myosin, caldesmon can be dephosphorylated by smooth muscle phosphatase.

Although the caldesmon mechanism of actomyosin ATPase regulation is supported by a wide variety of experimental evidence, it is not without its critics. Some suggest that this flip-flop mechanism is an oversimplification which cannot explain the relaxation produced by reductions in cytosolic calcium late in contraction. Others suggest that reactions such as the phosphorylation of caldesmon observed in homogenates, *in vitro*, have little bearing on the regulation of contraction, *in vivo*. And still others suggest the existence of a second, as yet unknown, calcium-binding regulatory protein. In any case, the combined data strongly suggest that, in addition to the phosphorylation of myosin, a second calcium-sensitive regulatory step exists.

## Calcium

### Calcium mobilization

Myoplasmic calcium concentration has long been recognized as the primary regulator of the contractile state in all muscle types. In smooth muscle, calcium concentration hovers near 0.1 μM during resting conditions and climbs an order of magnitude or more during excitation. This rise is transient and falls rapidly to a plateau value only two to three times that observed during resting conditions. The calcium transient slightly precedes the myosin phosphorylation transient mentioned above, but otherwise the two time courses are quite similar. Although these transients vary in magnitude with artery type and method of contraction, they are characteristic of both agonist-induced and potassium-induced contractions.

The calcium added to the myoplasm during excitation derives from both intracellular and extracellular sources (Fig. 3). Inside the cell, calcium is bound and buffered at many locations including the inner surface of the plasmalemma and the endoplasmic reticulum. However, the largest and most important intracellular calcium pool is the sarcoplasmic reticulum

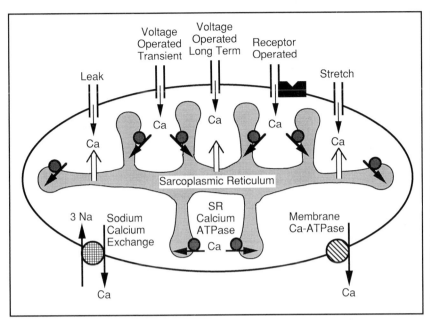

*Fig. 3. Calcium handling in vascular smooth muscle. The calcium added to the myoplasm during excitation derives from both intracellular and extracellular sources. Extracellular calcium enters the vascular smooth muscle cell through specialized channels, indicated in the figure as: (1) leak; (2) voltage-operated transient; (3) voltage-operated long term; (4) receptor operated; and (5) stretch-operated calcium channels. The main intracellular source of calcium is the sarcoplasmic reticulum (SR). Myoplasmic calcium is removed by the action of a calcium ATPase in the SR membrane, and by another different calcium ATPase in the surface membrane. Sodium–calcium exchange may also participate in calcium extrusion from vascular smooth muscle cells, although the importance of this mechanism is uncertain. See text for further details.*

(SR) which, as in skeletal muscle, is composed of a contiguous network of membrane-bound tubules. Many of these tubules are superficial, near the plasmalemma, and others are deep within the cell. Together, the SR occupies 2–7 per cent of the cell volume and smaller arteries generally contain a smaller volume of SR than larger arteries. By virtue of their content of calsequestrin, a specialized calcium-binding protein, the SR represent a high affinity, high capacity, and highly buffered intracellular calcium pool capable of releasing approximately 150–200 nmol of calcium per ml of cell water.

Calcium within the SR is released in response to inositol 1,4,5-triphos-phate ($IP_3$), which in turn is synthesized from membrane phosphatidyl inositol in a sequence of reactions initiated by broad variety of agonist–receptor interactions. The $IP_3$ thus formed acts on specific G-protein coupled receptors (see Chapter 4) on the surface of the SR to open cation channels in the SR membrane. In response to $IP_3$, superficial SR appear to release more of their calcium than deep SR, suggesting that the superficial SR either contain a higher density of $IP_3$ receptors, or that the $IP_3$ half-life is sufficiently short that only a small fraction of the amount synthesized can reach the deep SR before being degraded. In addition to the IP3 receptor, the SR appear to possess a receptor for caffeine, which also re-

leases calcium but from a different pool than that released by $IP_3$.

In other muscle types, myoplasmic calcium concentrations of 2–3 µM stimulate a further release of calcium from the SR. This so-called 'calcium-induced calcium release' apparently results from the activation of large cation channels in the SR. The calcium concentration at which this effect occurs is directly proportional to magnesium concentration and inversely proportional to intracellular cAMP concentration. Although this mechanism, which serves to amplify the effects of any stimulus raising intracellular calcium, has been demonstrated in certain types of smooth muscle, its physiological importance in cerebral arteries remains speculative.

Whereas the release of intracellular calcium is responsible for the initial rise in calcium, the sustained elevation in myoplasmic calcium late in contraction is maintained largely by the entry of extracellular calcium. The relative importance of extracellular calcium in the maintenance of contraction, however, appears to vary considerably between arteries of different size and in different vascular beds. Smaller arteries, by virtue of their smaller relative content of SR, are generally more dependent on extracellular calcium than larger arteries. In addition, arteries of the cerebral circulation are in general more dependent on extracellular calcium for the maintenance of tone than arteries of most other vascular beds.

### Calcium channels

Extracellular calcium traverses the plasmalemma by any of at least four different routes: (1) leak channels; (2) potential-operated channels; (3) receptor-operated channels; and (4) stretch-operated channels. Calcium entry through leak channels represents a continuous, constant but small contribution to myoplasmic calcium that is resistant to organic calcium channel blockers but can be blocked by lanthanum and cobalt. Potential-operated calcium channels exist in several different subtypes including the 'L-type' (long-lasting) and 'T-type' (transient) channels identified by electrophysiological studies. Both main types increase their conductance as the membrane becomes depolarized, but only the 'L-type' channels can be blocked by organic calcium channel blockers (see below: 'Cerebrovascular electrophysiology').

Receptor-operated calcium channels increase their calcium conductance in response to agonist binding at specific membrane receptors. Although the nature of the coupling between the receptors and the channels remains unclear, some evidence suggests that G-protein α-subunits may be involved. Receptor-operated channels are generally non-specific cation channels and can be distinguished from potential-operated channels on pharmacological grounds, by virtue of their relative insensitivity to calcium channel blockers. In addition, the uptake of isotopic calcium via receptor-operated channels is additive to that through potential-operated channels. Because current through receptor-operated calcium channels may in some cases be sufficient to slightly depolarize the membrane, certain agonists may activate both receptor-operated and potential-operated channels.

Stretch-operated calcium channels increase their conductance to calcium in response to mechanical deformation of the plasmalemma. An interesting feature of these channels is the latency of several seconds which separates the mechanical and ionic events. It suggests a biochemical transduction mechanism involving at least several different steps. Physiologically, this channel is of potentially great importance as a mechanism capable of explaining both myogenic and autoregulatory behaviour.

### Mechanisms returning myoplasmic calcium to basal levels

Following an increase in myoplasmic calcium, a variety of mechanisms are activated to return calcium toward resting levels. The same calcium-calmodulin complexes that activate MLCK and initiate contraction, also stimulate the membrane calcium ATPase and begin the extrusion of calcium. Indeed, the high activity and quick recruitment of the membrane calcium ATPase is credited for the finding that a significant fraction of all calcium released from internal stores ends up being extruded into the extracellular space. Although some studies suggest that this rapid extrusion may be transiently inhibited by certain agonist-receptor interactions, the physiological importance of such inhibition remains to be demonstrated in cerebral arteries.

Another possible membrane mechanism affecting calcium extrusion is the sodium–calcium exchange pump. This mechanism uses the large inward directed gradient in sodium to exchange extracellular sodium for intracellular calcium. Because three sodium ions are exchanged for each calcium ion, this mechanism has a depolarizing effect on membrane potential. Whereas sodium–calcium exchange is of undoubted importance in cardiac muscle and myometrium, its role in cerebrovascular smooth muscle remains to be elucidated.

A large fraction of the calcium released to initiate contraction is resequestered by the SR. The calcium ATPase of the SR is different than that of the plasmalemma, has a mass of approximately 105 kD, and requires Mg-ATP as its energy source. This ATPase not only returns myoplasmic calcium toward resting levels following the initiation of contraction, it also appears to buffer myoplasmic calcium concentrations. Owing to the extensive network of superficial SR, extracellular calcium traversing the membrane must also traverse these SR before it can bind with calmodulin and influence the contractile state. Consistent with this view, low concentrations of dihydropyridine calcium channel openers can significantly increase total cellular calcium content without initiating contraction.

Clearly, regulation of myoplasmic calcium concentration in vascular smooth muscle is a complex and dynamic integration of multiple processes operating at both the membrane and intracellular levels (Fig. 3). The precise balance of these mechanisms varies considerably, not only among arteries of different size and anatomical location, but also with age, dietary status, and species. Nonetheless, calcium homoeostasis of all arteries is maintained by an exact matching of the amount of calcium entering, and the amount being extruded from, the cell during each contraction cycle.

## Contraction cycle energetics and efficiency

The high degree of specialization characteristic of the mechanisms which govern contraction and calcium homoeostasis in vascular smooth muscle, is also characteristic of the metabolic pathways yielding ATP in this tissue. The total ATP and phosphocreatine stores in vascular smooth muscle are low and average only 1 µmol per gram. Thus, extra ATP synthesis is required just to reach peak contractile force. Approximately 30 per cent of the ATP synthesized in vascular smooth muscle derives from aerobic glycolysis and much of this latter ATP is preferentially utilized by the membrane sodium–potassium ATPase. Nonetheless, a key characteristic of vascular smooth muscle is that its oxygen consumption is proportional to its force generation.

Another key metabolic characteristic of vascular smooth muscle is its efficiency, which is more than 20 times that of skeletal muscle. Although the mechanisms responsible for this high efficiency remain uncertain, several interesting hypotheses have been proposed. The best supported of these is the so-called 'Latch' hypothesis of Murphy and colleagues (Murphy, 1989; Murphy *et al.*, 1990). According to this hypothesis, myosin can be dephosphorylated even when attached to actin during contraction. Such dephosphorylation produces an attached crossbridge which cycles and detaches very slowly, thus maintaining force with little or no ATP consumption. The hypothesis also proposes a second calcium-binding protein with a calcium affinity greater than that of calmodulin and which governs the maintenance of sustained tone. Given such a system, calcium could regulate the force and cycle rate independently, as has been observed. Despite considerable investigative effort, however, the identity of this second calcium binding protein remains to be determined.

An alternative to the Latch hypothesis is the possible existence of a separate fibrillar domain composed of the thin filament proteins filamin, actin, and desmin. In theory, this system could disassociate and reassociate to form cross-links capable of fixing cell length. Such a system could easily maintain force with little or no energy consumption. But as for the Latch hypothesis, the biochemical evidence necessary to fully support this hypothesis remains incomplete.

## Mechanisms of relaxation

It is clear from multiple independent lines of evidence that calcium is the primary regulator of the contractile state in vascular smooth muscle. In turn, mechanisms of relaxation can be conveniently divided into two broad categories: (1) those that reduce the concentration of myoplasmic calcium available to support contraction; and (2) those that reduce the sensitivity of the contractile proteins to calcium.

## Modulation of myoplasmic calcium concentration

The myoplasmic calcium concentration is the dynamic summation of the influences tending to increase calcium (SR calcium release and the entry of extracellular calcium) and those tending to decrease it (sequestration by the SR and extrusion by sarcolemmal calcium ATPase). Thus, relaxation can be achieved by: (1) decreasing the extracellular entry; (2) decreasing the intracellular release; (3) enhancing the intracellular sequestration; or (4) enhancing the sarcolemmal extrusion of calcium.

A common experimental approach to decrease the entry of extracellular calcium is to block the membrane channels permeable to calcium. This can be achieved with certain heavy ions such as lanthanum, and also with a large group of organic compounds. These drugs, known as calcium antagonists, comprise a structurally and functionally diverse group of compounds that includes three main families. The first and oldest is that of verapamil and its derivatives (D-600 for example). The second includes the dihydropyridine derivatives (nifedipine, nimodipine, nitrendipine, etc.). The third is the benzothiazepine family, which includes diltiazem. All these compounds inhibit calcium entry through L-type potential-operated calcium channels. Because the entry of extracellular calcium is particularly important for the maintenance of tone late in contraction, these compounds are effective inhibitors of sustained tone and have clinical value as antihypertensives. These compounds are also somewhat selective for certain tissues and vascular beds, an effect often attributed to variations in the calcium channel subtypes which populate different tissues. The calcium channel blocker most potent in the cerebral circulation appears to be nimodipine, a member of the dihydropyridine family.

A more physiological mechanism of reducing extracellular calcium entry is to hyperpolarize the smooth muscle membrane. This decreases the probability of activation for potential-operated calcium channels and thereby decreases calcium entry. Because vascular membrane potential is dominated by potassium permeability (see below, 'Cerebrovascular Electrophysiology'), any perturbation which increases potassium conductance, or increases the transmembrane potassium gradient, will hyperpolarize the membrane and produce relaxation.

Whereas the entry of extracellular calcium generally supports sustained contractile tension, the release of intracellular calcium is more closely involved in the initiation of contraction. Thus, inhibition of SR calcium release also inhibits the development of contractile tone. For example, ryanodine, a compound which promotes the slow depletion of SR calcium, blocks not only the release of SR calcium, but also the generation of force. In addition, interference with the synthesis of IP3, the second messenger which couples membrane-receptor activation to SR calcium release, attenuates force production.

Physiologically, the release of SR calcium is resistant to the actions of cyclic nucleotides and other second messengers of relaxation. In contrast, cAMP and cGMP are important stimulators of the membrane calcium ATPase, and thereby enhance calcium extrusion and promote relaxation. The cyclic nucleotides act through combination with specific kinases (A-kinase and G-kinase) which phosphorylate the calcium ATPase. Both nucleotides are synthesized in response to a wide variety of physiological stimuli by adenylate cyclase and guanylate cyclase, respectively, are degraded by specific phosphodiesterases, and are among the most important and ubiquitous of all second messengers in vascular physiology.

Several lines of evidence suggest that A-kinase may also phosphorylate a component of the SR, and thereby stimulate the sequestration of myoplasmic calcium by the SR. Evidence suggesting a similar role for G-kinase is generally weaker and more speculative than that for cAMP and A-kinase. Nonetheless, the data suggest that cyclic nucleotides stimulate not only the extrusion of myoplasmic calcium, but also its intracellular sequestration by the SR.

## Modulation of calcium sensitivity

A relatively new concept in vascular physiology is that the calcium sensitivity of the contractile proteins may be physiologically modulated to alter force production in vascular smooth muscle. In support of this concept are experiments in which the calcium–force relation, whether determined directly or indirectly in intact smooth muscle, varies with the conditions of contraction. In preparations where the smooth muscle plasmalemma has been treated to increase its permeability, the calcium–force relation can also be shown to vary in response to the addition of cyclic nucleotides to the bathing medium. Such variations illustrate that the calcium–force relation is not constant, but instead may vary with time, tissue, and physiological condition.

The primary mechanism underlying changes in calcium sensitivity appears to be selective phosphorylation of the contractile proteins. Myosin can be phosphorylated at multiple sites on both the light chains and the heavy tails. Some of these phosphorylations stimulate, and others inhibit, myosin ATPase activity *in vitro*. Sites of myosin phosphorylation vary considerably among the various myosins found in animals from different phylogenetic levels. The physiologic frequency of such phosphorylations and their regulator implications are topics of continuing and intensive biochemical investigation.

Another contractile protein that is a substrate for phosphorylation is MLCK. Activated A-kinase can phosphorylate a site on MLCK which reduces its affinity for calmodulin, and thus reduces its activity. Activated A-kinase can also phosphorylate a second 'inactive' site on MLCK, but this appears to have no effect on MLCK activity. Activated G-kinase can also add a phosphate to MLCK, but only at the 'inactive' site and thus the regulatory significance of this phosphorylation is doubtful. Other kinases, including protein kinase C, also appear capable of phosphorylating MLCK, but the significance of these reactions remains uncertain and under investigation.

As mentioned above, caldesmon can be phosphorylated by more than one enzyme, and this reduces its inhibitory influence on actomyosin ATPase. Again, the physiological significance of this phosphorylation remains in question, but it nonetheless demonstrates the great potential of selective phosphorylation as a possible modifier of the overall calcium sensitivity of the contractile apparatus of vascular smooth muscle.

In addition to the contractile proteins which serve as substrates for phosphorylation, the enzymes mediating this phosphorylation have also been extensively investigated. Central among these are the abundant A-kinase and G-kinase. Other kinases, including one activated by calcium–calmodulin, have also been implicated in the phosphorylation of contractile proteins. One of the most interesting of these kinases, however, is protein kinase C, a membrane-linked enzyme dependent on both phospholipid and calcium.

The numerous studies of protein kinase C (PKC) reveal a complex, multifaceted signalling enzyme. Perhaps the most important feature of PKC is its activation by diacylglycerol, the cleavage product of the action of phospholipase C on membrane-derived phosphatidylinositol. Thus, when membrane receptor activation triggers the breakdown of phosphatidylinositol and the synthesis of IP3, it also leads to the activation of PKC. Certain phorbol esters also activate PKC and although these agents were important in the early discovery and characterization of PKC, they probably play no role in the physiological activation of the enzyme.

In vascular smooth muscle, PKC has been reported to increase contractile protein calcium sensitivity. In contrast, PKC has also been reported to decrease myosin affinity for MLCK, decrease ATPase activity, and decrease the ability of MLCK to phosphorylate myosin. Additionally, it has been reported to decrease the affinity of myosin for actin by phosphorylating myosin at up to three different sites on both the light chains and the heavy tail. PKC has also been suggested to phosphorylate MLCK and decrease its affinity for calcium–cal-

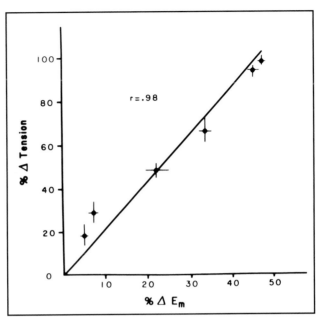

Fig. 4. Relation of tension to membrane potential. Shown here is the relation between change in membrane potential ($E_m$) and development in tension ($\Delta tension$) measured in isolated cat cerebral arteries. Activation was induced by increasing doses of serotonin beginning at $3 \times 10^{-9}$ M.

modulin. PKC appears capable of phosphorylating caldesmon and relieving its inhibition of actomyosin ATPase. Activators of PKC have also been associated with the phosphorylation of numerous unidentified low molecular weight proteins found in vascular smooth muscle. Clearly, the results of studies of PKC are inconsistent and its role in vascular contraction remains to be demonstrated conclusively. Nonetheless, PKC retains the potential of being an important modulator of the contractile process.

## Cerebrovascular electrophysiology

Electrophysiological events occurring at the plasma membrane regulate many cellular functions within excitable cells. With respect to arterial smooth muscle, one of the principal events regulated largely by changes in resting membrane potential in non-dividing differentiated cells is activa-

tion of contractile elements and cell shortening. Indeed, membrane electrical events and biochemically mediated mechanical events are very tightly coupled in cerebral arterial muscle. In both cat cerebral and canine carotid arteries, for example, a membrane depolarization of only a few mV will elicit a substantial increase in tension (Fig. 4). The precise influence of membrane electrophysiological events on activation of arterial muscle, however, varies depending upon the nature of the stimulus initiating the event.

## Determinants of membrane potential in cerebral arterial smooth muscle

The resting membrane potential ($E_m$) of cerebral arterial muscle is between –60 and –70 mV. This $E_m$ is established and maintained largely by the muscle cell permeability to $K^+$. The slope of the relationship between $E_m$ and extracellular $K^+$ approaches 60 mV per 10-fold change (decade) between 10 and 100 mM $[K]_o$, as predicted by the constant field theory for a $K^+$ selective membrane. The contributions of other ion species, namely, $Na^+$ and $Cl^-$, are relatively small.

It is difficult to determine the contribution of active transport to the $E_m$ in cerebral arterial muscle. Application of ouabain, a Na-K ATPase inhibitor, to cat basilar artery produces a 12 mV depolarization, consistent with inhibition of Na-K ATPase. However, the finding that ouabain may also inhibit potassium conductance ($g_K$) in arteriolar smooth muscle complicates this interpretation. Thus, electrogenic transport of $Na^+$ and $K^+$ in cerebral arterial muscle probably contributes to the $E_m$, although the size of this contribution relative to the reduction in $g_K$ is uncertain and probably small.

Cerebral arterial muscle is capable of generating action potentials in response to membrane depolarization. The primary ion species carrying the inward current is $Ca^{2+}$ and thus action potential amplitude is not affected by tetrodotoxin, and is markedly reduced by verapamil and inorganic $Ca^{2+}$ channel blockers.

## $K^+$ currents in cerebral arterial muscle

Given the central importance of potassium as a determinant of cerebrovascular membrane potentials, it is not surprising that potassium current is carried across the membrane by several different types of channels. These $K^+$ channels all carry current out of the cell, are voltage activated and have been characterized by their sensitivity to known blockers of $g_K$, namely, 4-

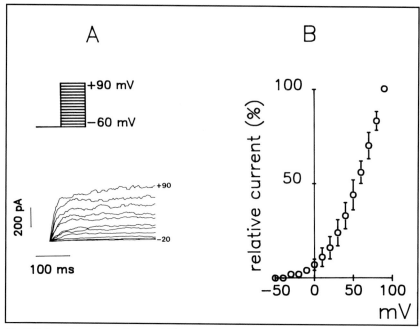

Fig. 5. Voltage dependence of outward $K^+$ current. Shown here is the voltage dependence of outward current activation as measured in isolated patch-clamped cat cerebral arterial muscle cells. These $K^+$ currents were obtained from cells with a holding potential of –60 mV. Initial currents were observed beginning at around –20 mV. The left panel (A) indicates the raw data as measured, and the right panel (B) indicates the voltage current relation (reproduced with permission from Bonnet et al., 1991).

aminopyridine (4-AP) and tetraethylammonium (TEA). To date, three physiologically relevant types of $K^+$ channels have been identified including a delayed rectifier $K^+$ channel, a $Ca^{2+}$-regulated outward current, and an ATP-sensitive $K^+$ channel.

The delayed rectifier $K^+$ channel has been demonstrated in cerebral arterial smooth muscle using the whole-cell voltage clamp technique. Outward current through this channel is activated at depolarizing voltages above –30 mV, and can be blocked by 4–AP (Fig. 5).

Outward current through the $Ca^{2+}$-sensitive $K^+$ channel is activated by increases in intracellular $Ca^{2+}$ concentration, but may be insensitive to changes in extracellular $Ca^{2+}$. These

channels have been identified in single channel recordings from whole-cell patch-clamped cat cerebral arterial muscle cells by a marked increase in open time probability ($P_o$) in the presence of $Ca^{2+}$ ionophore (Fig. 6). Such measurements indicate a peak unitary conductance of 82 ps, suggesting that these channels may carry the largest current of any of the three known $K^+$ channels.

The ATP-sensitive $K^+$ channel is the third known carrier of outward potassium current. Its ability to carry $K^+$ increases as intracellular ATP concentration falls, giving it a potentially important role in the coupling between metabolic and electrical events in the vascular smooth muscle cell. From this perspective, it is interesting to note that a variety of other intracellu-

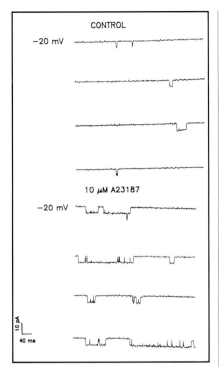

CONTROL

−20 mV

10 μM A23187

−20 mV

*Fig. 6. Opening behaviour of single $K^+$ channels. This figure indicates raw data obtained from cat cerebral arterial muscle using cell-attached patches. The holding potential was around −40 mV. The downward deflections, corresponding to outward current, indicate the opening of single $K^+$ channels during (A) control activity and (B) in the presence of the $Ca^{2+}$ ionophore A23187. This increased $K^+$ channel activity is most likely due to the ionophores' effect on increasing intracellular $Ca^{2+}$.*

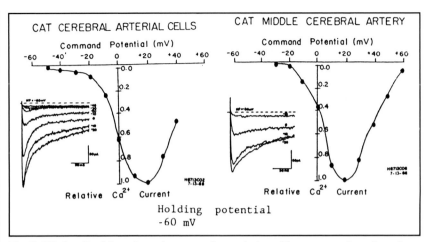

*Fig. 7. Whole cell calcium channel current-voltage relations. The current–voltage data shown here were obtained from patch-clamped cat cerebral arterial muscle cells recorded in the whole cell mode. The extracellular $Ca^{2+}$ was 2.5 mM and the holding potential was −60 mV. The left panel indicates data from individual cells, and the right panel indicates whole artery data. Activation of this inward current begins at between −30 and −40 mV.*

lar mediators, including pH and possibly $Ca^{2+}$, might also modulate $K^+$ current through this channel. Sulphonylurea derivatives are known to block current through these channels.

## $Ca^{2+}$ channels in cerebral arterial muscle

In contrast to $K^+$ channels which carry current out of the cell, $Ca^{2+}$ channels carry current into the cell by virtue of the large inward directed gradient in calcium across the plasma membrane.

Two general types of $Ca^{2+}$ channels have been described in arterial muscle, based upon the voltage ranges at which activation occurs and their sensitivity to dihydropyridines. The 'L-type' $Ca^{2+}$ channels are sensitive to dihydropyridines and are maintained in a conductive state during prolonged depolarization. The 'T-type' $Ca^{2+}$ channels are transient in nature, have a lower threshold for activation, and are insensitive to dihydropyridines.

Cerebral arterial smooth muscle appears to possess both main types of voltage-dependent $Ca^{2+}$ channels. Cell attached patches of rabbit cerebral arteries reveal two subtypes of 'T-type' calcium channels with unitary conductances of 9 and 5 ps, when barium is used as the charge carrier. Rabbit cerebral arteries also appear to possess another voltage-dependent $Ca^{2+}$ channel whose probability of opening in solutions containing 10 mM $Ca^{2+}$ is significantly reduced by membrane hyperpolarization beyond

−40 mV. The opening frequency and mean open time of these latter channels is markedly increased by serotonin. Studies of rat cerebral arteries with 80 mM $Ba^{2+}$ as the charge carrier suggest a voltage-dependent $Ca^{2+}$ channel with a conductance of 30 ps. Whole cell measurements in the guinea-pig basilar artery using 10 mM $Ba^{2+}$ as the charge carrier have identified an L-Type $Ca^{2+}$ current with a peak current of 292 pA at +15 mV. From these studies it is clear that voltage-operated $Ca^{2+}$ channels are present and probably play an important role in the physiological activation of cerebral arterial muscle (Fig. 7).

## Physiological modulation of membrane electrical events in cerebral arterial muscle

Stimuli which ultimately depolarize or hyperpolarize cerebral arterial muscle will activate or inhibit contractile tone, respectively. Such stimuli include a variety of neurotransmitters and circulating humoral agents which act on specific membrane receptors. Many of these receptors in turn activate 'receptor-operated' channels and increase membrane permeability to specific ions such as calcium or chloride. The type and distribution of these 'receptor-operated' channels varies widely among species and among arteries in the same species from different anatomical origins. The general features of the electrophysiological responses to many of the relevant neurotransmitters and humoral agents effecting arterial muscle have been recently reviewed by Hirst & Edwards (1989). Additional characteristics of neurotransmission unique to the cerebral circulation are reviewed elsewhere in this volume (see Chapters 3, 4, 10, 11 and 12).

A variety of other physiological stimuli, such as changes in transmural pressure and blood gases, also have important effects on cerebrovascular membrane potentials. Because these stimuli usually have multiple mechanisms of action, previous studies have often overlooked their electrophysiological effects. Nonetheless, recent evidence suggests that these electrophysiological effects are important components of the integrated multifactorial cerebrovascular responses to such perturbations.

## Transmural pressure and electromechanical coupling in cerebral arterial muscle

Increasing transmural pressure in isolated cerebral arterial segments depolarizes muscle cells, induces spontaneous action potentials and reduces internal diameter. In cat middle cerebral arteries, increasing transmural pressure from 20 to 160 mmHg decreases Em from –68 to around –36 mV. The slope relating the change in Em to the change in pressure increases as external $[Ca^{2+}]$ is elevated and decreases upon reduction of $[Ca^{2+}]$ below control levels. The frequency of spontaneous electrical spike activity in these isolated pressurized cerebral arteries also increases as transmural pressure is increased from 100 to 140 mmHg. These findings suggest that either pressure or the subsequent mechanical deformation of the vascular muscle wall activates arterial muscle by increasing $Ca^{2+}$ permeability and depolarizing the membrane. Similar phenomena have been documented in a variety of vascular beds.

The 'sensor' translating the mechanical stimulus into the observed biophysical processes in arterial muscle is not yet well defined. Although several laboratories have examined the role of the vascular endothelium in this response, controversy concerning the heterogeneity of endothelial function and the methods used to remove the endothelium obscures any consensus. For example, endothelin, a potent vasoactive peptide shown to open $Ca^{2+}$ channels, is a prime candidate as a pressure released endothelial contractile factor. However, it has not been shown to be released from pressurized cerebral arteries, and generally does not fit the profile necessary to act as a mediator in this regard. Thus, stretch may promote the release of some endothelium-derived contracting factor other than endothelin.

Another obvious possibility for an arterial 'pressure-sensor' is a 'stretch-activated' ion channel in either the vascular smooth muscle, the endothelium, or both. Such mechanosensitive ion channels have been described in both smooth muscle and aortic endothelial cells, but they are non-specific and exhibit conductance for multiple ion species with some selectivity for either cations or anions. Mechanosensitive channels are generally studied under patch-clamp conditions and usually conduct currents large enough to significantly affect $E_m$. Therefore, these channels remain plausible as mediators of the cerebrovascular smooth muscle depolarization and action potential generation observed upon elevations in transmural pressure, which certainly play a major role in the phenomenon of autoregulation (Chapter 9).

Inward current flow through pressure-sensitive channels may also make an important contribution to membrane potential, even under 'quiescent' conditions, *in vivo*. At a normal transmural pressure of 100 to 120 mmHg, the $E_m$ in a cerebral artery is approximately –40 to –45 mV, and thus is more depolarized than previously thought based on studies of non-pressurized arterial preparations and isolated single cells. This implies that at a normal arterial pressure, the $E_m$ *in vivo* is always at or near its threshold for mechanical activation of around –50 mV. The contribution from the pressure-sensitive channels also brings $E_m$ closer to the operational voltage ranges for the $K^+$ and $Ca^{2+}$ channels (approximately –30 mV and –40 mV, respectively). Thus, the higher the pressure, the greater the degree of depolarization and the more reactive the arterial muscle is to influences which activate ionic- or voltage-dependent mechanisms. Indeed, in cat cerebral arteries, the sensitivity to serotonin increases by an entire order of magnitude as transmural pressure is elevated from 60 to 140 mmHg.

## Effect of $pO_2$ on membrane properties of cerebral arterial muscle

Whereas the ability of hypoxia to relax isolated segments of smooth muscle *in vitro* has long been recognized, this response has only recently been demonstrated in isolated cerebral arteries. Because this response can be demonstrated in cerebral arteries from rabbits, sheep, dogs, and monkeys, it is apparently independent of species. The magnitude and time course of the direct effects of hypoxia, however, vary considerably among arteries from different vascular beds.

One factor contributing to this variability is the potential role of the vascular endothelium. *In vitro* studies of carotid and cerebral arteries from both rabbits and sheep suggest that

removal of the vascular endothelium attenuates hypoxic relaxation. Because the endothelium in some cases can release a factor which hyperpolarizes vascular smooth muscle, release of such a factor during hypoxia may be responsible for the vascular hyperpolarization observed in response to hypoxia in canine carotid arteries. However, hypoxia must also have a direct effect on cerebral arteries independent of the endothelium, because a significant hypoxic relaxation is still observed in cerebral arteries denuded of endothelium.

In cat cerebral arteries, hypoxia inhibits $Ca^{2+}$-dependent action potential generation. One possible explanation of this response is that hypoxia lowers intracellular ATP levels enough to activate the ATP sensitive $K^+$ channels present in cerebral arteries. This activation would increase outward $K^+$ current, hyperpolarize the membrane, and relax the artery.

This mechanism has been demonstrated in canine coronary arteries and also appears present in cerebral arteries of both the cat and the rabbit. In freshly dispersed cat cerebral arterial muscle cells, reducing $pO_2$ in the bathing solution significantly increases the probability of opening and the mean open time of an 82 ps $K^+$ channel when recorded in the cell attached mode. Addition of glibenclamide, an inhibitor of the ATP-sensitive $K^+$ channel, blocks the increase in $K^+$ channel activity observed under hypoxic conditions. Thus, hypoxia appears to have a direct influence on cellular events mediating membrane electrical phenomena in cerebral arterial muscle.

## Effects of $pCO_2$ and pH on electrophysiological properties of cerebral arterial muscle

Reduction of extracellular pH by increases in either $pCO_2$ or $[H^+]$ dilate cerebral arteries. In isolated cat cerebral arteries, this dilatation appears to be mediated by membrane hyperpolarization. In contrast, reduction of $pCO_2$ or elevation of pH while maintaining $pCO_2$ constant depolarizes and contracts cat cerebral arteries. Both effects appear to be mediated, at least in part, by changes in $g_K$. Elevation of either $pCO_2$ or $[H^+]$ increases the slope of the relation between $E_m$ and $[K]_o$ consistent with an increase in $g_K$ according to constant field theory. Conversely, reductions in either $pCO_2$ or $[H^+]$ decrease the slope of the relation between $E_m$ and $[K]_o$, indicating a decrease in $g_K$.

In patch-clamped isolated cerebral arterial muscle cells, reduction of extracellular pH from 7.43 to 7.20 increases the peak outward current by

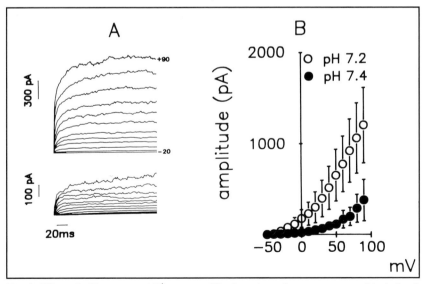

*Fig. 8. Effects of pH on outward $K^+$ currents. The data shown here were measured in isolated, patch-clamped cat cerebral arterial muscle cells. The left panel indicates the raw data as measured and the right indicates the voltage–current relations in relation to pH. The decrease in pH from 7.4 to 7.2 produced a dramatic increase in peak outward current (reproduced with permission from Bonnet et al., 1991).*

35 per cent in a majority of the cells (Fig. 8). Because this effect can be observed even when *intracellular* pH is heavily buffered by dialysis with 10 mM HEPES buffer, it remains possible that *extracellular* $H^+$ may have a direct effect on cerebrovascular $K^+$ channels. Alternatively, changes in pH on either side of the membrane may alter the transmembrane 'field potential' and thus modify the conformational state of the channel proteins, changing their conductance characteristics. In either event, it is clear that free hydrogen ion, a common byproduct of both vascular and parenchymal metabolism, is an important modulator of membrane electrical properties, and thus the reactivity, of the major arteries of the cerebral circulation.

## Cerebrovascular endothelium

In addition to their smooth muscle and connective tissue, cerebral arteries also have a lining of endothelial cells which constitute an interface between the circulating blood and the underlying media of the arteries. Following the discovery by Furchgott in late 1970s that endothelial cells can release chemical factors capable of altering vascular tone, the majority of endothelial research in recent years has focused on this tissue as a potential regulator of vascular resistance. In addition to its short-term effects on vascular tone, however, the endothelium also synthesizes and releases a wide range of other products including proteins, peptides, prostanoids, and other small, labile and diffusible compounds. This diversity of synthetic products enables the endothelium to serve, in addition to its acute influence on vascular tone, many additional functions including haemostasis, angiogenesis, growth and differentiation of the underlying vascular smooth muscle, and barrier function.

## The role of endothelium in vascular growth and development

### Embryonic and foetal development of endothelium

In the developing rat brain, by embryonic day 11–12 (total gestation ≈ 21 days) the dorsal neural groove is fused and endothelial cells, with immature nuclei and intracytoplasmic organelles, are located primarily around the neural tissues. Between embryonic days 13 and 16, the matrix cells in the neural tube begin to produce neuroblasts that migrate from the matrix layer and intraneural blood vessels and pericytes first begin to appear. In the earliest phases of vasculogenesis, tentacles appear to grow out from the distal end of the vascular cord and these cells later extend to form the afferent blood vessel. In the brain, this formation of new penetrating vascular trunks and intracortical capillary branchings is terminated before global brain size reaches a plateau.

While the vascular sprouts are developing, endothelial cells synthesize new proteins, increase their cytoplasmic thickness and generate many cell processes that protrude from both the luminal and abluminal sides. In human foetal endothelial cells, immunostaining can demonstrate the presence of factor VIII (von Willebrand factor) within the first four weeks of gestation. These early endothelial cells appear to protrude first into the lumen, and then with advancing gestational age, form a flat continuous sheet. To accommodate the growth of the underlying vascular layer, the endothelial cells become elongated and their thicknesses decrease with age from ≈0.6 to ≈0.2 μm. The rate of this process varies considerably, and is generally slower in cerebral arteries than in arteries from any other vascular bed. In maturing cerebral arteries, anionic sites appear gradually on both the luminal and abluminal surfaces of the endothelium and basement membrane, a development which correlates well with the early acquisition of blood brain barrier characteristics. By 15–18 weeks of gestational age, human cerebral microvessels begin to exhibit barrier function but the endothelial cells still do not yet contain many key enzymes such as choline acetyl transferase. By this age, glial foot processes begin to form, but these lack many of the ultrastructural features characteristic of mature astrocytes, suggesting that the endothelial cells and interendothelial junctions are not yet fully functional. At term, the degree of functionality of the endothelium, both in terms of barrier characteristics and of vasomotor influence, varies considerably among different brain regions. Not surprisingly, the endothelial cells of vessels of the germinal matrix are probably less functionally mature than in vessels of any other brain region. For example, the less mature germinal matrix microvessels often contain intraluminal microvilli and junctional complexes (primarily tight junctions) that are not continuous between the lateral endothelial cell walls.

In newborns, the blood–brain barrier

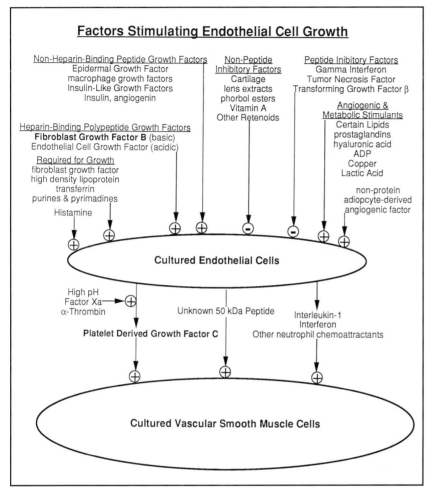

*Fig. 9. Endothelial growth factors. Summarized here are the major factors released from cultured endothelial cells which influence growth in cultured vascular smooth cells. Also shown are a variety of factors, organized by class, which influence the growth of endothelial cells in culture. This summary is not meant to be exhaustive, but rather to indicate the complex nature of the mechanisms governing growth and proliferation of both endothelium and vascular smooth muscle. See text for details.*

is up to ten times more permeable to small polar non-electrolytes such as mannitol, than in adults. In addition, newborn endothelial cells also have a higher water content, contain more prostacyclin, and exhibit more negatively charged vesicles than adult endothelial cells. Nonetheless, all of the saturable transport systems are operative at birth including those for purines, monocarboxylic acids, and glucose. For the glucose transporter, however, the half-saturation concentration of glucose is about threefold higher in newborns than in adults. In the immediate postnatal period, the carrying capacity for glucose, as for many other substrates, increases progressively with age. Simultaneously, brain-to-blood transport mechanisms that were inoperative at birth gradually become functional and the permeability to most compounds, including horse radish peroxidase, declines markedly. In addition, the vasoactive capacity of the maturing endothelial cells improves markedly during the first weeks of life. Thus at birth, the cerebrovascular endothelium is functionally immature, but this immaturity quickly disappears so that very early in life the endothelium can begin to exert its diverse influences on vascular tone, haemostasis, angiogenesis, barrier function, and the growth and differentiation of vascular smooth muscle.

## Endothelial factors influencing vascular cell growth and division

Throughout vascular development, there is a delicate interplay between the endothelial and underlying vascular layers. Although the exact nature of this interaction is still largely unknown, a growing number of factors have been identified in cultured endothelial cells which stimulate vascular cell cultures, including interleukin-1, interferon, and other neutrophil chemoattractants (see Fig. 9). In terms of vascular growth, however, the most important factors identified to date are the peptide mitogens, which include platelet-derived growth factor C (PDGFc). This peptide factor is basic, appears to bind to specific receptors on both smooth muscle cells and fibroblasts, and may be an A-chain homodimer of regular platelet-derived growth factor which has both A and B chains. Production of PDGFc is stimulated by alkaline media, by coagulation factor Xa, and alpha-thrombin. In non-cerebral endothelial cells, PDGFc production is inhibited by acetyl-LDL but not by native LDL or cholesterol and this effect requires uptake of acetyl-LDL by a receptor which is absent in cerebral endothelia. Agents that kill endothelial cells, such as high concentrations of cholesterol, phorbol esters or hydrogen peroxide, generally first stimulate a burst of PDGFc release.

Endothelial cells also contain another, as yet unidentified, growth factor which may account for more than one-half of the total growth promoting activity present in endothelial cell cultures. This unidentified factor is a peptide distinct from regular platelet derived growth factor, basic fibroblast growth factor, transferrin, bovine fibronectin and epidermal growth factor. When isolated from the conditioned medium of bovine cerebral capillary endothelium, it promotes DNA synthesis in astrocytes and pericytes, but not in oligodendrocytes or endothelial cells *in vitro*. The peptide's molecular weight is ≈50 kD and it has been proposed that it is involved in the local signaling between cell types that control new vessel formation during development, in regeneration after brain tissue injury, or in tumour formation.

In addition to its effects on the growth and development of vascular cells in culture, the conditioned medium from endothelial cell cultures also stimulates vascular collagen synthesis and secretion. Of particular interest is the finding that different types of endothelial cells stimulate the synthesis of different kinds of collagen. The factors which mediate this response are at present unknown and remain topics of intensive investigation.

## Factors influencing endothelial cell growth and division

Just as vascular growth and differentiation is influenced by factors re-

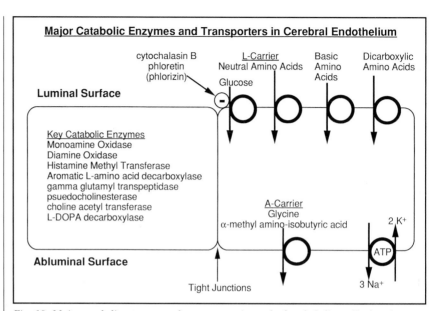

Fig. 10. Major catabolic enzymes and transporters in cerebral endothelium. To function as a blood–brain barrier, cerebral endothelial cells contain a number of catabolic enzymes, as shown here, which are unique to cerebral arteries. In addition, cerebral endothelium also possesses a facilitated diffusion glucose carrier, and spearate carriers for neutral, basic, and dicarboxylic amino acids. The abluminal membrane apparently contains the A-carrier and a uniquely high concentration of the Na-K ATPase pump.

leased by the endothelium, the growth and differentiation of the endothelium, in turn, is influenced by numerous factors external to the endothelial cell (see Fig. 10). Although endothelial cell turnover in adults is typically very low, less than 0.1 per cent per day, it can increase dramatically during neovascularization, such as that which occurs during wound healing and repair. The process usually begins with the breakdown of venular basement membranes followed by endothelial cell proliferation and migration. This is followed by the formation of a new vascular tube, deposition of a new basement membrane, and differentiation. Given the key role of endothelium in this process, and the importance of neovascularization in recovery from organ transplants and ischaemic insults, interest in the processes governing en-

dothelial cell growth and division has grown dramatically in recent years.

What we now know of the factors that stimulate endothelial cell growth and division has been learned primarily from tissue culture studies. For example, we know that human endothelial cultures require fibroblast growth factor (see below), transferrin, and high-density lipoprotein (HDL), but do not respond to platelet-derived growth factor. Some of these requirements appear angiogenic but are not mitogenic and probably serve nutritional needs. In this category are certain lipids, prostaglandins, hyaluronic acid, and HDL, whose role may be to activate a key enzyme of cholesterol metabolism, 3-hydroxy-3-methylglutaryl coenzyme A. Other factors in this category, such as ADP, copper, or lactic acid, may act either metabolically or indirectly through leukocyte activation.

Clearly, the most important factors governing endothelial cell growth and division are the mitogenic factors which include the polypeptides, polyamines, the purine and pyrimadine nucleosides (which probably serve nutritional requirements), nonprotein adipocyte-derived angiogenic factor, and histamine. Given the mitogenic effects of histamine, which appear to involve H1 receptors, a role for mast cells in endothelial growth has also been proposed. However, the most important single class of factors influencing endothelial cell growth is the polypeptide family of growth factors.

One of the best recognized of the peptide endothelial growth factors is fibroblast growth factor (FGF), a ubiquitously distributed basic polypeptide which is both mitogenic and chemotactic. FGF contains 154 amino acids, binds heparin, and acts via a specific receptor. This factor may also preferentially bind and interact with components of the extracellular matrix. FGF is generally not secreted by cells in culture, but can be secreted by isolated macrophages.

Another important growth factor is endothelial cell growth factor (ECGF) also known as acidic fibroblast growth factor. ECGF is closely related to the basic FGF described above: both molecules contain the same number of amino acids with greater than 50 per cent homology, suggesting a common ancestral origin. ECGF is an acidic (anionic) single chain polypeptide produced in the hypothalamus and other neural tissues. It is biologically active in the nanogram per ml range, but is not detectable in plasma or serum. ECGF can be isolated in either an alpha form ($\approx$17 kD) or a beta form ($\approx$20 kD), both of which contain methionine double sulfide bonds. Based on digestion studies, it has been proposed that the beta form may be a precursor of the alpha form. ECGF acts on a specific receptor that has the characteristics of a single chain polypeptide ($\approx$150 kD) that is internalized upon ligand binding. The receptor appears to be quite labile, and can be rapidly up-regulated or down-regulated. The biological activity of ECGF requires the presence of serum and its effects are potentiated by glycosaminoglycan heparin; indeed heparin may be a cofactor for ECGF. ECGF delays premature senescence of human endothelial cells and is a potent chemoattractant for endothelial cells, *in vitro*.

Aside from ECGF and FGF, there are other peptide growth factors that do not bind heparin. In this group are epidermal growth factor and angiogenin, a 14.4 kD cationic protein secreted by cells in culture which is 35 per cent homologous with pancreatic ribonuclease but has no enzymatic activity of its own. Another group of non-heparin binding peptide growth factors are the macrophage growth factors, whose release is stimulated by hypoxia or lactate. Insulin and insulin-like factors also fall into this group. Although insulin receptors are found on many endothelial cell types, as are receptors for insulin-growth factor I and II, the exact function of these receptors is as yet unclear. Similarly, endothelial cell cultures contain thrombin binding sites, but the importance and role of these sites remain controversial.

In addition to the peptide growth factors are peptide inhibitory factors. These include gamma interferon and tumour necrosis factor, which may be responsible for the leaky endothelium found in tumours. Transforming growth factor beta, a 25 kD peptide found in platelets, also inhibits proliferation *in vitro* but paradoxically stimulates angiogenesis *in vivo*. Non-peptide inhibitory factors include cartilage and lens extracts, phorbol esters (by virtue of PKC activation), vitamin A and other retnoids. Clearly, the regulation of endothelial growth and differentiation is a complex and multifactorial process which is only just beginning to be understood.

## Barrier function

Another major function served by the cerebral endothelium is that of a barrier between the circulating blood and the brain parenchyma. Although the properties of this barrier vary considerably between the cerebrum and cerebellum, between large arteries and small, with age, and among different species, certain barrier characteristics of all cerebral endothelium are relatively uniform. All cerebral endothelial cells can be characterized by tight junctions and close astrocytic relationships, high concentrations of degradative enzymes, uniquely adapted mechanisms of selective transport, and pronounced vulnerability to hypertensive and haemorrhagic insults.

## Cerebral endothelium as a physical barrier

Recent advances in tissue culture techniques have revolutionized the study of the blood–brain barrier (BBB) properties of cerebral endothelia. Primary cultures of bovine microvessel endothelial cells cultured onto polycarbonate membranes can now be used as an *in vitro* model system for estimating the potential permeability of a solute through the BBB, *in vivo*. These cells retain, up to the fiftieth generation, numerous characteristics of the BBB, including tight junctions, sparse pinocytotic vesicles, and monoamine oxidase activity. In addition, primary cultures of microvascular endothelial cells from rat brain are capable of synthesizing all their lipids, including neutral lipids, phospholipids, and glycolipids, from the water-soluble compounds glucose, acetate, acetoacetate and β-hydroxybutyrate. In these cultures the ketone bodies, especially acetoacetate, are the preferred substrates for lipid synthesis. This ability of cultured cerebral endothelial cells to take up lipids, principally cholesterol, contained in the serum lipoproteins indicates that these cultures are powerful tools for the study of both the metabolic and the physical characteristics of the cells making up the BBB.

From such studies, it is clear that a key characteristic of cerebral endothelial cells, whether studied *in vivo* or *in vitro*, is the presence of tight junctions between adjacent cells. In addition, brain endothelial cells contain few pinocytotic vesicles and demonstrate a high electrical resistance which is ≈1000 times that found in endothelial cells of mesenteric origin, and ≈100 times that found in endothelia of skeletal muscle origin. This tight physical barrier restricts not only the movement of solutes into and out of the brain, but also of water; attempts to identify a paracellular pathway for water across the intact BBB have been

generally unsuccessful. The physical barrier is not absolute, however, as it has been demonstrated in mice and rats that blood-borne molecules taken into the cerebral endothelium by adsorptive endocytosis and conveyed to the Golgi complex can, either by themselves or as vehicles for other molecules excluded from the brain, undergo transcytosis through the BBB without compromising the integrity of the barrier.

Another key finding to come from studies of cerebral endothelial cells is their close relationship with astrocytes. In general, cerebral endothelial cells are in contact with numerous astrocytic foot processes which form a 'glial sheath'. The processes are probably not involved in endothelial nutrition, but they may be involved in differentiation of cerebral endothelial cells and maintenance of BBB properties. Maturation of these astrocytic processes usually parallels the acquisition of BBB characteristics, and when the barrier breaks down, the astrocytic processes often appear to be abnormal. When non-brain tissue is transplanted into the brain, the new vessels take on the characteristics of BBB vessels. These observations suggest that some feature of the intracranial environment predisposes to the development of BBB characteristics and, at present, the glia remain as possible candidates as the mediators of this effect.

## Cerebral endothelium as a biochemical barrier

The enzymatic profile of cerebral endothelial cells is distinct from that of endothelial cells of extracranial origin. Aromatic L-amino acid decar-

boxylase, gamma glutamyl transpeptidase, pseudocholinesterase, choline acetyl transferase, and L-DOPA decarboxylase are all largely unique to cerebral endothelial cells. Alkaline phosphatase is also largely specific to cerebral endothelium and can be stained with black lead sulphide salt to render the brain microvasculature visible by both light microscopy and microradiography. Glutamic acid decarboxylase immunoreactivity has been reported in cerebral endothelial cells, suggesting that a non-neuronal GABA system may be present. Cerebral endothelial cells also contain quantities of diamine oxidase and histamine methyl transferase capable of quickly degrading any histamine which enters the cells. Other penetrant amines can be quickly degraded by monoamine oxidase. And in support of this substantial and diverse catabolic capacity, cerebral endothelial cells have a high density of mitochondria, up to six times that found in endothelia of skeletal muscle origin. Together, these characteristics enable cerebral endothelia to function as a highly efficient biochemical barrier to circulating neurohormones and transmitters.

Cerebral endothelial cells also express a unique profile of non-enzymatic antigens. In addition to the von-Willebrand factor (VIII) found in most endothelial types, cerebral endothelial cells contain the distinctive proteins alpha-globin, histone 2A, histone 2B, and histone 3. The most abundant protein of cerebral endothelial cells, however, appears to be a cytoplasmic actin of ≈46 kD which is localized in the endothelial plasma membrane and may play a role in the regulation of BBB permeability. Neurothelin, a protein with a molecular mass of $\approx 43$ kD, is found in the hatched chick on the luminal surface of cerebral, but not extracranial, endothelial cells. Consistent with this localization, neurothelin has been proposed to play a role in the interaction between vascularization and neuronal differentiation. Other unique, as yet unidentified, antigens have been found in cerebral endothelial cells, including one found in both the cytoplasm and on the luminal surface, one found mainly in the basement membrane of vessels smaller than 20 μm in diameter, and a third 46 kD antigen specific for brain capillaries. Myelin basic protein has been observed in cultured guinea-pig cerebrovascular endothelial cells, and it has been suggested that the presence of this protein may contribute to the pathogenesis of chronic relapsing experimental allergic encephalomyelitis. Thus, in spite of potential species differences, cerebral endothelial cells express a unique and specific antigenic identity. This identity, in turn, is most probably responsible for the patterns of adhesion to lymphocytes, neutrophils, tumour cells and viruses characteristic of cerebral endothelial cells, *in vivo*.

## Selective transport

To effect cerebral homoeostasis, cerebral endothelial cells must not only exclude unnecessary and potentially deleterious blood-borne compounds, they must also facilitate the transfer of essential nutrients and fuels from the blood into the cerebral interstitium. To this end, cerebral endothelial cells possess a highly specialized system of transport proteins. Among these, the carrier with the highest capacity is that for glucose, which is probably of the facilitated diffusion type. This carrier exhibits a strong homology to the glucose carrier of erythrocytes and is strongly inhibited by cytochalasin B and phloretin, and less strongly inhibited by phlorizin. The mRNA for this carrier (2.9 kB) is expressed only in cerebral endothelial cells. Interestingly, the glucose permeability of the BBB usually appears closely linked to the rate of cerebral phosphorylation, but this effect is probably due to an increased surface area for exchange (increased PS product) resulting from an increased rate of cerebral perfusion. In contrast, the apparent shrinkage of the cerebral L-glucose space in chronic hyperglycaemia may indeed be due to changes in the BBB permeability to L-glucose, possibly indicating a decrease in carrier density. In Alzheimer's disease, there is a marked decrease in the cerebral glucose transporter in microvessels, and this decrease is most pronounced in the cerebral neocortex and hippocampus, regions that are most affected in Alzheimer disease.

Neutral amino acids, basic amino acids, and dicarboxylic amino acids all traverse the BBB on separate carriers. The neutral amino acids enter via the L-carrier system, as defined by Christensen (1979). However, in contrast to the L-carrier system in extracranial endothelia, in cerebral endothelial it appears subject to adaptive regulation. The A-carrier, which transports glycine and alpha-methyl amino-isobutyric acid, appears to be absent at the luminal membrane of cerebral endothelial cells, but may be present at the abluminal membrane thus providing a way for cerebral endothelia to maintain low levels of these compounds in the cerebrospinal fluid. The monocarboxylic acids acetate, butyrate, lactate, and pyruvate all compete for and are carried by a single carrier. Cultured bovine cerebral endothelial cells are also able to incorporate tritiated choline by a carrier-mediated mechanism, and the characteristics of this carrier are modulated by the presence of choline in the incubating medium.

Given the observation that cerebral endothelial cells appear to transport potassium from the brain interstitium to blood via a ouabain-sensitive mechanism, BBB endothelia are thought to contain unusually high concentrations of sodium–potassium ATPase in their abluminal membrane. This mechanism would help keep the cerebrospinal fluid concentrations low enough to maintain normal neuronal excitability and, equally important, may play a role in the fluid and volume balance of the intracranial compartment. Another ion transported from the brain to the blood is lead, and this flux compensates for any $PbOH^+$ which in the rat has been reported to passively enter the endothelial cells from the bloodstream.

Certain peptides also appear to be selectively transported from blood to brain by the cerebral endothelia. For example, a high affinity saturable mechanism has been described for the transport of delta-sleep-inducing peptide across the BBB. Although IgG molecules are normally excluded from entering the brain, cationization of IgG molecules appears to greatly facilitate the transport of these plasma proteins through the BBB *in vivo*. Such protein transport systems, however, appear to be quite heterogeneous in their distribution throughout the cerebral endothelia. In cats, the endothelia of certain well-defined loci including the hypothalamic median eminence, subfornical organ, and area postrema are relatively permeable to endogenous albumin, as is the endothelium of the arteries of the circle of Willis. In contrast, the other smaller intracerebral arteries of the feline brain are almost completely impermeable to labelled albumin.

The BBB's permeability to many molecules appears subject to active regulation. The most abundant protein in brain capillaries now appears to be actin, suggesting a possible role for contractile proteins in modulating BBB permeability changes. In addition, cerebral endothelial cells contain high concentrations of adenylate cyclase and elevated levels of cyclic AMP have been associated with increased overall BBB permeability, increased rates of pinocytosis and formation of brain oedema. This adenylate cyclase appears coupled to cell surface receptors for histamine and other nonadrenergic agonists. The effects of cAMP within cerebral endothelial cells are mediated by cAMP-dependent kinase, which is present in significant concentrations. Also present in cerebral endothelia are important quantities of calmodulin-dependent protein kinase and cGMP-dependent protein kinase (G-kinase). The presence of the G-kinase in cerebral endothelia suggests a potential role for guanylate cyclase in the regulation of BBB permeability, a suggestion borne out by the observation that dibutyryl cGMP in the rat cerebral microvessels increases both pinocytotic activity and the permeability to albumin in a dose-dependent manner. Thus, as for many processes of biochemical regulation, BBB permeability appears to be regulated by the phosphorylation profile of the proteins within the cerebral endothelial cells.

A variety of influences external to the cerebral endothelia modulate BBB permeability. Although atrial natriuretic factor, vasopressin and endothelin (see below) have highly variable effects, other substances such as arachidonic acid can induce a reversible alteration of cerebral endothelial permeability to trypan blue albumin in cultured rat cerebral endothelial cells. Similarly, a stable synthetic analogue of thromboxane A2 has been reported to increase endothelial permeability in the major cerebral arteries of the rabbit to Evans Blue dye and horseradish peroxidase. In cultured mouse endothelial cells, hydraulic conductivity can be decreased by methyl-prednisolone and increased by hydrocortisone. In feline cerebral arteries, bradykinin causes brain oedema and selectively increases BBB permeability to small molecules such as sodium–fluorescein. In addition, the BBB permeability to ammonia is strongly affected by blood pH, and ammonia passes more freely in alkaline than in acidic conditions. Of greatest interest, however, is the suggestion that BBB permeability may be directly influenced by neuronal input. Cerebral parenchymal capillaries appear to receive an adrenergic innervation from the locus coeruleus and stimulation of this innervation increases BBB permeability to small molecules and water. Serotonin at physiological doses can also increase BBB permeability to Evan's Blue and sodium in rats, and this effect appears dependent on increases in endothelial cAMP levels. Thus, the permeability of the BBB is dynamic and probably under tight and complex physiological regulation.

## Modulation of vascular tone

Complementing their abilities to influence vascular growth and regulate the exchange of materials between blood and the cerebral interstitum, cerebral endothelial cells also have the capacity to modulate the contractile tone of cerebral arteries (Fig. 11). This modulation can result in either decreased tone, mediated through the release of endothelium-derived relaxing factors (EDRF), or can result in increased tone, mediated through the release of endothelium derived contracting factors (EDCF). Certain characteristics of this modulation are identical in all vascular beds, but others are unique to the cerebral circulation and vary not only among species, but also among cerebral arteries of different sizes and ages.

### EDRF and its mechanism of action

One feature of endothelium dependent relaxation common to all vascular beds is that it is mediated by activation of soluble vascular guanylate cyclase (MW $\approx$ 150–230 kD). Guanylate cyclase, which catalyses the formation of cyclic guanosine monophosphate (cGMP) from guanosine triphosphate (GTP), is also present in cerebral arteries as a particulate enzyme (MW $\approx$300–400 kD), but this form plays no major role in endothelium-dependent vasodilatation. Soluble guanylate cyclase is a haem-containing protein with a Km for GTP of $\approx$20–100 $\mu$M. It can be inhibited by either methylene blue or by Lily compound #83583. High concentrations of both ATP and

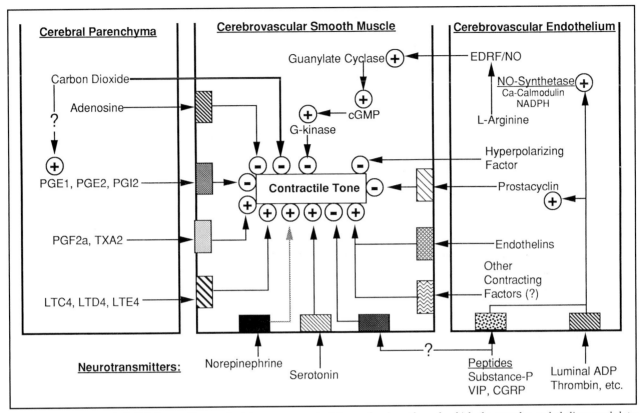

*Fig. 11. Vasoactive endothelial factors. Summarized here are several of the mechanisms through which the vascular endothelium modulates cerebrovascular tone. As discussed in the text, the endothelium releases EDRF, now known to be nitric oxide, which promotes vasodilation through stimulation of vascular soluble guanylate cyclase. Endothelium also releases a variety of other vasoactive factors including prostacyclin, endothelins, and other as yet unidentified contracting factors. Endothelial cells respond to a wide variety of factors, depending on the type and number of receptors populating the outer endothelial cell membrane. For comparison, vasoactive factors released from the cerebral parenchyma are also shown.*

arachidonate have also been reported to inhibit guanylate cyclase, although these effects remain somewhat controversial. Activation of soluble guanylate cyclase occurs in response to nitric oxide, and has also been reported in response to certain lipids such as lysophosphatidyl choline and oleic acid.

The mechanisms whereby the cGMP synthesized by guanylate cyclase produces relaxation remain uncertain. Although it is clear that cGMP exerts its effects through activation of a cGMP-dependent protein kinase (G-kinase), the substrates of this kinase

remain under intensive investigation. Several lines of evidence suggest that G-kinase phosphorylates a sarcolemmal calcium transport protein, which enhances calcium extrusion from the smooth muscle cell. Other evidence suggests that G-kinase phosphorylates a component of the sarcoplasmic reticulum membrane to stimulate sequestration of calcium within the cell. It is also possible that G-kinase may alter contractile force by phosphorylating a contractile protein, possibly myosin light chain kinase. Whatever the substrate, it is now clear that cGMP produces relaxation by altering the phosphorylation profile of the

proteins in the vascular smooth muscle cells.

Given that nitric oxide can activate guanylate cyclase, and that it can be easily derived from nitrogenous vasodilators such as nitroglycerine and nitroprusside, nitric oxide was proposed in 1986 to be the main chemical signal released by the endothelium to initiate vascular relaxation. Since that time, a large and growing body of evidence has appeared which supports the view that nitric oxide is the physiologically most important EDRF. In the canine basilar artery, for example, the relaxations produced by both native EDRF and nitric oxide are significantly augmented in the presence of superoxide dismutase and catalase, are not affected by indomethacin, but are reduced in the presence of haemoglobin which binds nitric oxide. Furthermore, both nitric oxide and native EDRF have similar half-lives of approximately 6–10 s. Observations such as these have now been made in the cerebral arteries of many species including mouse, cat, dog, pig, monkey and human, reinforcing the view that EDRF is not a free radical or prostaglandin derivative, and instead is probably nitric oxide or a closely related compound.

Notwithstanding the bulk of evidence supporting the hypothesis that EDRF is nitric oxide, some evidence has been obtained suggesting that there may be more than one EDRF in the cerebral circulation. For example, endothelium-mediated relaxation often involves hyperpolarization of the smooth muscle and nitric oxide is unable to reproduce this effect in rabbit middle cerebral arteries. In cat pial arteries, nitro blue tetrazolium can block relaxation responses to nitric oxide, nitroprusside and nitroglycerine, but does not inhibit endothelium-dependent relaxations to acetylcholine. In the mouse pial circulation, endothelium dependent relaxations to bradykinin have been reported to be blocked by superoxide dismutase, a free radical scavenger. These and other similar data lend support to the concept that, at least in the cerebral circulation, endothelium-dependent relaxation involves the release of more than just nitric oxide. Candidates for these controversial secondary EDRFs include endothelium-derived hyperpolarizing factors,

free radicals, and possibly other as yet unidentified low molecular weight compounds.

## Regulation of EDRF synthesis and release

### Nitric oxide synthase and its regulation

The prevailing view that nitric oxide constitutes the main EDRF in the cardiovascular system as a whole has stimulated significant effort toward elucidating the mechanisms whereby nitric oxide can be synthesized *in vivo*. These efforts have now convincingly demonstrated that nitric oxide is generated by the enzyme nitric oxide synthetase. As shown in cultured bovine aortic endothelial cells, the enzyme is particulate and requires both calcium-calmodulin and NADPH for activity. The substrate for nitric oxide synthetase is L-arginine, and its products are nitric oxide and citrulline. Derivatives of L-arginine, such as L-N monomethylarginine (L-NMMA), nitroarginine, and L-nitroarginine methyl ester, have been shown to competitively inhibit endothelium dependent relaxations in cerebral arteries from the mouse, rat, cat, sheep, cow, and human.

As revealed by *in situ* hybridization studies, the distribution of nitric oxide synthetase is most interesting. The enzyme is found not only in the endothelium as expected, but an inducible form is also found in the medial layer of many cerebral arteries. Equally important, the enzyme is abundant in certain nerve terminals and this finding has fueled the hypothesis that nitric oxide may act as a neuronal transmitter within the central nervous system. Other evidence

suggests that nitric oxide synthetase may be involved in the non-adrenergic, non-cholinergic neurogenic vasodilatation characteristic of certain cerebral arteries. In addition, nitric oxide synthetase is abundant in macrophages which apparently produce nitric oxide as part of the host defense system. Thus, the pathway for *in vivo* synthesis of nitric oxide is now well established.

One key to understanding the activation of endothelial nitric oxide synthetase derives from its dependence on calcium–calmodulin. Given this requirement, any influence which increases cytosolic calcium concentration within the endothelial cell will increase the probability for its synthesis of nitric oxide. In this manner, many of the same mechanisms that govern cytosolic calcium in smooth muscle also govern cytosolic calcium and thus nitric oxide production in endothelial cells. For example, the entry of extracellular calcium through membrane calcium channels plays a key role in the synthesis and release of nitric oxide. Some membrane calcium channels in endothelial cells are potential sensitive, and thus the relative abundance and activity of potassium channels (of which five endothelial types have already been identified) in the endothelial cell membrane ultimately affect nitric oxide production and release. On the other hand, many endothelial cell membrane calcium channels appear to be receptor operated. Clearly, there is still much to learn about stimulus-response coupling in endothelial cells, but all initial findings suggest that these cells follow the same principles of electrophysiological and pharmacological coupling established in other cell types.

Under resting conditions, endothelial cells continuously synthesize and release significant quantities of nitric oxide. When this basal release is inhibited by the systemic administration of an arginine analogue such as L-NMMA, arterial pressure increases markedly. Topical administration of arginine analogues to rat basilar or mouse pial arteries reduces arterial diameter up to 12 per cent, suggesting a continuous basal release of nitric oxide in cerebral arteries. Similar evidence of basal release of EDRF has also been obtained in sheep cerebral arteries. Such findings reinforce the view that the vascular endothelium imparts a major tonic influence on vascular tone throughout the cerebrovascular tree. A corollary of this view is that any influence which tends to reduce the basal release of nitric oxide would also tend to increase basal cerebrovascular resistance.

One important pharmacological tool used to probe endothelium-mediated vasorelaxation is the calcium ionophore, A23187. Because this agent acts to enhance the transmembrane entry of extracellular calcium into the endothelial cell, it is receptor-independent and therefore can provide an estimate of maximal endothelial vasodilator capacity. Why A23187 acts preferentially on endothelial cells and not on vascular smooth muscle cells remains a mystery that is perhaps partially explained by differences in the lipid composition of endothelial and vascular smooth muscle cell membranes. Alternatively, some non-ionophoretic property of A23187 may also be involved. In isolated perfused rabbit cerebral arteries, A23187 produces vasorelaxation only when per-

fused intraluminally, and extraluminal application has little effect unless high concentrations are used, in which case constriction is observed. Nonetheless, endothelium-dependent relaxation to A23187 has been reported in the cerebral arteries of many species including mouse, dog, sheep, monkey and human. In general, this relaxation is unaffected by cyclooxygenase and lipoxygenase inhibitors, although a few reports to the contrary have been published. More importantly, the relaxation to A23187 is often more consistent and of greater magnitude than that produced by most receptor-dependent relaxants.

### Agonist-induced release of EDRF

The majority of endothelium-dependent relaxants mediate their effects through interaction with specific receptors on the endothelial cell membrane. Among these relaxants are neurohormones such as acetylcholine, histamine, and serotonin, and also peptides such as angiotensin II, bradykinin, substance-P, and vasopressin. For all of these agents, the extent of endothelium-dependent relaxation attained is a function of not only the local concentration of the relaxant, but also of the membrane concentration of receptors and the mechanisms coupling those receptors to EDRF synthesis and release. Because all receptor populations are labile, subject to turnover, and heterogeneously distributed, endothelium mediated relaxation to receptor-dependent relaxants also exhibits these characteristics.

One of the best studied of all endothelium-dependent receptor-mediated

relaxants is the neurohormone acetylcholine. It was the relaxant first shown by Furchgott and colleagues in the late 1970s to be endothelium-dependent. Since that time, acetylcholine has been shown to produce endothelium-dependent relaxation in cerebral arteries of many species including the mouse, rat, rabbit, cat, dog, pig, piglet, and human. Several species, however, are excluded from this list including monkeys and sheep which have cerebral arteries that relax poorly to acetylcholine. Of those species that respond well, acetylcholine-induced relaxation has been demonstrated in cerebral arteries of all sizes, from the larger arteries of pial circulation and circle of Willis, to the much smaller arterioles of the pial circulation. Typically, these relaxations are mediated by muscarinic receptors located predominantly on the luminal surface, although the exact receptor subtype involved remains controversial and may vary among species. In general, acetylcholine-induced relaxation of cerebral arteries is not affected by superoxide dismutase or catalase, is resistant to indomethacin, and is sensitive to both methylene blue and oxyhaemoglobin. Acetylcholine-induced relaxations, however, are also associated with hyperpolarizations up to 20 mV in rabbit and cat middle cerebral arteries, and can be attenuated by simultaneous application of bradykinin, suggesting that acetylcholine stimulates more that the release of just nitric oxide from cerebral endothelial cells.

Another neurohormone that stimulates endothelium-dependent relaxation of cerebral arteries is histamine. In the middle cerebral arteries of humans and monkeys, this relaxation appears to be mediated by endothelial $H_1$ histamine receptors. In contrast, endothelium-dependent relaxation to histamine in rat middle cerebral arteries has been attributed to endothelial $H_2$ receptors. In rabbit cerebral arteries, both $H_1$ and $H_2$ receptors appear to be present on the endothelium. Interpretation of these findings is complicated, however, by the presence of both $H_1$ and $H_2$ receptors on the underlying smooth muscle; 'pure' endothelial or vascular effects of histamine are unlikely. Similarly, serotonin has been reported to stimulate endothelium-dependent relaxation in some cerebral arteries, but interpretation of these findings is uncertain owing to the probable existence of serotonin receptors (5HT1 and/or 5HT2) in the vascular smooth muscle of most cerebral arteries.

Of the peptides that produce endothelium-dependent vasodilatation, bradykinin is probably the most widely studied. In cerebral arteries, bradykinin has now been reported to produce relaxation in the mouse, rat, cat, and human. In most cases, the relaxation induced by bradykinin appears to be sensitive to methylene blue but not indomethacin, suggesting a role for guanylate cyclase typical of that in most endothelium-dependent relaxation. In mouse and cat pial arterioles, however, the free radical scavengers superoxide dismutase and catalase inhibit bradykinin-induced relaxation, suggesting an additional role for the hydroxyl radical in these responses. Although the physiological importance of the free radical component of the endothelial response to bradykinin remains controversial, it now seems clear that the

relaxation responses to bradykinin involve multiple mechanisms in the cerebral circulation.

Substance-P is another peptide capable of stimulating endothelium-dependent relaxations in cerebral arteries. This response has been reported in cerebral arteries from the cat, dog, sheep, pig, and human and in some cases can be elicited by sub-nanomolar concentrations of the peptide. In dog cerebral arteries, the relaxations to substance-P appear to be mediated primarily by the endothelial release of nitric oxide. As for many peptidergic responses, endothelium-dependent relaxation to substance-P is rapid, short lasting, and exhibits a marked tachyphylaxis.

In dog cerebral arteries, the peptide vasopressin has been reported to stimulate endothelium-dependent relaxation via action on endothelial V1 receptors. As for serotonin and histamine, however, characterization of this response is complicated by the simultaneous presence of receptors on both the endothelium and smooth muscle of canine cerebral arteries. In other species, such as the sheep, vasopressin has no endothelium-dependent effect in cerebral arteries.

Angiotensin II is one of the most interesting of the peptides reported to stimulate endothelium-dependent relaxation in cerebral arteries. This response, which has been observed in both rat and rabbit cerebral arterioles, typically involves the release of nitric oxide, an increase in cyclic GMP, and is sensitive to methylene blue. The nitric oxide, however, can apparently be liberated from the terminal arginine residue of the angiotensin II molecule. Equally unique, the terminal ar-

ginine of angiotensin II can be liberated under certain conditions which may require cyclooxygenase activity. Although this response is both complicated and controversial, it is a novel suggestion that a receptor-mediated endothelium-dependent relaxant may also donate the substrate for nitric oxide formation.

Aside from the neurohormones and peptides, many other agents stimulate specific endothelial receptors to promote the release of EDRF. Among these are the purines ADP and ATP. Rabbit, cat, and human cerebral arteries all exhibit endothelium dependent relaxation to these relaxants, presumably through interaction with endothelial $P_2$ receptors. In cat and rabbit middle cerebral arteries, ADP also produces smooth muscle hyperpolarization, suggesting the endothelial release of some factor (presumably endothelium-derived hyperpolarizing factor) in addition to EDRF. As for histamine, serotonin, and vasopressin, receptors for ADP and ATP can be present on both the endothelium and smooth muscle of cerebral arteries, making characterization of endothelial ADP and ATP responses difficult. To complicate matters further, cerebral artery and endothelial $P_2$ receptors are also subject to tachyphylaxis.

Some members of the pyrimidine family, namely UDP and UTP, also appear to be capable of promoting endothelium-dependent relaxation. In human pial arteries, this response is rapid, transient and obtains between 0.1 and 10 μM. At higher concentrations, or in endothelium-denuded preparations, both UDP and UTP produce cerebral artery contractions. Once again, receptors for these agents are present on both the endothelium and vascular smooth muscle of cerebral arteries and the two populations of receptors produce opposite effects on cerebral artery tone.

In monkey cerebral arteries, low concentrations of prostaglandin F2-alpha have been reported to produce endothelium-dependent relaxations that can be inhibited by pretreatment with either arginine derivatives (L-NMMA) or oxyhaemoglobin. In addition, relaxation-inducing prostanoid receptors have been reported in feline basilar and middle and cerebral arteries. In guinea-pig basilar artery, the leukotriene LTD4 has been reported to stimulate the release of EDRF. Although all these effects of prostanoids on EDRF release remain controversial, they further illustrate the range and diversity of receptors which may be present on cerebrovascular endothelial cells.

Clearly, the endothelial cells of cerebral arteries can promote vascular relaxation in response to a wide variety of stimuli. From the large thrombin molecule, for which endothelial receptors have been reported in rabbit, dog, and human basilar arteries, to the very small magnesium ion for which concentration decreases in the physiological range (1.2 to 0.8 mM) promote endothelium-dependent relaxation in feline cerebral arteries, many different physiological signals are continuously integrated by cerebral endothelial cells to determine their output of endothelium-derived relaxing factors. By virtue of the diversity of these signals, the release of endothelium-derived relaxing factors must be labile, heterogeneous, and under some precise form of coordinated regulation. The exact nature of this regulation, however, remains unknown.

## Endothelin and its mechanism of action

In contrast to its capacity to promote the relaxation of contractile tone, the endothelium can also release factors which enhance contractile tone in cerebral arteries. A wide variety of endothelial products capable of producing vasoconstriction have now been identified, the most prominent of which is the peptide endothelin. Endothelin exists in three closely related forms, creatively named endothelins 1, 2, and 3, and distinct genes for each of the three forms have been identified. Endothelin-1 is first synthesized as a 203 amino acid chain which is then subsequently digested into a 39 amino acid sequence (Big Endothelin) and finally into the active 21 amino acid form which contains two disulphide bonds. Endothelin, in either its precursor or final form, is not stored intracellularly and thus appears to be synthesized 'on demand' in response to a variety of factors including thrombin, epinephrine, and phorbol esters. Circulating concentrations of endothelin are normally quite low ($\approx 1$ pg/ml) and the peptide degrades very slowly if at all. Endothelin appears to be removed from the circulation primarily by binding to high affinity sites which are most prominent in the lung, kidney, and liver.

Endothelin promotes vasoconstriction by binding directly to specific vascular smooth muscle receptors, of which subtypes may exist for each of the three endothelins. Interestingly, in both cat and dog basilar arteries these receptors appear to be preferentially located on the abluminal surface of the smooth muscle cells. The binding of endothelin to its receptor causes activation of vascular phospholipase-C which in turn promotes the synthesis of inositol triphosphate, a subsequent increase in cytosolic calcium, and vasoconstriction. In most arteries, including those of the cerebral circulation, endothelin-induced vasoconstriction is slow in onset, requires the presence of extracellular calcium, and can be attenuated but not blocked by nitric oxide-releasing vasodilators such as nitroglycerine or nitroprusside. In cat middle cerebral arteries, contractions produced by endothelin are associated with small depolarizations of up to 6 mV.

In some arteries, endothelin also appears to activate phospholipase A2 which then leads to liberation of arachidonic acid and synthesis of prostanoids. This effect may help to explain why cyclooxygenase inhibitors have been reported to either enhance (cat middle cerebral artery) or attenuate (newborn piglet pial arteries) endothelin-induced contractions. Exogenous endothelin may also stimulate the synthesis of vasodilator prostanoids within the endothelial cell, as suggested by the finding in newborn piglets that low concentrations of endothelin produce modest endothelium-dependent cerebral vasodilatation which can be blocked by indomethacin. In general, however, endothelin-induced contractions cannot be blocked by inhibitors of either cyclooxygenase or lipoxygenase.

Receptors for endothelin appear to be present in most arteries and many veins. In cerebral arteries, endothelin receptors have been demonstrated in the rat, rabbit, cat, goat, newborn and adult pig, cow, and human. In addition, endothelin receptors appear to be distributed among cerebral arteries of all sizes including the basilar, middle cerebral, pial arteries and arterioles. Most interestingly, relatively high densities of endothelin receptors are found in the rabbit choroid plexus, where endothelin has profound effects on choroid plexus blood flow at concentrations too low to elicit vasoconstriction in other cerebral arteries.

## Non-endothelin endothelium-derived contracting factors

In addition to endothelin, endothelial cells can also synthesize thromboxane $A_2$, the major vasoconstrictor product of the cyclooxygenase pathway. Evidence that cerebral endothelial cells may release this thromboxane has been obtained in dog cerebral arteries, where angiotensin I and A23187 have both been reported to stimulate endothelium-dependent contractions that could be blocked by inhibitors of thromboxane $A_2$ synthetase. Although the physiological stimuli which promote the release of these vasoconstrictor prostanoids remains unclear, endothelial receptors for acetylcholine, ADP, ATP, angiotensin, leukotriene $C_4$, oxyhaemoglobin, and prostaglandin $H_2$, have all been proposed to be coupled to the release of vasoconstrictor prostaglandins in various cerebral arteries.

Endothelial cells are also capable of generating superoxide anions, primarily through the cyclooxygenase pathway. Evidence that these anions may contribute to endothelium-dependent contraction has been ob-tained in dog cerebral arteries, where contractile responses to A23187 could be attenuated by either endothelium removal, cyclooxygenase, or superoxide dismutase, a superoxide anion scavenger. Because neither catalase, a scavenger of hydrogen peroxide, or deferoxamine, a scavenger of hydroxyl radicals, blocked the vasoconstrictor response, neither of these free radicals appears to be involved.

## Physiological stimuli promoting release of EDRFs and EDCFs

Not long after the discovery that endothelial cells could modulate vascular smooth muscle tone, it was proposed that the vasodilatation observed in response to acute hypoxia involved the release of EDRF, and observations consistent with this hypothesis were obtained in canine coronary artery preparations. In rabbit cerebral arteries, however, hypoxia apparently promotes the simultaneous release of both relaxing and contracting factors from the endothelium, and the ratio of these opposing influences varies among different cerebral arteries. In dog basilar arteries, for example, acute hypoxia produces only an endothelium-dependent vasoconstriction that can be blocked by organic calcium entry blockers. In sheep middle cerebral arteries, a similar response has also been reported which was insensitive to either superoxide dismutase or catalase, suggesting that free radicals are not involved.

Owing to the rapid time course and reversible nature of these contractions, endothelin release was probably also not involved. In both dog and rabbit cerebral arteries, however, indomethacin has been reported to attenuate or block endothelium-de-pendent hypoxic vasoconstriction, suggesting that thromboxane $A_2$ may be the endothelium-derived contracting factor released in response to acute hypoxia. A direct test of this hypothesis using thromboxane $A_2$ synthetase inhibitors, however, has yet to be performed.

Another physiological stimulus proposed to release endothelial vasoactive factors is stretching of the cerebral artery wall. In both canine basilar and cat middle cerebral arteries, stretching causes the development of active tone which can be abolished by endothelium removal. This active tone is associated with depolarizations of up to 30 mV and appears to be the result of the active endothelial release of a contracting factor, rather than the inhibition of release of an endothelium-derived relaxing factor. As stated above (see section on 'Transmural pressure and electromechanical coupling in cerebral arterial muscle' earlier), this putative contracting factor is probably not endothelin. Although the chemical identity of the contracting factor released remains unknown, it produces an increase in vascular cytosolic calcium (as shown by Fura II imaging) that can be blocked by organic calcium entry blockers.

Similar to stretch is the contractile response to increased luminal flow. In rabbit pial arteries, the magnitude of this response is graded with the rate of flow. The contraction is dependent on extracellular calcium and is thought to result from the shear stresses arising between the luminal fluid and the luminal surfaces of the endothelial cells. Although endothelium impairment has been reported to attenuate the response, the exact role of the endothelium remains controversial.

## Endothelial pathophysiology

Given the many important functions of the cerebral endothelium, it is obvious that any endothelial damage could compromise cerebrovascular, and possibly also cerebral, homoeostasis. For this reason, significant effort has been directed toward elucidating endothelial pathophysiology in a variety of different disease states. First among these is intracranial haemorrhage. Because haemoglobin can inhibit endothelium mediated relaxation, and basal release of EDRF is substantial, it is conceivable that the haemoglobin released by lysed erythrocytes could contribute to the cerebral vasospasm that so often accompanies intracranial haemorrhage. Consistent with this hypothesis, subarachnoid haemorrhage is associated with blood–brain barrier trauma characterized by swelling and thickening of the intima, vacuolization of endothelial cells, and disruption of tight junctions. Subsequent to the haemorrhage are ultrastructural vascular changes, subendothelial oedema, and medial necrosis. Correspondingly, most if not all endothelium mediated relaxation responses are depressed in the region of haemorrhage. Although a causal role for the cerebral endothelium in the vasospasm characteristic of post-haemorrhagic sequela has not yet been firmly established, it is clear that endothelial function is profoundly compromised by this insult.

Another common cerebrovascular trauma is that imposed by ischaemia and/or prolonged hypoxia. In rats, 6 h of cerebral ischaemia produce endothelial denudation in the small cerebral arteries which in turn may precipitate subsequent thrombus and oedema formation. After 12 h of middle cerebral artery occlusion in the rat, endothelial albumin and water permeability are increased for up to 3 days. During this period, blood–brain barrier permeability to small molecules increases gradually, as does the accumulation of oedematous fluid. With sustained hypoxia alone, the changes are less severe, but relaxation responses to most endothelium-dependent relaxants are still depressed. Most purely hypoxic changes in cerebral endothelial function, however, appear to be largely reversible.

Atherosclerosis also negatively impacts cerebral endothelial function. Cerebral endothelial cells appear less vulnerable to high cholesterol diets than larger arteries such as the aorta, but fatty streaks do form in cerebral arteries and the extent of these formations is well correlated with decreased responsiveness to endothelium dependent vasodilators. In squirrel monkey cerebral arteries, the endothelium appears preferentially permeable to low-density lipoprotein (LDL) and concentrations of injected LDL are highest in the luminal endothelium and lowest in the arterial medial layer. Low density lipoprotein receptors have also been observed in bovine cerebral endothelium.

Systemic hypertension is another disease state with ramifications for the cerebral endothelium. It has long been recognized that acute hypertension can disrupt the blood–brain barrier and this disruption now appears to be most prevalent in the cerebral veins and venules where it is characterized by an increased permeability to molecules of many different sizes, as is typical of enhanced vesicular transport. Chronic hypertension, on the other hand, is associated with both intimal and medial cerebral arterial abnormalities including functional changes in endothelial permeability resulting from increased pinocytotic activity, increased endothelial cell turnover, and distinct changes in endothelial cell morphology. In spontaneously hypertensive rats, the onset of hypertension is associated with transient depressions of endothelial sodium–potassium ATPase activity and reductions in calcium ATPase content. Compared to the cerebral arteries of normotensive rats, those from spontaneously hypertensive rats are less responsive to endothelium-dependent vasodilators such as acetylcholine and bradykinin. However, cerebrovascular responses to endothelium-independent vasodilators such as nitric oxide, nitroglycerine, and adenosine, are equivalent in normotensive and spontaneously hypertensive rats, suggesting that chronic hypertension induces a primary functional lesion in cerebral endothelial cells. In the extreme case, a variety of findings now indicate that sustained dilation of cerebral arteries with disruption of the blood–brain barrier, and not vasospasm, is the precipitating factor in the aetiology of hypertensive encephalopathy.

Malignant astrocytoma cells also effect an increase in cerebral blood–brain barrier permeability. This permeability increase, and its associated cerebral oedema, appears to be mediated by a protein (molecular weight > 10 kD) and possibly other as yet unknown factors that combine to enhance the endothelial abundance of sodium–potassium ATPase. Unfortunately, irradiation of these gliomas can further increase endothelial permeability and induce a delayed radiation necrosis characterized by decreased endothelial cellularity, impoverishment of endothelial nuclei, and increased pinocytotic activity.

Many other disease processes also affect the cerebral endothelium. HIV infections are associated with increased blood–brain barrier permeability. Cold injury induces increased pinocytotic activity, oedema and swelling of both cerebral endothelial cells and adjacent astrocytic foot processes. Similarly, fluid percussion injury and head trauma are both associated with increased blood–brain barrier permeability and loss of endothelium-dependent relaxation responses. Diabetes, a disease which has relatively little effect on cerebrovascular function, still produces a modest increase in the number of cytoplasmic vesicles in cerebral endothelial cells. Thus, the cerebral endothelium can be viewed as a relatively fragile interface between the circulating blood and cerebral vasculature, which is often functionally compromised in many diseases.

Even in the absence of disease, the normal processes of ageing effect many endothelial changes. In rats, advanced age is associated with approximately a 50 per cent decrease in the responses of cerebral arterioles to endothelium-dependent vasodilators. Again, this decrease occurs in the ab-

sence of any change in responses to endothelium-independent vasodilators such as nitroglycerine, suggesting that a primary change in endothelial function is part of the normal ageing process. Age-related decreases in endothelium-mediated relaxation responses to thrombin have also been observed in human cerebral arteries, and these have been attributed to decreases in the density, but not the type, of endothelial thrombin receptors. Similarly, the cerebral arteries from foetal and newborn lambs exhibit greater reactivity to ADP and acetylcholine, but less reactivity to nitric oxide and nitroglycerine, than corresponding arteries from adult sheep; development also is associated with important changes in cerebral endothelial function. In addition, adaptation to high altitude hypoxia is associated with enhancement of endothelium-mediated vasodilator responses to acetylcholine and substance-P in the ovine cerebral circulation. Thus overall, the cerebral endothelium is a functionally and structurally complex tissue, and by virtue of this complexity, its characteristics are labile and subject to the many influences of environment, maturation, and disease.

## Summary

In this chapter, we have seen that cerebral arteries are the complex functional units of the cerebrovascular circulation responsible for integrating a wide variety of physiological signals and effecting a tight regulation of both cerebral perfusion and the transfer of materials between the cerebral interstitum and the vascular compartment. The integrated physiological signals include short-term inputs such as changes in the local ionic environment, oxygen and carbon dioxide tensions, and neurohormone concentrations, as well as more complex stimuli such as wall stretching, transmural pressure, and luminal flow. Longer term inputs include hormonal influences acting directly on the vascular smooth muscle, directly on the vascular endothelium, or indirectly on the vascular smooth muscle through the stimulated release of endothelial growth factors. As effectors for cerebral perfusion, cerebral arteries possess the complex biochemical machinery necessary for contraction, and this is controlled as in most arteries through regulation of cytoplasmic calcium concentration and calcium sensitivity of the contractile proteins. As regulators of the exchange of materials between blood and the cerebral interstitum, cerebral arteries possess a highly specialized vascular endothelium with unique structural and biochemical characteristics.

Given the complexity inherent in the structure and function of cerebral arteries, it should not be surprising that the characteristics of these arteries are labile and change markedly in response to many different stimuli including changes in environment, ageing, and diet. Although our appreciation of the basic mechanisms responsible for cerebral artery function has advanced impressively in the past decade, we are only now coming to understand how these many mechanisms might be coordinated and governed as a whole. It is the understanding of this overall coordination of adaptive responses that will undoubtedly remain a goal of many future studies.

## References for further reading

### The biochemistry of vascular smooth muscle contraction

Adelstein, R.S. & Sellers, J.R. (1987): Effects of calcium on vascular smooth muscle contraction. *Am. J. Cardiol.* **59,**(3) 4B–10B.

Bell, D.R., Webb, R.C. & Bohr, D.F. (1985): Functional bases for individualities among vascular smooth muscles. *J. Cardiovasc. Pharmacol.* **7,** (Suppl. 3) S1–11.

Berridge, M.J. & Irvine, R.F.(1989): Inositol phosphates and cell signalling. *Nature* **341,** 197–205.

Grover, A.K. & Daniel, E.E (eds.) (1985): *Calcium and contractility*, pp. 1–487. Clifton, NJ: Humana Press.

Hai, C.M. & Murphy, R.A. (1989): $Ca^{2+}$, crossbridge phosphorylation, and contraction. *Annu. Rev. Physiol.* **51,** 285–298.

Itoh, T., Fujiwara, T., Kubota, Y., Nishiye, E. & Kuriyama, H. (1990): Roles of protein kinase C on the mechanical activity of vascular smooth muscles. *Am. J. Hypertens.* **3,** (8 Pt 2) 216S–219S.

Itoh, T., Ueno, H. & Kuriyama, H. (1995): Calcium-induced calcium release mechanism in vascular smooth muscles assessments based on contractions evoked in intact and saponin-treated skinned muscles. *Experientia* **41,**(8) 989–996.

Karaki H. (1990): The intracellular calcium-force relationship in vascular smooth muscle. Time- and stimulus-dependent dissociation. *Am. J. Hypertens.* **3,** (8 Pt 2) 253S–256S.

Khalil, R., Lodge, N., Saida, K. & van Breemen, C. (1987): Mechanism of calcium activation in vascular smooth muscle. *J. Hypertens.* Suppl. **5,**(4) S5–15.

Laher, I. & Bevan, J.A. (1989): Stretch of vascular smooth muscle activates tone and $45Ca^{2+}$ influx. *J. Hypertens. Suppl.* **7,**(4) S17–20.

Marin, J. (1988): Vascular effects of calcium antagonists. Uses in some cerebrovascular disorders. *Gen. Pharmacol.* **19,** (3) 295–306.

Morgan, K.G. (1990): The role of calcium in the control of vascular tone as assessed by the $Ca^{2+}$ indicator aequorin. *Cardiovasc. Drugs Ther.* **4,** 1355–1362.

Murphy, R.A., Rembold, C.M. & Hai, C.M. (1990): Contraction in smooth muscle: what is latch? *Prog. Clin. Biol. Res.* **327,** 39–50.

Murphy, R.A. (1989): Contraction in smooth muscle cells. *Annu. Rev. Physiol.* **51,** 275–283.

Ohnishi, S.T. & Endo, M. (eds.) (1981): *The mechanism of gated calcium transport across biological membranes*, pp. 1–325 New York: Academic Press.

Siegmen, M.J., Somlyo, A.P. & Stephens, N.L. (eds.) (1986): *Regulation and contraction of smooth muscle*, pp. 1–507. New York: Alan R. Liss.

Silver, P.J. (1985): Regulation of contractile activity in vascular smooth muscle by protein kinases. *Rev. Clin. Basic Pharm.* **5,** (3–4) pp. 341–395.

Somlyo, A.P. & B. Himpens. (1989): Cell calcium and its regulation in smooth muscle. *FASEB J.* **3,** 2266–2276.

Sperelakis, N. & Ohya, Y. (1990): Cyclic nucleotide regulation of $Ca^{2+}$ slow channels and neurotransmitter release in vascular muscle. *Prog. Clin. Biol. Res.* **327,** 277–298.

van Breemen, C. & Saida, K. (1989): Cellular mechanisms regulating $[Ca^{2+}]$ in smooth muscle. *Annu. Rev. Physiol.* **51,** 315–329.

Vanhoutte, P.M., Paoletti, R. & Govoni, S. (eds.) (1988): *Calcium antagonists: pharmacology and clinical research*, pp. 1–802. New York: New York Academy of Sciences.

Crass, M.F. & Barnes, C.D. (1982): *Vascular smooth muscle: metabolic, ionic, and contractile mechanisms*, pp. 1–205. New York: Academic Press.

### Cerebrovascular electrophysiology

Bonnet, P., Rusch, K.J. & Harder, D.R. (1991): Characterisation of an onward current in freshly dispersed cerebral artery muscle cells. *Pflügers Arch.* **418,** 292–296.

Christensen, H.N. (1979): Developments in amino acid transport, illustrated for the blood–brain barrier. *Biochem. Pharmacol.* **28,** 1989–1992.

Harder, D.R. & Madden, J.A. (1986): Membrane ionic mechanisms controlling activation of cerebral arterieal muscle. In: *Neural regulation of brain circulation*, eds C. Owman & J.E. Hardebo, pp. 81–91. New York: Elsevier.

Hille, B. (1984): *Ionic channels of excitable membranes*, pp. 1–426. Sunderland, MA: Sinauer Associates.

Hirst, G.D. & Edwards, F.R. (1989): Sympathetic neuroeffector transmission in arteries and arterioles. *Physiol. Rev.* **69**, 546–604.

Kitamura, K., Suzuki, H., Ito, Y. & Kuriyama, H. (1990): Similarity and diversity of electrical activity in vascular smooth muscles. *Prog. Clin. Biol. Res.* **327**, 257–276.

Kuriyama, H. & Kitamura, K. (1985): Electrophysiological aspects of regulation of precapillary vessel tone in smooth muscles of vascular tissues. *J. Cardiovasc. Pharmacol.* **7**, (Suppl. 3.) S119–128.

Nelson, M.T., Patlak, J.B., Worley, J.F. & Standen, N.B. (1990): Calcium channels, potassium channels, and voltage dependence of arterial smooth muscle tone. *Am. J. Physiol.* **259**, (1 Pt 1) C3–18.

Rusch, N.J. & Hermsmeyer, K. (1988): Measurement of whole-cell calcium current in voltage-clamped vascular muscle cells. *Mol. Cell. Biochem.* **80**, (1–2) 87–93.

## Cerebrovascular endothelium

Adams, D.J., Barakeh, J., Laskey, R. & C. Van Breemen. (1989): Ion channels and regulation of intracellular calcium in vascular endothelial cells. *FASEB J.* **3**, 2389–2400.

Bohme, E. & Schmidt, H.H. (1989): Nitric oxide and cytosolic guanylate cyclase: components of an intercellular signalling system. *Z. Kardiologie* **78**, (Suppl. 6) 75–79.

Bradbury, M.W. (1984): The structure and function of the blood–brain barrier. *Fed. Proc.* **43**, 186–190.

Busse, R., Pohl, U. & Luckhoff, A. (1989): Mechanisms controlling the production of endothelial autacoids. *Z. Kardiologie* **78**, (Suppl. 6) 64–69.

Busse, R., Trogisch, G. & Bassenge, E. (1985): The role of endothelium in the control of vascular tone. *Basic Res. Cardiol.* **80**, 475–490.

Forstermann, U. (1986): Properties and mechanisms of production and action of endothelium-derived relaxing factor. *J. Cardiovasc. Pharmacol.* **8**, (Suppl. 10) S45–S51.

Furchgott, R.F. (1990): The 1989 Ulf von Euler lecture. Studies on endothelium-dependent vasodilation and the endothelium-derived relaxing factor. *Acta Physiol. Scand.* **139**, 257–270.

Furchgott, R.F. & Vanhoutte, P.M. (1989): Endothelium-derived relaxing and contracting factors. *FASEB J* **3**, 2007–2018.

Ignarro, L.J. (1989): Biological actions and properties of endothelium-derived nitric formed and released from artery and vein. *Circ. Res.* 65: 1–21, 1989.

Ignarro, L.J. (1989): Endothelium-derived nitric oxide: actions and properties. *FASEB J.* **3**, 31–36.

Ignarro, L.J. (1989): Endothelium-derived nitric oxide: pharmacology and relationship to the actions of organic nitrate esters. *Pharmaceut. Res.* **6**, 651–659.

Ignarro, L.J. (1989): Heme-dependent activation of soluble guanylate cyclase by nitric oxide: regulation of enzyme activity by porphyrins and metalloporphyrins. *Semin. Hematol.* **26**, 63–76.

Ignarro, L.J. & Kadowitz, P.J. (1985): The pharmacological and physiological role of cyclic GMP in vascular smooth muscle relaxation. *Annu. Rev. Pharmacol. Toxicol.* **25**, 171–191.

Le Monnier de Gouville, A.C., Lippton, H.L., Cavero, I., Summer, W.R. & Hyman, A.L. (1989): Endothelin – a new family of endothelium-derived peptides with widespread biological properties. *Life Sci.* **45**, 1499–1513.

Long, C.J. & Berkowitz, B.A. (1989): What is the relationship between the endothelium derived relaxant factor and nitric oxide? *Life Sci.* **45**, 1–14.

Luscher, T.F. & Vanhoutte, P.M. (1990): *The endothelium: modulator of cardiovascular function*, p. 228. Boston: CRC Press.

Moncada, S., Palmer, R.M. & Higgs, E.A. (1989): Biosynthesis of nitric oxide from L-arginine. A pathway for the regulation of cell function and communication. *Biochem. Pharmacol.* **38**, 1709–1715.

Moncada, S., Palmer, R.M. & Higgs, E. (1988): The discovery of nitric oxide as the endogenous nitrovasodilator. *Hypertension* **12**, 365–372.

Murad, F. (1986): Cyclic guanosine monophosphate as a mediator of vasodilation. *J. Clin. Invest.* **78**, 1–5.

Rapoport, R.M. & Murad, F. (1983): Endothelium-dependent and nitrovasodilator-induced relaxation of vascular smooth muscle: role of cyclic GMP. *J. Cyclic Nucleotide Protein Phosphor. Res.* **9**, 281–296.

Ryan, U.S. (ed). (1990): *Endothelial cells*, Vol. II, p. 282. Boston, CRC Press.

Vane, J.R., Anggard, E.E. & Botting, R.M. (1990): Regulatory functions of the vascular endothelium. *N. Engl. J. Med.* **323,** 27–36.

Vanhoutte, P.M. (1989): Endothelium and control of vascular function. State of the Art lecture. *Hypertension* **13,** 658–667.

Vanhoutte, P.M., Auch-Schwelk, W., Boulanger, C., Janssen, P.A., Katusic, Z.S., Komori, K., Miller, V.M., Schini, V.B. & Vidal, M. (1989): Does endothelin-1 mediate endothelium-dependent contractions during anoxia? *J. Cardiovasc. Pharmacol.* **13,** (Suppl. 5) S124–S128.

Vanhoutte, P.M., Rubanyi, G.M., Miller, V.M. & Houston, D.S. (1986): Modulation of vascular smooth muscle contraction by the endothelium. *Annu. Rev. Physiol.* **48,** 307–320.

# Chapter 7

# Metabolism of the central nervous system

## Astrid Nehlig

*INSERM U398, Université Louis Pasteur, Faculté de Médecine, 11 rue Humann, 67000 Strasbourg, France*

## Introduction

A closed system in equilibrium neither requires nor supplies any kind of energy. The functionality of a system is highly dependent on its relative disequilibrium and its dynamics to reach equilibrium. This happens in living organisms which can only function continuously as open systems. Maintenance of these so-called 'metastable' conditions needs a continuous supply of energy. All living organisms, except in very short transient conditions, can be considered as steady-state systems in which the rate of energy production equals the rate of energy utilization.

The brain converts fuel energy into various kinds of work. How can brain work be defined, or, in other words, what are the mechanisms which might consume energy in the brain? First, brain work and functional activity mainly result from the maintenance of electrical activity and the restitution of ionic gradients after electrical discharge. This activity requires concentrating and pumping systems which are able to displace ions and other substances against electro-chemical gradients. Regular functional activity also involves synthetic processes, which manufacture various molecules required for normal function, structural integrity and turnover of various molecules.

The brain can be considered as one of the most metabolically active organs of the body and its specific nutritional requirements have been described in detail. The purpose of this review is to discuss brain energy metabolism, but mostly at the cellular compartment level, as other chapters of this book will be devoted to cerebral circulation and coupling between flow and metabolism in the brain. We will discuss first how the brain gets and transforms energy and what the cerebral energy consumption represents. Then, the cellular heterogeneity as well as brain function in normal energy states will be considered.

## How does brain get and transform energy?

### Specificities of brain energy metabolism

#### Brain oxygen consumption is high and is a limiting factor

The brain can be considered as one of the most active organs in the whole body. This quite high activity is reflected by a very high rate of oxygen uptake and consumption. Thus,

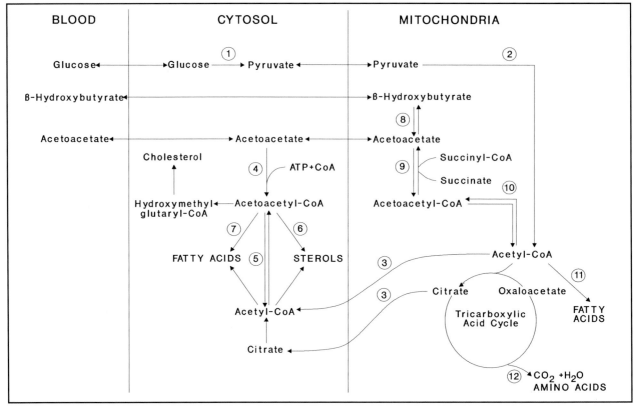

*Fig. 1. Cytosolic and mitochondrial pathways of glucose and ketone body metabolism.*
*1. Glycolysis. 2. Pyruvate dehydrogenase. 3. Transport of acetyl groups from the mitochondria to the cytosol (i.e. as citrate, acetyl-L-carnitine or N-acetylaspartate). 4. Acetoacetyl-CoA synthetase. 5. Cytosolic acetoacetyl-CoA thiolase. 6. Direct use of acetoacetyl-CoA for sterol biosynthesis. 7. Direct use of acetoacetyl-CoA for fatty acid biosynthesis. 8. β-Hydroxybutyrate dehydrogenase. 9. 3-Ketoacid-CoA transferase. 10. Mitochondrial acetoacetyl-CoA thiolase. 11. Mitochondrial fatty acid elongation system. 12. Mitochondrial amino acid biosynthesis.*

oxygen consumption of the brain represents about 3.5 ml/100 g/min, or 49 ml/min for the whole brain. An adult man weighing about 70 kg consumes approximately 250 ml oxygen/min. Thus, the brain oxygen requirement in adult man can account for about 20 per cent of that of the whole body for a relative weight of only 2 per cent and is highest as compared to other organs. Brain oxygen uptake is even higher in children in the first decade of life, where it reaches up to 50 per cent of the total oxygen supply to the body. In adults, in normal physiological conditions,

the brain takes up about 30 to 40 per cent of the total oxygen content of arterial blood flowing through it and the average daily brain oxygen consumption represents about 74 litres. The brain is thus considered as being tightly dependent upon circulation especially because of its high oxygen needs.

Oxygen is utilized in the brain almost exclusively for oxidation of glucose. The adult brain has an energy requirement of approximately 17 cal/100 g/min. For an adult brain weighing about 1500 g, the total en-

ergy requirement represents then about 250 cal/min or 20 watts, whereas the energy requirement of the whole body amounts to approximately 1275 cal/min. Whatever the energy is used for, since ATP is the universal medium for energy exchange, cells and especially brain cells derive their energy for any of their activities from ATP, storage of interconvertible and rapidly usable energy. ATP is derived from glucose oxidation. It has been calculated that the entire brain generates 38–48 mmol/min of ATP under non-stimulated conditions.

## Brain energy metabolism is dependent on glucose

Unlike many other organs, which are able to use different kinds of substrates for their energy metabolism and biosynthetic needs, the adult brain in normal conditions is almost exclusively dependent on glucose as substrate for its energy metabolism. Under normal conditions, no positive cerebral arteriovenous difference has been found for any other energy-yielding substance, and no other substrate has been shown to be able to reverse efficiently the effects of hypoglycaemia.

This almost exclusive use of glucose by the adult brain for oxidative metabolism under normal conditions is translated by a respiratory quotient of 1.0. Indeed, the normal conscious human brain consumes oxygen at a rate of 156 $\mu$mol/100 g tissue/min and produces carbon dioxide at the same rate. Thus, considering that for 1 $\mu$m of glucose completely oxidized to $CO_2$ and $H_2O$, 6 $\mu$mol of $O_2$ are consumed and 6 $\mu$mol of $CO_2$ produced, $O_2$ consumption and $CO_2$ production can account for a mean rate of glucose utilization of 26 $\mu$mol/100 g tissue/min. Measurements of glucose utilization lead to a rate of 31 $\mu$mol/100 g/min, which is even in excess by 5 $\mu$mol/100g/min over the amount needed for the total $O_2$ consumption recorded. The fate of the excess glucose is not precisely known, but is probably distributed in lactate, pyruvate and other metabolic intermediates and also used for biosynthetic processes.

Thus, approximately 7 per cent of the glucose taken up by the brain is converted to lactate and transferred into the lactate pool of the whole body. The functional importance of this lactate pool is not yet known exactly. About 30 per cent is degraded via the oxidative citric acid cycle for energy. The remaining 60 per cent are converted into amino acids and other neurotransmitters (see section: 'Interrelationships between brain energy metabolism, neurotransmitters and other neurochemical compounds').

Glucose appears then to be the fundamental and almost exclusive substrate of adult brain energy metabolism in normal conditions. However, one must notice that glucose is usually supplied to the brain in large excess. Indeed, in basic conditions, only 10 per cent of the arterial blood glucose content is taken up by the brain.

Nevertheless, because of its relatively low carbohydrate stores, brain activity relies on using very rapidly the glucose supplied by the arterial bloodstream. Indeed, the levels of true glycogen, 1.5–3.0 $\mu$mol/g, are quite small for a tissue so dependent on its sources of energy. The levels of intermediates of glucose metabolism are also quite low and there seems to be little or no intracellular glucose under normal conditions.

## Cerebral utilization of ketone bodies

It is well known that marked hypoglycaemia very rapidly leads to disturbances in cerebral function and even to coma. However, the brain, like many other organs, is metabolically adaptable, even in the choice of its energy metabolism substrates. Although it has never been demonstrated that the brain may be able to function without glucose, the brain has the capacity to utilize other substrates in specific conditions when glucose supply is limited.

Indeed, it was shown more than 20 years ago that, in human patients treated for severe obesity by complete fasting for several weeks, ketone bodies ($\beta$-hydroxybutyrate and acetoacetate), originating in hepatic fatty acid degradation, can also be utilized as energy sources in the neuronal and glial cells of the central nervous system. The consequence of this ketone body utilization is a large drop in the respiratory quotient. In this fasting situation, the rates of utilization of ketone bodies by the brain account for more than 50 per cent of the total oxygen consumption. The activity of the enzymes for metabolising ketone bodies in the adult brain is sufficient to oxidize $\beta$-hydroxybutyrate and acetoacetate to acetyl-CoA at the rate at which they are taken up from the blood.

Cerebral utilization of ketone bodies is directly related to their circulating level. Thus, in the adult brain in normal conditions, ketone levels in the blood and their use by the brain are almost negligible. However, in ketotic situations, such as starvation, fat-feeding or ketogenic diets, diabetes, uraemia, and hepatic insufficiency, cerebral utilization of ketone bodies is largely increased.

Ketone bodies are also actively used by the immature brain. In the rat, the newborn animal is transiently hypoglycaemic, and becomes rapidly ketotic with the onset of suckling, as a result of the high fat content of maternal milk. Ketosis lasts until weaning. Ketone bodies are taken up by the brain at rates 3 to 4 times higher in the

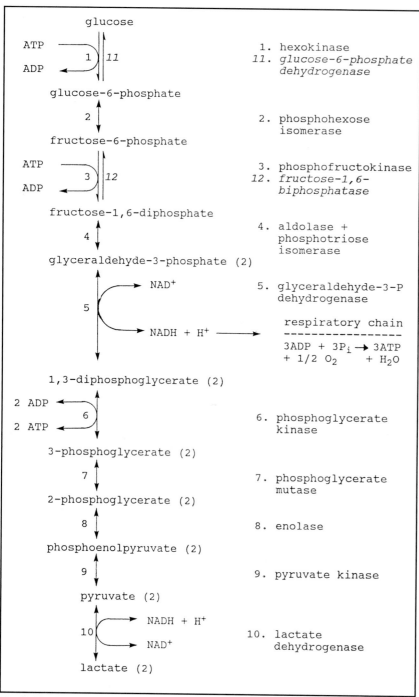

*Fig. 2. Glycolysis and gluconeogenesis (steps of gluconeogenesis are shown in italics).*

balance. In addition, the activity of enzymes of ketone body metabolism is well correlated with their active use during suckling and remains high enough to face ketosis in adulthood. The subcellular distribution of the different enzymes of ketone body utilization as well as the fate of these metabolic substrates in the brain are shown in Fig. 1.

In the developing brain, ketone body utilization appears then to be a physiological mechanism, whereas in the adult brain it may be considered as a 'survival mechanism' rather than a normal metabolic situation. It should however be stressed that β-hydroxybutyrate and acetoacetate alone are not able to maintain normal brain function. Thus, although ketone bodies can partially replace glucose, they cannot fully satisfy brain energy requirements in the total absence of glucose. Indeed, after weeks of starvation, the human brain reduces its normal daily glucose consumption (100–115 g), but still uses about 50 g of glucose.

## Cerebral metabolism of glucose and biosynthesis of energy

In this section, we will discuss the main metabolic routes of glucose in the brain, glycolysis and gluconeogenesis, the tricarboxylic acid cycle and the transport of electrons along the respiratory chain as well as the pentose phosphate pathway, glycogenesis and glycogenolysis.

The routes of glucose metabolism in the brain are similar to those which are generally operating in all mammalian cells. Utilization of glucose by the brains of large mammals (cat, dog or man) *in vivo* normally proceeds at a

immature rat or in the newborn human infant than in the adult and may represent as much as 30 to 70 per cent of the total energy metabolism

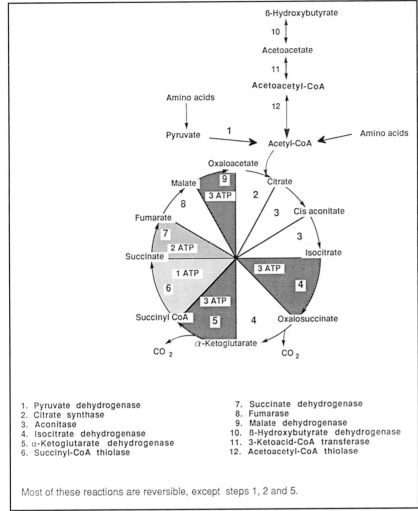

1. Pyruvate dehydrogenase
2. Citrate synthase
3. Aconitase
4. Isocitrate dehydrogenase
5. α-Ketoglutarate dehydrogenase
6. Succinyl-CoA thiolase
7. Succinate dehydrogenase
8. Fumarase
9. Malate dehydrogenase
10. ß-Hydroxybutyrate dehydrogenase
11. 3-Ketoacid-CoA transferase
12. Acetoacetyl-CoA thiolase

Most of these reactions are reversible, except steps 1, 2 and 5.

*Fig. 3. The tricarboxylic acid cycle: biosynthesis of ATP.*

rate ranging from 10 to 41.5 μmol/100 g/min, with a mean value of 30 μmol/100 g/min. About 5 mol/100 g of glucose are usually converted to 10 μmol/100 g lactate in each minute. In rodents, the rate of glucose utilization is about twice as high as in larger species and ranges from 50 to 80 μmol/100 g/min.

## Glycolysis

The first step of glucose utilization is glycolysis which occurs in the cytosol, and leads, from 1 mol of glucose, to the formation of 2 mol of pyruvate (in aerobic conditions) or 2 mol of lactate (in anaerobic situations) (Fig. 2). Glycolysis normally occurs at a rate of 20 μmol of glucose utilized per g tissue per h, but may under maximal stimulation reach values as high as 1500–2000 μmol of lactate produced per h, equivalent to 1000 μmol glucose used per g tissue per h. Such rates can only be maintained for very short periods and imply that glycolytic enzymes, which normally react at only a low percentage of their maximal activity, operate then close to their maximal capacities.

The rate of glycolysis is controlled at three enzymatic steps, at the level of hexokinase, phosphofructokinase and pyruvate kinase, the major step of regulation being the one catalysed by phosphofructokinase. The regulation of hexokinase and phosphofructokinase is coupled to and partly mediated by glucose-6-phosphate and by the interactions with the adenine nucleotides which they share. The activity of pyruvate kinase is also inhibited by ATP, as are the two other enzymes.

The result of glycolytic reactions leading to the synthesis of pyruvate from glucose can be summarized as follows:

$$Glucose + 2P_i + 2ADP + 2NAD^+ \Rightarrow$$
$$2\,Pyruvate + 2\,ATP + 2NADH$$

At this level, the anaerobic and aerobic pathways diverge. In the presence of oxygen, pyruvate enters the tricarboxylic acid cycle and is further oxidized to $CO_2$ and $H_2O$. In anaerobic conditions, pyruvate is reduced to lactate by lactate dehydrogenase and the net result of glycolysis in the absence of oxygen is then:

$$Glucose + 2ADP + 2P_i \Rightarrow$$
$$Lactate + 2ATP$$

$$\Delta G^{\circ\prime} = -47.0\ kcal/mol$$

Lactate production of the brain under aerobic conditions accounts for only 0 to 4 per cent of the whole glucose metabolism. $\Delta G^{\circ\prime}$ represents the free energy change occurring during a reaction. The maximal work performance of a reaction under standard conditions is given by its $\Delta G^{\circ\prime}$ value. For

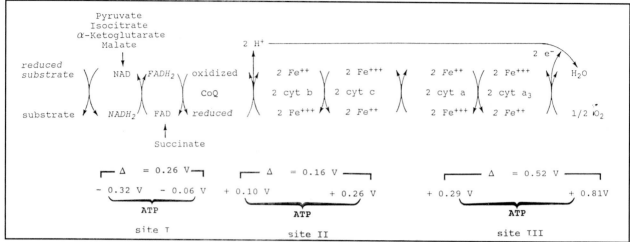

*Fig. 4. Mitochondrial respiratory chain (reduced forms are in italic).*

The following sequence affects the passage of hydrogen ions removed from the substrate to oxygen: NAD dehydrogenase, FAD flavoprotein, coenzyme Q (CoQ), also called ubiquinone, followed by the cytochrome (cyt) chain, in sequence cytochromes b, c, a and $a_3$. The latter transfers two electrons to oxygen.

Each potential increase of about 0.16 volt liberates enough energy to synthesize a high-energy bond and form a molecule of ATP from ADP and phosphate. Thus, three ATP are formed during the transfer of hydrogen and electrons between NAD and oxygen, two ATP between flavoproteins and oxygen, and one ATP between cytochromes and oxygen. The amount of energy liberated from NADH oxidation would be sufficient to synthesize 7 mol of ATP, but the actual yield of ATP is only 3 mol of ATP formed per mol of NADH oxidized. The three segments of the respiratory chain that provide energy to generate ATP by oxidative phosphorylation are called energy-conserving sites, sites I, II and III respectively.

example, the oxidation of glucose to lactate is an exergonic reaction, i.e. it proceeds with a decrease in free energy.

## Gluconeogenesis

In most tissues, such as the liver for example, the glycolytic pathway may be reversed and then easily lead back to the formation of glucose from lactate. However, the gluconeogenesis pathway relies on the activity of two key enzymes, glucose-6-phosphatase and fructose-1,6-biphosphatase, which allow the two unidirectional reactions of glycolysis, catalysed by h e x o k i n a s e          a n d phosphofructokinase, to be bypassed (Fig. 2).

For quite a long time, gluconeogenesis was thought not to occur in mam-

malian brain cells, due to the very low activity of glucose-6-phosphatase and to the apparent absence of fructose-1,6-diphosphatase. However, some gluconeogenesis has been shown to occur in the brain, although at a very slow rate as compared to glucose consumption. Indeed, pyruvate is incorporated into cerebral glycogen at rates equal to about 3 per cent of those by which glucose is incorporated into glycogen. Brain fructose-1,6-diphosphatase activity has also been demonstrated and characterized. However, the maximal activity of gluconeogenesis reaches a value of 1 to 4 $\mu mol/g/h$ *in vitro*, which is very low compared to the activity of glycolytic enzymes which ranges from about 350 to 4500 $\mu mol/g/h$.

## The tricarboxylic acid cycle

The second sequence of reactions leading to complete glucose oxidation is the tricarboxylic acid cycle which fully oxidizes, in the mitochondria, the pyruvate molecule formed in the cytosol to $CO_2$ and $H_2O$ (Fig. 3).

The first stage of the tricarboxylic acid cycle is catalysed by a cluster of enzymes called the pyruvate dehydrogenase complex. By this reaction, pyruvate undergoes oxidative decarboxylation, a dehydrogenation process in which the carboxyl group disappears as $CO_2$ and the acetyl group appears as acetyl-CoA.

$$\text{Pyruvate} + \text{NAD}^+ + \text{CoA-SH} \Rightarrow$$
$$\text{Acetyl-CoA} + \text{NADH} + CO_2$$

$$\Delta G^{\circ\prime} = -8.0 \text{ kcal/mol}$$

**Table 1. States of respiratory control**

|  | Conditions limiting the rate of respiration |
| --- | --- |
| State 1 | Availability of ADP and substrate |
| State 2 | Availability of substrate only |
| State 3 | The capacity of the respiratory chain itself in the presence of saturating amounts of substrates and components |
| State 4 | Availability of ADP only |
| State 5 | Availability of oyygen only |

This reaction is irreversible in mammalian tissues. Acetyl-CoA is a high-energy compound with a free energy of hydrolysis equivalent to that of ATP which is probably used to make the condensation of acetyl-CoA and oxaloacetate irreversible in the forward direction.

As shown in Fig. 3, not only pyruvate but also amino acids and ketone bodies can enter the tricarboxylic acid cycle via acetyl-CoA. β-Hydroxybutyrate and acetoacetate are interconvertible through a reaction catalysed by the mitochondrial enzyme β-hydroxybutyrate dehydrogenase. Then acetoacetate is converted to acetoacetyl-CoA in the presence of succinyl-CoA by 3-ketoacid-CoA transferase. In the last reaction of this sequence, two molecules of acetyl-CoA are formed from acetoacetyl-CoA and CoA (see Fig. 1).

In contrast to glycolysis which takes place as a linear sequence, the tricarboxylic acid cycle functions in a cyclic manner. At the end of each cycle, oxaloacetate is regenerated and is ready to react again with another mol of acetyl-CoA to start a new cycle. In each turn of the cycle, one mol of acetyl-CoA enters and two mol of $CO_2$ leave. The whole sequence of the tricarboxylic acid cycle can be summarized as follows:

$$Pyruvate + 4NAD^+ + FAD + GDP + P_i$$
$$\Rightarrow 3CO_2 + 4NADH + FADH_2 + GTP$$

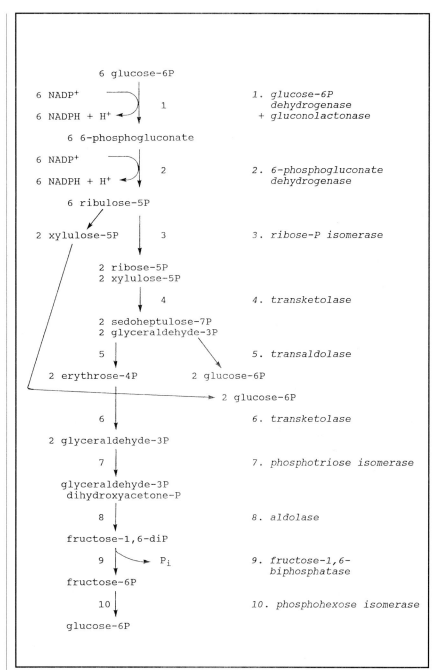

*Fig. 5. Pentose phosphate activity.*

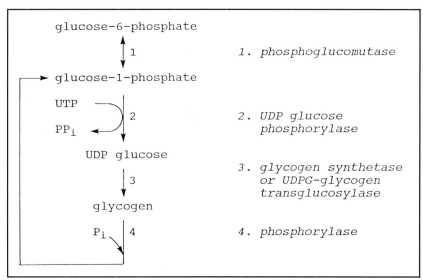

*Fig. 6. Glycogenesis and glycogenolysis.*

The availability of pyruvate as a source of acetyl-CoA as well as that of oxaloacetate and of the coenzymes in the correct redox state represent major limitations of overall rates of the tricarboxylic acid cycle. Likely control points of the cycle are also located at the isocitrate, α–ketoglutarate and succinate dehydrogenase steps.

## The respiratory chain and oxidative phosphorylation

As a result of oxidations catalysed by the dehydrogenases of the tricarboxylic acid cycle, 3 mol of NADH and one of $FADH_2$ are produced for each mol of acetyl-CoA oxidized. Oxidation of these reduced nucleotides back to their original form is of great importance because it occurs via a chain of reactions through which the energy of the oxidations is retained in utilizable form. During passage along the chain, reducing equivalents from NADH will generate three high-energy phosphate bonds by the esterification of ADP to ATP in the process of oxidative phosphorylation. However,

$FADH_2$ produces only two high-energy phosphate bonds, because it transfers its reducing power to ubiquinone or coenzyme Q, thus bypassing the first site for oxidative phosphorylation in the respiratory chain (Fig. 4). A further high-energy phosphate is generated at the level of the cycle itself during the conversion of succinyl-CoA to succinate. Thus, 12 new high-energy phosphate bonds are generated for each turn of the cycle.

The conservation of the energy of mitochondrial oxidations occurs at the level of the respiratory chain (Fig. 4). It is made possible by the large decrease of free energy during the oxidation of NADH and $FADH_2$. For example, oxidation of NADH can be summarized as follows:

$$NADH + 1/2\, O_2 + H^+ \Rightarrow NAD^+ + H_2O$$

$$\Delta G^{\circ\prime} = -51.3\ \text{kcal/mol}$$

This large free energy change is the expression of the fact that NADH has a very low and oxygen a very high affinity for electrons. The amount of en-

ergy liberated from NADH oxidation would be sufficient to synthesize 7 mol of ATP (ATP synthesis corresponds to a $\Delta G^{\circ\prime}$ of 7.3 kcal), but the yield is only 3 mol ATP formed per mol NADH oxidized.

During glycolysis there is a net formation of two high-energy phosphate bonds, equivalent to approximately 17.7 kcal/mol of glucose. Conversely, the complete oxidation of one molecule of glucose yields about 687 kcal. All the phosphorylations occur at the substrate level. In intact respiring mitochondria, oxidation of substrates via an NAD-linked dehydrogenase leads to the synthesis of 3 mol of ATP from 3 mol of ADP per 1/2 mol of $O_2$ consumed, i.e. the P:O ratio = 3. When substrates are oxidized via a flavoprotein-linked dehydrogenase, 2 mol of ATP are formed and the P:O ratio = 2. These reactions are known as oxidative phosphorylation at the level of the respiratory chain. Thus, it can be considered that about 46 per cent of the free energy resulting from the complete oxidation of glucose is captured as high-energy phosphate. The remaining free energy is liberated as heat and contributes to the maintenance of body temperature. There must be a redox potential of about 0.2 volts or a free energy change of approximately 8.9 kcal between components of the respiratory chain at the site of formation of each mol of ATP.

The rate of respiration of mitochondria is controlled by the level of ADP which is, with oxygen, an essential component of the phosphorylation process. Three sites of phosphorylation have been identified metabolically on the respiratory chain which coincide with the sites identified on thermodynamic grounds. These have been called sites I, II and III. Site I is the NADH–coenzyme Q reductase complex; site II is the reduced coenzyme Q-cytochrome c reductase complex; site III is the cytochrome c oxidase complex (Fig. 4).

Oxidation and phosphorylation are tightly coupled in mitochondria and respiration cannot occur via the respiratory chain without the concomitant phosphorylation of ADP. Five different states of respiratory control have been defined by Chance and Williams (1956) which control the rate of respiration in mitochondria (Table 1).

Usually, in the resting state, cells are in state 4 in which respiration is controlled by the availability of ADP. When work is performed, ATP is converted to ADP, allowing more respiration to occur which in turn leads to the synthesis of ATP.

## Mechanisms of oxidative phosphorylation: the chemiosmotic theory

Two main hypotheses have been advanced to explain the coupling between respiration and phosphorylation.

The chemical hypothesis postulates direct chemical coupling at all stages of the process, as in the reactions that generate ATP during glucose oxidation. This hypothesis postulates the existence of a high-energy intermediate (I~X) linking oxidation with phosphorylation. But this intermediate has never been isolated and this theory is no longer considered.

The chemiosmotic theory of Mitchell (1979) postulates that oxidation of components in the mitochondrial respiratory chain generates hydrogen ions which are ejected at the outside of the mitochondrial membrane. The electrochemical difference of potential resulting from the asymmetric distribution of the hydrogen ions (both a chemical potential originating in the difference in pH and an electrical potential) is used to drive a membrane-located ATP synthetase which in the presence of ADP and $P_i$ forms ATP.

## Pentose phosphate pathway

At the level of glucose-6-phosphate, there are three diverging pathways. The most prominent one in the brain is glycolysis which represents a very high proportion of all the glucose used. The other two ones are the pentose phosphate pathway (Fig. 5) and glycogenesis (Fig. 6).

In the pentose phosphate pathway, three molecules of glucose-6-phosphate give rise to three molecules of $CO_2$ and three molecules of ribulose-5-phosphate. The latter are rearranged to regenerate two molecules of glucose-6-phosphate and one molecule of glyceraldehyde-3-phosphate by a series of reactions involving mainly two enzymes, transketolase and transaldolase (Fig. 5). As in glycolysis, oxidation is achieved by dehydrogenations, but in the presence of NADP as hydrogen acceptor.

The result of the reactions of the pentose phosphate pathway can be summarized as in the equation below:

$$3 \text{ Glucose-6P} + 6 \text{ NADP}^+ \Rightarrow 3 CO_2 + 2 \text{ Glucose-6P} + \text{Glyceraldehyde-3P} + 6 \text{ NADPH} + 6 H^+$$

The enzymes of this metabolic pathway are located in the cytoplasm of the cell. The major functions of the pentose phosphate pathway are to provide the cell with both ribose for nucleotide and nucleic acid biosynthesis and NADPH for reductive syntheses outside the mitochondria. Indeed, it is generally assumed that this pathway is the main source of NADPH. This NADPH would be mainly used for lipid biosynthesis and for the reduction of glutathione. Glutathione is needed in its reduced form to maintain the structural integrity of many membrane proteins and enzymes.

Under basal conditions, at least 5 to 8 per cent of brain glucose is likely to be metabolized by the pentose phosphate pathway in the adult monkey and 2 to 3 per cent in the adult rat. The pathway is however much more active during development, reaching a peak during myelination, certainly because of its major contribution to synthesis of NADPH required for membrane lipid biosynthesis.

Flux through the pentose phosphate pathway is apparently regulated by the concentrations of glucose-6-phosphate, glyceraldehyde-3-phosphate, fructose-6-phosphate and NADPH. Turnover rates of this pathway decrease in situations of increased energy requirement, for example during and after high rates of stimulation.

## Glycogenesis and glycogenolysis

From glucose-6-phosphate, another alternative metabolic route is glycogen synthesis (Fig. 6). The first step of this pathway is the conversion of glucose-6-phosphate to glucose-1-phosphate by phosphoglucomutase. The speed of this reaction allows an

**Table 2. Yield of ATP from various substrates under aerobic and anaerobic conditions**

| Substrate | Conditions | ATP yield (mol) per mol of substrate used |
|---|---|---|
| Glucose | Aerobic, complete oxidation | 38 |
| Glucose | Anaerobic, conversion to lactate | 2 |
| Glycogen | Aerobic, complete oxidation | 39 |
| Glycogen | Anaerobic, conversion to lactate | 3 |
| Acetoacetate | Aerobic, complete oxidation | 24 |

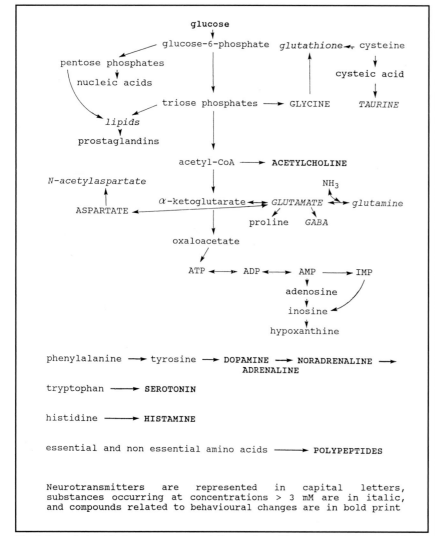

Neurotransmitters are represented in capital letters, substances occurring at concentrations > 3 mM are in italic, and compounds related to behavioural changes are in bold print

*Fig. 7. Interrelationships between energy metabolism, amino acids, neurotransmitters and adenine nucleotides.*

equilibrium to be maintained between the two phosphorylated hexoses in which glucose-6-phosphate is preponderant and represents 95 per cent. Then, glycogen synthesis occurs by means of two non-equilibrium reactions (Fig. 6). Since in this chain of reactions, UTP is synthesized at the expense of ATP, the incorporation of one molecule of glucose into glycogen costs two molecules of ATP.

Glycogen breakdown is initiated by the action of the enzyme phosphorylase which induces successive ruptures of the –1,4– linkages of glycogen to yield glucose-1-phosphate units.

Under normal circumstances, the concentration of glycogen within the brain remains constant and low, but seems to undergo quite a fast turnover. The glycogen level within the brain ranges from 2 to 4 µmol/g whereas it reaches 20–25 µmol/g in muscle and about 250 µmol/g in the liver. Although glycogen stores in the brain can only support a very short period of activity, cerebral glycogen metabolism is of great importance. Indeed, glycogen represents a store of glucose to be used, e.g. in situations where transport of glucose from blood to tissue becomes limiting. Glycogen stores are also very sensitive to various endocrine, pharmacological or pathological influences.

## High-energy phosphates and energy supply to the brain

### ATP production from glucose oxidation

The conversion of 1 mol of glucose to pyruvate is associated with the net production of 2 mol of ATP. Two mol of NADH are produced during the conversion of one mol of glucose to pyruvate, which will result in the synthesis of 6 mol of ATP. In the conversion of 2 mol of acetyl-CoA, 2 mol of NADH are produced, which lead to the production of 6 mol of ATP by oxidation. Glucose produces 2 mol of acetyl-CoA, via glycolysis and the pyruvate dehydrogenase reaction, and the oxidation of acetyl-CoA via the tricarboxylic acid cycle produces 24 mol of ATP.

Thus, the complete oxidation of 1 mol of glucose under aerobic conditions produces 38 mol of ATP. If glycogen is used as the substrate for glycolysis, 39 mol of ATP will be produced for the complete oxidation of every glucose equivalent. The oxidation pathways appear then to be 19 times as efficient for the production of ATP than anaerobic glycolysis (Table 2).

However, the yield of ATP presented in Table 2 represents a maximum. Indeed, several reactions contribute to a decrease in net ATP production from glucose oxidation. In man, 1 to 4 per cent of the glucose is used in the brain via anaerobic glycolysis and appears as a net output of lactate. Moreover, some glucose enters the pentose phosphate pathway and from the tricarboxylic acid cycle, there is a very efficient biosynthesis of amino acids. Thus, it can be calculated that the true ATP yield from the oxidation of 1 mol of glucose would be rather 30–35 mol, of which 1 per cent at the most is provided by glycolysis.

Between different organs of the body, the situation may vary widely. For example, isolated perfused rat heart consumes 6–9 μmol $O_2$/g wet weight per min under moderate work and derives about 3 per cent of its ATP from glycolysis. In the kidney, $O_2$ uptake is 3–6 μmol/g/min but glycolysis represents only 5 per cent of the energy synthesized. Conversely, vascular smooth muscle uses little $O_2$, only about 0.10 μmol/g/min, but derives around 30 per cent of the total ATP produced from glycolysis.

## Phosphocreatine acts as an energy reserve

The amide phosphate of creatine is an important form of storage of high energy phosphate in the brain. The levels of phosphocreatine (PCr) range from 3.0 to 5.0 μmol/g wet weight and fall rapidly during ischaemia *in vivo* or at the expense of ATP during electrical stimulation *in vitro*. PCr is more labile than ATP and acts as an energy reserve for it. Energy exchange between the ATP and PCr is mediated by ATP-creatine phosphotransferase, also called creatine kinase, which catalyses the following reaction:

$$PCr + ADP + H^+ \rightleftharpoons creatine + ATP$$

As shown in the last decade, PCr and creatine kinase may play quite an important role in energy transfer through cells. Indeed, the total creatine content of excitable tissues, both as creatine (Cr) and phosphocreatine (PCr), is generally at least three- to fourfold greater than that of the total adenylates (ATP + ADP + AMP). The enzyme is distributed equally between cytosol and mitochondria with only minor differences in kinetic and physical properties of both forms. At physiological pH, the equilibrium of the reaction is in favor of ATP which means that any ADP formed will be immediately rephosphorylated to ATP. Indeed, PCr has a $\Delta G^{\circ\prime}$ of hydrolysis of –10.3 kcal/mol, somewhat greater than the $\Delta G^{\circ\prime}$ for ATP hydrolysis. Thus, ADP and AMP levels in the brain increase appreciably only after the PCr reserve has decreased to a low level.

In conclusion, glycogen, glucose, ATP and PCr represent the primary total energy stores of the brain. Total ATP levels (about 3 μmol/g) would be depleted in about 6 s at an estimated turnover rate of 0.5 μmol/s/g. However, the presence of adenylate kinase in brain tissue enables two ADP molecules to generate one additional ATP, which would bring the maximal ATP availability to 9 s. PCr, in the presence of creatine kinase, could regenerate about 5 μmol/g ATP, so that total ATP supplies would last about 20 s in the absence of glucose and glycogen supply.

## Interrelationships between brain energy metabolism, neurotransmitters and other neurochemical compounds

### Amino acids

It was shown more than 30 years ago that the fundamental characteristic of mature brain energy metabolism *in vivo* is the active and fast incorporation of glucose carbon into amino acids, mainly those of the glutamate group, glutamate, glutamine, aspartate, and GABA. Indeed, as shown in Fig. 7, brain composition differs from that of other organs because of the high concentration of these four amino acids, about 10 μmol/g wet weight for glutamate, 5 μmol/g for glutamine and 2.5 μmol/g for both aspartate and GABA. They are derived from intermediates of the tricarboxylic acid cycle via the reactions catalysed by aspartate and alanine aminotransferases. They play an important role as neurotransmitters and precursors of other cerebral amino acids.

Glutamate metabolism in the brain has also been shown to occur in at least two different compartments. In fact, two distinct tricarboxylic acid cycles have been shown to operate, each linked to a separate pool of glutamate. The large pool of glutamate, located in neurons and nerve endings, leads to the synthesis of a small pool of glutamine and to GABA and is related to brain energy metabolism itself. The small pool of glutamate, located in glial cells, leads to the synthesis of a large active pool of glutamine which diffuses back to the neurons, which are impermeable to glutamate, and acts as a precursor pool for glutamate and GABA and thus would rather be related to neurotransmission.

Arising from the metabolism of the essential amino acid methionine, and related to glycine, are cysteine and taurine. Glycine and taurine are both inhibitory neurotransmitters.

## Neurotransmitters

The link between glucose metabolism and acetylcholine biosynthesis is indirect, glucose acting as a source of energy-rich substances, of which acetyl-CoA is, with choline, the immediate precursor of acetylcholine (Fig. 7). This reaction is catalysed by choline acetyl transferase. The average cerebral content of acetylcholine is about 7 nmol/g in the corpus callosum up to 60–80 nmol/g wet weight in the caudate nucleus.

Catecholamines, serotonin and his-

tamine can be related not only because of their chemical structure associated with neurotransmission, but also because of their occurrence in distinctive cell types and pathways in the brain. These different substances are all formed from three essential aromatic amino acids, phenylalanine for dopamine, noradrenalin and adrenalin, tryptophan for serotonin and histidine for histamine (Fig. 7). These compounds are neurotransmitters and vasoactive agents. They are present in the brain in nanomolar concentrations in most regions, except for dopamine in caudate nucleus and noradrenalin in hypothalamus where they both reach micromolar amounts.

## Adenosine and related substances

It is often considered that the adenine nucleotide system represents a pool of constant size composed of ATP + ADP + AMP that can be charged by oxidative phosphorylation or discharged during performance of work. However, it has been shown that during stimulation, adenine nucleotides are degraded and their breakdown occurs at the level of AMP, which leads to the formation of adenosine, inosine and hypoxanthine in the respective proportions of 54, 14 and 13 per cent (Fig. 7). Adenosine has been shown to act as a neuromodulator, inhibiting the release of most neurotransmitters and acting on specific well-characterized receptors, both neuronal and vascular.

## Techniques for measurement of brain energy metabolism

A wide variety of methods has been used to study brain energy metabolism. We will only recall some of them and describe them.

## Biochemical techniques

In these methods, determinations of changes in carbohydrate metabolites and energy cofactor stores are assessed in microdissected samples in animals exposed to specific experimental conditions. They assume that the rates of change reflect those before ischaemia. This approach has permitted the study of intermediate metabolism such as the rates of flux of various pathways and the identification of control points and has allowed the study of changes in energy metabolism in pathological situations.

## Incorporation of radioactivity from substrates into metabolites

In these techniques, at a given time after the administration of a radioactive precursor to the animal, the brain is removed and the precursor as well as the products of interest are isolated and assayed for concentration and radioactivity. These techniques allow the identification of metabolic pathways and rates of flux through different steps of the pathway. They have been used for the study of neurotransmitter, lipid and amino acid metabolism and biosynthesis as well as protein synthesis.

## Measurement of brain oxygen utilization

Brain oxygen consumption has specifically been measured by means of an oxygen electrode. Changes in local partial pressure of oxygen ($pO_2$) during pathological states can be measured continuously by means of an oxygen electrode applied to the surface of the exposed brain cortex.

## Arteriovenous differences of various substances across the brain

During brain function, substrates are taken up by the brain from the blood and some products are released by the brain into the circulation. Since arterial blood has essentially the same composition within the whole body, arterial blood can be sampled from any artery or even from the left ventricle of the heart. Venous blood must be representative of brain metabolism and can be readily obtained, with only about 3 per cent extracerebral contamination, from the superior bulb of the internal jugular vein in humans, which can be punctured under local anaesthesia to allow measurements in fully conscious subjects. For many laboratory animals, since there is communication between cerebral and extracerebral blood, venous blood is sampled at the confluence of the saggital sinuses, even though it is not representative of some parts of the brainstem. These techniques are quite simple but give no quantitative measurement of the rates of utilization or production of a given substance, since arteriovenous differences depend both on the rates of utilization or production and on blood flow.

Indeed, when a substance is exchanged between brain and blood, its rate of uptake by the brain from the arterial blood or release to the venous blood in a steady state is equal to its utilization or production by the brain. This can be written as follows:

$$CMR = CBF\,(C_A - C_V)$$

where CMR is the cerebral metabolic rate of utilization or production of the substance, CBF the cerebral blood flow and $(C_A - C_V)$ the cerebral arteriovenous difference. If the rate of cerebral blood flow and the arteriovenous difference are known, the net rate of utilization or production of a substance by the brain can be calculated. This method has been the basis of many quantitative studies of brain metabolism *in vivo* and has been widely applied for the determination of cerebral blood flow by the inert gas method of Kety & Schmidt (1948). Chapter 8 describes the techniques of cerebral blood flow measurements in more detail.

## Measurement of local cerebral metabolic rates with quantitative autoradiographic techniques

Several quantitative autoradiographic techniques have been developed recently to study brain metabolism, blood flow and protein synthesis at the regional level by means of specific labelled markers. Indeed, the brain is a very heterogeneous organ and all the techniques described above allow only average measurements in the brain as a whole. Thus, metabolic changes can only be observed when they concern the whole brain or rather large regions. The quantitative autoradiographic techniques, the [$^{14}$C]2-deoxyglucose method of Sokoloff *et al.* (1977) for the assessment of local cerebral glucose utilization, the [$^{14}$C]iodoantipyrine technique of Sakurada *et al.* (1978) for the measurement of local cerebral blood flow, and several techniques using a labelled amino acid for the determination of local cerebral brain protein synthesis, have been set up recently to visualize subtle changes in discrete brain areas (see Chapter 8 for more details).

## Histochemical assessment of the activity of enzymes of brain energy metabolism

As for autoradiographic techniques, histo-, cyto- and immunohistochemical techniques allow the regional detection of various enzymes of brain energy metabolism at regional, cellular and even electron microscopic levels. Studies have been most frequently carried on lactate dehydrogenase, cytochrome oxidase, glucose-6-phosphate dehydrogenase and β-hydroxybutyrate dehydrogenase. Levels of enzyme activity change with altered neuronal activity. However, such changes take hours or even days to occur. Thus, regional enzyme activity correlates rather with long-term changes in metabolism whereas the deoxyglucose method for glucose utilization reflects acute changes in activity.

## Nuclear magnetic resonance

Nuclear magnetic resonance (NMR) has become quite a widely used technique. NMR relies on the principle that atomic nuclei with a odd number of nucleons (protons and neutrons) have an intrinsic magnetism that makes each a magnetic dipole, in essence, a bar magnet. They orient either in the direction of the field (low-energy state) or opposite to the direction of the field (high-energy state). The absorption of radiation of a radio-frequency specific to each nucleus causes the nucleus to reorient from the low- to the high-energy state by a process called resonance. The resonance absorption induces a signal in a detector coil.

The nuclei used for NMR studies include $^1$H, which is the nucleus of 99.98

per cent of all hydrogen atoms occurring in nature, $^{13}$C, which is the nucleus of 1.1 per cent of all carbon atoms, and $^{31}$P, which is the nucleus of all phosphorus atoms. With $^{31}$P, magnetic resonance spectroscopy allows the assessment of cellular respiration and the separate quantification of the adenylate phosphates, creatine phosphate, sugar phosphates, and phospholipids, although these molecules cannot be used yet for metabolic mapping. With $^{13}$C, some measurements on glucose metabolism substrates and intermediates have already been performed. $^1$H NMR allows the assay of a wide range of cerebral metabolites such as glucose, lactate, amino acids and phosphocreatine/creatine pools. Other detectable nucleides of most interest in these studies are $^{15}$N, $^{17}$O, $^{19}$F, $^{23}$Na, $^{25}$Mg, $^{39}$K and $^{43}$Ca.

## Cerebral energy consumption

The main mode of function of the central nervous system is through excitation and conduction of nervous influxes along axons and dendrites and the transfer of information across synapses. The energy synthesized in the brain almost exclusively from glucose is then used for active ion pumping, maintenance of electrochemical gradients, fast axoplasmic transport, synthesis of macromolecules and neurotransmitters and also maintenance of structural integrity.

## Energy consumption for neurotransmission, ion transport and maintenance of electrochemical gradients

### Spike activity

Estimates on heat production of un-myelinated nerve fibres give values which range from $14 \times 10^{-6}$ cal/g/impulse to $2.8 \times 10^{-3}$ cal/g of neuronal surface material/impulse. Nerve cell density per unit volume of cerebral cortex varies considerably between different areas and even within a given area. The number of nerve cells may be estimated as $2 \times 10^6$ and $6$–$15 \times 10^6$ for 1 cm$^2$ of the human and the cat cerebral cortex, respectively. Assuming that a neuron generates about 10 impulses per s for a weight of 50 ng, the energy consumption due to spike activity alone can be calculated as $7 \times 10^{-5}$ cal/g/s and $2$–$5 \times 10^{-5}$ cal/g/s for the human and cat cerebral cortex respectively. In man, the brain uptake of oxygen represents 3.5 ml O$_2$/100 g/min, which means also $5.5 \times 10^{-4}$ m l O$_2$/g cortex/s, and corresponds to $2.75 \times 10^{-3}$ cal/g cortex/s. The same calculation for glucose leads to the value of $36 \times 10^{-3}$ cal/g cortex/s. These estimates indicate that only about 0.3–3 per cent, or even less, of the cortical energy consumption can be accounted for by spike activity or, in other words, that a large amount of energy expenses is for purposes other than action potentials. Indeed, it has been estimated that a tenfold increase in net spike activity would only lead to a 3–25 per cent increase in metabolic demand.

### Ion transport

In fact, a large proportion (more than 50 per cent) of the energy consumption of the brain is used for pumping processes, in other words for the active transport of ions to sustain and restore the membrane potentials discharged during the process of excitation and conduction. These ionic pumps are mainly Na$^+$-K$^+$ ATPase and the Ca$^{2+}$ pump. For example, Na$^+$-K$^+$ ATPase moves 3 Na$^+$ from inside to outside against 2 K$^+$ in the opposite direction, with a concomitant hydrolysis of 1 ATP. Moreover, even apparently quiescent neurons consume energy, partly because of the constant leakage of ions through membranes which must be counteracted by active transport needing a constant supply of ATP.

Indeed, the normal transmembrane Na$^+$ and K$^+$ gradients are restored after a transient depolarization by the transport of Na$^+$ out of and K$^+$ into the cell at the expense of ATP. Transmembrane fluxes of Na$^+$ and K$^+$ are intense, especially in the cerebral cortex, and the extracellular concentration of K$^+$ has been shown to be quickly responsive to changes in metabolism. Thus, even a transient interruption of O$_2$ supply is reflected by a significant increase in the extracellular concentration of K$^+$.

Similarly, relatively direct coupling between neuronal metabolism and excitability appears to be mediated by intracellular free Ca$^{2+}$ and the influx of Ca$^{2+}$ into presynaptic endings is a prerequisite for transmitter release. Restoration of normal Ca$^{2+}$ gradients requires energy, since external concentrations (about $10^{-3}$ mol/l) are higher than internal levels of Ca$^{2+}$ (about $10^{-7}$ mol/l) and the negative membrane potential leads to the entry of Ca$^{2+}$ into the cell. Extrusion of Ca$^{2+}$ out of the cell occurs by both Ca$^{2+}$/Na$^+$ exchange related to the Na$^+$ gradient resulting from the operation of Na$^+$-K$^+$ ATPase and the Ca$^{2+}$ pump responsible for net extrusion of calcium.

Inhibition in the nervous system is mediated by an increase in Cl$^-$ per-

Table 3. Representative values for local cerebral metabolic rates for glucose in the conscious rat and monkey, expressed as mol/100g/min (means ± SEM)[a]

| Brain structure | Rat (n = 10) | Monkey (n = 7) |
|---|---|---|
| *Auditory system:* | | |
| Auditory cortex | 162 ± 5 | 79 ± 4 |
| Medial geniculate body | 131 ± 5 | 65 ± 3 |
| Inferior colliculus | 197 ± 10 | 106 ± 6 |
| Superior olive | 133 ± 7 | 63 ± 4 |
| Cochlear nucleus | 113 ± 7 | 51 ± 3 |
| *Visual system:* | | |
| Visual cortex | 107 ± 6 | 59 ± 2 |
| Lateral geniculate body | 96 ± 5 | 39 ± 1 |
| Superior colliculus | 95 ± 5 | 55 ± 4 |
| *Limbic system:* | | |
| Parietal cortex | 112 ± 5 | 47 ± 4 |
| Nucleus accumbens | 82 ± 3 | 36 ± 2 |
| Amygdala | 52 ± 2 | 25 ± 2 |
| Hippocampus | 79 ± 3 | 39 ± 2 |
| Hypothalamus | 54 ± 2 | 25 ± 1 |
| Mammillary body | 121 ± 5 | 57 ± 3 |
| Pontine grey matter | 62 ± 3 | 28 ± 1 |
| *Motor system:* | | |
| Sensory-motor cortex | 120 ± 5 | 44 ± 3 |
| Caudate-putamen | 110 ± 4 | 52 ± 3 |
| Globus pallidus | 58 ± 2 | 26 ± 2 |
| Substantia nigra | 58 ± 3 | 29 ± 2 |
| Cerebellar cortex | 57 ± 2 | 31 ± 2 |
| Cerebellar nuclei | 100 ± 4 | 45 ± 2 |
| *White matter:* | | |
| Corpus callosum | 40 ± 2 | 11 ± 1 |
| Internal capsule | 33 ± 2 | 13 ± 1 |
| Cerebellar white matter | 37 ± 2 | 12 ± 1 |

[a] Rat values from Sokoloff *et al.* (1977) are reproduced with permission from *J. Neurochem.* (©Raven Press). Monkey values from Kennedy *et al.* (1978) are reproduced with permission *Ann. Neurol.* (copyright ©The American Neurological Association).

meability. The normal hyperpolarizing inhibitory postsynaptic potential (IPSP) depends on a steep electrochemical gradient of $Cl^-$ across neuronal membranes, which must be maintained by a $Cl^-$ pump, and thus requires energy. Indeed, if the gradient is reduced or inversed, the inhibition may loose some of its effectiveness.

It then appears that, under normal conditions, the maintenance of ionic gradients requires 50–60 per cent of the total $O_2$ consumption, i.e. 12–16 µmol ATP/g wet weight/min, of which the major fraction may be accounted for by $Na^+$-$K^+$ ATPase.

## Neurotransmitter reuptake

The $Na^+$ gradient is predominantly used for osmotic equilibration compensating for the reaccumulation of neurotransmitters released during nervous activity. Thus, the uptake of amino acid neurotransmitters by brain cells is an energy-consuming process and is to an important extent dependent on both an adequate supply of energy and the presence of $Na^+$. However, it is not certain whether the active transport is coupled to metabolism directly, or via the transmembrane gradient of $Na^+$ concentration maintained by the operation of $Na^+$-$K^+$ ATPase.

## Synthesis of macromolecules and neurotransmitters

It should be recalled that not all of the oxygen consumption is used for energy metabolism. Indeed, the brain contains different oxidases and hydrolases which are involved in the synthesis and metabolism of various neurotransmitters. For example, tyrosine hydroxylase, dopamine-β-hydroxylase and tryptophan hydroxylase operate in the biosynthesis of dopamine, noradrenaline and serotonin respectively. These oxygenases incorporate molecular oxygen in the reaction products. These catecholamines are also deaminated to their respective aldehydes by monoamine oxidase in the presence of oxygen. However, if one takes into account the maximal rates of all these oxidases, the total amount of oxygen used in the turnover of these neurotransmitters can only account for quite a small proportion of the whole oxygen consumed by the brain. Indeed, it represents only 1–3 per cent of the amount of ATP consumed in cation transport. However, this amount only represents ATP and oxygen used in monoamine synthesis, but since the major part of the energy may be used in reuptake and storage of the transmitters, the energy cost may be somewhat larger.

## Maintenance of cellular structural integrity

The rate of degradation and replacement of cellular macromolecules is not accurately known. It has however been estimated that a cell body may resynthesize up to 2000 mitochondria and replace its whole population of macromolecules within a day. Likewise, it has also been postulated that synaptic vesicles are used only a few times before being degraded. According to these estimates, it appears likely that these synthetic processes may consume quite a large proportion of the total brain energy needs.

For example, it is well known that the activity of some transmembrane proteins, such as ATPases, adenylate cyclases and cytochrome oxidases, depends on their phospholipid environment. Some other enzymes are responsible for the interconversion of phospholipids which in turn may alter membrane fluidity and influence ionic channels and enzymatic activities. Indeed, interaction between an agonist and its receptor leads to a whole cascade of molecular events which includes the turnover of membrane structure, $Ca^{2+}$ transmembrane movements and protein phosphorylation which are all energy-consuming processes.

## Cellular heterogeneity of brain energy metabolism

Brain energy needs appear to be heterogeneous at different levels, according both to the region, the type of cell considered, neuron or glia, and also within a single cell.

## Regional heterogeneity of brain energy metabolism

### Oxygen and glucose utilization

Since the work of Kety and Sokoloff, it has been clearly established, both in humans and in animals, that local cerebral metabolic rates for oxygen ($LCMRO_2$) and for glucose ($LCMRglc$) are tightly coupled and are quite heterogeneous within the brain. $LCMRO_2$ and $LCMRglc$ are lowest and quite homogeneous in white matter (Table 3). They are higher in grey matter and the ratio between $LCMRglc$ in the most (inferior colliculus) and least active brain structures (hypothalamus and amygdala) reaches a value of about 4.0. The distribution of metabolic activity within the different brain areas is also typical of the species considered, $LCMRglc$ being about 2.0 to 4.0 times higher in the rat than in man and monkey (Table 3). The regions with the highest glucose consumption are usually the auditory areas, followed by cortical areas and posterior areas, such as brainstem and cerebellar nuclei. The less energy-demanding structures are hypothalamus, most limbic and motor regions. It has also been shown that regions rich in synaptic contacts such as the molecular layer of the hippocampus exhibit high metabolic activities.

### Enzyme regional distribution

Regional heterogeneity of brain energy metabolism is also apparent at the level of the enzymes. Cytochrome oxidase activity is, as for $LCMRO_2$ and $LCMRglc$, rather higher in grey than in white matter. Likewise, functionally active areas which have elevated $LCMRglc$ are also characterized by high cytochrome oxidase activity. Thus, regions which exhibit enhanced levels of spontaneous and synaptic activities such as brainstem auditory nuclei and the $CA_3$ area of hippocampus are rich in cytochrome oxidase. Physiologically active regions, such as the basal ganglia, thalamus, brainstem and spinal cord exhibit an elevated cytochrome oxidase activity. Within some structures, areas of higher enzymatic activity can be seen, such as zones involved in colour processing in the primate visual cortex and laminar patterns of activity within the cerebral cortex. Regarding regional distribution of other enzymes, the specific activities of pyruvate dehydrogenase, NAD-isocitrate dehydrogenase and hexokinase are higher in the cerebral cortex than in the striatum, pons and medulla whereas NADP-isocitrate dehydrogenase activity is highest in the pons and medulla, intermediate in the cerebral cortex and lowest in the striatum. Conversely the regional distribution of other enzymes, such as lactate dehydrogenase, citrate synthase, NAD-malate dehydrogenase and β-hydroxybutyrate dehydrogenase, is rather homogeneous throughout the brain. This latter homogeneity of enzyme distribution is attributed to the fact that most of these enzymes only work at a very low proportion of their maximal rates and are able to cope with quite large metabolic changes.

Concerning enzymes directly related to the maintenance of energy stores, phosphocreatine kinase does not show large differences between various brain regions. However, some data indicate that phosphocreatine kinase activity may be increased in regions rich in synaptic contacts, such as the molecular layer of the cerebellum. Moreover, it also appears that high levels of phosphocreatine and phosphocreatine kinase are characteristic of regions in which energy expenditure for processes such as ion pumping may be large.

## Metabolic heterogeneity between neurons and glia

The various types of glial cells (astroglia, microglia and oligodendroglia) have been estimated to occupy altogether about 50 per cent of the whole brain volume. Astrocytes separate neuronal cells from each other and from blood vessels and regulate the composition of the neuronal environment. They form a syncytium, allowing the spread of electric potentials and the diffusion of material over quite large distances and may function as $K^+$ buffers since they act nearly as perfect $K^+$ electrodes. They also act as a buffering system by taking up the neurotransmitters released from neurons during synaptic transmission. This is particularly obvious in the compartmentation of brain amino acid metabolism. Glial cells take up GABA which is converted *in situ* to glutamine. Glutamine is then transported back to the neuron where it becomes a precursor for glutamate and GABA. Glial cells also contain most of the cerebral glycogen reserves, about 2 μmol/g. This glial distribution is related to the preferential location of glycogen synthetase within glial cells. This glycogen store may act as a glucose reserve for neighbouring neurons in high-demand periods.

The contribution of glial cells to the overall metabolic rate is not really clear yet. For some authors, the metabolic rate in glial cells seems to be at least as high as in neurons. However, it appears rather likely that, judging by their lower mitochondrial density, the metabolic activity in glial cells would be lower than in neurons. This hypothesis is in accordance with the fact that neurons die from hypoxia or hypoglycaemia whereas glial cells not only survive but still proliferate. Moreover, neurons have also a greater oxidative metabolic activity, as assessed by cytochrome oxidase histochemistry, than glial or endothelial cells.

Concerning enzyme distribution within neurons and glial cells, hexokinase is likely to be found in most cell types. The activities of lactate dehydrogenase and of the pentose phosphate pathway enzymes are higher in astrocytes than in neurons. Moreover, the activity of glucose-6-phosphate dehydrogenase, a rate-limiting enzyme of the pentose phosphate pathway, is high in myelinated fibres and tends to vary with the degree of myelination. Regarding the tricarboxylic acid cycle, enzyme activities are highest in neurons, intermediate in astrocytes and lowest in oligodendrocytes. Finally, the rate of metabolism of ketone bodies in neurons and oligodendrocytes is about three times as efficient as in astrocytes.

There is also some metabolic heterogeneity between different neuronal types. Functional attributes, rather than size, are dominant modulators of metabolic activity in neurons. Thus, the levels of enzyme activity are coupled with the level of spontaneous activity and with the degree and type of synaptic evoked activity. For example, in the cat geniculate nucleus, Y-cells sustain a higher rate of discharge than X-cells and the activity of cytochrome oxidase is higher in the former than in the latter.

## Heterogeneity of energy metabolism within the neuron

It has been shown that during the stimulation of nerves, axon terminals rather than cell bodies are characterized by high levels of metabolic activity. Likewise, the distribution of cytochrome oxidase varies according to the portion of the cell considered, for example between dendrites and soma of the same cell, different segments of the same dendrite and also between different classes of axon terminals. The axonal trunk usually has low cytochrome oxidase activity and axon terminals possess various levels of enzyme activity. The highest cytochrome oxidase activity is seen in dendrites and synaptic terminals which receive excitatory inputs. The level of cytochrome oxidase is however not consistently related to the type of neurotransmitter present in the neuron. It is also not correlated with the distribution of hexokinase, or glutamate dehydrogenase, but coincides with that of succinate dehydrogenase. Finally, a parallel distribution of cytochrome oxidase and $Na^+$-$K^+$ ATPase has been reported within neurons, providing evidence that energy is actively synthesized just at the site where high rates of ion transport and ATPase activity occur. Indeed, it seems that localized pools of ATP are associated with each sodium pump. The size of these pools has been estimated to range from 100 to 400 mol of ATP per molecule of ATPase, although the exact location and mechanisms maintaining these pools are not known yet. The ATP in these pools is thought to be produced by a group of tightly associated enzymes and the pumps appear to use this locally-formed energy to the exclusion of any other source of extrinsic ATP.

## Homoeostasis of cerebral energy metabolism

Homoeostasis of cerebral energy metabolism at the synapse requires close coupling between physiological activity expressed in terms of energy use, blood flow which represents energy supply, and finally metabolism or energy production. Capillaries play an active role in bringing the amount of substrate necessary closer to its site of utilization.

## Sites of regulation of cerebral energy production

The systems responsible for cerebral energy homoeostasis coexist at three consecutive sites along the pathway followed by energy substrates in the brain. These are cerebral circulation, the interface between the cerebral parenchyma and the cerebral circulation, and finally the cerebral parenchyma itself.

At the first site of regulation, cerebral blood flow remains constant in basic physiological conditions and varies as a function of functional activation or during variations of the cerebral energy environment, e.g. changes in tissue glucose levels. This adaptation takes place either at the structural level through changes in the vascular supply or at the physiological level through vasodilation of the cerebral vascular bed.

The second site of regulation modulates the exchanges of the main substrates of cerebral energy metabolism, oxygen and glucose, in response to a functional activation or to a decrease in the arterial supply of energy substrates. At that site, the release of some products of energy metabolism also takes place, such as $CO_2$ in basic physiological conditions and lactate in pathophysiological situations.

At the third site or intrinsic site of homoeostasis in the cerebral parenchyma, the rate of the different metabolic pathways is oriented and modulated according to the energy needs during functional activation or deficiency in basic energy substrates.

## Mechanisms involved in cerebral energy homoeostasis

Regulatory mechanisms involved in cerebral energy homoeostasis in various physiological or pathophysiological conditions remain a matter of controversy. Two main theories attempt to explain the phenomena underlying brain energy homoeostasis: the metabolic and the neurogenic theory.

In the metabolic theory, changes in blood flow are considered to be secondary to changes in metabolism. Indeed, physiological activation in the brain induces ion movements and release of neurotransmitters. These changes cost no or little energy, but restoration of homoeostasis at the level of the nerve terminals requires ion pumping and uptake of neurotransmitters which are both energy-consuming processes. The amount of ATP used during these processes is resynthesized from glucose and oxygen. In this classical model, the

products of metabolic activity induce vasodilation and hence increased perfusion within the areas of augmented activity.

More recently, a direct effect of physiological activity on blood flow has been stressed. In the neurogenic theory, both metabolism and blood flow can be globally modulated by activity in ascending neuronal pathways and blood flow may be changed locally through neuronal circuits. In this theory, blood flow levels are controlled both locally and globally by physiological activity and vary simultaneously with metabolic activity resulting in rapid changes both in the delivery of substrates and in the removal of metabolic products.

## Brain energy homoeostasis at the cellular level

Energy homoeostasis at the cellular level is regulated by a close correlation between functional demand and the capacity of neuronal cells for oxidative metabolism. During transient changes in cellular energy metabolism, when oxygen availability is normal, the rate of ATP synthesis is controlled by two variables. These are the phosphorylation state ratio, $[ATP]/[ADP][P_i]$, and the redox state of intramitochondrial pyridine nucleotides $[NAD^+]/[NADH]$ which is related to the activities of various rate-controlling dehydrogenases. In situations of limited oxygen availability, the rate of energy metabolism is mainly regulated at the level of phosphofructokinase. Thus, when energy expenditure is increased, the concentrations of the activators of the enzyme (ADP, AMP, $P_i$) increase, whereas the levels of the inhibitors of the enzyme (ATP, PCr) decrease, enhancing the flux through the glycolytic pathway. However, when energy needs are increased over prolonged periods, the cells respond by increasing their number of mitochondria or the quantity of the respiratory chain enzymes or even both.

The homoeostasis of the pool of high-energy intermediates and mainly of ATP depends on the fast exchanges between the ATP/ADP couple on the one hand and phosphocreatine/creatine couple on the other hand. By operation of the creatine kinase shuttle, most of the ATP produced in the mitochondria is immediately used to phosphorylate creatine between the inner and the outer mitochondrial membrane. Phosphocreatine then leaves the mitochondria to act as the primary energy source at the sites where ADP needs to be phosphorylated. Then, creatine diffuses back to the mitochondria to be rephosphorylated. The operation of the creatine kinase shuttle is very efficient for the maintenance of brain energy homoeostasis. Indeed, the total creatine content (PCr + Cr) of brain tissue is several-fold higher than that of the total adenylates (ATP + ADP + AMP), which provides a greater potential for energy transfer per unit of time by means of the phosphocreatine shuttle than by the diffusion of ATP and ADP.

## Conclusion

Glucose represents the almost exclusive substrate for brain energy metabolism. However, glucose is provided to the brain in large excess since only 10 per cent of the arterial glucose supply is taken up by the brain in basic physiological conditions. Conversely, the brain takes up 30 to 40 per cent of the oxygen supply from the blood, which may easily represent a limiting factor.

However, in specific conditions of reduction in glucose availability, the brain is somehow flexible in its choice of substrates. Indeed, during development or in ketotic states in the adult, the brain is able to sustain part of its energy needs with ketone bodies.

The brain is one of the organs in which energy homoeostasis is best controlled. In particular, ATP plays a very important role in the maintenance of cerebral function. Under basic physiological conditions, ATP is mainly synthesized by mitochondrial oxidative phosphorylation. Moreover, the operation of creatine phosphokinase and adenylate kinase contribute to the maintenance of a constant ATP level. It seems that more than 50 per cent of the energy produced from oxygen and glucose is utilized for the maintenance of ionic gradients and the operation of ionic pumps.

Most of the mechanisms underlying the control of homoeostasis of brain energy metabolism are now better understood in basic physiological conditions. However, the exact contribution of these different mechanisms in various pathophysiological situations remains to be further explored.

**Acknowledgements:** The author wishes to thank Dr B. Barrere for helpful discussions.

## Summary

This chapter summarizes the main characteristics of energy metabolism in the brain. Glucose is the predominant fuel of energy metabolism in the adult brain. Glucose is mainly oxidized via glycolysis and the tricarboxylic acid cycle. The complete oxidation of glucose through these pathways leads to the formation of 38 ATP per mol of glucose and accounts for the greater part of the oxygen used by the brain. The latter represents about 20 per cent of the oxygen consumption of the whole body for a relative weight of the brain of only 2 per cent. A quite small proportion of the glucose entering the brain (just a few per cent) is utilized by alternative routes, the pentose phosphate pathway and glycogenesis. In addition to ATP, high energy phosphate is also stored in the brain as phosphocreatine which acts as an energy reserve for ATP. Moreover, the developing brain under normal conditions or the adult brain in situations of fasting or ketotic diabetes is able to use ketone bodies in addition to glucose for its energy metabolism and biosyntheses. In the normal adult brain, glucose is a very active precursor of various amino acids and other neurotransmitters. In addition to neurotransmitter synthesis and reuptake, the energy synthesized in the brain is also used for spike activity, ion transport, and maintenance of electrochemical gradients and of cellular integrity. Brain energy metabolism is characterized by multiple levels of heterogeneity. Indeed, at the regional level, glucose and oxygen utilization as well as enzyme distribution are heterogeneous. Moreover, there is a metabolic heterogeneity between neurons and glia and also within the neuron itself. Finally, this chapter discusses the different levels of homoeostatic control of brain energy metabolism.

## List of abbreviations

ADP:      Adenosine 5'-diphosphate;
AMP:      Adenosine 5'-monophosphate;
ATP:      Adenosine 5'-triphosphate;
ATPase:   Adenosine 5'-triphosphatase;
$C_A, C_V$:   Arterial and venous concentration;
CBF:      Cerebral blood flow;
CMR:      Cerebral metabolic rate;
CoA:      Coenzyme A;
CoQ:      Coenzyme Q;
$\Delta G^{o'}$:     Free energy change;

FAD:      Flavin adenine dinucleotide, oxidized form;
$FADH_2$:   Flavin adenine dinucleotide, reduced form;
GABA:     Gamma-aminobutyrate;
GDP:      Guanosine diphosphate;
GTP:      Guanosine triphosphate;
LCMRglc:  Local cerebral metabolic rates for glucose;
$LCMRO_2$:  Local cerebral metabolic rates for oxygen;
$NAD^+$:    Nicotinamide adenine dinucleotide, oxidized form;

NADH:     Nicotinamide adenine dinucleotide, reduced form;
$NADP^+$:   Nicotinamide adenine dinucleotide phosphate, oxidized form;
NADPH:    Nicotinamide adenine dinucleotide phosphate, reduced form;
PCr:      Phosphocreatine;
$P_i$:       Inorganic phosphate;
$PP_i$:      Inorganic pyrophosphate;
UDP:      Uridine diphosphate;
UTP:      Uridine triphosphate.

## References

Boulton, A.A., Baker, G.B. & Butterworth, R.F. (1989): *Neuromethods,* Vol. 11, *Carbohydrates and energy metabolism.* Humana Press, Clifton.

Chance, B. & Williams, G.R. (1956): The respiratory chain and oxidative phosphyorylation. *Adv. Enzymol.* **17**, 65–134.

Collins, R.C. (1987): Physiology–metabolism–blood-flow couples in brain. In: *Cerebrovascular diseases*, eds. M.E. Raichle & W.J. Powers, pp 149–163. New York: Raven Press.

Erecinska, M. & Silver, I.A. (1989): ATP and brain function. *J. Cereb. Blood Flow Metab.* **9**, 2–19.

Kennedy, C., Sakurada, O., Shinohara, M., Jehle, J. & Sokoloff, L. (1978): Local cerebral glucose utilization in the normal conscious Macaque monkey. *Ann. Neurol.* **4**, 293–301.

Kety, S.S. (1957): The general metabolism of the brain *in vivo*. In: *Metabolism of the nervous system*, ed. D. Richter, pp. 221–237. London: Pergamon Press.

Kety, S.S. & Schmidt, C.F. (1948): The nitrous oxide method for the quantitative determination of cerebral blood flow in man: theory, procedure and normal values. *J. Clin. Invest.* **27**, 476–483.

McIlwain, H.S. & Bachelard, H.S. (1985): *Biochemistry and the central nervous system*, 5th edn. Edinburgh: Churchill Livingstone.

Mitchell, P. (1979): Keilin's respiratory chain concept and its chemiosmotic consequences. *Science* **206**, 1148–1159. (Mitchell reviews the evolution of the chomiosmotic concept in his Nobel lecture.)

Sakurada, O., Kennedy, C., Jehle, J., Brown, J.D., Carbin G.L. & Sokoloff, L. (1978): Measurement of local cerebral blood flow with iodo[$^{14}$C]antipyrine. *Am. J. Physiol.* **234,** H59–H66.

Siesjö, B.K. (1978): *Brain energy metabolism.* New York: John Wiley and Sons.

Smith, C.B. & Sokoloff, L. (1981): The energy metabolism of the brain. In: *Molecular basis of neuropathology*, eds. A.N. Davison & R.H.S. Thompson, pp. 104–131. London: Edward Arnold.

Sokoloff, L. (1973): Metabolism of ketone bodies by the brain. *Ann. Rev. Med.* **24**, 271–280.

Sokoloff, L. (1989): Circulation and energy metabolism of the brain. In: *Basic neurochemistry: molecular, cellular and medical aspects*, 4th edn, eds. G.J. Siegel, B.W. Agranoff, R.W. Albers & P.B. Molinoff, pp. 565–590. New York: Raven Press.

Sokoloff, L., Reivich, M., Kennedy, C., Des Rosiers, M.H., Patlak, C.S., Pettigrew, K.D., Sakurada, O. & Shinohara, M. (1977): The [$^{14}$C]deoxyglucose method for the measurement of local cerebral glucose utilization: theory, procedure and normal values in conscious and anesthetized albino rat. *J. Neurochem.* **28**, 897–916.

Wong-Riley, M.T.T. (1989): Cytochrome oxidase: an endogenous marker for neuronal activity. *Trends Neurosci.* **12**, 94–101.

# Neurophysiological Basis of Cerebral Blood Flow Control: An Introduction

**PART 2**

## Chapter 8

# Methods of investigation of cerebral circulation and energy metabolism

**Pierre Lacombe[1] and Mirko Diksic[2]**

[1]*Laboratoire de Recherches Cérébrovasculaires, CNRS UA 641, University Paris VII, France;* [2]*Montreal Neurological Institute. 3801 University Street, Montreal, Quebec, Canada H3A 2B4*

## Introduction

### Differences between the techniques used in man and animals

This chapter is divided into two parts. One describes the methods used in animals and the other one those used in man. The main reason for this distinction is the different objectives of the methodological developments, towards medical investigation associated with non-invasiveness on the one hand, and towards the best spatial and/or temporal resolution to attain the finest level of investigation in cerebral tissue on the other hand. The brain size of the species considered is, indeed, the major technical conditioning factor. In primates, including man, several extracranial detectors of radioemitting tracers can be positioned around the brain. In small laboratory species, the brain size is adequate for the autoradiographic techniques performed on histological slides. The highest degree of tissue localization is obtained in the latter case, but autoradiography requires the sacrifice of the animal after a single measurement, whereas repeated measurements are possible with the former.

### Variety of variables involved

Several variables can be explored for the investigation of the cerebral circulation and metabolism. For studies on the circulation, first the pial vessel re-activity or the intracranial pressure, indirectly reflecting the brain blood circulation have been studied, then followed by measurement of the blood in large cerebral vessels or the blood flow perfusing through the tissue. Most investigations have favoured the tissue perfusion, undoubtedly because this variable is clinically relevant. However, tissue perfusion brings into play not only blood flow but also cerebral capillary permeability, which may vary in some physiological or pathophysiological situations.

For studies on metabolism, the choice of the variable to be explored mainly amounts to the choice of the metabolic substrate investigated. As for the tissue perfusion measurement, brain extraction of the metabolic tracer through the blood–brain barrier (BBB) should be taken into account.

## Interest of quantification and radioactive tracers

Quantification of a functionally relevant variable provides reference values for comparisons between normality and pathology in humans and can thus be useful as a diagnostic marker. The control values obtained in animals can be compared to those of other studies, and values obtained in defined experimental situations can likewise be compared with each other. Moreover, absolute values are useful data for large scale statistical analysis including the major statistical parameters indicative of the variability and distribution of the values.

In all the techniques described below designed for measuring cerebral blood flow (CBF), cerebral blood volume (CBV), BBB permeability and energy metabolic rate, reliable

quantification has been a prominent methodological objective. It has been achieved much more readily (for a given principle) by measuring the radioactivity of labelled tracers than by chemical assays of non-radioactive substances. Nonetheless, non-radioactive indicators have not been completely abandoned. For instance, dyes are still used for qualitative assessment of BBB disruption.

The use of radioactivity requires labelling of the tracer whereas natural substances can be readily used as tracer provided their assay is technically available. However, the development of radiolabelling technology has yielded a huge variety of biological tracers and a correspondingly wide field of investigation. In man, for instance, the introduction of γ-emitting tracers has allowed extracranial counting and thus non-invasive exploration.

## Part A: Methods used in animals (P. Lacombe)

## Generalities

### Species

An excessive variety of species has been used for the investigation of cerebral blood flow and metabolism. This diversity has resulted in major difficulties in comparing studies, and putative species differences have been frequently claimed to explain discrepant results.

The objective of the proposed experimental study has mostly guided the choice of the animal species. For

instance, studies on ischaemia were first performed on the monkey or gerbil because of either similarity to humans in the organization of their brain blood supply or a particular experimental advantage (incompleteness of the circle of Willis). Conversely, cats and dogs have been avoided because of the complex anastomotic vascular network supplying their brain (see Chapter 5), whereas the cat was the favourite species for neurophysiological studies. Another factor has been the traditional practice in a given experimental field, for instance, cardiovascular experimentors prefer

dogs or piglets.

During the last decade, there has been a general trend to a widespread use of the rat, thanks to the transposition of experimental models to this species and also to neurochemical studies mainly performed on the rat. Another reason is the smaller size of the animal which requires a reduced amount of expensive radiolabelled tracer and enables easier sectioning of the brain for histological work. However, other species, such as the rabbit, still represent a good compromise for experimentation, and it remains the favourite model for certain specific purposes such as studies related to immunological reactions (endotoxaemia, fever, etc.).

## Experimental protocols and animal preparations

### Types of animal preparation

Whatever the technique used for *in vivo* cerebrovascular experimentation, two main types of experimental protocols are to be distinguished: conscious and anaesthetized animals.

Conscious preparations have the considerable advantage of avoiding general anaesthetic agents at the time of measurement, but require special equipment or implantation, introduced either chronically or acutely with a brief recovery period under local anaesthesia. However, in a test situation, conscious animals exhibit an integrated response, including behavioural and respiratory changes, etc., which may complicate the interpretation of the results.

By contrast, acute preparations require general anaesthesia, often associated with a curarizing agent to enable artificial respiration. This preparation authorizes a wide range of experimental approaches, such as stereotaxic or other delicate manoeuvres on the brain of the immobilized animal. Moreover, this kind of approach provides more specific data on the phenomenon studied, since behavioural or respiratory changes do not perturb the response, e.g. during selective, localized intracerebral stimulation.

An experimental drawback commonly encountered in acute prepara-

tions, although often disregarded, is the cortical spreading depression. Such a phenomenon is elicited practically every time the cortical mantle is perforated (electrode, EEG screw, cannula, etc.), and sometimes even following a supradural intervention (skull drill, etc.). The inherent susceptibility is highest in the cortex of lissencephalic species, and under anaesthesia. This phenomenon has been shown to perturb the cerebral circulation and energy metabolism for as long as several hours (for review, see Seylaz, 1968; Hansen and Lauritzen, 1988; Lacombe *et al.*, 1992; Mraovitch *et al.*, 1992).

### Conscious animal preparations

#### Practical implications

Conscious animal preparations necessitate a chronic implantation in the brain and usually in the femoral circulation. The cerebral implantation may consist of probes inserted in the brain or the skull and sealed on the skull surface (for details see clearance techniques and laser-Doppler flowmetry). This is usually performed under barbiturate anaesthesia. A recovery time of 10–15 days is needed for any possible inflammation or oedema to be resorbed in the case of intracerebral implantation. The femoral implantation consists of venous and arterial catheters to measure blood pressure, withdraw blood samples and inject tracers, etc. (see sections 'Tissue perfusion techniques using a diffusible tracer' and 'Non-diffusible tracer techniques'). Since this is a light surgical procedure, volatile anesthetics such as halothane or isoflurane are generally adequate for a preparation period which does not ex-

ceed 30–45 min. The incisions are sutured and infiltrated with local anaesthetic (lidocaine) and the animal is then allowed to recover from the short-duration general anaesthesia for at least 2 h.

This last procedure is widely used, probably because it is easier and more reliable than a chronic arterial and venous catheter implantation. However, some investigators have developed special chambers for measurements in freely moving, unstressed rats (see for review Bryan, 1990) particularly relevant for some investigations, and requiring chronic vascular implantations.

#### *Immobilization of the animal*

This point is particularly critical for the preparations involving conscious animals. The freely moving preparation, which seems at first sight the most satisfactory, is in fact difficult to put into practice, especially when serial arterial blood samples are to be regularly collected at fixed short intervals. This difficulty reduces the number of successful experiments. It has been overcome by various restraining devices. The most commonly used is the 'slight' or 'gentle restraint' first developed by Sokoloff's group for cats and rats (Reivich *et al.*, 1969; Sokoloff *et al.*, 1977; Sakurada *et al.*, 1978; Ohno *et al.*, 1979). It involves immobilization of the hindquarters by plaster cast, which is the minimal procedure of restraint to secure the femoral catheters.

A comparative study of the cerebrovascular effects of restraint in the rat was conducted by Lasbennes and colleagues (1986), using as reference the freely moving rat with chronically implanted catheters. By measuring the plasma corticosterone, these authors showed that the degree of stress found in conscious, gently restrained rats was significantly higher than in freely moving rats. However, this difference had almost no repercussion on the cerebral circulation. Moreover, the gentle immobilization stress was only minimally reduced by the diazepam used at anxiolytic doses (Lasbennes and Seylaz, 1986). Thus, despite the stress inherent in the preparation, the classical experimental protocol can be considered convenient for most investigations, including measurement of cerebral glucose utilization (Soncrant et al., 1988; and see for review Bryan, 1990).

It can be further improved by the use of, for instance, a hammock instead of a rigid plaster cast (Seylaz et al., 1983). It is of some importance to associate any preparation on the conscious animal with sham sessions in the experimental conditions, prior to the real experiment(s). The unfamiliar surroundings and the imposed restraint may prevent the animal from responding normally to environmental and/or pharmacological stimuli. The substantial effect of habituation and handling on both the level and reproducibility of lCBF values of chronically implanted animals was shown early by Haining et al. (1968). The value of habituation was shown to be even greater when the restraint involved the head.

## Anaesthetized animal preparations

### Convenience of anaesthetics
This kind of preparation is utilized when very invasive surgical proce-

dures such as craniotomy (for intracerebral exploration) or cardiac catheter insertion (for the radiolabelled microsphere technique) are required prior to the cerebrovascular investigation planned. It presents the advantages, of first, a more specific approach of the phenomenon studied than in the conscious animal, avoiding sudden unwanted behavioural ventilatory changes and a general stress reaction, either of which could obscure the phenomenon studied, and second a better reproducibility of the physiological state of the animal can be expected thereby enabling a more homogeneous series of animal preparations.

The choice of the agent inducing general anaesthesia is always a matter of debate in this field of research, since all compounds responsible for central narcosis can be suspected to bias the cerebrovascular reactivity. Its practical characteristics should be (i) to induce a long-lasting anaesthesia, adapted to the degree of invasiveness of the surgical procedure envisaged; (ii) to maintain a fairly stable and reproducible anaesthetic and physiological state of the animal at the time of measurement; (iii) to be devoid of marked effects on the circulation or the metabolic rate of the brain. To fulfil these conditions as far as possible many investigators have used sequential association of two agents. The first one, employed for the surgical procedure, should give a deep anaesthesia (stage III) but have a limited duration of effect, short enough to be completely dissipated at the time of measurement. The choice of the agent subsequently administered is more debatable, since its possible cerebrovascular effects prevail at the time of

measurement. Many volatile anaesthetics have been discarded because of their vasodilatory effect on the brain, barbiturates because of their long-lasting depression of the metabolic activity, and ketamine (a phenyclidine derivative found to be a noncompetitive antagonist of the NMDA receptor) because of its possible local stimulation of both the cerebral metabolic rate and blood flow (see for review and references, Siesjö, 1978; Lacombe et al., 1980; Cavazutti et al., 1987; Gumbleton et al., 1989; Archer et al., 1990).

To compensate for any respiratory depression, paralysing agents are often administered to enable artificial ventilation and adjust the blood gases to the narrow range of physiological values, improving thereby the reproducibility of the results obtained. Curare is also used for experiments which require complete immobilization of the head, such as stereotaxic interventions or acute flowmetry using a laser probe. However, several studies have demonstrated a stress reaction altering the cerebral blood flow and metabolism under pharmacological paralysis while anaesthesia was dissipating (see for review Bryan, 1990).

### Anaesthetic protocols commonly used
The volatile anaesthetic, halothane, was found to properly meet the conditions defined above for the first agent (Siesjö, 1978) and has been widely employed. It is tending now to be progressively replaced by isoflurane.

Nitrous oxide (70 per cent), giving a superficial (Stage II) anaesthesia, has been used in the rat. By virtue of its analgesic properties, it was not co-administered with another anaesthetic but with a neuromuscular blocker (d-tubocurarine) to allow artificial respiration. Nitrous oxide has been shown to have minor cerebrovascular effects. Alternatively, these effects may be considerably altered or obscured by the following factors: (i) the body temperature, rapidly decreasing in an animal ventilated with a gas mixture including a volatile agent; (ii) immobilization by neuromuscular blockade for artificial ventilation; and (iii) the stressed state of the animal as evidenced by the significant influence of adrenalectomy related to partial restraint or external stimuli. Only local measurement of CBF and CGU revealed such changes and their particular cerebral distribution (Dahlgren *et al.*, 1981; Ingvar and Siesjö, 1982). It is worth noting that the use of nitrous oxide can potentiate the action of other anaesthetics such as diazepam, or blunt the metabolic depression of pentobarbital (Hoffman *et al.* 1986).

Althesin, a mixture of steroids (alphaxalone and alphadolone) shown to have cerebrovascular effects similar to those of barbiturates, but with a considerably shorter dissipation time, was successfully used in the rabbit (Lacombe *et al.*, 1985). The protocol involved i.v. perfusion at low dose, and the CBF measurements were performed at a low level of anaesthesia.

Chloralose (α isomer) is commonly used for neurophysiological and cardiovascular studies. Under α-chloralose, the blood pressure remains elevated and the activity of the autonomic nervous system is believed to be unaffected. Its anAesthetic effects resemble that of chloralhydrate, although the pharmacologically active molecules are different. The α-chloralose solution (1 per cent in saline) should be prepared fresh immediately before use by heating to 60 °C, and infused at 35–40 °C before crystallizing (for anaesthesia guide, see Stoelting and Miller, 1989).

Chloralose is associated with urethane or halothane to provide suitable anaesthesia for the surgical step. In conjunction with halothane (1 per cent) for femoral catheter insertion, it has been extensively used by the group of D.J. Reis for cerebral vascular and metabolic investigations in the artificially ventilated rat and recently validated (Bonvento *et al.*, 1994). Compared to the conscious animal, α-chloralose or chloral hydrate have been shown to have a moderate general depressant action on the cerebral metabolic rate (Dudley *et al.*, 1982), most notably in the neocortex and thalamic relay nuclei (Grome and McCulloch, 1983).

In summary, the anaesthetic protocols retained for cerebrovascular investigations are designed to minimally affect the nervous system at the time of measurement. This can be indirectly tested by the stability of the arterial blood pressure and heart rate during the surgical procedure or painful stimuli, and in EEG monitoring the occurrence of trains of low voltage activity attests to impending recovery of consciousness (see 'Notice to Contributors' of the *Journal of Physiology* 445, pp. xii and xiii, 1992). In fact, most types of animal preparation are conditioned by the technique of measurement, and the critical point is to obtain (i) proper control conditions, i.e. to adjust the control conditions as closely as possible to the experimental situation, and (ii) reproducible physiological conditions.

## Evaluation of a technique

### Quantification and variability

The normal range of absolute flow values given by a quantitative CBF technique is certainly indicative of the accuracy of the measurement, but it is far from sufficient to evaluate a technique. Indeed, a technique may provide accurate blood flow values in resting conditions, and similar values in test conditions, which means that the technique is totally insensitive. Also, the mean value may appear correct, but mask a considerable scattering of the values.

Two other types of averaging processes must be considered. Values obtained from local measurements, e.g. by autoradiogram analysis of, at most, a few milligrams of tissue, cannot be readily compared with values obtained from regional measurements (several tens of milligrams of tissue), or global measurements (whole brain or one hemisphere, representing grams of tissue). Each is the average of a given volume of tissue and thus has a specific spatial resolution. Likewise, the time constant of a measurement conditions the temporal resolution of the technique and thereby the quantification of each single value. The clearance technique, for example, gives more accurate values at elevated flow rates, with an acquisition period shorter than at the resting flow rate. Spatial and temporal resolution therefore both define the level of the averaging processes, which considerably influences the values. These problems are further addressed below, with the description of each technique.

### Sensitivity and spatial resolution

The sensitivity of a CBF technique is

commonly tested by the reactivity to inhalation of a gas mixture containing 4–8 per cent carbon dioxide ($CO_2$). The reactivity observed around normal CBF values ranges from 2 per cent flow change per mm Hg of $CO_2$ for global measurement (Lacombe *et al.*, 1980), to 3–5 per cent for regional measurements, and up to 8–15 per cent according to the brain structure using the autoradiographic technique of local measurement (Iadecola *et al.* 1987) or laser Doppler flowmetry (Bonvento *et al.*, 1994). An equivalent reference test is not available for cerebral metabolic rate (CMR). Only the absolute values for CMR of glucose or oxygen obtained in control conditions (usually the unanaesthetized rat) give an indication of the accuracy of the technique of CMR measurement.

Another criterion of sensitivity (or spatial resolution) for both CBF or CMR measurements when multiregional techniques are employed is the ratio of values obtained in grey matter versus white matter, in resting conditions. This ratio ranges from 1.3 to 6.0 considering all CBF tracer techniques together (Lacombe *et al.*, 1980). It is usually between 1.7 and 5.0 for the diffusible tracer techniques. Within this well-represented group, it has been possible to distinguish two procedures for measuring tracer concentration in the brain: brain dissection and autoradiography (Table 1). This table provides quantitative evidence of the better spatial resolution obtained with the autoradiographic technique (higher ratios). It also demonstrates the significance of the experimental paradigm followed. The use of anaesthetic considerably reduces the ratio, probably through metabolic depression mainly affecting

**Table 1.** Ratios ($R_1$ or $R_2$) of cerebral blood flow (CBF) or glucose utilization (CGU) in the cortical grey matter to that in white matter, using either brain dissection or autoradiography. Two experimental paradigms are considered: anaesthetized or conscious animals (values calculated from studies on the rat).

| Brain dissection | | Anaesthetized rats | Conscious rats | Variable |
|---|---|---|---|---|
| Brain dissection | $R_1$ | 1.7–1.9 | 2.0–2.2 | } CBF using |
| Autoradiography | $R_2$ | 2.3–3.1 | 3.6–5.0 | [$^{14}$C]iodo-antipyrine |
| Autoradiography | $R_2$ | 2.0–2.5 | 2.8–5.2 | CGU using [$^{14}$C]2-deoxyglucose |

$R_1$ is the ratio of the values in the parietal cortex to those in the corpus callosum.
$R_2$ is the ratio of the values in the somato-motor or somato-sensory cortical area to those in the corpus callosum (genu) or the internal capsule.

the neocortex (this influence is twofold: direct on the metabolic activity, and indirect through behavioural suppression).

Table 1 also reports the equivalent ratios for cerebral glucose utilization (CGU) using autoradiography which shows a similar difference in range between anaesthetized and conscious rat preparations. The wider range of values found in the conscious animal reflects the better contrast observed by all investigators in CGU autoradiograms compared to CBF autoradiograms. The difference may be attributable to (i) a finer local control of the cerebral tissue energy metabolism (cytoarchitectural organization) than CBF, since the latter is conditioned by the pattern of local vasculature (columnar organization in cortex, see Chapter 5), and (ii) the CGU tracer is trapped within cells, whereas the CBF tracer is diffusible unless the brain tissue is instantly frozen and kept at low temperature.

## Reproducibility and temporal resolution

Like the sensitivity and the spatial resolution, reproducibility is associated with the temporal resolution, since it

is closely related to the time constant of a technique. The duration of a measurement represents the averaging time of the measurement.

The variability coefficient (standard deviation expressed as a percentage of the mean value, SD per cent) is the most widely available distribution parameter expressing intragroup variability, an integral part of the reproducibility. Since it takes into account both technical fluctuations due to measurement errors (in weight, volume, activity, etc.) and temporal, biological fluctuations, the variability coefficient allows a valuable estimation of the precision of a technique. However, its interpretation is dependent upon the size of the group considered. Groups of less than six individuals should be considered with caution.

Differences in the variability coefficients of results obtained by different techniques cannot be directly interpreted. High values (exceeding 20 per cent) may indicate either a limited or variable technical accuracy or a lack of reproducibility in the experimental protocol. Low values in resting conditions (5–10 per cent) may indicate either an excellent methodological reproducibility (both in the technique and the experimental protocol), or a poor temporal resolution, or even a lack of sensitivity. Thus, this index of global reproducibility should necessarily be complemented by a test of technical sensitivity, e.g. a change in systemic capnia (see above).

Clearly, the technical criteria mentioned above have to be considered altogether to properly evaluate the accuracy of a technique, and this makes extensive comparative studies very complex. Moreover, comparison of absolute flow values should take into account the type of preparation of the animal and its physiological state. For instance, stressful situations markedly increase the CBF level in artificially ventilated rats and rabbits, and have no or only a moderate effect in spontaneously breathing animals. The CGU is altered in both situations (for references, see Lasbennes *et al.*, 1986; Bryan, 1990). As already noticed, anaesthetics often also significantly influence the cerebral circulation and metabolism.

In conclusion, the accuracy of a technique can be evaluated in practice from its absolute values, its sensitivity and its variability coefficient. The latter parameter also indirectly reflects its temporal and spatial resolution.

## Investigation of the cerebral blood circulation

## Generalities and classification of the techniques

Starting from the early direct observations of pial vessels through a cranial window by Forbes and Wolf in 1928,

and the first application of the Fick principle for CBF measurement by Kety and Schmidt, in 1945, a huge variety of techniques of investigation of the cerebral circulation has been developed (for references see Purves, 1972; Edvinsson and MacKenzie, 1977).

The aim of this section is to help the investigator choose the technique the most convenient for his or her experimental purpose. The first question concerns the variable to measure, among the following which have been explored:

- local vasoreactivity, *in situ* on pial vessels or *in vitro* on isolated vessels;

- intracranial pressure, which indirectly reflects global cerebral circulatory changes;

- blood flow in large vessels supplying or draining the brain;

- pial intravascular blood pressure;

- local or regional tissue perfusion (expressed as blood volume per unit mass of tissue per unit time);

- thermal or electrical conductivity of brain tissue;

- velocimetric measurement of tissue perfusion by laser-Doppler flowmetry;

- electromagnetic or ultrasonic blood velocimetry.

Because of its physiological and pathophysiological relevance, tissue perfusion has been preferred by most investigators, but the study of local vasoreactivity remains a necessary complement. For a simplified classification, we have considered three main families of techniques:

(1) those 'directly' measuring tissue perfusion using a tracer or probes implanted within the cerebral tissue;

(2) those 'indirectly' measuring tissue perfusion using flowmetry techniques in the large cerebral or cranial vessels or by the intracranial or intravascular pressure;

(3) those measuring local vasoreactivity, either *in vivo* or *in vitro*.

Table 2 summarizes these techniques and their main features. It can be seen that within a given family, the techniques have paired features. For instance, the tissue blood perfusion techniques are either quantitative and discontinuous, or qualitative and continuous.

## Tissue perfusion techniques using a diffusible tracer

### Fick principle

As reported in Table 2, these techniques are quantitative and discontinuous. Most of them require a tracer which is diffusible, so that the Fick principle is applicable. This principle states that the amount of tracer $Q_i$ carried by the blood supplying the compartment i per unit of time is equal to the product of the blood flow $F_i$ perfusing this compartment and the arterio-venous difference of the tracer concentration $(C_a - C_v)$.

$$dQ_i/dt = F_i (C_a - C_v) \qquad (1)$$

This equation assumes that the tracer is perfectly diffusible and that the blood flow is stable (in steady state) during the measurement period. The tracer should also fulfill all the usual conditions of indicators: inertness,

**Table 2. Classification of the techniques of investigation of the cerebral circulation**

| | CBF quantification (ml/100 g.min) | Number of measurements | Degree of brain regionalization |
|---|---|---|---|
| **A. Measurement of tissue blood perfusion** | | | |
| I.  Diffusible tracer techniques | | | |
| – clearance techniques: | | | |
| arterio-venous difference | + | n | 0 |
| intracerebral probes | + | n | + |
| extracranial counting | + | n | ++ |
| – saturation techniques: | | | |
| brain dissection | + | 1 | ++ |
| autoradiography | + | 1 | +++ |
| II.  Non-diffusible tracer techniques | | | |
| microspheres | + | 1–6 | ++ |
| molecular microspheres | + | 1 | +++ |
| III. Non-tracer tissue perfusion techniques | | | |
| – conductibility measurement | ± | continuous | + |
| – velocimetry measurement (laser-Doppler flowmetry) | ± | continuous | + |
| **B. Indirect measurement of tissue perfusion** | | | |
| – large vascular trunk flowmetry | 0 | continuous | 0 |
| – variations of CBV and ICP | 0 | continuous | 0 |
| **C. Measurements of local vasomotor activity** | | | |
| – *In situ* vessels | 0? | continuous | + |
| vessel calibre | | | |
| pial arterial pressure | | | |
| – *In vitro* vessels | 0 | continuous | + |

CBF quantification: +, CBF in ml/100 g.min; ±, CBF changes in percentage or in arbitrary units; 0, qualitative evaluation.
Number of measurements: n, repeatable without limitation.
Degree of regionalization: 0, global measurement; +, one or few cerebral structures; ++, multiregional; +++, multilocal measurements.

detectability at tracer dose, etc. (see Lacombe *et al.*, 1980; Kety, 1985, for more details and references). Equation 1 has been applied in two different ways, either to a desaturation process (clearance) or to a saturation process.

## Clearance techniques

As mentioned in Table 2, clearance techniques are global or pauciregional and allow multiple measurements in the same animal. They require a first phase in which the brain is loaded with the tracer (intracarotid or intravenous infusion, or inhalation) and a second phase in which tracer administration is stopped inducing a clearance. CBF is calculated from the time-concentration clearance curve of the tracer, using one of three different means for measuring the tracer concentration: the arterio-venous difference, intracerebral probes and extracranial counting.

The technique involving measurement of the arterio-venous concentrations is a direct application of equation 1: tracer concentrations are assayed in serial arterial and venous samples. Siesjö's group widely used this technique in the 1970s, but these investigators also provided evidence that extracerebral contamination of the venous blood was difficult to avoid, the tracer being inhaled (Nilsson and Siesjö, 1976). Since technological developments tended to favour multiregional and local CBF investigation, this technique has been largely supplanted by others.

### Measurement by intracerebral probes

This technique has become the most common of the clearance techniques in the animal. Because of the practical impossibility of simultaneously measuring the tissue content $Q_j$, arterial $C_a$ and venous $C_v$ concentrations

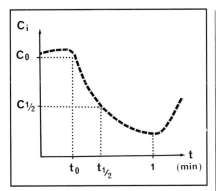

*Fig. 1A. Example of a clearance curve as obtained using hydrogen or helium as tracer added to the inhaled gas mixture (about 10–20 per cent). The clearance is obtained by replacing the tracer by a similar amount of nitrogen. Every second the brain tracer concentration ($C_i$) is fed into a computer, and the regional flow is calculated on line according to equation 5. $C_0$ is the initial tracer concentration and $C_{1/2}$ is half this value; $t_0$ is the beginning of the acquisition period and $t_{1/2}$ corresponds to $C_{1/2}$. (From Lacombe, 1992, with permission.)*

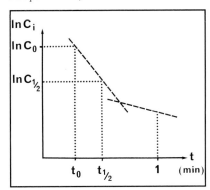

*Fig. 1B. Same clearance curve where the concentration axis is expressed in naperian logarithms. The curve is then expressed as a straight line if the clearance is monoexponential (one homogeneous compartment) or as two straight lines if biexponential (rapid and slow component corresponding to the grey and white matter flow respectively).*

of the tracer, equation 1 is modified to use $C_i$, the tracer concentration in the compartment i:

$$C_i = Q_i / M_i$$

where $M_i$ is the mass of tissue. To avoid performing venous sampling, the following relationship between $C_i$ and $C_v$ is used:

$$C_v = C_i / \lambda \qquad (2)$$

where $\lambda$ is the tissue–blood partition coefficient between venous blood and brain tissue. This equation again assumes that the tracer is perfectly (instantaneously) diffusible. The $\lambda$ value is generally close to unity, ranging from 0.7 to 1.2. It must be determined for each species and each tracer in white and grey matter by separate experiments.

In a desaturation process, $C_a(t)$ rapidly tends to zero so that:

$$C_i(t) = C_0 \, e^{-K_i t} \qquad (3)$$

where $C_0$ is the initial tracer concentration (at $t_0$, time zero) and $K_i$ the 'apparent flow' related to $F_i$ by the equation:

$$F_i = \lambda \, K_i \qquad (4)$$

During the clearance, $C_i(t)$ follows a decreasing exponential curve from $C_0$. Equation 3 is solved at time $t_{1/2}$ corresponding to $C_0/2$ (see Fig. 1A). Then:

$$F_i = \lambda \ln 2 / t_{1/2} \qquad (5)$$

which is the operational equation of the clearance technique. This equation assumes that the desaturation curve is monoexponential, i.e. it reflects the clearance of only one compartment.

Hydrogen is commonly used in the animal as tracer for the clearance technique, which requires platinum electrodes chronically implanted in the brain. These probes are polarized by hydrogen and give a signal proportional to its tissue concentration,

measured between platinum and a silver reference electrode placed under the skin (Haining *et al.*, 1968; Pasztor *et al.*, 1973; and see Lacombe *et al.*, 1980 for other references). Helium has also been successfully used, associated with mass spectrometric analysis of the tracer concentration (Seylaz *et al.*, 1983).

*Advantages*

(a) The best temporal resolution among the quantitative technique of CBF measurement. At flow rates around 100 ml/100 g.min, a 20 s clearance may be sufficient to obtain reliable values.

(b) Practically unlimited, repeated measurements in the same animal (e.g. every 5 min at the above flow rate).

(c) Easy calculation of the CBF values.

(d) Well adapted to chronic investigation in conscious animals, including associated behavioural studies.

*Disadvantages/limitations*

(a) Invasive for the brain. This problem mainly concerns acute implantation experiments (see also the phenomenon of cortical spreading depression, in section 'Conscious animal preparations'), since healing is satisfactory in chronic implantation experiments if properly performed.

(b) Spatial resolution somewhat limited. It depends on many parameters (denuded length of the electrode tip, excessive tracer diffusibility, blood flow level, etc.) and is not easy to estimate.

The brain volume 'sensed' by the electrode may be roughly estimated to be of the same order of magnitude as the brain samples dissected in the tissue sampling technique, based on the overall similarities of the CBF values obtained, i.e. 30–60 mg of tissue, corresponding to a sphere of 4–6 mm in diameter. Such a volume is inhomogeneous in rat brain, so that the flow value obtained reflects both rapid grey matter flow and low white matter flow. Flow inhomogeneity was recognized early on by Fieschi and colleagues as responsible for underestimation of the grey matter flow (for references see Lacombe *et al.*, 1979). This factor will have less importance in species with larger brains (more homogeneity in the same volume of cerebral tissue). Thus, values found in the rabbit, the cat or the monkey should be more precise. Moreover, flow inhomogeneity can be experimentally detected using a semilog plot of the clearance curve (see Fig. 1B). A biexponential clearance curve will give two straight lines. The investigator should normally select the straight part of the first one, which corresponds to the fast component flow (grey matter).

(c) Technical difficulties in recording the very low current intensity (μA) due to electrical parasites. Thus, the use of thinner electrodes with short tips in order to improve spatial resolution is limited because it reduces the signal-to-noise ratio and leads to parasited recordings.

(d) Low degree of multiregionality in hydrogen clearance measure-

ments (1–4 electrodes in the rat). It should be noted that each electrode requires its own recording channel (multiplexing is not possible because the polarization curves are different).

### Measurement by extracranial counting

This clearance technique is the most widespread in man, because it allows repeated measurements which are not invasive for the brain. It has also been developed in the monkey, especially in the baboon by Harper's group (Harper and MacKenzie, 1977).

Its principle is based on the use of radioactive tracers, mainly γ-emitting, that can be detected outside the skull (Hoedt-Rasmussen *et al.*, 1966). Further, increasing the number of detectors allows multiregional measurements to be made in relation to the brain size of the species investigated.

The clearance curves have been exploited according to two types of analysis:

(1) The compartmental analysis based on equation 3 above, considering that two functionally different compartments contribute to the clearance curve. A mathematical treatment using a computer allows us to distinguish the two exponential curves (see Part B of this chapter).

(2) The stochastic analysis based on the principle of the mean transit time (t), which is equal to the ratio of the area A delimited by the desaturation curve between $t_0$ and infinity, to the initial height of the curve $H_0$ (for references see Purves, 1972, and Lacombe *et al.*, 1980):

$$\bar{t} = A/H0 \qquad (6)$$

This method assumes that the distribution of the transit time of any molecule of tracer in an organ is a random variable, and this variable is related to the blood flow F (tissue perfusion) as follows:

$$\bar{t} = Vd/F \qquad (7)$$

where Vd is the volume of distribution of the tracer. Practically, the following relation is used for CBF calculation:

$$F = \lambda\, H_0/A \qquad (8)$$

(for details, see Purves, 1972; Harper and MacKenzie, 1977). Only the mean blood flow is calculated by this analysis the flow in the grey and white cerebral matter cannot be distinguished.

The use of external counting implies that extracerebral tissues in the detector field are not contaminated. In primates, this condition is respected by administering the tracer through the main route of blood supply to brain, i.e. the internal carotid. Additional precautions or adaptations are necessary for other species or other routes of tracer administration (intravenous or inhalation). (For details about species see Chapter 5 of this book; Purves, 1972; Edvinsson & MacKenzie, 1977; and Part B of this chapter for humans.)

The main drawback of extracranial detection is that the brain volume investigated is not easy to localize and delimit. If the recorded radiation energy is high (500 keV) the γ rays are little absorbed and the signals detected can originate from both cortical layers and deep cerebral structures, including from the opposite hemisphere. For weaker γ radiation energy, such as the tracer [133]Xenon (80 keV), a better localization is obtained by appropriate collimation of the detector cone area. However, such a radiation energy is associated with radiation scattering (through the Compton effect), and a substantial part (up to 50 per cent) of the activity recorded may originate from outside of the collimation cone (Meric and Seylaz, 1977). Thus, this factor makes the use of γ-emitters somewhat illusory for regional CBF measurement in the animal.

Tremendous improvements toward both non-invasiveness and better cerebral localization have later resulted from two major methodological developments:

(1) Single photon emission computed tomography (SPECT), which led to a significant increase in the degree of regionalization;

(2) Positron emission tomography (PET), which led to considerable improvement of the spatial resolution (see Part B of this chapter).

However, these techniques are not commonly used for measurements in the animal (see section B for references from Raichle's group; and also Kiyosawa *et al.*, 1989; Pinard *et al.*, 1993).

*Advantages*
(a) Repeated measurements, non-invasive for the brain.

(b) Although limited, multiregional measurements can be performed.

(c) Good temporal resolution.

*Disadvantages/limitations*
(a) Limited spatial resolution so

that only large-brain-size species (primates) can be conveniently investigated.

(b) The brain volume under investigation is not precisely defined. Improvement in this respect requires an advanced technological investment.

(d) Intracarotid tracer administration (to avoid extracerebral tissue contamination) in the animal implies the use of anaesthetics.

## Saturation techniques

This group of techniques is very homogeneous, the techniques being distinguished only by the tracer used. They are termed saturation techniques because the CBF measurement is performed during the tracer administration, while the brain tracer concentration is increasing, i.e. in the course of a saturation process, although saturation is not actually reached. At time T, the brain circulation is arrested and the tracer distribution is fixed until determination of brain tracer concentration. CBF is calculated according to equation 1, integrated under the particular conditions of a saturation process, between times $t_0$ and T, the beginning and the end of measurement corresponding to tracer perfusion:

$$C_i(T) = \lambda \, K_i \int_0^T C_a(t) \, e^{-K_i(T-t)} \, dt \quad (9)$$

where the flow $F_i$ is related to $K_i$ by equation 4. This is the operational equation of the saturation techniques, first used by Kety's group in 1955. Since it comprises a deconvolution, its resolution in practice requires a computer. The two experimental data are: $C_i(T)$, the local tracer concentration at time T, determined *post mortem*, and $C_a(t)$, the arterial tracer contamina-

tion-time curve for $0 < t < T$, the period of tracer administration (see Fig. 2).

The technique currently used results from several methodological developments beginning with the studies of Kety's group (see Reivich *et al.*, 1969; Sakurada *et al.*, 1978, for references). The main points were the choice of tracer which has to be (i) not diffusion-limited for valid quantification and (ii) convenient for autoradiographic studies (non-volatile). Insufficient diffusibility was found for antipyrine (Eklöf *et al.*, 1974; Sakurada *et al.*, 1978; Ohno *et al.*, 1979; and see below 'Tracer diffusibility'). Antipyrine was replaced by ethanol (Eklöf *et al.*, 1974; Lacombe *et al.*, 1979) or iodoantipyrine (Sakurada *et al.*, 1978; Ohno *et al.*, 1979). The latter is now commonly used for CBF autoradiographic studies in the rat, and [14C] labelling was found most appropriate for radiation energy (Lear, 1988). However, it was observed that iodoantipyrine was not convenient in the rabbit because of a significant uptake by red blood cells of this species. (For more details about the conditions that a tracer should fulfil, see Lacombe *et al.*, 1980.)

Practically, the tracer is intravenously infused at a constant rate for 30–40 s, while serial arterial blood samples are collected every 2–4 s. At the end of the infusion period circulatory arrest is provoked and the brain is removed. It is then either immediately dissected on a cool plate (0 °C) in the tissue sampling technique, or frozen (–40 °C) for later processing in the autoradiographic technique. The brain is then sectioned in a cryostat (–20 °C) and the slices are put in contact with sensitive X-ray film (for details, see Lear, 1988).

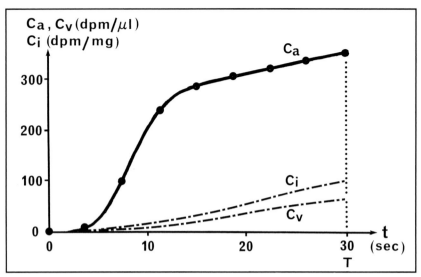

*Fig. 2. CBF measurement using the saturation technique involves the determination of $C_a(t)$, the curve of arterial tracer concentration. Arterial blood is serially sampled (8–10 values in 30 s). T is the time of circulatory arrest. The simultaneous evolution of $C_i(T)$ and $C_v(t)$ has been computed. (From Lacombe et al., 1979, with permission.) Only $C_i(T)$ and $C_a(t)$ are necessary for CBF calculation.*

Several studies have examined and estimated the causes of error of these techniques. Among the practical conditions to respect, the time lag between the $C_a(t)$ curve obtained by sampling arterial blood and the one corresponding to the brain perfusion should be rigorously allowed for (Lacombe *et al.*, 1979; Bryan, 1986; Lasbennes *et al.*, 1986). The importance of preventing post mortem tracer diffusion in brain has also been pointed out (Lacombe *et al.*, 1979; Hatakeyama *et al.*, 1992). In the case of high CBF values (e.g. from 300–500 ml/100g.min), as found with the autoradiographic technique under some activated conditions, especially in the conscious rat, a ramp $C_a(t)$ curve is strongly recommended. Such an arterial input function reduces the overestimation errors in $C_i(T)$ which might lead to excessively high CBF values, by moving the steep exponential part of the CBF-$C_i(T)$ relationship (see equa-

tion 9) towards higher $C_i(T)$ values. Technically, a ramp $C_a(t)$ curve requires a progressively increasing rate of infusion, which can only be obtained with a programmable syringe pump. Theoretical limitations common to all diffusible tracer techniques are considered in 'Limitations common to all diffusible tracer techniques' below.

*Advantages*
(a) Cerebral mapping of CBF, either multiregional (tissue sampling) or multilocal (autoradiography), the latter giving the best spatial resolution especially with the $^{14}$C label.

(b) Convenient for many species and many experimental paradigms, and allowing various reliable comparisons to be made.

(c) Accuracy thoroughly established, and all causes of error

quantified (Lacombe *et al.*, 1979). It is thus reliable provided that certain methodological conditions are respected.

*Disadvantages/limitations*
(a) The mapping obtained is instantaneous and at a single moment in the same animal. Consequently, this technique requires the use of groups of animals (control, test) for statistical treatment, and the homogeneity of each group must be ensured with respect to the physiological parameters and the reactivity tested.

(b) CBF calculation requires a computer.

(c) The autoradiographic technique takes a long time to achieve a rigorous quantification: each CBF value is an average of 4–6 optical density values measured, and the fit of the standards curve should have a very high and reproducible correlation coefficient.

## Limitations common to all diffusible tracer techniques

The diffusible tracer techniques share certain theoretical assumptions, which underly practical limitations.

### Tracer diffusibility
Excessive and insufficient tracer diffusion have both been suspected. Excessive diffusibility will induce non-perfusional (direct intercompartmental) tracer transport, i.e. independent of tissue perfusion, that will lead to overestimation of blood flow values. This occurs with gaseous tracers such as hydrogen, which diffuse directly from arteriolar or arterial blood to the brain parenchyma.

Limited diffusibility was shown for antipyrine and water by Siesjö's and Sokoloff's group (Eklöf *et al.*, 1974; Sakurada *et al.*, 1978). More diffusible tracers such as iodoantipyrine, ethanol, butanol or nicotine have since been selected for CBF studies (Lacombe *et al.*, 1980). However, they are still probably diffusion-limited. A simple model developed by Kety in 1951 (see Kety, 1985) allows evaluation of the influence of limited diffusion on CBF values, according to the equation:

$$m = 1 - e^{-PS/F} \text{ for } 0 < m < 1 \quad (10)$$

where m is an extraction coefficient reflecting the degree of equilibration of tracer between blood and tissue. The coefficient m can be introduced as a function of F in equation 1. PS is the product of tracer permeability (P) and capillary exchange surface (S). The use of this model provided evidence that limited tracer diffusion does not really influence CBF values below 250–300 ml/100 g.min (Lacombe *et al.*, 1979). Above this range, the model has to be adapted with corresponding PS values.

### *Partition coefficient λ (lambda)*
This coefficient is characteristic of a tracer, since it represents a ratio of solubility between venous blood and brain tissue (equation 2). For a given tracer, it has to be determined in separate experiments for each species, each cerebral tissue (grey or white matter), and eventually for certain particular or extreme conditions such as oedema, ischaemia or tumour, in which the chemical composition of brain tissue or venous blood may be considerably altered, thereby changing the tracer solubility.

Eklöf *et al.* (1974) showed that if the λ values were near unity (0.80 > λ > 1.20) a 20 per cent difference in this value caused negligible changes in the calculated CBF values for flow levels up to 200 ml/100 g. min. Thus, the lower the flow values, the more insignificant becomes the error introduced by an inaccurate λ value. For this reason, the same grey matter λ value is utilized for all cerebral structures, both white and grey.

It is worth mentioning that for given experimental data, increasing the λ value results in a decrease in the calculated CBF. This can be inferred from equation 5, and also equation 9, (although less readily in the latter case, because λ is present in two places considering $K_i = F_i \lambda /$, equation 4). This can be used as a test of accuracy of the computer calculation program.

### *Flow inhomogeneity*

This factor of underestimation has already been considered (see 'Measurement by intracerebral probes', on clearance techniques limitation, b). The influence is all the lower as brain samples become smaller (negligible for autoradiographic studies). The presence of two functional compartments of non-homogeneous flow can be detected in clearance techniques (see above) and in autoradiograms but not in dissected brain samples. In the latter case, if particular care is taken to distinguish at least white and grey matter and to dissect small samples, this cause of underestimation remains limited (Lacombe *et al.*, 1979).

## Non-diffusible tracer techniques

### Radioactive microspheres

This technique occupies a particular place among the quantitative tissue perfusion techniques, because it resembles the saturation (diffusible tracer) technique with respect to the experimental protocol, but it uses particulate tracers, microspheres, and allows several (up to six) successive CBF measurements in the same animal. It therefore seems very attractive to the investigator since paired control and test measurements can be obtained in the same animal.

In practice, a bolus of radiolabelled microspheres is infused into the arterial circulation, and these are trapped at the precapillary level. The CBF is assessed according to the principle of indicator fractionation of Sapirstein (1958), which assumes that the number of microspheres trapped in each region of interest i or j is proportional to the blood flow $F_i$ or $F_j$ supplying this region:

$$\frac{F_i}{A_i} = \frac{F_j}{A_j} = \frac{F_{ref}}{A_{ref}} \quad (11)$$

where A is the local radioactivity, and $F_{ref}$ and $A_{ref}$ represent respectively the reference flow and reference activity. The flow is calculated from the reference arterial sample (ref) obtained by quantitative sampling during a period covering the microsphere arterial infusion and distribution. Several CBF measurements (usually 2–4) can be performed by successive infusions of microspheres differently labelled with γ-emitting nuclides: $^{85}$Sr, $^{141}$Ce, $^{46}$Sc, $^{125}$I, $^{51}$Cr, $^{113}$Su, etc.

The following theoretical conditions must be met to ensure validity and accuracy:

(1) total and stable trapping of the microspheres on the first pass through the capillaries;

(2) uniform mixing within the arterial blood at all levels of the cerebral arterial tree as well as in the reference sample;

(3) uniform distribution so that they are trapped in direct proportion to the blood flow in all the regions investigated.

The optimal diameter of the microspheres, that which best meets these conditions, has been determined to be 15 μm, and they should be suspended in a saline solution and ultrasonicated just before being infused into the left atrium or ventricle of the heart. The latter point implies specific invasive surgery (thoracotomy) usually involving a chronic implantation.

*Advantages*

(a) Repeated measurements (2–4) in the same animal.

(b) Easy quantification and CBF calculation.

(c) Multiregional CBF values (brain dissection).

(d) Excellent temporal resolution.

*Disadvantages/limitations*

(a) Invasiveness of the cardiac catheter implantation.

(b) Degree of localization dependent on the quantity of microspheres reaching the given brain samples. To obtain sufficient accuracy, regional CBF measurement requires the infusion of large number of microspheres, which may not be innocuous, especially if repeated several times. This is why the use of microspheres for regional CBF measurement is more adapted to large species (e.g. cat, pig) than to small laboratory animals such as the rat.

(c) Spatial resolution inferior to that of the saturation technique using brain dissection. Autoradiography cannot be applied with γ-emitting nuclides (too long mean free path of the ray).

(d) Absence of proper verification of theoretical conditions 2 and 3 (see above). The validity of the microsphere technique has thus been questioned, especially for blood flow measurement in the brain. Haemorheological phenomena, such as plasma skimming due to axial streaming of the microspheres and right-angle bifurcations (penetrating arteries), tend to limit the validity of the measurement. Unfortunately, the influence of these factors on the validity of the technique cannot be readily quantified in the brain.

Finally, caution must be used for comparison of non-diffusible tracer CBF data with those obtained with diffusible tracer techniques. Indeed, interpretation of blood flow results based on precapillary-trapped particles and tissue perfusion differs, the former probably reflecting essentially arteriolar vasoreactivity and the latter flow through the capillaries. These two vascular compartments are probably subjected to different regulatory mechanisms, a debated issue.

## Molecular microspheres

Because of its attractive temporal resolution, the indicator fractionation technique of Sapirstein (1958) was also applied using a variety of tracers, including at first diffusible ones such as iodoantipyrine (Gjedde *et al.*, 1980; Van Uitert *et al.*, 1981; and for references see Lacombe *et al.*, 1980). Then, this technique in fact belonged to the saturation techniques presented above. In the 1980s, another application of this principle was developed, based on the use of 'molecular microspheres'. As for particulate microspheres, these molecules are presumed to be completely extracted by the brain during a single pass and sequestered for several minutes (by high-capacity amine-binding sites). This was first observed with the iodoamphetamine family of compounds, mainly the isopropyl-iodoamphetamine (Lear *et al.*, 1982).

Compared to the microsphere technique, this tracer has the advantage of considerably improving the spatial resolution, allowing autoradiographic investigation. However, only one measurement per animal can be performed. Other kinds of 'molecular microspheres' have been subsequently proposed, and HMPAO (hexamethyl-propyleneamine-oxime) labelled with $^{99}$Technetium has already been extensively employed for clinical applications thanks to its convenient biological half life (see Part B of this chapter).

## Non-tracer tissue perfusion techniques

These techniques require the use of probes and have the unique advantage of reflecting practically the continuous, dynamic aspect of the circulatory variable explored in 'real time'.

Unfortunately, this feature is associated with poor absolute quantification due to the difficulty in performing rigorous *in vivo* calibration.

The flow measurement is based on the principle of a relationship between tissue perfusion and the variable investigated, i.e. thermal or electrical conductivity, or erythrocyte velocimetry. Conductivity probes are usually implanted within the brain tissue (small diameter probes) or on the brain surface, whereas velocimetry probes are usually placed over the cortex. Both types of probe provide a local exploration of the brain circulation in one or two regions simultaneously investigated.

## Electrical and thermal conductivity measurements

### Electrical conductivity
Electrical conductivity (changes in resistance between two electrodes) was soon shown to be of low specificity with regard to local blood flow. It also reflects local ionic changes and electrolytic phenomena (especially oxidative) between the electrode and the brain tissue. Its development has thus remained limited.

### Thermal conductivity
On the contrary, thermal conductivity has been shown to be reliable and specific with regard to local blood flow changes. This technique is based on the fact that heat exchange in the tissue is mainly dependent on blood convection, so that overall the thermal conductivity increment is closely related to changes in tissue perfusion (see Carter, 1991, for review).

Since the early 1960s, two major developments have resulted from this principle, one based on the use of thermocouples, and the other based on the use of thermistors. The latter has been exploited more, because it was demonstrated that by using paired thermistors implanted in homologous brain regions, a linear relationship between the signal increments and tissue perfusion changes could be obtained over a wide range of flow levels (Seylaz, 1968). In this method, one thermistor is slightly heated (+0.3 to 0.5 °C) and, instead of measuring the change in resistance, the signal corresponds to the current necessary to heat the thermistor to maintain a constant the difference in temperature between the two probes (hence it was named a thermal clearance technique.)

The main criticism addressed to these techniques concerns the brain invasiveness of the probes, as for the hydrogen clearance technique, and the poorly defined volume of brain tissue 'sensed', possibly including functionally different compartments (white matter, grey matter, cerebrospinal fluid). Recent developments have been directed towards non-invasive, surface, cortical probes for medical purposes, often in association with laser-Doppler flowmetry (Carter, 1991).

## Velocimetry measurements (laser-Doppler flowmetry)

Velocimetry measurements can be applied either to the tissue surface to measure local tissue perfusion (using a laser beam) or to large vascular trunks (using an ultrasonic source) (see below 'Large vascular trunk flowmetry').

Laser-Doppler flowmetry consists of delivering a monochromatic infrared light source (laser beam) to a tissue, and analysing the reflected light received by a photodetector. The light emitted is scattered by both static and mobile (erythrocytes) elements of the tissue, but only the erythrocytes induce a Doppler frequency shift, whatever their direction of movement. Analysis of the backscattered light yields the fraction that is shifted (proportional to blood volume), together with the frequency of the Doppler shift (proportional to velocity). Since the two relationships are linear, the blood flow is the product of the blood volume explored and the red cell velocity (principle of mean transit time, see above 'Clearance techniques' equation 7, and Purves, 1972). It is continuously measured, and expressed in arbitrary units (volume not quantified). (For details and references, see Dirnagl *et al.*, 1989; Carter, 1991).

In practice, a flexible optical fibre transports the laser beam to and from the cortical surface. The tissue volume explored has been estimated to be approximately 1 mm$^3$ (Dinargl *et al.*, 1989). So far, except in a recent study (Bonvento *et al.*, 1994), the applications to CBF investigations have been performed on anaesthetized and immobilized (curarized) animals to avoid artifacts due to movement. One difficulty resides in the possible occurrence of cortical spreading depression if the cortical surface is touched or otherwise disturbed. This problem is circumvented by emitting and receiving the light through a thinned-down layer of skull or through a translucent, closed cranial window, chronically implanted, the dura remaining intact in both cases. Artifactual values are often obtained if large vessels (monodirectional flux) are in the field of the probe, or if changes in haematocrit occur.

Although recent, this technique shows a great deal of promise for a variety of cerebral circulatory investigations, because of its quantitative and continuous features. Its accuracy and reliability with respect to CBF investigation have already been tested (Skarphedinsson et al., 1988.; Dirnagl et al., 1989), however it is too early to give a definitive opinion about its use in the different paradigms of brain experiments. Its sensitivity seems similar to that of the microsphere technique, probably because vascular particle velocity is involved in both techniques.

## Vascular flowmetry and blood volume techniques

These techniques belong to the family of 'indirect' techniques of CBF measurement because they do not directly reflect tissue perfusion. Their value is in offering a continuous measurement of the variable investigated, without requiring invasive brain tissue implantation.

Two types of techniques can be distinguished: the quantitative ones aimed at investigating a large vascular trunk, and the qualitative ones aimed at investigating global cerebral circulatory changes. Their development has been slowed down by the introduction of the laser-Doppler flowmetry presented above, which has the additional advantages of being quantitative and measuring local tissue perfusion.

## Large vascular trunk flowmetry

Large vascular trunk flowmetry includes either the venous outflow techniques or flow velocimetry in arteries

and veins. Their use is now very limited for cerebrovascular investigations, and their interest has become mainly historical.

Cerebral venous outflow was first thoroughly investigated in the dog by Rapela and co-workers in 1964 (see Purves, 1972 or Lacombe et al., 1980 for references), using direct volumetry. This method raised the problem of the isolation of a purely cerebral vascular bed, i.e. how to avoid contamination by extracerebral blood due to the extra-intracranial anastomoses (see Chapter 5). This problem is particularly difficult to solve in carnivorous animals which have an extremely complex system of anastomoses in their cerebral vascular bed and thus requires extensive surgical preparation.

Velocimetric measurements on large vascular trunk are carried out using electromagnetic or ultrasonic flowmeter probes placed around or attached close to the vessel. Measurements can be quantified by probe calibration prior to implantation or *post mortem*. Owing to the complexity of the cerebrovascular bed and the small size of the cerebral vessels, the multiple developments of the quantitative pulsed Doppler technique in peripheral vascular beds have only rare counterparts in investigation of the cerebral circulation (Gotoh et al., 1982).

## Variations of cerebral blood volume (CBV) and intracranial pressure (ICP)

Variations of CBV and ICP have been postulated to reflect variations of CBF. This principle is based on the fact that the brain is enclosed in a rigid skull so that changes in brain blood

volume will induce ICP changes (Monro–Kellie theory). However, the relationship which links ICP and CBV is complex and still unclear. Since CBV amounts to 3–4 per cent of the total brain volume, comprising approximately 10 per cent of blood in arteries, the arterial blood volume which can be assumed to more directly reflect the vasoreactivity controlling tissue blood supply is therefore very low, and interpretation of ICP or CBV changes are thus subject to uncertainty. Moreover, other factors such as CSF production and circulation, and possible rapid changes in capillary permeability, may interfere with these variables.

*Advantages*
(a) Continuous investigation of the variable. This is particularly convenient for studying brief vascular events such as myogenic or neurogenic responses.

(b) Non-invasive for the brain.

(c) Simple technology.

*Disadvantages/limitations*

(a) Qualitative investigation, since the relationship to brain tissue perfusion is indeterminate.

(b) Investigation of the brain circulation is either global, or regional but of poorly-defined cerebral origin.

## Measurement of local vasomotor activity

The techniques used to investigate the local vasomotricity mostly address the question of the reactivity of vascular segments of the pial surface. Despite this spatially limited exploration, these techniques have led to very fruitful results, probably because the pial vessels studied (mostly arteries and arterioles) are a key site of regulation of downsteam cerebral tissue perfusion. Such studies can be carried out either *in situ* on the brain surface or *in vitro* on isolated cerebral vessels.

## *In situ* vessels

### *Measurement of the vessel calibre*

The modern version of the direct observation of pial vessels through a cranial window first developed by Forbes and Wolff as early as 1928 and Fog in 1939 is the TV image-splitting introduced by Wahl and Kuschinsky in the 1970s (for references, see Kuschinsky and Wahl, 1978). This technique continuously and quantitatively measures the calibre of a given pial vessel down to a diameter of approximately 50 μm.

The main interest of such *in situ* investigation is that the local CSF environment can be altered (e.g. pH, $K^+$, adenosine, etc.), and pharmacological agents can be applied locally (in the subarachnoid space) to the vessel studied, avoiding the blood–brain barrier. The action of neurotransmitters released by perivascular nerves is mimicked or, during stimulation, can be locally modulated. Differences between various vascular segments can be identified, which is of great value in studying the pial vascular motricity with respect with some physiological factors such as arterial blood pressure, $PaCO_2$ etc. Many variants of this technique exist, including multiple vessel diameter measurement (Auer and Haydn, 1979), blood velocimetry (Gotoh *et al.*, 1982) and superfusion of the arachnoid inside a closed window

(Levasseur *et al.*, 1975).

One difficulty common to all open skull preparations is that the cortex environment has to be maintained in normal physiological conditions, i.e. to be locally protected especially from loss of heat and $CO_2$. This is obtained by continuously flowing artificial cerebrospinal fluid of normalized composition over the exposed cortical area.

Recently, combined developments of fluorescence microscopy and computerized imaging have led to the appearance of the technique of confocal laser microscopy with which the cerebral microcirculation may be studied *in vivo* (Villringer *et al.*, 1989; Dirnagl *et al.*, 1992). Dynamic observations of all sizes of pial and also intracortical blood vessels (to a depth of about 200 μm) can be made, with quantification of the events observed.

### *Measurement of pial arterial blood pressure*

This other approach of the local vasomotor activity of pial arteries was developed by Heistad's group in the early 1980s (see for references Faraci and Heistad, 1990). This technique involves the measurement of arterial pressure with a glass micropipette that has a sharp bevelled tip of 3–6 μm diameter, connected to a servo-null pressure device. The micropipette filled with a NaCl solution is inserted into the lumen of pial arteries of rabbits and cats using a micromanipulator. This investigation is usually completed by the measurement of the arterial diameter using an electronic micrometer coupled to a TV camera and a microscope. The diameter of the 'small' pial arteries studied have been 100–150 μm in the rabbit and

200–300 μm in the cat (Baumbach and Heistad, 1983).

Although this technique is not a direct measurement of the pial vasomotor activity, its major value is that it allows comparison of the segmental resistance of different parts of the cerebrovascular bed, including between species. The cerebral vascular resistance (CVR) is determined from the classical relationship:

$$CVR = (Pa_i - Pa_j)/F_{ij} \qquad (12)$$

where $F_{ij}$ represents the blood flow within the whole brain volume perfused by the pial vessel investigated, with the upstream blood pressure $Pa_i$ and the downstream blood pressure, $Pa_j$.

The resistance of 'large' cerebral arteries is evaluated by the difference between the aortic and the pial arterial pressure. The blood flow value is measured in separate experiments, considering either regional or global CBF values.

This method to determine segmental cerebral vascular resistance is based on three assumptions:

(1) the presence of the micropipette should neither affect the arterial wall responsiveness nor the intravascular flow and downstream regional CBF;

(2) the open skull preparation should adequately maintain the cortical area exposed in normal physiological conditions as mentioned above;

(3) the CBF value used to assess the vascular resistance should spatially correspond rigorously to the single pial artery studied.

This last assumption is difficult to test directly, and more generally, the issue of the correlation between pial vascular motricity and the underlying parenchymal blood perfusion has not yet been satisfactorily resolved. The reason is probably the difficulty of approaching the intraparenchymal vascular events. It is likely that the mechanisms controlling these two variables are, at least partly, different. The issue could now be reconsidered using *in vivo* confocal laser microscopy mentioned above, which enables investigation both at the pial surface and within the superficial cortical layers.

### In vitro vessels

The study of isolated cerebral vessels has been applied to all vascular segments that can be sampled, from large trunks to small pial vessels (down to a diameter of ≈ 100 µm), and also to perforating arteries and arterioles (down to 20–30 µm). For reasons of size, large laboratory species such as cats, rabbits and dogs have often been preferred to rats.

The most common technique consists of measuring the tension exerted by a vascular segment mounted between two L-shaped rods or wires placed in a bath (e.g. Högestätt *et al.*, 1982). Changes in isometric tension of the vascular wall are recorded under the influence of various vasomotor agents. The comparison between vessels of different origin (brain, myocardium, limb) is readily made.

A more sophisticated technique consists of perfusing the vascular segments considered and of measuring the upstream perfusion pressure (e.g. Sercombe *et al.*, 1985). This has the advantage of distinguishing between the effects of drugs applied perivascu-larly from drugs administered intravascularly with the perfusing solution. The disadvantage is that, in practice, its application is limited to vessels with a sufficient resistance to perfusion for easily measured changes in pressure to be obtained without excessively high flow. Variants of this technique include measurement of vessel diameter instead of pressure (e.g. Dacey and Duling, 1984), and the preferential determination of the vasoreactivity of perforating arterioles (without individual cannulation) by ligating the major trunk and certain branches of the vessel.

*Advantages*

(a) Quantitative, continuous investigation of the vessel vasomotricity, allowing detailed pharmacological investigation to be performed, e.g. definition of the functional receptors (type and subtypes).

(b) Comparisons with vessels of different vascular beds of the organism and different species, and between veins and arteries.

(c) Very convenient for mechanistic studies on cerebral vessel functioning.

*Disadvantages/limitations*

(a) Pial vasomotricity represents only one aspect of the regulation of the cerebral circulation.

(b) Experimental difficulties increase as the size of the vessels studied decreases, and also with the invasiveness due to cannulation and difficulties of isolation of the vessel. Comparisons between homologous vessels from different species (different diameter) should be prudently drawn with regard to response amplitude.

---

### Investigation of cerebral capillary permeability and blood volume

Exploration of the cerebral circulation cannot neglect one of the variables regulating the supply of nutrients to the brain, namely the capillary permeability. Although the BBB restricts the passage of blood elements, proteins and small lipid-insoluble molecules from blood to brain tissue, physiologically vital exchanges occur which ensure brain tissue homoeostasis.

Cerebral blood volume (CBV) measurement must be considered in this section because this variable has to be determined in order to calculate capillary permeability. Indeed, all these techniques are based on estimates of the distribution of a tracer: first that within the intravascular space, and second, depending on the specific permeability of the tracer involved with respect to the BBB, distribution within the surrounding extracellular space and parenchyma. The permeability assessment must therefore take into account the amount of tracer remaining inside the vascular space by substracting it from the total brain concentration or activity (see below, equations 18 and 19).

### Measurement of cerebral blood volume (CBV)

As already mentioned, CBV has been little used as an index of the cerebral circulation in animal studies (see

'Variations of cerebral blood volume and intracranial pressure'). However, when quantitatively measured in conjunction with CBF measurement, it may become relevant in some situation of human pathology. The ratio CBF/CBV can be considered as an index of reserve of the cerebral perfusion. Its main value in the animal resides in the necessary determination of tracer intravascular blood content for BBB permeability measurement, as noted above.

Various types of tracer are employed for CBV investigation, with different carriers in blood. These may be either plasma proteins (e.g. serum albumin with bound Evans blue), erythrocytes (labelled with $^{51}$Cr) or plasma solutes such as $^{14}$C- or $^{3}$H-labelled hydrophilic molecules (sucrose, dextran, inulin, mannitol) or amino acid ($\alpha$-aminoisobutyric acid). The latter molecules are used for BBB permeability measurements as well.

Following i.v. infusion, the tracer distributes into the vascular space during several minutes, after which the circulation is arrested (time T). CBV is calculated according to the equation:

$$CBV = C(T)_{brain} / C(T)_{blood} \quad (13)$$

where $C_{brain}$ is the concentration at time T in dpm/100 g of brain tissue and $C_{blood}$ in dpm/100 ml of blood.

Equation 13 assumes that the tracer occupies entirely, and only, the intravascular space. The distribution time is chosen so that this assumption will be respected. $C_{blood}$ can be measured either from an arterial blood sample or from a venous blood sample draining the brain. After several minutes of circulation, $C_a$ and $C_v$ tend to become equal. CBV is expressed in ml/100 g

of brain tissue or as a percentage. (The units of equation 13 are not homogeneous due to the difference in density of brain tissue and blood.)

Several causes of error affect the CBV values. Each tracer gives CBV values 'smeared' by specific molecular properties related to its carrier or its BBB permeability. Thus, it is essential to state the origin of the CBV data: erythrocyte space, sucrose space, etc. Non-uniform distribution of erythrocytes may occur in the brain (see section 'Radioactive microspheres' for details about the microsphere technique), and there exists a minimal net flux of hydrophilic molecules across the blood–brain barrier. Plasma protein-bound tracers or dextran thus seem more reliable for accurate CBV measurement. For instance, global post mortem CBV values reported from the rat range from 0.5 per cent using labelled erythrocytes to 1 per cent using radioiodinated serum albumin (RISA) and to 1.5–3.5 per cent using hydrophilic molecules (5–10 min distribution period).

Importantly, multiregional measurements provide evidence of a non-uniform intracerebral distribution of blood volume. The cortical grey matter value is generally 2–3-fold higher than the corpus callosum white matter, and non-neural tissues such as the choroid plexus or the pia-arachnoid can be strongly labelled (Blasberg *et al.*, 1983), as well as structures devoid of BBB (the pineal and the pituitary glands, the circumventricular organs), according to the tracer used.

Under *in vivo* conditions, it has been evaluated in man that CBV comprises the following compartments: arterial bed $\approx$ 10 per cent, capillaries $\approx$ 20 per

cent and venous blood $\approx$ 70 per cent. It is thus clear that the circulatory state at the time of CBV measurement intervenes. Decapitation or circulatory arrest are associated with lower CBV values, due to loss of cephalic blood, than *in vivo* estimation.

## Measurement of cerebral capillary permeability

Two different aims can be distinguished in measuring cerebral capillary permeability. One is the study of the functional properties of the BBB in physiological or pathological conditions (transport of nutrients, hormones or drugs to the brain), and the other is the study of the functional integrity of the BBB (changes induced by experimental disruption, injury, etc.). The first aim concerns 'specific' BBB permeability, explored with certain molecules presenting a specific physiological (or pathological) interest, whereas the second aim concerns 'unspecific' BBB permeability which can be explored by a family of substances.

### Specific BBB permeability

Quantitative measurement of BBB permeability to certain molecules was introduced by Crone as the 'indicator diffusion method' in 1963 and then modified by Yudilevich and colleagues (1977). It requires the use of two tracers, one being non-diffusible (sucrose, inulin) to estimate the intravascular space, the other a test molecule of unknown diffusibility. A mixture of the two is infused as a bolus in carotid artery, and serial venous samples are then withdrawn following a single passage through the brain microcirculation.

Extraction E of the test molecule is calculated as follows:

$$E = 1 - R_{test} / R_{ref} \quad (14)$$

where R is the ratio of the concentration of a given tracer (test or ref) in venous plasma to that in the infused mixture. The permeability P is estimated according to equation 10 (see 'Tracer diffusibility' above):

$$\ln(1 - E) = PS / F \quad (15)$$

where S is the capillary exchange surface and F the blood flow, two variables that must also be determined.

The main advantage of this technique is that it allows repeated measurements in the same animal. However, it is global and the question of possible extracerebral contamination of the venous samples should be seriously considered. There have been further developments in man, but it has been largely replaced in the animal by the 'brain uptake index' (BUI) technique of Oldendorf (1971).

Measurement of the BUI also requires infusion into the carotid artery of a mixture of two tracers, but in this case, the reference is a standard of high diffusibility, assumed to be completely extracted by the brain tissue. The mixture of differently labelled ($^{14}$C, $^{3}$H) tracers is allowed to distribute in the brain during a single passage, i.e. $\approx$ 15 s in the rat.

The uptake by the brain is calculated as follows:

$$BUI = R_{brain} / R_{mixture} \cdot 100 \quad (16)$$

where $R_{brain}$ is the concentration ratio of the test molecule to the reference molecule in the brain, and $R_{mixture}$ is the same ratio in the mixture infused. This index thus expresses in percent-

age the change of the original mixture when it perfuses brain and is extracted by the parenchyma. The usual reference molecules are water, ethanol or butanol.

*Advantages*

(a) Quantitative expression of BBB permeability, relative to a standard; BUI can be multi-regional.

(b) Accuracy related to the fact that the tracers are infused together, with necessarily the same distribution (temporal and spatial).

(c) Many different molecules can be studied and compared.

*Disadvantages/limitations*

(a) These techniques are not independent of blood flow. At high flow rates, the capillary transit time may become too short for complete diffusion from blood to brain tissue.

(b) BUI does not directly reflect a physiological variable.

(c) Global results (indicator diffusion method) of questionable cerebral origin; multiregional results (BUI) limited to the vascular bed supplied by the artery used for tracer infusion.

## Non-specific BBB permeability

The techniques which measure non-specific permeability are based on the use of only one tracer, which is a small, biochemically inert, hydrophilic molecule such as inulin, sucrose, dextran, mannitol or $\alpha$-aminoisobutyric

acid (AIB). These tracers are poorly permeable through the BBB, and thus particularly suitable for demonstrating BBB disruptions or leakage.

In contrast to the techniques exposed above ('Specific BBB permeability'), the tracers are intravenously infused, and allowed to distribute for 10 min within the brain. Thus, the results obtained can be multiregional. Rapoport's group first developed this technique in the rat in the late 1970s. The capillary permeability-surface product (PS as defined in equation 17 below) is the variable calculated. PS is expressed by the ratio:

$$PS = C_i(T) / \int_0^T C_{plasma}(t) \, dt \quad (17)$$

where $C_{plasma}(t)$ is the time-related plasma concentration and $C_i(T)$ the brain parenchymal tracer concentration at time T. $C_i(T)$ is calculated from the measured brain concentration $C_{brain}$ in dpm/mg, after substracting the residual tracer contained in the intravascular blood. The intravascular blood content is in fact the CBV as defined in equation 13 above:

$$C_{brain} = (100 - V).C_i + V.C_{plasma} \quad (18)$$

where V in percentage is usually negligible in comparison to 100 giving:

$$C_i = (C_{brain} - V) . C_{plasma} \quad (19)$$

The intravascular tracer content V is calculated in additional experiments using the same tracer with diffusion times of 2–4 and 30–40 min (Ohno *et al.*, 1978).

It should be emphasized that the vascular space measured *post mortem* in the animal does not represent the total CBV, but rather the volume of blood remaining in the dissected brain after removal of the dural sinuses, choroid plexus, and meninges. In principle, it is assumed that $C_{plasma}$ always largely exceeds $C_{brain}$, so that back diffusion of the tracer from brain to plasma is negligible. This theoretical prerequisite is easily respected thanks to the low diffusibility of the tracer. (A high plasma-to-brain gradient yields significant regional counting in the brain.) However, the drawback is the very high plasma concentration at time T (up to 100-fold higher than in brain parenchyma), so that a high degree of accuracy in determining V is crucial. This difficulty has been partly overcome by the use of α-aminoisobutyric acid (AIB), a synthetic, inert, neutral amino acid of smaller molecular weight (104 daltons versus 342 daltons for sucrose).

The technique involving [$^{14}$C]AIB, developed in 1983 by Blasberg and colleagues, has the following advantages: AIB is transferred across the BBB by a carrier-mediated mechanism which is the rate-determining step for its accumulation in brain tissue; brain-to-blood backflux of AIB is minimized by cellular accumulation, so that the tissue activity reflects the magnitude of its unidirectional transport; tissue labelling obtained with AIB is greater than with other hydrophilic molecules (higher permeability, parenchymal uptake), and the autoradiographic technique can be applied. However, one drawback is the necessary use of another tracer to estimate the intravascular space.

*Advantages*

(a) Technique less invasive than the BUI technique (i.v. instead of intracarotid tracer infusion), so that it can be more easily performed on unanaesthetized animals;

(b) Multiregional measurements in the whole brain, thanks to the i.v. infusion of the tracer.

(c) Autoradiography can be used with tracer molecules of intermediate diffusibility, such as AIB.

*Difficulties/limitations*

(a) Measurement of the intravascular tracer content in brain is required. This value is not easy to measure accurately.

(b) Interpretations of changes in PS can be ascribed to changes, in either CBV or S, the capillary surface, or P, the capillary permeability. The influence of each of these variables has to be estimated (a change in CBF alters both CBV and S) and deducted from the effect observed.

## Measuremement of cerebral energy metabolism

### Generalities, objectives and classification of the techniques

As implied by the title of this section, only the techniques related to cerebral energy metabolism are considered, as opposed to, e.g. protein synthesis or long-term homoeostatic metabolic changes. Chapter 7 deals with the metabolic pathways followed and the intermediary metabolites found in various experimental conditions, and the metabolic fate of molecules within the brain. Our concern here is restricted to the techniques of quantification of the metabolism, i.e. the cerebral metabolic rate (CMR), measured by consumption of a substrate or production of a metabolite.

As for the cerebral circulation, a variety of techniques have been employed to quantify brain energy metabolism. However, the feature which distinguishes them is the tracer used, rather than the methodological principle. Since only one variable is measured (consumption of a metabolite), choosing a technique usually implies choosing a substrate as tracer.

In addition to the relevance of the tracer as a physiological or pathophysiological index in itself (see section B), the results obtained by these techniques allow important comparisons to be made with both the functional activity of the brain for anatomofunctional studies, and the local CBF. Comparison between the local cerebral distributions of metabolic activity and blood flow provides information on the coupling of these two variables, a result of considerable physiological significance (see 'Correlating CBF and CGU measurements' below and Chapter 12).

Table 3 recapitulates the different techniques of measurement of energy metabolism considered in this chapter. They are classified according to two characteristics: the degree of regionalization and the continuity or non-continuity of the measurement. As for the techniques of investigation of the cerebral circulation, the techniques of investigation of cerebral metabolic activity have associated features. Repeated measurement is associated with a low degree of regionalization, whereas multiregional meas-

Table 3. Classification of the techniques of measurement of cerebral energy metabolism

| | CMR quantification ($\mu$mol/100 g.min) | Number of measurements | Degree of brain regionalization |
|---|---|---|---|
| I. Global measurements by arteriovenous difference | | | |
| – oxygen and glucose consumption | + | n | + |
| – lactate and pyruvate production | + | n | + |
| II. Local multiregional measurement of glucose consumption | | | |
| – [$^{14}$C]2-deoxyglucose technique | + | 1 | ++ or +++ |
| – [$^{14}$C]glucose techniques | + | 1 | (dissec. or autorad.) |
| III. Indices of local metabolism | | | |
| – polarography and spectrometry of $O_2$ | ± | continuous | + |
| – spectrofluorimetry of NAD/NADH | ± | continuous | + |
| IV. Global measurement by NMR spectroscopy | | | |
| – $^{31}$P nucleus: phosphorylated compounds (energy state and intracellular pH) | 0 or + | n | + |
| – $^{1}$H nucleus: lactate production | ± | n | + |

CMR quantification: +, CMR in $\mu$mol/100 g.min; ±, per cent changes; 0, qualitative evaluation.
Number of measurements: n, repeatable without limitation.
Degree of regionalization: +, few cerebral structures ++, multiregional; +++, multilocal measurements.

urement (brain dissection or autoradiography) is associated with a single measurement. Nuclear magnetic resonance (NMR) spectrometry occupies a special place because this recent technique is still undergoing rapid development, and its technical features (degree of regionalization, sensitivity, etc.) are in constant progress.

## Global measurement by arteriovenous difference

### Principle

Metabolic substrates are taken up from blood by the functioning brain tissue and catabolites are released into the circulation. Considering the substrate to be a tracer, the Fick principle can be applied to estimate a net utilization (see equation 1; 'Fick principle', above). Assuming that the substrate taken up is metabolized, the CMR is given by:

$$CMR = CBF (C_a - C_v) \qquad (20)$$

where $C_a - C_v$ is the cerebral arteriovenous difference in tracer concentration. This principle is valid for both the utilization or the production of a given substance. The above equation assumes that the measurement takes place in a steady state period (from $t_0$ to T). Thus, CMR determination is indirect and requires the simultaneous measurement of two variables (see also CBF measurement by clearance in 'Clearance techniques' above).

### Oxygen and glucose consumption

Considering oxygen and glucose as tracers, the above equation allows calculation of their consumption by the brain in ml or $\mu$mol/100 g.min. This technique has been employed by Kety and co-workers in man and extensively used in the rat by Siesjö's group in the 1970s (Nilsson and Siesjö, 1976). The tracer used for CBF measurement was $^{133}$Xenon and the

venous blood was withdrawn from the superior sagittal sinus (for review, see Siesjö, 1978).

Since many substances can be measured in the arterial and venous samples, the respiratory quotient RQ (ratio between the arteriovenous difference for carbon dioxide and oxygen) can be determined. It was found close to unity, indicating that carbohydrates were the main substrate for the *in vivo* brain.

Further technological development has allowed measurement, quantitatively and multiregionally, of both glucose consumption (see Local measurements of glucose consumption below for laboratory species, and Part B for man), and oxygen consumption in man, using positron emission tomography (PET) (see Part B of this chapter).

### Lactate and pyruvate production

By additional measurements of arteriovenous difference of lactate and pyruvate, two useful indices can be calculated: the oxygen/glucose index and the lactate/glucose index (Siesjö, 1978). These ratios give quantitative data on the balance between the aerobic and anaerobic pathways of glucose metabolism (the former being the major fraction, $\approx 90$ per cent in normal conditions). However, the question of possible extracerebral contamination of the venous samples is particularly relevant for lactate, which is produced by extracranial muscles.

*Advantages*

(a) Relatively easy to set up, yielding quantitative CMR data for several metabolites.

(b) Measurements can be repeated in various experimental conditions in the same animal.

(c) Useful for indicating coupling between the global CBF and CMR for several substances.

(d) Brief temporal resolution.

*Disadvantages/limitations*

(a) Global results (white and grey matter not distinguished).

(b) Purely cerebral origin of venous blood is not certain in many animal species (but less risk of extracerebral contamination in primates).

(c) Limited accuracy of the results due to the combined measurement errors of two variables.

(d) Difficult to apply in the unanaesthetized animal.

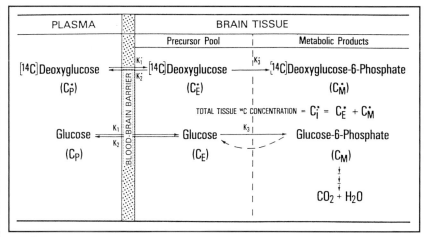

Fig. 3A. Diagrammatic representation of the theoretical model of the deoxyglucose method of Sokoloff. $C^*_i$ represents the total $^{14}C$ concentration in a single homogeneous tissue of the brain. $C^*_P$ and $C_P$ represent the concentrations of $[^{14}C]2$-deoxyglucose ($[^{14}C]2$-DG) and glucose in the arterial plasma, respectively; $C^*_E$ and $C_E$ represent their respective concentrations in the tissue pools that serve as substrates for hexokinase. $C^*_M$ represents the concentration of $[^{14}C]2$-deoxyglucose-6-phosphate in the tissue. The constants $k^*_1$, $k^*_2$ and $k^*_3$ represent the rate constants for carrier-mediated transport of $[^{14}C]2$-DG from plasma to tissue, for carrier-mediated transport back from tissue to plasma, and for phosphorylation by hexokinase, respectively. The constants $k_1$, $k_2$, and $k_3$ are the equivalent rate constants for glucose. $[^{14}C]2$-DG and glucose share and compete for the carrier that transports them between plasma and tissue and for hexokinase which phosphorylates them to their respective hexose-6-phosphates. The dashed arrow represents the possibility of glucose-6-phosphate hydrolysis by glucose-6-phosphatase activity, if any. (From Sokoloff et al., 1977, with permission.)

## Local measurements of glucose consumption

These techniques are based on the fact that glucose is the main energy substrate of the brain. This assumption is respected in normal physiological conditions and in many, though not all, other experimental conditions. During development or a long-lasting fast, ketone bodies support cerebral energy production (see Chapter 7).

In the late 1970s two different groups developed a technique to quantify glucose consumption in brain, one using $[1\text{-}^{14}C]2$-deoxy-D-glucose as tracer (Sokoloff and co-workers), and the other using $[^{14}C]$glucose as tracer (Hawkins and co-workers). Both techniques can be applied by autoradio-

graphy, giving results spatially comparable to those obtained for CBF measures using $[^{14}C]$iodoantipyrine.

## The $[^{14}C]2$-deoxyglucose technique (2DG)

The principle of this technique is based on the structural analogy of 2DG and natural glucose (Sokoloff *et al.*, 1977). At an initial step of the glycolysis, the 2DG-6-phosphate (2DG-6P) formed is metabolically inert, i.e. it is not recognized by the following enzyme in the metabolic pathway, and thus remains trapped intracellularly (see Fig. 3A).

Practically, a tracer dose of $[^{14}C]2$-DG is intravenously infused and the arterial plasma contamination is measured by serial arterial blood sampling. The protocol is therefore similar to that of

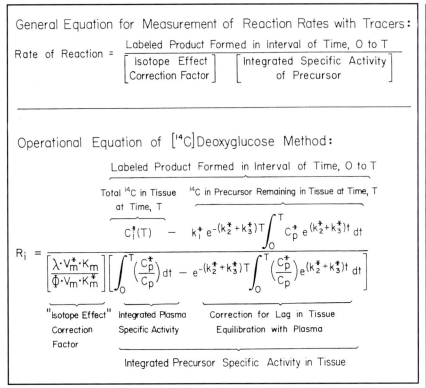

*Fig. 3B. Initial operational equation of radioactive deoxyglucose method and its functional anatomy. T represents the time at the termination of the experimental period; λ equals the ratio of the distribution space of deoxyglucose in the tissue to that of glucose; Φ equals the fraction of glucose which, once phosphorylated, continues down the glycolytic pathway; and $K^*_m$ and $V^*_m$ and $K_m$ and $V_m$ represent the familiar Michaelis–Menten kinetic constants of hexokinase for deoxyglucose and glucose, respectively. The other symbols are the same as those defined in Fig. 3 A. (From Sokoloff, 1978, with permission.) The 'isotope effect' correction factor was later termed 'lumped constant' (LC). A more complex equation, corresponding to a model less restrictive about the constancy of arterial plasma glucose concentration was also developed (Sokoloff et al., 1983), as well as a model taking into account $k^*_4$ (see Part B of this chapter).*

the CBF saturation techniques except that (i) the non-radioactive glucose concentration in plasma has to be measured in the same arterial samples, and (ii) the measurement period lasts 45 min instead of 30–40 s for CBF measurement. At time T (45 min) the animal is sacrificed and the brain tracer concentration $C_i(T)$ is measured either in the dissected brain or autoradiographically. Figures 3A and B show the model of the 2DG method and the operational equation

defining the $CMR_{Glucose}$ per unit mass in tissue i. This variable is usually called local cerebral glucose utilization (lCGU) and the values are expressed in $\mu mol/100\,g.min$.

The main theoretical conditions are the following:

(1) steady state during the measurement period of cerebral function, CGU, and also blood flow and BBB transport;

(2) functional homogeneity of the tissue compartment (as for tissue saturation CBF measurement);

(3) the measurement period must be at least 30 min, so that the amount of 2DG-6P accumulated is sufficient to be accurately measured and the unreacted 2DG (term subtracted in the denominator in Fig. 3B) is low, but it must also be less than 60 min, so that the loss of 2DG-6P (due to the glucose-6-phosphatase activity) and $^{14}CO_2$ production remains negligible and the error due to inhomogeneity of the rate constants is minimal. A period of 30–45 min has been found to be optimal for normal rats (Mori et al., 1990). This technique thus has a long measurement period, but it should be borne in mind that the first 10–20 min of measurement are the most important (high gradient of 2DG from blood to brain) for determining the final value, and this time, 'the impregnation time', is all the shorter when the metabolic rate is high.

The operational equation first given by Sokoloff et al. (1977) assumed that the arterial plasma glucose concentration remains constant during the measurement period. Since this constraint has proved to be cumbersome in many experimental conditions, another operational equation which is more versatile with respect to this assumption has been derived (Sokoloff et al., 1983; and Part B of this chapter), which authorizes 20 per cent variations in glycaemia from the baseline level.

The lumped constant (LC) and the rate constants $k_1$, $k_2$ and $k_3$ have to be determined in separate experiments, for each species and, ideally, for the particular experimental conditions considered. The procedure has been described by Sokoloff *et al.* (1977). Differences between rate constants from one structure to another, especially between grey and white matter, is a cause of error. This error is minimized as the declining curve of plasma 2DG approaches zero at time T, which justifies the 45 min measurement period.

This technique quickly proved very successful, and the most reliable for quantitative measurement of CGU. Moreover, since data on local CGU yields important information on local brain activity, it was clear that it represented a major advance in this field of neuroscience. It has been adapted, using [$^{18}$F]2-DG as tracer and positron emission tomography (PET) to the monkey (Kiyosawa *et al.*, 1989) and man (see Part B of this chapter). Refinements have also been made on the initial model, such as for instance taking into account the rate constant $k^*_4$ (see Part B of this chapter).

## The [$^{14}$C]glucose techniques

Several investigators have attempted to develop techniques for measuring glucose consumption over a short duration, i.e. 2–10 min (see Gjedde *et al.*, 1980; and Hawkins *et al.*, 1979, for references). Since these techniques use [$^{14}$C]glucose, whereby the $^{14}$C follows multiple pathways (see Chapter 7), they require simplifying hypotheses and related assumptions and approximations.

The calculations first proposed by Hawkins and colleagues in 1974 using [2-$^{14}$C]glucose are based on the fact that, once within the metabolite pool, $^{14}$C is incorporated into many intermediary metabolites representing a large amount relative to the [$^{14}$C]glucose taken up. At a given moment, the rate at which $^{14}$C is incorporated is given by the product of CGU and the specific activity of the available glucose (see Hawkins *et al.*, 1979, for details and references).

The first assumption is that the loss of $^{14}$C via $CO_2$ formed is negligible because $^{14}$C is temporarily trapped in intermediary metabolites. This assumption requires a measurement period as short as 2–10 min. Another assumption is that, the exchange of glucose between blood and brain being very rapid, the integral of brain tissue glucose specific activity can be assimilated with that of blood plasma after a sufficient time (a few minutes) (Hawkins *et al.*, 1979).

Later versions of the method allowed for the loss of $^{14}$C through metabolism to $^{14}CO_2$ by considering it was proportional to the total amount of $^{14}$C in the metabolite pool and CGU; the measurement period was reduced from 10 to 5 min, and [1-$^{14}$C]glucose was found more convenient for investigation of energy metabolism than [2-$^{14}$C]glucose. The premise of specific activity equivalence has been replaced by separate measurements of the constants (see Hawkins *et al.*, 1988). Despite the considerable advantage of a shorter temporal resolution, the [$^{14}$C]glucose method has been less widely accepted than the [$^{14}$C]2-deoxyglucose method (see Part B of this chapter for the use in man).

*Advantages*

(a) Quantitative mapping of the cerebral glucose utilization.

(b) Measurement independent of the CBF.

(c) 2DG technique reliable: it has become a widely accepted technique which yields reference values.

(d) Can be applied in the conscious animal.

*Disadvantages/limitations*

(a) A single measurement per animal, thus necessitating use of groups of homogeneous animals.

(b) Long measurement period (especially 2DG technique) giving poor time resolution and difficulties in correlating results with CBF values (see 'Correlating CBF and CGU measurements' below).

(c) Overall execution of the technique is time-consuming (see 'CBF measurement by autoradiography' in 'Saturation techniques' above).

## Correlating CBF and CGU measurements

### Principles and limitations

Parallel measurements of CBF using iodoantipyrine and CGU using 2-deoxyglucose in the rat have been performed by many groups, sometimes by using double-tracer labelling in the same animal (see 'Multiple tracer measurements' below). The autoradiographic technique is particularly suitable for these studies, because of its optimal spatial resolution for both variables.

One difficulty is the considerable difference in the temporal resolution of the two measurements, 30–45 s versus 30–45 min. When both the vascular and the metabolic events studied are very stable (steady state ensured for both variables for 45 min), this is not critical. However, this is not the case in many experimental situations in which the vascular events are brief or more transient than the metabolic events, thereby precluding a direct interpretation. There is no completely satisfactory solution to this problem. One way to limit the difficulty is to adjust as far as possible the conditions and timing. For instance, not only the activation or stimulation process should be similar for both measurements, but the CBF measurement period should take place at a time equivalent to 10–20 min after 2DG administration. In fact, the 'impregnation time' for the 2DG measurement is approximately 15–20 min in control conditions and as little as 10 min in activated conditions (see 'The $[^{14}C]2$-deoxyglucose technique' above), so that such a comparison between these two variables has a certain foundation. The stability of the circulatory response can also be verified by measurement at, e.g. 5 and 15 min, but this is not a necessary prerequisite, since a circulatory response may be physiologically relevant although transient.

The reader should be aware of the fact that, because CBF and CGU differ in their sensitivity to physiological parameters (capnia, temperature, arterial blood pH), investigation of their coupling requires that particular care be taken to obtain a comparable state for the two types of measurement. For instance, CBF is very sensitive to changes in systemic $pCO_2$, even moderate ones, whereas CGU is not.

If one uses an anaesthetic which induces a respiratory depression and hence hypercapnia, CBF will increase whereas CGU will decrease due to the metabolic depression (Cavazzuti *et al.*, 1987). The coupling pattern of this anaesthetic will therefore appear to be that of a vasodilator agent. Similarly, certain central activations in the conscious animal induce hyperventilation and hence hypocapnia which affects the cerebrovascular response. Thus, the true pattern of coupling can be interpreted only in the animal in a strictly normal or normalized (by artificial ventilation) state of capnia.

## Multiple tracer measurements

As already mentioned, the relationship between CBF and CGU can be studied by double-tracer measurements which, by the use of only one series of animals instead of paired series, circumvents the above difficulties. This also eliminates inter-animal variability and reduces other experimental errors such as variations in tissue thickness and brain structure location in autoradiographic measurements. However, multiple radionuclide studies involve other sources of error: (i) separation of exposures from the different tracers, (ii) densitometric measurement, and (iii) data comparison between the autoradiographs (for details, see Lear, 1988).

Two main strategies have been proposed for the simultaneous measurement of CBF and CGU in the same animal. One exploits the difference in the characteristics of the radionuclides used for labelling the two tracers (half-life or emission energy), and the other exploits the difference in solubility of the two tracers in an organic solvent.

Triple tracer autoradiography has also been developed, complementing CBF and CGU measurement with another variable such as protein synthesis (Mies *et al.*, 1986) or recently measurement of pH, a particularly relevant variable for studies of ischaemia (Nakai *et al.*, 1988).

## Correlations observed and interpretation

A tight correlation between CBF and CGU values was observed early on in the rat (see for review McCulloch, 1982, and Chapter 9) based on linear regression of the values of the two variables for all the cerebral structures analysed. This analysis usually provides a correlation coefficient close to unity indicating that the levels of blood flow and glucose utilization present a similar hierarchy of distribution within the brain. However, several assumptions limit this statistical analysis.

First, as reported by McCulloch (1982), a test of the homogeneity of the CBF/CGU ratios must first be performed in both control and test situations. Only the results corresponding to cerebral structures for which this homogeneity is confirmed can be interpreted. Second, usually a large number of regions are analysed using a small number of animals (6–8) per group, which is a limiting factor for this test (Ford *et al.*, 1991).

The interpretation of 'uncoupling' is proposed when CBF and CGU do not vary alike between the test and control situation. Either blood flow increases whereas glucose use remains unchanged, or blood flow decreases whereas glucose use is unchanged or even increased. Such dissociations of

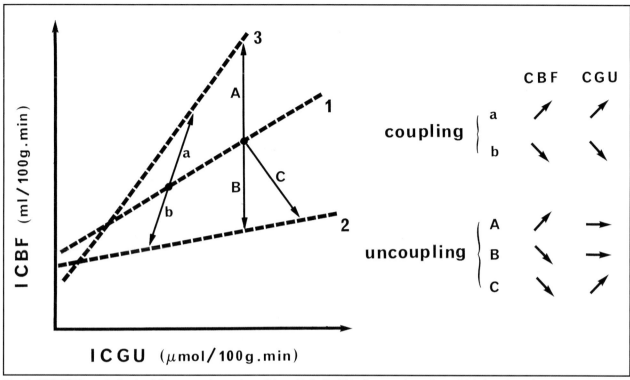

*Fig. 4. CBF-CGU correlation in different experimental conditions (1, 2, 3). This diagram shows the multiple hypothetical situations of coupling and uncoupling experimentally observed between the local cerebral blood flow (lCBF) and the corresponding local cerebral glucose utilization (lCGU) found in a variety of cerebral structures, as expressed in absolute values.*

the changes in the vascular and metabolic activities are not easy to establish statistically. A modification of the slope of the regression line is an equivocal indicator: the constancy of the slope can be obtained in association with either coupling or uncoupling (see Fig. 4). Thus, some investigators use the term 'resetting' to characterize the flow-metabolism relationship when the CBF/CGU ratios are not constant (Kuschinsky *et al.*, 1983; Bonvento *et al.*, 1991; and see Chapter 9 for details and references). Likewise, the use of relative changes (e.g. per cent) in flow or glucose use can be criticized because it masks the baseline values. It should be emphasized that the behaviour of individual cerebral structures is the most

meaningful variable. This is demonstrated by the lines connecting the identical structures in the two situations studied, control and test (see Fig. 4).

The major interest in uncoupling is that it may constitute a rare functional argument in favour of a separate control of one variable (blood flow), which may be independent of the other (metabolic activity), thereby suggesting a possible neurogenic or hormonal control in addition to the metabolic control (see Chapter 12). This may occur either globally or regionally in certain structures only, the latter case being more demonstrative since it provides evidence of heterogeneity of coupling mechanisms in the

same animals.

Whatever the technical sophistication, the investigator should keep in mind that the spatial resolution of different variables may not be identical. For instance, flow measurement has poorer spatial resolution than glucose utilization for physiological reasons: the former is organized at the cytoarchitectonic level whereas the latter is organized at the cellular level.

### Indices of local metabolism (probe techniques)

These techniques use probes to continuously measure one or more of the metabolites of energy metabolism. It can be expected that future developments will favour techniques which are quantitative, which reflect one independent variable or simultaneously several related variables, and which explore the intracellular compartment. Below we briefly describe some existing techniques which present significant advantages. They share the usual disadvantages of probe techniques (see CBF measurements, in 'Measurement by intracerebral probes' and 'Non-tracer tissue perfusion techniques' above).

## Oxygen polarography

The technique measuring tissue oxygen levels by polarography was the first developed (see for review Leniger-Follert, 1985). It requires either platinum electrodes, the same as those used for hydrogen clearance (see 'Measurement by intracerebral probes', above), with opposite polarity, or glass electrodes. The advantages are that it is easily set up, and can continuously reflect local tissue oxygen partial pressure. Its drawbacks are those described for the hydrogen clearance technique, plus its dependence on blood flow and the difficulty of quantifying in absolute values.

## Mass spectrometry measurement of tissue pO$_2$ and pCO$_2$

This technique allows continuous and quantitative measurement of tissue partial pressure of oxygen (pO$_2$) and carbon dioxide (pCO$_2$) by gas microsampling through an implanted cannula covered with a polyethylene membrane at the extremity (Pinard *et al.*, 1978). These measurements can be associated with repeated CBF measurements by helium clearance (Seylaz *et al.*, 1983). Thus, data related to O$_2$ consumption together with CO$_2$ production and blood flow can be ob-

tained simultaneously from the same brain region, and their interactions can be studied. The major drawback of this technique is its technical complexity and the high cost of the mass spectrometer.

## Redox state of NAD/NADH and cytochrome a,a$_3$

These studies have attempted to estimate the cellular free energy production which in the brain almost totally originates from oxidative metabolism in the mitochondrial respiratory chain (see Chapter 7). The behaviour of the respiratory chain is determined by *in situ* detection of the redox steady state of intramitochondrial nicotinamide adenine dinucleotide (NAD) or cytochromes, without penetration of the brain.

Using surface spectrofluororeflectometry, it is possible to locally evaluate in cortical tissue the ratio of NAD/NADH which is related to the cellular redox state. The technique consists in delivering UV light to the tissue surface and in analysing the reflected fluorescence emitted by NADH (Jöbsis *et al.*, 1977; Dora, 1984).

Its major drawback stems from the results obtained. It seems that the NAD/NADH ratio is very stable in most experimental situations and that it varies only in pathological conditions such as ischaemia (Tanaka *et al.*, 1986). One explanation is that the reduction of this coenzyme, closely associated with dehydrogenation of Krebs cycle intermediates, is sensitive to changes in the supply of these substances. This reduction occurs in the first step in the transport of reducing equivalents from substrates to O$_2$ and is thus sensitive to changes in their

concentrations (Jöbsis *et al.*, 1977). Another explanation could be limited sensitivity of the detector.

Compared to NAD/NADH, the reaction involving cytochrome a,a$_3$ (cytochrome oxidase) is at the terminal part of the respiratory chain and directly implicates O$_2$. Thus, O$_2$ utilization is proportional to the rate of electron transfer from cytochrome a,a$_3$ to molecular O$_2$, and *in vivo* changes in its redox state are directly indicative of cellular oxygen sufficiency.

There are several difficulties with these techniques. First, interpretation of the signals in terms of the cellular redox state is hampered by the interaction of many factors, either metabolic in the case of NADH, or local blood volume and haemoglobin oxygenation in the case of cytochromes (Jöbsis *et al.*, 1977). Moreover, in contrast to the two other types of technique presented above (see 'Principles and limitations', and 'Multiple tracer measurements' above), the measurements of redox state have not been successfully applied in the unanaesthetized, chronic animal preparation, probably because signal detection requires complete immobilization of the animal.

## NMR spectroscopy

### Principles and development of *in vivo* NMR spectroscopy

Despite wide utilization for 20 years for *in vitro* studies in tissues and isolated organs, *in vivo* nuclear magnetic resonance (NMR) spectroscopy is a recent development, later than NMR imaging. Both techniques are based on the same principles. For a nucleus of a given atomic species, the phenomenon of NMR is observed when the sample, placed in a strong magnetic field, receives an electromagnetic pulse with a frequency characteristic of the nuclear species. The loss of excitation of the resonant nuclei, after the end of the pulse, induces the emission of a transient signal named free induction decay. The spectrum of the frequencies contained in this signal constitutes the data used in NMR spectroscopy and imaging. For a given nuclear species, this spectrum should contain only one peak at the resonance frequency of the nucleus. In fact, in living tissues, the chemical environment of the nucleus studied modifies the characteristics of the resonance signal, inducing a frequency shift named chemical shift. The NMR spectrum may thus contain several peaks which represent different chemical environments for a given nucleus, i.e. its presence in a given chemical species.

Theoretically, all nuclei which have a magnetic moment of spin different from zero can give rise to a NMR signal and be used for NMR spectroscopy, e.g. phosphorus ($^{31}$P), proton ($^{1}$H), carbon ($^{13}$C), fluorine ($^{19}$F), sodium ($^{23}$Na), potassium ($^{39}$K), etc. However, *in vivo* applications of these techniques have been hampered because for some of these nuclei the natural abundance in living tissues and their intrinsic sensitivity is very low compared to the water proton used in imaging.

In the early 1980s the development of surface coil techniques for NMR spectroscopy allowed the first *in vivo* spectrum of the brain to be obtained (Ackerman *et al.*, 1980). A surface coil for excitation pulse emission induces a non-homogeneous excitation pulse

in a volume beneath the coil which covers approximately a half-sphere with a radius equal to the coil radius, thus allowing a relative spatial selection. The advent of such surface coils has given a tremendous acceleration to *in vivo* explorations of cerebral tissues during the last few years (see for review Meric, 1991).

More recently, combination of the technique of NMR imaging with the methods of NMR spectroscopy has enabled local NMR measurement with a spatial resolution suitable for individual spectroscopic investigation of the main cerebral structures, both in man and animals (Fisher *et al.*, 1992; Fernandez *et al.*, 1992).

## $^{31}$P spectroscopy

$^{31}$P spectroscopy is, at the present time, one of the most widely used for *in vivo* cerebral tissue exploration. This is due to its relatively good sensitivity and the high concentration of phosphorus-containing molecules in living systems. Although the phosphorus sensitivity is about 15 times lower than that of the proton, it is relatively easy to obtain a spectrum of brain tissue with an acquisition time compatible with the time-course of cerebral metabolic events. Another asset lies in the nature of the molecules detected *in vivo*: phosphomonoesters (phosphorylcholine and phosphorylethanolamine), inorganic phosphate, phosphodiesters (glycero-

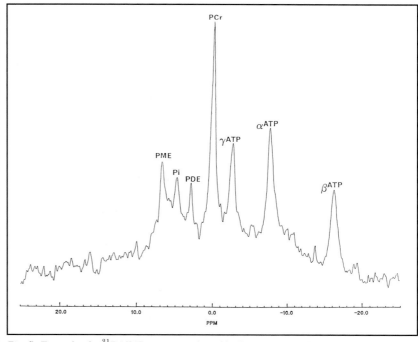

*Fig. 5. Example of a $^{31}$P NMR spectrum of a rabbit brain in normal physiological conditions. Each peak corresponds to a phosphorylated compound. The unit in abscissa is the relative frequency expressed in PPM (parts per million), with phosphocreatine (PCr) peak as reference. This spectrum was obtained from the Fourier transform of the sum of 96 signals (free induction decays) accumulated during 5 min, to improve the signal to noise ratio. (From P. Roucher, with permission.) Abbreviations: PME: phosphomonoesters; Pi: inorganic phosphate; PDE: phosphodiesters; ATP: adenosine triphosphate, with the three phosphates γ, α and β.*

phosphorylcholine and glycerophos-phorylethanolamine), phosphocre-atine and the three phosphates ($\gamma$, $\alpha$, $\beta$) of nucleoside triphosphates which are essentially ATP in the brain (see Fig. 5). There is another broad reson-ance, superimposed on the others, which originates from membrane phospholipids and cranial bone phos-phates.

The position of the various peaks in the spectrum provides information on the intracellular pH and com-plexed $Mg^{2+}$ ion available. Inorganic phosphate is present in cells in two forms ($H_2PO_4^-$) and $HPO_4^{2-}$) with relative concentrations and a rate of interchange dependent of H+ concen-tration. This rate of interchange is suf-ficiently high for only one 'averaged' inorganic phosphate peak to be dis-cerned. Its position is calculated with respect to the phosphocreatine peak position which is stable for a pH range larger than the physiological range. It is thus possible to establish a titration curve linking intracellular pH with the chemical shift of the inorganic phos-phate peak. It is also possible to calcu-late the level of magnesium ion bind-ing, which is given by a shift of the ATP resonance peaks.

Many studies have already shown that $^{31}$P-NMR spectroscopy is a useful tool for investigation of the metabolic ef-fects of a drug before, during and after a reversible pathology such as hypoxia or ischaemia. Above all, this technique is the first to provide quan-titative data on the metabolic energy state of the brain cells.

## $^1$H spectroscopy

The proton is the nucleus with the highest NMR sensitivity, but the res-onances of the different detectable metabolites are masked by the overlap of the resonance of the protons of water. Techniques of water signal sup-pression are necessary to observe sig-nals from metabolites other than water. An *in vivo* $^1$H-NMR spectrum of the brain, after water signal sup-pression, exhibits resonances which originate mainly from inositol, choline, the creatine–phosphocre-atine pool, the glutamate–glutamine pool, N-acetyl-aspartate (NAA) and lactate. In most studies, $^1$H-NMR has been used to detect only lactate and NAA because the other peaks are diffi-cult to resolve and their reproduci-bility is then very poor. The explora-tion of lactate variations correlated with intracellular acidosis during pa-thological states has proven very fruit-ful.

Recent studies have shown that it is possible to exploit all the data con-tained in an *in vivo* $^1$H spectrum using the technique of 2D spectroscopy, by which the information is dispatched in two dimensions. For example, these 'maps' enable detection of several aminoacids (GABA, taurine, alanine) and inositols, which are not visible on a classic one-dimension spectrum.

## Other nuclei of interest

### $^{13}$C spectroscopy

Carbon spectroscopy, while of consid-erable interest for investigating meta-bolic pathways, remains relatively un-used because of its very low sensitivity (0.1 per cent of proton sensitivity) and the low natural abundance of $^{13}$C (1.1 per cent), factors which lead to very long acquisition times. Consequently, only metabolites with a concentration higher than 0.1 M can theoretically be detected *in vivo*. However, it is possible to use exogenous metabolites containing one or more carbons sub-stituted by a $^{13}$C atom. $^{13}$C-glucose has often been used in this way. $^{13}$C-NMR spectroscopy allows a truly quantita-tive and kinetic analysis of fluxes in the cycles of intermediate metabo-lism. However, the high quantities-needed *in vivo* and the cost of syn-thesis of these labelled molecules has so far limited the application of this technique to humans or animals such as monkeys and rabbits.

In an attempt to overcome this limita-tion, it has been recently proposed to use 'heteronuclear methods', i.e. spec-troscopic methods which exploit the resonance signals of the protons asso-ciated with the $^{13}$C nuclei of a labelled compound. These methods, by com-bining the specific detection of la-belled metabolites with the high sensi-tivity of proton spectroscopy, have considerably advanced *in vivo* $^{13}$C brain spectroscopy (Fitzpatrick *et al.*, 1990).

### $^{23}$Na and $^{39}$K spectroscopies

These nuclei have been investigated because of the interest in knowing the intracellular content of sodium or potassium ions. The sensitivity and the concentrations of these nuclei is sufficient for *in vivo* $^{23}$Na- and $^{39}$K-NMR spectroscopy to be technically feasible. However, their spectra ex-hibit only one peak which corre-sponds both to the intracellular pool and the extracellular pool. Several at-tempts, especially in liver and heart, have used shift reagents which do not diffuse into cells, so that one peak is exhibited at the normal position, corresponding to the intracellular pool, and another at a shifted position, corresponding to the extracellular pool. The acute toxicity of these shift reagents and their low BBB permeability limit cerebral investigations, especially in humans.

## $^{19}F$ spectroscopy

Fluorine is a nucleus with a high NMR sensitivity and it is nearly absent from living tissues. These properties make $^{19}F$ a good marker for exogenous molecules. Indeed, it is possible to study the metabolism of drugs such as fluorinated anaesthetics, anticarcinogen molecules (e.g. 5-fluorouracile), etc. $^{19}F$-NMR spectroscopy has been extensively used for measurement of cerebral blood flow and glucose metabolism. These techniques are the transposition to NMR spectroscopy of the methods described in the previous sections:

(1) Evaluation of the local cerebral metabolic rate of glucose using fluoro-deoxyglucose according to Sokoloff's method (see 'The [$^{14}C$]2-deoxyglucose technique' above). The use of this technique is still limited by the need for very high concentrations of fluoro-deoxyglucose which are not devoid of pharmacological effects.

(2) Measurement of cerebral blood flow based on the clearance technique (see 'Clearance techniques' above). Most of these studies have involved inhalation of a tracer gas (trifluoromethane, Freon-23). Such measurements have similar sensitivity and reproducibility compared to classical radioisotopic methods. Toxicological studies are needed before these methods can be widely applied.

*Advantages*

(a) Quantitative and repeatable measurements.

(b) Non-invasive for the brain.

(c) *In vivo* intracellular investigations of the state of energy metabolism, pH, etc.

(d) Technique still in progress, with continuous extension of its possibilities, especially toward localization in brain.

*Disadvantages/limitations*

(a) Technique requiring an expensive and elaborate technology to compensate for its limited sensitivity.

(b) Investigations at present limited to one cerebral region, not precisely defined.

## Conclusions: how to choose a technique?

In the face of such diversity of techniques for studying cerebral circulation and energy metabolism, the choice of the investigator could be based on the following criteria.

## Choice of the variables

Measurement of one variable gives only one aspect of the phenomenon studied. For instance, measurement of blood tissue perfusion may be meaningless under a hypertension exceeding the upper limit of autoregulation because the tracer will invade the tissue at all sites of BBB breakdown to a degree unrelated to the local blood flow. In this case, measurement of the

cerebral capillary permeability is a necessary complement in order to determine whether or not the BBB is disrupted. Another example is that it is often necessary to have some index of metabolic changes as well as flow to evaluate the site of action of a pharmacological agent (vascular, parenchymal).

## Choice of the technique

Tables 2 and 3 which report the main features of the techniques available will help in this choice. Crucial features are the spatial resolution and degree of multiregionality which are to be balanced against the dynamic aspect of the variable (temporal resolution) and the possibility of repeated or continuous measurements. At this step, technical limitations should be considered in relation to the experimental protocol envisaged. Pilot experiments should be profitable to determine secondary effects. For instance, hypocapnia resulting from hyperventilation induced in the test situation may hamper interpretation of CBF changes, so that artificial respiration should then be envisaged.

Conclusive experimental arguments are usually obtained by complementary approaches. The necessity of complementarity may explain the considerable diversity of the techniques used in animals for investigating cerebral blood flow and metabolism.

**Acknowledgements:** The author wishes to thank Dr Philippe Meric for valuable discussion and precious advice about many points of the manuscript, and Dr Pascal Roucher who wrote the section on NMR spectrometry. The author's own studies were supported by grants from the CNRS (UA 641), the INSERM (U182), and the Université de Paris VII.

## Part B: Methods used in man (M. Diksic)

### Cerebral blood flow measurement in humans

There are several distinct methods for the estimation of regional cerebral blood flow (rCBF) in human brain. Here we will limit our discussion to the methods that measure brain perfusion flow, those using radionuclides as tracers and the tomographic approach in image reconstruction. These methods can be distinguished on the basis of the physical characteristics of the radionuclides used as tracers into those using low energy photon-emitting radionuclides and those using coincidence radiation (positron-emitting radionuclides which, after annihilation, produce two 511 keV $\tau$-rays). On the basics of these emission characteristics the methods are called single photon emission tomography (SPECT) and positron emission tomography (PET). Both imaging methods permit regional measurement which can be quantified in PET and be semi-quantitative in SPECT. In this section we will concentrate only on the compounds presently used or those which have shown exceptionally promising results. We will also place more emphasis on the bolus PET measurements in contrast to the equilibrium method. The reason for this is that the latter is rarely used these days. Quite a large spectrum of labelled compounds have been investigated for cerebral blood flow measurements. These days the compounds most often used are $^{15}$O-labelled water (Raichle *et al.*, 1983; Meyer, 1989, 1991; Talbot *et al.*, 1991; Sergent *et al.*, 1992) and butanol for PET measure-

ments (Raichle *et al.*, 1976) and $^{99m}$Tc-labelled HMPAO (Troutner *et al.*, 1984; Neirinckx *et al.*, 1987), $^{123}$I-labelled iodoamphetamines and their N-substituted derivatives (Nishizawa *et al.*, 1989) and $^{133}$Xe (Obrist *et al.*, 1975, 1979; Lassen, 1985) for SPECT measurements.

### PET measurements

#### Dynamic equilibrium method

The measurement of brain blood flow using PET was developed by using $^{15}$O-labelled water 'synthesized' *in vivo* in the lungs with the help of carbonic anhydrase and an equilibrium approach. The subject breathed $^{15}$O-labelled carbon dioxide delivered in the air for about 10 min before data acquisition started and delivery was continued throughout the imaging procedure. The change in tissue radioactivity with time, including correction for radioactive decay, is:

$$\frac{dQ_i}{dt} = F_i(C_{a_t} - C_{v_t}) - \lambda Q \qquad (1)$$

where $\lambda$ is the constant of radioactive decay ($\lambda = \ln2/2$ expressed in min$^{-1}$), and other symbols are as described in section A, 'Fick principle', of this chapter. After inserting $dQ_i/dt = 0$ (dynamic equilibrium) and converting amounts of radioactivity into concentrations and rearranging (Jones *et al.*, 1976), we obtained an equation for the rCBF (ml/100 g min):

$$rCBF = \frac{\lambda \cdot 100}{\dfrac{C_a^{eq}}{C_i^a} - \dfrac{1}{p}} \qquad (2)$$

where $C_a^{eq}$ and $C_i^{av}$ are respectively equilibrium arterial plasma and average brain tissue (averaged over acquisition time) concentrations of $[^{15}O]H_2O$, and p is the tissue–blood partition coefficient for water (p = 0.9 ml/g). This method has been criticized by several authors on the grounds that the water extraction fraction is not a linear function for the entire range of the blood flows found in the human brain, as well as the propagation of error due to changes in $C_a^{eq}$ not being linear. For example, a 10 per cent change in $C_a^{eq}$ results in an error of 20–30 per cent in rCBF (Herscovitch and Raichle, 1983; Meyer and Yamamoto, 1984).

#### Bolus method

The method generally used these days for the rCBF estimates is the bolus approach developed by Raichle and associates (1983). The rCBF is estimated on the basis of an image collected during 30–60 s after the bolus reaches the brain. From the equations above, an equation for estimation of the rCBF (represented by F in the equation) can be derived (see equation 3 below, where $C^{av}$ is an average tissue radioactivity concentration over the $T_1$–$T_2$ time interval and C(t) is the tissue radioactivity concentration as a function of time during this interval, $C_a(\tau)$ is the arterial plasma tracer concentra-

$$C^{av} = \int_{T_1}^{T_2} C(t) \cdot dt = \int_{T_1}^{T_2} F\, e^{-k_2 \cdot t} \int_0^T C_a(\tau) \cdot e^{k_2 \cdot \tau} \cdot d\tau \cdot dt \qquad (3)$$

tion as a function of time ($\tau$) and $k_2$ (min$^{-1}$) is the constant for the egress of the tracer from the tissue. By solving this equation using one of the standard methods, the value of F (blood flow) is obtained. The [$^{15}$O]H$_2$O bolus method also has some drawbacks because of the diffusion limitation of water. The rCBF above approximately 65 (ml/100 g min) is progressively underestimated as a result of the water diffusion limitation. This problem can be overcome by using butanol labelled with $^{15}$O- or $^{11}$C which is a freely diffusible tracer and has an extraction fraction of one. The bolus water method has been systematically evaluated and it has been shown by several investigators that there is a small delay between the bolus reaching the brain and the time at which it reaches the output of the radial artery catheter (the location from which the brain input function is generally sampled). To obtain accurate estimates of the rCBF one must account for this delay in the calculation (Herscovitch *et al.*, 1983; Meyer, 1989). Another set of corrections required is the difference in dispersion of the input function between the heart and brain on one hand, and the point of sampling of the brain arterial input function on the other. There have been several approaches to correct for this dispersion (Meyer *et al.*, 1987). The bolus method described above uses only one image in the calculation of rCBF acquired during 30–60 s.

Recently there was another approach suggested, which uses a sequential set of images acquired after a bolus injection. This approach permits estimation of the parameter, rCBF, p, and delay time using a curve fitting method (Meyer, 1989). This methodology allows us a pixel-by-pixel estimation of these parameters. Of course, this approach is limited by the count rate obtained in different regions of each frame. Probably the greatest benefits of these refinements can be used in different pathologies (e.g. tumours, arteriovenous malformations), though unfortunately there has been very little information available on its use in these pathological situations. An example of PET images obtained with a Scanditronix PC-2048B scanner are shown in Fig. 6.

Quantitative estimates of rCBF are very important in studies where blood flow or oxygen utilization are evaluated as variables. However, lately there have been some very interesting results on the neurological organization and processing of different neuronal inputs (Posner *et al.*, 1988; Talbot *et al.*, 1991; Sergent *et al.*, 1992) using additive qualitative PET imaging and bolus injected [$^{15}$O]H$_2$O. An example of the use of this methodology is shown in Fig. 7 where very specific brain activations are shown in a group of piano players.

## SPECT measurement

In general, a radiopharmaceutical must have some basic characteristics before it can be considered good for the SPECT measurement of the perfusion rCBF. It must be lipophilic in order to easily cross the blood–brain barrier. Ideally its brain extraction must be close to one (see 'Tracer diffusibility' and 'Molecular microspheres' in Section A), and it must stay in the brain without being re-distributed, long enough to permit SPECT imaging to take place. The radiopharma-

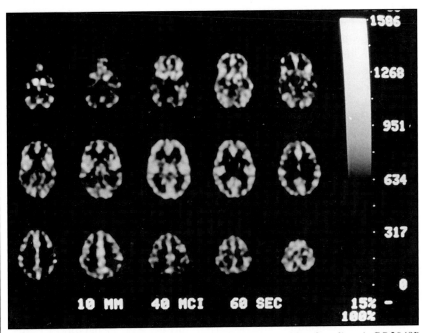

*Fig. 6. A set of brain images obtained in a normal volunteer with a Scanditronix PC-2048B scanner. The scanner has eight rings producing 15 slices of brain with a resolution of about 6.8 mm FWHM (courtesy of Dr E. Meyer).*

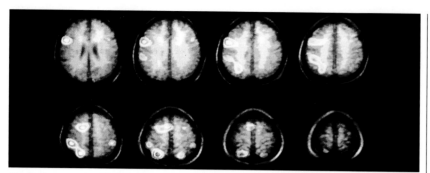

*Fig. 7. An example of PET images showing cortical activations during sight-reading and piano playing; (The figure is reproduced from Talbot et al. (1991), with permission; courtesy of Dr. J. Sergent.) (see colour section)*

ceutical should be available in a kit form to permit easy distribution and general availability.

## 99mTc HMPAO and isopropyl-[123I]iodoamphetamine

The brain perfusion blood flow measurements with SPECT have been greatly enhanced and accelerated with the development of a $^{99m}$Tc HMPAO radiopharmaceutical by Amersham International (Neirinckx *et al.*, 1987; and see 'Molecular microspheres' in Section A). This development was based on the pioneering work of researchers at the University of Missouri, Columbia (Troutner *et al.*, 1984). The kit CERETEC™ marketed by Amersham International has about 6 h of chemical stability after being reconstituted. This prolonged stability makes it very convenient for clinical use, especially for studies of epileptic episodes. The principle underlying the use of this $^{99m}$Tc radiopharmaceutical is based on the chemical characteristics of the original compound (e.g. high lipophilicity) delivered by plasma, and that it is metabolically altered in the brain into a hydrophilic compound. The original compound is highly lipid soluble and its distribution in the brain is

roughly proportional to the regional blood flow. Once the compound is in the brain tissue (extravascular space) it decomposes (chemical change) into highly hydrophilic $^{99m}$Tc-labelled products which are trapped in the brain for an extended period of time (Nakamura *et al.*, 1989). The original $^{99m}$Tc-labelled compound has a high brain extraction, but the extraction fraction (equation 10 in Section A) is significantly less than one (Murase *et al.*, 1991). Because of this, there is no linear relation between the regional brain uptake and rCBF. However, as shown by several investigators (Inugami *et al.*, 1988; Lassen *et al.*, 1988) it is possible to introduce a simple correction to obtain an almost perfect relation between the regional brain uptake of $^{99m}$Tc-HMPAO and rCBF.

*N*-isopropyl-*p*-[$^{123}$] iodoamphetamine (Winchell *et al.*, 1980) in $^{123}$I-labelled IMP is most often used as a tracer for SPECT imaging of the brain perfusion blood flow. Under normal physiological conditions, at early times after injection IMP distributes in the brain proportionally to the cerebral blood flow. However, in some pathological conditions (e.g. ischaemia) the distribution might not be directly related to the blood flow (Kuhl *et al.*, 1982).

The proportionality between the regional cerebral blood flow and the uptake of radioactivity after i.v. injection of IMP holds up to about one hour after injection (Homan *et al.*, 1984). Clearance from the grey matter is substantially greater and after about 24 h there is more radioactivity in the white matter than in the grey matter (Nishizawa *et al.*, 1989).

## The $^{133}$Xenon method

There has been a great need to measure rCBF in patients recovering from brain injuries with the hope of obtaining a relationship between rCBF and the patient's recovery. Very valuable clinical evaluation of the brain perfusion has been done by using the $^{133}$Xe method. This method has been developed into a method that permits bed side measurement of brain perfusion (Obrist *et al.*, 1975, 1979). The method is based on the analysis of the tissue washout curve and it can be completely noninvasive (the patient breathes a mixture of $^{133}$Xe and air for a few minutes) (Obrist *et al.*, 1967 and 'Measurement by extracranial counting' in Section A). The method can also be an intravenous (Austin *et al.*, 1972) or intraarterial injection (Kohlmeyer, 1985). The re-breathing approach has a disadvantage; it requires complete cooperation from the patient during the re-breathing period. This could be of special concern when measurements are done in some pathological conditions (e.g. stroke; post-operative assessment). For these reasons, the completely noninvasive method might be of limited value in those patients who would benefit the most from it. The intravenous approach for tracer injection is of special interest because it does not require any patient cooperation and it can be done on patients who are anaesthetized or unconscious.

The $^{133}$Xe method has also been developed into the SPECT tomographic method (Lassen, 1985). The $^{133}$Xe method is clinically very attractive because the isotope is widely available, requires minimal technical expertise, and in its simplest form can be used at the patient's bedside (Obrist *et al.*, 1979). The latter could be of special importance when measurements on trauma or stroke patients are contemplated. In addition, each examination is rather inexpensive (compared to other nuclear medicine methods for the same objective) and it can be repeated in the same patient every 30 min if required. This is in contrast to [$^{123}$I]IMP which requires about 24 h between measurements. Of course, the method has some deficiencies, mainly differential solubility of $^{133}$Xe in the white and grey matter as well as the rather low γ-ray energy emitted by $^{133}$Xe. The lower energy of γ radiation results in a large scatter fraction being recorded from the different areas of the brain, especially areas with low radioactivity/flow (see 'Measurement by extracranial counting' in Section A). The latter makes measurements in the area of very low flow (e.g. infarcts) very unreliable.

## Oxygen utilization measurements

### Dynamic equilibrium method

The brain's oxygen utilization is measured using $^{15}$O-labelled oxygen gas and a PET scanner. There is no method for measuring brain oxygen

$$\frac{dC}{dt} = F \cdot C_a^{H_2O}(t) + F \cdot E_O \cdot C_a^{O_2}(t) - \left(\frac{F}{p} + \lambda\right) \cdot C(t) \tag{4}$$

utilization with SPECT. Oxygen labelled with $^{15}$O is inhaled over a period of 10 min to obtain a steady-state (equilibrium method), then scanning is done similarly to that described above for CBF measurement using the dynamic equilibrium method. Actually, during [$^{15}$O]O$_2$ inhalation one measures the regional oxygen extraction ratio in the brain. During [$^{15}$O]O$_2$ inhalation, only about 30–40 per cent of the oxygen is extracted by the brain tissue leaving some 60–70 per cent of labelled oxygen in the cerebral veins. This radioactivity is limited to the brain vascular compartment (haemoglobin bound oxygen), and by independently measuring cerebral blood volume

with [$^{11}$C]CO or [$^{15}$O]CO, correction of the tissue signal can be made for the vascular contribution. Tissue oxygen is promptly used in several brain metabolic processes of which glycolysis is the most prominent. These metabolic processes result in the release of [$^{15}$O]H$_2$O (metabolic byproduct) which contaminates the tissue signal and arterial input function, as arterial plasma is also contaminated with $^{15}$O-labelled metabolic water. The latter contributes 20–30 per cent to the total plasma (Jones *et al.*, 1976) radioactivity and requires appropriate corrections to be made. In short, the tissue radioactivity measured during [$^{15}$O]O$_2$ inhalation must be corrected for recirculating labelled water

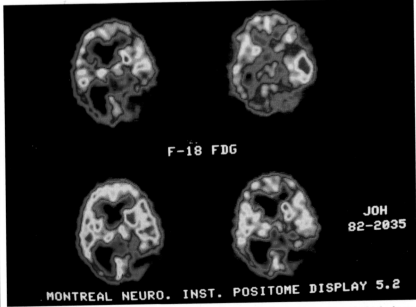

*Fig. 10. A set of images obtained in a patient with a recurrent brain tumor. Note the high glucose uptake in the active tumor tissue located next to the excision cavity. These images were obtained with a three-slice PET scanner (POSITOME IIIp) with an axial resolution of about 1.1 cm and a slice thickness of approximately 1.5 cm (Thompson et al., 1986). (see colour section)*

($C_aH_2O$ in equation 4 below) and un-extracted $[^{15}O]O_2$ in the venous blood ($C_aO_2(t)$ in equation 4). Analogous to the above description of rCBF measurement, for the change in the tissue radioactivity with time ($dC/dt$), we can write equation 4 (above):

where F, $\lambda$, and p are described above, $C_B(t)$ is the whole blood tracer concentration, and $E_O$ is the oxygen extraction ratio. Inserting $dC/dt = 0$ (equilibrium method), and assuming that $C_aH_2O(t)$, $C(t)$, and $C_aO_2(t)$ are constant (they do not change with respect to time during acquisition; in equation 5 they are represented with a bar over a symbol), and solving for $E_o$, we have:

$$E_o = \overline{C}_B \cdot \left(\frac{F}{p} + \lambda\right) - F \cdot \frac{\overline{C}_a^{H_2O}}{F \cdot \overline{C}_a^{H_2O}} \quad (5)$$

Here F represents the blood flow during $[^{15}O]CO_2$ inhalation, assumed to be the same as that during flow measurement. In other words, one can obtain the regional oxygen extraction by knowing the regional blood flow from an independent measurement. Then the regional oxygen utilization according to the dynamic equilibrium method can be estimated by the following equation:

$$rCMRO_2 = E_o \cdot F \cdot P_aO_2 \quad (6)$$

where $P_aO_2$ is the arterial blood oxygen content.

## Bolus method

This method is analogous to the bolus method for rCBF measurement described above. However, in this method $[^{15}O]O_2$ is inhaled with air in a deep breath ($\approx 70$ mCi) which is held for some 15 s to facilitate the exchange of $^{15}O$-labelled $O_2$ from the inhaled air and haemoglobin-bound oxygen (Mintun et al., 1984). Soon after this, PET scanning is initiated and acquisition of the brain data is typically done for 40 s. With the start of scanning, the arterial blood sampling begins and is carried out during the entire time of PET scanning. The movement of oxygen from blood to tissue can be presented in the form of a two-compartment model. In a differential form the movement of the tracer through the second compartment (tissue) is as in equation 7, below, which is analogous to equation 4, but

without the radioactive decay constant. The radioactive decay constant is not used because tissue radioactivity is collected (integrated) during 40 s and it is corrected during the reconstruction phase (Raichle et al., 1983) by introducing an average correction factor equal to $(1 - e^{-\lambda T})/\lambda T$. After integration of equation 7 we get equation 8.

Note that $k_2 = F/p$ and $C_2(t)$ is the concentration of $^{15}O$ in compartment two (tissue compartment) as a function of time. Using the measured arterial blood radioactivity concentration and the brain $O_2$ extraction, the radioactivity present in the vascular compartment ($C_1$; compartment one) can be estimated as in equation 9 (Meyer, 1991).

Radioactive events detected (after correction for scatter) in the brain tissue between times $T_1$ and $T_2$ ($C_{PET}$) are equal to the sum of the radioactivity in the two compartments. After adding these two terms (equations 8 and 9) and solving for the oxygen extraction $E_O$ we obtain (equation 10):

Here R is equal to the small-to-large vessel haematocrit ratio (Meyer, 1991). Assuming that the plasma radioactivity is derived from $^{15}O$ in the form of $H_2O$ and that all $[^{15}O]O_2$ is bound to haemoglobin, then:

$$C_a^{H_2O}(t) = C_p(t) \cdot 0.8/0.92 \quad (11)$$

Here 0.8 and 0.92 represent the water content of blood and plasma, respectively (Davis et al., 1953). The oxygen utilization is again estimated by equation 5. Of course, for the latter method the arterial blood should be properly sampled in order to give us a very good representation of the input function.

$$\frac{dC_2}{dt} = F \cdot C_a^{H_2O}(t) + F \cdot E \cdot C_a^{O_2}(t) - \frac{F}{p} \cdot C_2(t) \quad (7)$$

$$C_{2(t)} = F \cdot C_a^{H_2O}(t) * e^{-k_2 \cdot t} + F \cdot E \cdot C_a^{O_2}(t) * e^{-k_2 \cdot t} \quad (8)$$

$$C_1(t) = CBV \cdot R \cdot C_a^{O_2}(t) \cdot (1 - 0.835 \cdot E_0) \quad (9)$$

$$E_0 = \frac{C_{PET} - F \int_{T_1}^{T_2} C_a^{H_2O}(t) * e^{\frac{F_t}{p}} \cdot dt - CBV \cdot R \cdot \int_{T_1}^{T_2} C_a^{O_2}(t) \cdot dt}{F \cdot \int_{T_1}^{T_2} C_a^{O_2}(t) * e^{\frac{F_t}{p}} \cdot dt - CBV \cdot R \cdot 0.835 \cdot \int_{T_1}^{T_2} C_a^{O_2}(t) \cdot dt} \quad (10)$$

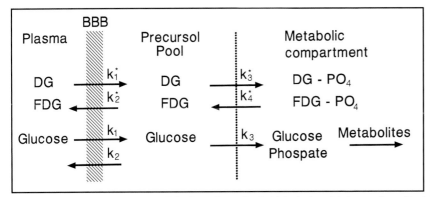

*Fig. 8. A schematic representation of the deoxyglucose brain biological model. It must be realized that the model is a crude approximation of the biological reality. The rate constants are first order rate constants ($min^{-1}$) describing the transfers of the tracer (with asterisk) and tracee (without asterisk) between the different compartments. Note that the constant related to the phosphatase activity is disregarded for glucose, as there is no biochemical data which would support its action onto glucose-6-phosphate. The division of the tissue into different distinct compartments is to aid us in solving this model. It should be noted that the precursor pool, for example, could be in part intracellular.*

## Brain glucose utilization measurement with PET

### Glucose or deoxyglucose as tracer?

There have been two basically different approaches to measuring brain regional glucose utilization with PET. The basic difference stems from the fact that one method uses radioactively labelled D-glucose (tracer and tracee are chemically the same) whereas the second approach uses a radioactively labelled glucose analogue, 2-deoxy-D-glucose ($^{11}$C-labelled) or 2-deoxy-2-fluoro-D-glucose labelled with $^{18}$F.

The principles on which the deoxyglucose method is based are discussed above in Section A. In principle, the biological model for glucose and deoxyglucose could be presented as a three-compartmental model (Fig. 8). The compartments are: plasma, precursor, and metabolic pool. In both cases the first two pools have at least

one divider, the blood–brain barrier. However, there is no change in the chemical form of tracer between them. The metabolic pool has a great distinction in the chemical composition of the products formed when glucose is compared to deoxyglucose. In the case of $^{11}$C-labelled glucose as tracer $^{11}$C-labelled $CO_2$, the product of glucose metabolism gets removed from the brain and for measurements longer than a few minutes, the tissue must be corrected for this loss. $^{11}$C-Labelled glucose as a tracer was introduced by Raichle *et al.* (1975), and recently the model and its use in PET imaging has been investigated in greater detail (Blomqvist *et al.*, 1985; Mintun *et al.*, 1985). The use of labelled glucose has one advantage; since it is chemically identical to natural glucose (tracee), there is no need for the lumped constant required in the analysis of deoxyglucose tissue uptake data. This could be advantageous, especially in pathological conditions. However, there are several disadvantages arising from the need

to use uniformly labelled glucose as well as making the proper correction for the loss of [$^{11}$C]$CO_2$ from the brain soon after glucose is injected. It is probably very hard to synthesize uniformly labelled glucose because of the biological methods used for its synthesis (Ehrin *et al.*, 1980). The delivery of $^{11}$C-labelled $CO_2$ to the biological system starts at the time when there are some fragments of the glucose molecule already synthesized, resulting in a non-uniform distribution of the label. For the reasons mentioned above, the method using uniformly labelled glucose has not been widely used in PET measurements of brain glucose utilization.

## The deoxyglucose method

### Theoretical limitations

The method most often used for the measurement of brain glucose utilization is the deoxyglucose method and the most widely used tracer is $^{18}$F-labelled 2-deoxy-2-fluoro-D-glucose (in medical publications referred to as 2-fluoro-2-deoxy-D-glucose or 2-FDG). Labelled deoxyglucose (used here to mean both [$^{11}$C]2-deoxyglucose and 2-[$^{18}$F]FDG) has an advantage over glucose, as once it gets trapped in the brain as the 6-phosphate, the label stays in the brain for quite some time, permitting ample time for the images to be acquired. However, a noticeable loss from the brain tissue starts to occur about 60–80 min after injection. There is still quite a large controversy over whether or not glucose phosphatase, which is in a different cellular compartment (Schmidt *et al.*, 1989), starts to hydrolyse deoxyglucose 6-phosphate sometime soon after it is synthesized (Redies and Diksic, 1989a). This point is of little relevance as long as the objective is to measure glucose utilization (Redies and Diksic, 1989b). Certainly there will be some delay in the hydrolysis as the 6-phosphate must be relocated to the other side of the membrane before it can serve as a substrate for the glucose phosphatase. However, a delay of a few minutes before the substrate for dephosphatase arrives into the appropriate cellular compartment, would not alter the general conclusion that the loss of label from the brain is a consequence of phosphatase hydrolysis. One should also bear in mind that all rate constants describing tracer movement between different compartments are highly correlated. One can obtain a different combination fitting the same data equally well, though the rate of unidirectional trapping will always be the same.

## Equations

The set of differential equations describing movement of the tracer between different compartments, as shown in Fig. 8, is (Phelps *et al.*, 1979; Redies and Diksic, 1989b):

$$\frac{dC_E^*(t)}{dt} = K_1^* \cdot C_p^*(t) - (k_2^* + k_3^*) \, C_E^*(t) \quad (12)$$

$$\frac{dC_M^*(t)}{dt} = k_3^* \cdot C_E^*(t) - k_4^* \, C_M^*(t) \quad (13)$$

Here $C_E^*(t)$ and $C_m^*(t)$ are the concentration (nCi/g of brain) of the label in the precursor and metabolic pool, respectively, as a function of time, $C_p^*(t)$ is the plasma concentration (nCi/ml) as a function of time, and $k_2^*$, $k_3^*$, and $k_4^*$ are the first order rate constants (min$^{-1}$) described in the caption to Fig. 8. $K_1^*$ (ml/g.min) is actually the first order rate constant $k_1^*$ (min$^{-1}$) multiplied by the fractional volume of the vascular compartment in the unit weight of the brain (ml/g).

This set of differential equations can be solved for $C_E^*(t)$ and $C_m^*(t)$ by some standard methods. The tracer concentration in the tissue volume $C_T^*(T)$ by the PET scanner is the sum of $C_E^*(T)$, $C_M^*(T)$, and the radioactivity present in the vascular compartment. The solution yields equations 14–17 below (Phelps *et al.*, 1979; Redies *et al.*, 1987).

From a curve fit (least square or some other optimization procedure) of the

tissue measured radioactivity to the tissue calculated radioactivity (equation 16), the rate constants can be estimated. The same glucose utilization rates are obtained no matter which model (k$_3$- or k$_4$-model; synonymous with the number of rate constants used in the model) is used as long as consistency is used throughout the analysis. However, if data acquired more than 60–70 min after injection is used in the fitting, then there is a need to include the rate constant (k$_4^*$) to correct properly for the loss of radioactivity from the metabolic compartment (Redies *et al.*, 1987).

After the rate constants are estimated the rate of glucose utilization (R) in the k$_4$-model is calculated as (Phelps *et al.*, 1979; Kato *et al.*, 1984):

$$R = \frac{C_p}{LC} \cdot \left( \frac{C_T^*(T) - C_E^*(T)}{C_M^*(T)} \right) \cdot \frac{K_1^* \cdot k_3^*}{k_2^* + k_3^*} \quad (18)$$

After substituting values for $C_E^*(T)$ and $C_M^*(T)$ from equations above, this equation becomes:

$$R = \frac{C_p}{LC} \cdot \frac{C_T^*(T) - \frac{K_1^*}{\alpha_2 - \alpha_1}\left[(k_4^* - \alpha_1)e^{\alpha_1 \cdot t} + (\alpha_2 - k_4^*)e^{-\alpha_2 \cdot t}\right] \otimes C_p^*(t)}{\frac{k_2^* + k_3^*}{\alpha_2 - \alpha_1} \cdot \left[e^{-\alpha_2 \cdot t} - e^{-\alpha_2 \cdot t}\right] \otimes C_p^*(t)} \quad (19)$$

$$C_E^*(T) = \frac{K_1^*}{(\alpha_2 - \alpha_1)}\left[(k_4^* - \alpha_1) \cdot e^{-\alpha_1 \cdot t} + (\alpha_2 - k_4^*) \, e^{-\alpha_2 \cdot t}\right] \otimes C_p^*(t) \quad (14)$$

$$C_M^*(T) = \frac{K_1^* k_3^*}{(\alpha_2 - \alpha_1)}\left[e^{-\alpha_1 \cdot t} - e^{-\alpha_2 \cdot t}\right] \otimes C_p^*(t) \quad (15)$$

$$C_i^*(T) = C_E^*(T) + C_M^*(T) = \frac{K_1^*}{(\alpha_2 - \alpha_1)}\left[(k_3^* + k_4^* - \alpha_1) \cdot e^{-\alpha_1 \cdot t} - (k_3^* + k_4^* - \alpha_2) \cdot e^{-\alpha_2 \cdot t}\right] \otimes C_p^*(t) \quad (16)$$

where:

$$\alpha_{2,1} = \frac{1}{2}\left[(k_2^* + k_3^* + k_4^*)\right] \pm \sqrt{(k_2^* + k_3^* + k_4^*)^2 - 4k_2^* \cdot k_4^*} \quad (17)$$

and $\otimes$ is the symbol for the operation of convolution.

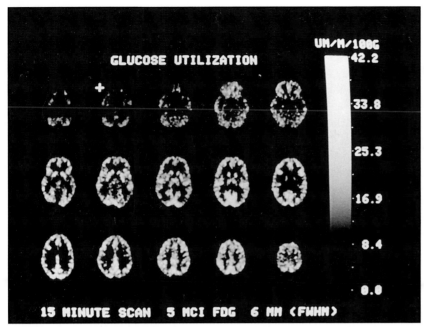

*Fig. 9. A set of glucose utilization images obtained in a normal volunteer using the scanner mentioned in Fig. 6.*

where $C_p$ and LC are the plasma glucose concentration (nmol/ml) and lumped constant, respectively. The lumped constant is the ratio of several other constants for glucose and deoxyglucose and the phosphorylation ratio (the ratio between glucose being phosphorylated and that proceeding further through the metabolic pathway; generally accepted to be one for glucose) and as such does not have units. Being the ratio between different constants for glucose and deoxyglucose generally one would expect that it would not vary between the different regions (See section A; and Phelps *et al.*, 1979). However, it was recently shown that in some pathologies (e.g. brain tumour) there is a difference between the affinities of different isoenzymes for glucose and deoxyglucose (Spence *et al.*, 1990). This could make it very difficult if not impossible to predict LC in brain tu-

mours. The lumped constant was measured in human brain and was estimated to be 0.52 and 0.56 for deoxy- and 2-fluoro-2-deoxy-D-glucose (Reivich *et al.*, 1985).

## The rate constant $k_4$* (dephosphorylation)

There have been several papers discussing the presence of $k_4$* (dephosphorylation) and with what delay, if any, it should be included in the model (Redies *et al.*, 1987; Redies and Diksic, 1989a; Schmidt *et al.*, 1989, 1992). Many of these discussions are academic as we must accept that all these model representations are approximations of the biological reality. Provided we keep that in mind, it is irrelevant at what time after injection $k_4$* is included in the calculation, as long as the time period is approximately the same for the experiments

in which the rate constants are estimated and those used in the glucose utilization measurements, and is somewhere between zero and 15 min. We must also realize that it is impossible to have a 'pure' (unique) biological compartment in any brain region. There are always going to be groups of neurons which use glucose at different rates making any group of neurons a heterogeneous compartment. It was recently shown (Schmidt *et al.*, 1992) by a computer simulation that the tissue heterogenous compartment (grey–white matter mixture) can give rise to an apparent $k_4$*. In conclusion, we can state that as long as the analysis is carried out in a consistent manner (see above), the outcome of the calculations will always be the same: the same glucose utilization rate.

An example of glucose utilization estimated after injection of 2-[$^{18}$F]FDG is shown in Fig. 9 for a normal brain and in Fig. 10 for a recurrent brain tumour.

Methods for *in vivo* estimation of brain glucose and oxygen utilization, and regional blood flow have been applied to the normal brain and different brain pathologies (Diksic & Reba, 1990). A review of their usefulness in the clinical assessment of different pathological states is beyond the scope of this section which deals only with a description of the methods.

*Abbreviations:* BBB, blood–brain barrier; CBF, cerebral blood flow; CBV, cerebral blood volume; CGU, cerebral glucose utilization; CMR, cerebral metabolic rate; EEG, electroencephalogram; NMR, nuclear magnetic resonance; PME: phosphomonoesters; $P_i$: inorganic phosphate; PDE: phosphodiesters; ATP: adenosine triphosphate, with the three phosphates $\gamma$, $\alpha$, $\beta$.

## References – Methods used in animals

Ackerman, J.J.H., Grove, T.H., Wong, G.G., Gadian, D.G. & Radda, G.K. (1980): Mapping of metabolites in whole animals by $^{31}$P NMR using surface coils. *Nature* **283**, 167–170.

Archer, D.P., Elphinstone, M.G. & Pappius, H.M. (1990): The effect of pentobarbital and isoflurane on glucose metabolism in thermally injured rat brain. *J. Cereb. Blood Flow Metab.* **10**, 624–630.

Auer, L.M. & Haydn, F. (1979): Multichannel videoangiometry for continuous measurement of pial microvessels. *Acta Neurol. Scand.* **60**, (Suppl. 72) 208–209.

Baumbach, G.L. & Heistad, D.D. (1983): Effects of sympathetic stimulation and changes in arterial pressure on segmental resistance of cerebral vessels in rabbits and cats. *Circ. Res.* **52**, 527–533.

Blasberg, R.G., Fenstermacher, J.D. & Patlak, C.S. (1983): Transport of $\alpha$-aminoisobutyric acid across brain capillary and cellular membranes. *J. Cereb. Blood Flow Metab.* **3**, 8–32.

Bonvento, G., Lacombe, P., MacKenzie, E.T. & Seylaz, J. (1991): Effects of dorsal raphe stimulation on cerebral glucose utilization in the anaesthetized rat. *Brain Res.* **567**, 325–327.

Bonvento, G., Charbonné, R., Corrèze, J.L., Borredon, J., Seylaz, J. & Lacombe, P. (1994): Is $\alpha$-chloralose plus halothane induction a suitable anesthetic regimen for cerebrovascular research? *Brain Res.* **665**, 213–221.

Bryan, R.M. Jr. (1986): A method for measuring regional cerebral blood flow in freely moving, unstressed rats. *J. Neurosci. Methods* **17**, 311–322.

Bryan, R.M. Jr. (1990): Cerebral blood flow and energy metabolism during stress. *Am. J. Physiol.* **259**, H269–H280.

Carter, L.P. (1991): Surface monitoring of cerebral cortical blood flow. *Cerebrovasc. Brain Metab. Rev.* **3**, 246–261.

Cavazzuti, M., Porro, C.A., Biral, G.P., Benassi, C. & Barbieri G.C. (1987): Ketamine effects on local cerebral blood flow and metabolism in the rat. *J. Cereb. Blood Flow Metab.* **7**, 806–811.

Dacey, R.G. Jr., Duling, B.R. (1984): Effect of norepinephrine on penetrating arterioles of rat cerebral cortex. *Am. J. Physiol.* **246**, H380–H385.

Dahlgren, N., Ingvar, M., Yokoyama, H. & Siesjö, B.K. (1981): Influence of nitrous oxide on local cerebral blood flow in awake, minimally restrained rats. *J. Cereb. Blood Flow Metab.* **1**, 211–218.

Dirnagl, U., Kaplan, B., Jacewicz, M. & Pulsinelli, W. (1989): Continuous measurement of Cerebral cortical blood flow by laser-Doppler flowmetry in a rat stroke model. *J. Cereb. Blood Flow Metab.* **9**, 589–596.

Dirnagl, U., Villringer, A. & Einhäpl, K.M. (1992): *In-vivo* confocal scanning laser microscopy of the cerebral microcirculation. *J. Microsc* **165**, 147–157.

Dora, E. (1984): A simple cranial window technique for optical monitoring of cerebrocortical microcirculation and NAD/NADH redox state. Effect of mitochondrial electron transport inhibitors and anoxic anoxia. *J. Neurochem.* **42**, 101–108.

Dudley, R.E., Nelson, S.R. & Samson, F. (1982): Influence of chloralose on brain regional glucose utilization. *Brain Res.* **233**, 173–180.

Edvinsson, L. & MacKenzie, E.T. (1977): Amine mechanisms in the cerebral circulation. *Pharmacol. Rev.* **28**, 275–348.

Eklöf, B., Lassen, N.A., Nilsson, L., Norberg, K., Siesjö, B.K. & Torlöf, P. (1974): Regional cerebral blood flow in the rat measured by the tissue sampling technique; a critical evaluation using four indicators C$^{14}$-antipyrine, C$^{14}$-ethanol, H$^{3}$-water and xenon$^{133}$. *Acta Physiol. Scand.* **91**, 1–10.

Faraci, F.M. & Heistad, D.D. (1990): Regulation of large cerebral arteries and cerebral microvascular pressure. *Circ. Res.* **66**, 8–17.

Fernandez, E.J., Maudsley, A.A., Higuchi T. & Weiner, M.W. (1992): $^1$H spectroscopic imaging of rat brain at 7 tesla. *Magn. Reson. Med.* **25**, 107–119.

Fisher, M., Sotak, C.H., Minematsu, K. & Li, L. (1992): New magnetic resonance techniques for evaluating cerebrovascular disease. *Ann. Neurol.* **32**, 115–122.

Fitzpatrick, S.M., Hetherington, H.P., Behar, K.L. & Shulman, R.G. (1990): The flux from glucose to glutamate in the rat brain in *vivo* as determined by $^1$H-observed, $^{13}$C-edited NMR spectroscopy. *J. Cereb. Blood Flow Metab.* **10**, 170–179.

Ford, I., McColl, J.H., McCormack, A.G. & McCrory, S.J. (1991): Statistical issues in the analysis of neuroimages. *J. Cereb. Blood Flow Metab.* **11**, A89–A95.

Gjedde, A., Hansen, A.J. & Siemkowicz, E. (1980): Rapid simultaneous determination of regional blood flow and blood–brain glucose transfer in brain of rat. *Acta Physiol. Scand.* **108**, 321–330.

Gotoh, F., Muramatsu, F., Fukuuchi, Y., Okayasu, H., Tanaka, K., Suzuki, N. & Kobari, M. (1982): Video camera method for simultaneous measurement of blood flow velocity and pial vessel diameter. *J. Cereb. Blood Flow Metab.* **2**, 421–428.

Grome, J.J. & McCulloch, J. (1983): The effects of apomorphine upon local cerebral glucose utilization in conscious rats and in rats anesthetized with chloral hydrate. *J. Neurochem.* **40**, 569–576.

Gumbleton, M., Taylor, G., Sewell, R.D.E. & Nicholls, P.J. (1989): Anaesthetic influences on brain haemodynamics in the rat and their significance to biochemical, neuropharmacological and drug disposition studies. *Biochem. Pharmacol.* **38**, 2745–2748.

Haining, J.L., Turner, M.D. & Pantal, R.M. (1968): Measurement of local cerebral blood flow in the unanesthetized rat using a hydrogen clearance method. *Circ. Res.* **23**, 313–324.

Hansen, A.J. & Lauritzen, M. (1988): Spreading depression of Leão. In: *Basic mechanisms of headache*, eds. J. Olesen & L. Edvinsson, pp. 99–107. Amsterdam: Elsevier.

Harper, A.M. & MacKenzie, E.T. (1977): Cerebral circulatory and metabolic effects of 5-hydroxytryptamine in anaesthetized baboons. *J. Physiol.* **271**, 721–733.

Hatakeyama, T., Sakaki, S., Nakamura, K., Furuta, S. & Matsuoka, K. (1992): Improvement in local cerebral blood flow measurement in gerbil brains by prevention of postmortem diffusion of [$^{14}$C]iodoantipyrine. *J. Cereb. Blood Flow Metab.* **12**, 296–300.

Hawkins, R., Hass, W.K. & Ransohoff, J. (1979): Measurement of regional brain glucose utilization *in vivo* using [2-$^{14}$C] glucose. *Stroke* **10**, 690–703.

Hawkins, R.A., Mans, A.M., Davis, D.W. & DeJoseph, M. R. (1988): Comparison of [$^{14}$C]glucose and [$^{14}$C]deoxyglucose as tracers of brain glucose use. *Am. J. Physiol.* **254**, E310–E317.

Hoedt-Rasmussen, K., Sveinsdottir, E. & Lassen, N.A. (1966): Regional cerebral blood flow in man determined by the intra-arterial injection of radioactive inert gas. *Circ. Res.* **18**, 237–247.

Hoffman, W.E., Miletich, D.J. & Albrecht, R.F. (1986): The effects of midazolam on cerebral blood flow and oxygen consumption and its interaction with nitrous oxide. *Anesth. Analg.* **65**, 729–733.

Högestätt, E.D., Andersson, K.E. & Edvinsson, L. (1982): Effects of nifedipine on potassium-induced contraction and noradrenaline release in cerebral and extracranial arteries from rabbits. *Acta Physiol. Scand.* **114**, 283–296.

Iadecola, C., Arneric, S.P., Baker, H.D., Tucker, L.W. & Reis, D.J. (1987): Role of local neurons in cerebrocortical vasodilation elicited from cerebellum. *Am. J. Physiol.* **252**, R1082–R1091.

Ingvar, M. & Siesjö, B.K. (1982): Effect of nitrous oxide on local cerebral glucose utilization in rats. *J. Cereb. Blood Flow Metab.* **2**, 481–486.

Jöbsis, F.F., Keizer, J.H., LaManna, J.C. & Rosenthal, M. (1977): Reflectance spectrophotometry of cytochrome *aa$_3$* in vivo. *J. Appl. Physiol.* **43**, 858–872.

Kety, S.S. (1985): Regional cerebral blood flow: estimation by means of nonmetabolized diffusible tracers – an overview. *Sem. Nucl. Med.* **XV(4)**, 324–328.

Kiyosawa, M., Baron, J.C., Hamel, E., Pappata, S., Duverger, D., Riche, D., Mazoyer, B., Naquet, R. & MacKenzie, E.T. (1989): Time course of effects of unilateral lesions of the nucleus basalis of Meynert on glucose utilization by the cerebral cortex. Positron tomography in baboons. *Brain* **112**, 435–455.

Kuschinsky, W., Suda, S., Bünger, R., Yaffe, S. & Sokoloff, L. (1983): The effects of intravenous norepinephrine on the local coupling between glucose utilization and blood flow in the rat brain. *Pflügers Arch.* **398**, 134–138.

Kuschinsky, W. & Wahl, M. (1978): Local chemical and neurogenic regulation of cerebral vascular resistance. *Physiol. Rev.* **58**, 656–689.

Lacombe, P. (1992): Approches expérimentales: circulation et métabolisme du cerveau. Chapter III. 2. In *Pharmacologie Cardiovasculaire et respiratoire*, ed. C. Advenier & P. Meyer, pp. 301–314. Paris: Hermann.

Lacombe, P., Meric, P., Reynier-Rebuffel, A.M. & Seylaz, J. (1979): Critical evaluation of cerebral blood flow measurements made with $^{14}$C-ethanol. *Med. Biol. Eng. Comput.* **17**, 602–618.

Lacombe, P., Meric, P. & Seylaz, J. (1980): Validity of cerebral blood flow measurements obtained with quantitative tracer techniques. *Brain Res. Rev.* **2**, 105–169.

Lacombe, P., Miller, M.C. & Seylaz, J. (1985): Sympathetic regulation of cerebral blood flow during reflex hypertension. *Am. J. Physiol.* **249**, H672–H680.

Lacombe, P., Sercombe, R., Correze, J.L., Springhetti, V. & Seylaz, J. (1992): Spreading depression induces prolonged reduction of cortical blood flow reactivity in the rat. *Exp. Neurol.* **117**, 278–286.

Lasbennes, F., Lestage, P., Bobillier, P. & Seylaz, J. (1986): Stress and local cerebral blood flow: studies on restrained and unrestrained rats. *Exp. Brain Res.* **63**, 163–168.

Lasbennes, F. & Seylaz, J. (1986): Local cerebral blood flow in gently restrained rats: effects of propranolol and diazepam. *Exp. Brain Res.* **63**, 169–172.

Lear, J.L. (1988): Quantitative multiple tracer autoradiography, considerations in optimizing precision and accuracy. *J. Cereb. Blood Flow Metab.* **8**, 443–448.

Lear, J.L., Ackermann, R.F., Kameyama, M. & Kuhl, D.E. (1982): Evaluation of [$^{123}$I]isopropyliodoamphetamine as a tracer for local cerebral blood flow using direct autoradiographic comparison. *J. Cereb. Blood Flow Metab.* **2**, 179–185.

Leniger-Follert, E. (1985): Oxygen supply and microcirculation of the brain cortex. *Adv. Exp. Med. Biol.* **191**, 3–19.

Levasseur, J.E., Wei, E.P., Raper, A.J., Kontos, H.A. & Patterson, J.L. Jr. (1975): Detailed description of a cranial window technique for acute and chronic experiments. *Stroke* **6**, 309–317.

McCulloch, J. (1982): Mapping functional alterations in the CNS with [$^{14}$C]deoxyglucose. In: *Handbook of psychopharmacology*, Vol. 15. eds. L.L. Iversen, S.D. Iversen & S.H. Snyder, pp. 321–409. New York: Plenum.

Méric, P. (1991): Etude par résonance magnétique nucléaire (RMN) *in vivo* du métabolisme cérébral. *Circ. Metab. Cerveau* **8**, 53–70.

Méric, P. & Seylaz, J. (1977): Radiation scattering and the determination of regional cerebral blood flow by radioisotope clearance. *Med. Progr. Technol.* **5**, 41–46.

Mies, G., Bodsch, W., Paschen, W. & Hossmann, K.A. (1986): Triple-tracer autoradiography of cerebral blood flow, glucose utilization, and protein synthesis in rat brain. *J. Cereb. Blood Flow Metab.* **6**, 59–70.

Mori, K., Schmidt, K., Jay, T., Palombo, E., Nelson, T., Lucignani, G., Pettigrew, K., Kennedy, C. & Sokoloff, L. (1990): Optimal duration of experimental period in measurement of local cerebral glucose utilization with the deoxyglucose method. *J. Neurochem.* **54**, 307–319.

Mraovitch, S., Calando, Y., Goadsby, P.J. & Seylaz, J. (1992): Subcortical cerebral blood flow and metabolic changes elicited by cortical spreading depression in rat. *Cephalalgia* **12**, 137–141.

Nakai, H., Diksic, M. & Yamamoto, Y.L. (1988): Validation of the triple-tracer autoradiographic method in rats. *Stroke* **19**,: 758–763.

Nilsson, B. & Siesjö, B.K. (1976): A method for determining blood flow and oxygen consumption in the rat brain. *Acta Physiol. Scand.* **96**, 72–82.

Ohno, K., Pettigrew, K.D. & Rapoport, S.I. (1978): Lower limits of cerebrovascular permeability to nonelectrolytes in the conscious rat. *Am. J. Physiol.* **345**, H299–H307.

Ohno, K., Pettigrew, K.D. & Rapoport, S.I. (1979): Local cerebral blood flow in the conscious rat as measured with $^{14}$C-antipyrine, $^{14}$C-iodoantipyrine and $^{3}$H-nicotine. *Stroke* **10**, 62–67.

Oldendorf, W.H. (1971): Brain uptake of radiolabeled amino acids, amines, and hexoses after arterial injection. *Am. J. Physiol.* **221**, 1629–1639.

Pasztor, E., Symon, L., Dorsch, N.W.C. & Branston, N.M. (1973): The hydrogen clearance method in assessment of blood flow in cortex white matter and deep nuclei of baboons. *Stroke* **4**, 556–576.

Pinard, E., Mazoyer, B., Verrey, B., Pappata, S. & Crouzel, C. (1993): Rapid measurement of regional cerebral blood flow in the baboon using $^{15}$O-labelled water and dynamic positron emission tomography.. *Med. Biol. Eng. Comput.* **31**, 495–502..

Pinard, E., Seylaz, J. & Mamo, H. (1978): Quantitative continuous measurement of pO$_2$ and pCO$_2$ in artery and vein. *Med. Biol. Eng. Comput.* **16**, 59–64.

Purves, M.J. (1972): *The physiology of the cerebral circulation*, 420 p. London: Cambridge University Press.

Reivich, M., Jehle, J., Sokoloff, L. & Kety, S.S. (1969): Measurement of regional cerebral blood flow with antipyrine-$^{14}$C in awake cats.

*J. Appl. Physiol.* **27**, 296–300.

Sakurada, O., Kennedy, C., Jehle, J., Brown, J.D., Carbin, G.L. & Sokoloff, L. (1978): Measurement of local cerebral blood flow with iodo[$^{14}$C]antipyrine. *Am. J. Physiol.* **234**, H59–H66.

Sapirstein, L.A. (1958): Regional blood flow by fractional distribution of indicators. *Am. J. Physiol.* **193**, 161–168.

Sercombe, R., Verrecchia, C., Oudart, N., Dimitriadou, V. & Seylaz, J. (1985): Pial artery responses to norepinephrine potentiated by endothelium removal. *J. Cereb. Blood Flow Metab.* **5**, 312–317.

Seylaz, J. (1968): Contribution à l'étude du mécanisme de la régulation du débit sanguin cérébral. *Helv. Physiol. Acta* **26**, 33–61.

Seylaz, J., Pinard, E., Meric, P. & Correze, J.L. (1983): Local cerebral pO$_2$, pCO$_2$, and blood flow measurements by mass spectrometry. *Am. J. Physiol.* **245**, H513–H518.

Siesjö, B.K. (1978): *Brain energy metabolism*, 607 p. Chichester: John Wiley & Sons.

Skarphedinsson, J.O., Harding, H. & Thoren, P. (1988): Repeated measurements of cerebral blood flow in rats. Comparisons between the hydrogen clearance method and laser Doppler flowmetry. *Acta Physiol. Scand.* **134**, 133–142.

Sokoloff, L. (1978): Mapping cerebral functional activity with radioactive deoxyglucose. *Trends Neurosci.* **1**, 75–79.

Sokoloff, L., Kennedy, C. & Smith, C.B. (1983): Metabolic mapping of functional activity in the central nervous system by measurement of local glucose utilization with radioactive deoxyglucose, Chapter X. In: *Handbook of chemical neuroanatomy*, Vol. 1. *Methods in chemical neuroanatomy*, eds. A. Björklund & T. Hökfelt, pp. 416–441. Elsevier.

Sokoloff, L., Reivich, M., Kennedy, C., Des Rosiers, M.H., Patlak, C.S., Pettigrew, K.D., Sakurada, O. & Shinohara, M. (1977): The [$^{14}$C]deoxyglucose method for the measurement of local cerebral glucose utilization: theory, procedure, and normal values in the conscious and anesthetized albino rat. *J. Neurochem.* **28**, 897–916.

Soncrant, T.T., Holloway, H.W., Stipetic, M. & Rapoport, S.I. (1988): Cerebral glucose utilization in rats is not altered by hindlimb restraint or by femoral artery and vein cannulation. *J. Cereb. Blood Flow Metab.* **8**, 720–726.

Stoelting, R.K. & Miller, R.D. (1989): *Basics of anesthesia*, 2nd edn. New York: Churchill Livingstone.

Tanaka, K., Dora, E., Greenberg, J.H. & Reivich, M. (1986): Cerebral glucose metabolism during the recovery period after ischaemia – its relationship to NADH-fluorescence, blood flow, EcoG and histology. *Stroke* **17**, 994–1004.

Van Uitert, R.L., Sage, J.I., Levy, D.E. & Duffy, T.E. (1981): Comparison of radio-labeled butanol and iodoantipyrine as cerebral blood flow markers. *Brain Res.* **222**, 365–372.

Villringer, A., Haberl, R.L., Dirnagl, U., Anneser, F., Verst, M. & Einhupl, K.M. (1989): Confocal laser microscopy to study microcirculation on the rat brain surface *in vivo*. *Brain Res.* **504**, 159–160.

Yudilevich, D. & Barry, D.I. (1977): Indicator diffusion and other non-destructive methods for the study of experimental modifications of the blood–brain barrier. *Exp. Eye Res.* **Suppl.,** 511–521.


## References – Methods used in man

Austin, G., Horn, N., Rouke, S. & Hayward, W. (1972): Description and early results on an intravenous radioisotope technique of measuring regional cerebral blood flow in man. In: *Cerebral blood flow and intracranial pressure*, Proc. 5th Int. Symp, Roma, Siena 1971, part II, Europ Neurol **8**, 43–51.

Blomqvist, G., Bergstrom, K., Bergstrom, M., Ehrin, E., Eriksson, L., Garmelius, B., Lindberg, B., Lilja, A., Litton, J.-E., Lundmark, L., Lunqvist, H., Malnborg, P., Mostrom, U., Nilsson, L., Stone-Elander, S. & Widen, L. (1985) Models for 11C-glucose. In: *The metabolism of the human brain studied with positron emission tomography*, eds. T. Greitz , D.H. Ingvar, & L. Widen, pp. 185–194. New York: Raven Press.

Davis, F.E., Kenyon, K. & Kirk, J. (1953): A rapid method for determining water content of human blood. *Science* **118**, 276–277.

Diksic, M. & Reba, R.C. Eds. (1990): *Radiopharmaceuticals and brain pathology studied with PET and SPECT*. Boca Raton, FL: CRC Press.

Ehrin, E., Westman, E., Nilson, S.O., Nilsson, J.L.G., Widén, L., Gretiz, T., Larsson, C.M., Tillberg, J.E. & Malmborg, P.A. (1980): A convenient method for production of $^{11}$C-glucose. *J. Label Comp. Radiopharmac.* **17**, 453–461.

Herscovitch, P. & Raichle, M.E. (1983): Effect of tissue heterogeneity on the measurement of cerebral blood flow with the equilibrium C$^{15}$O$_2$ inhalation technique. *J. Cereb. Blood Flow Metab.* **3**, 407–415.

Herscovitch, P., Markham, J. & Raichle, M.E. (1983): Brain blood flow measured with intravenous H$_2$O [$^{15}$O]. I. Theory and error analysis. *J. Nucl. Med.* **24**, 782–789.

Homan, B.L., Lee, R.G.L., Hill, T.C., Lovett, R.D. & Lister-James, J. (1984): A comparison of two cerebral perfusion tracers: isopropyl-[$^{123}$I] *p*-iodoamphetamine and [$^{123}$I]HIPDM in the human. *J. Nucl. Med.* **25**, 25–30.

Inugami, A., Kanno, I., Uemura, K., Shishido, F., Murakami, M., Tomura, N., Fujita, H. & Higamo, S. (1988): Linearization correction of $^{99m}$Tc-labelled hexamethyl-propylene amine oxime (HM-PAO) image in terms of regional CBF distribution: Comparison to C$^{15}$O$_2$ inhalation steady-state method measured by positron emission tomography. *J. Cereb. Blood Flow Metab.* **8**, S52–S60.

Jones, T., Chesssler, D.A. & Ter-Pogossian, M.M. (1976): The continuous inhalation of oxygen-15 for assessing regional oxygen extractions in the brain of man. *Br. J. Radiol.* **49**, 339–343.

Kato, A., Diksic, M., Yamamoto, Y.L., Strother, S.C. & Feindel, W. (1984): An improved approach for measurement of regional cerebral rate constants in the deoxyglucose method with positron emission tomography. *J. Cereb. Blood Flow Metab.* **4**, 555–563.

Kohlmeyer, K. (1985): The intraarterial xenon 133 method: principles and clinical application. In: *Cerebral blood flow and metabolism measurement*, eds. A. Hartmann & S. Hayer, pp. 1–18. Berlin: Springer-Verlag.

Kuhl, D.E., Barrio, J.R., Huang, S.-C., Selin, C., Ackermann, R.F., Lear, J.L., Wu, J.L., Liu, T.H. & Phelps, M.E. (1982): Quantifying local cerebral blood flow by *N*-isopropyl-[$^{123}$I]iodoamphetamine (IMP) tomography. *J. Nucl. Med.* **23**, 196–203.

Lassen, N.A. (1985): Cerebral blood flow tomography using Xenon-133 inhalation-methodological considerations. In: *Cerebral blood flow and metabolism measurement*, eds. A. Hartmann & S. Hayer, pp. 224–233. Berlin: Springer-Verlag.

Lassen, N.A., Andersen, A.R., Fribert, L. & Paulson, O.B. (1988): The retention of $^{99m}$Tc-d,l-HMPAO in the human brain after intra-arterial bolus injection. A kinetic analysis. *J. Cereb. Blood Flow Metab.* **8**, S13–S22.

Meyer, E. & Yamamoto, Y.L. (1984): The requirement for constant arterial radioactivity in the C$^{15}$O$_2$ steady-state blood flow model. *J. Nucl. Med.* **25**, 455–460.

Meyer, E., Tyler, J.L., Thompson, C.J. & Hakim, A.M. (1987): The time dependence of the $^{15}$O bolus model for CMRO$_2$ measurement by PET with respect to dispersion of the input function. *J. Nucl. Med.* **28**, 699.

Meyer, E. (1989): Simultaneous correction for tracer arrival delay and deispersion in CBF measurements by H$_2$ $^{15}$O autoradiographic method and dynamic PET. *J. Nucl. Med.* **30**, 1069–1078.

Meyer, E. (1991): $^{15}$O Studies with PET. In: *Radiopharmaceuticals and brain pathology studied with PET and SPECT*, eds. M. Diksic & R.C. Reba, pp. 165–197. Boca Raton, FL: CRC Press.

Mintun, M.A., Raichle, M.E., Martin, W.R.W & Herscovitch, P. (1984): Brain oxygen utilization measured with $^{15}$O radiotracers and positron emission tomography. *J. Nucl. Med.* **25**, 177–187.

Mintun, M.A., Raichle, M.E., Welch, M.J. & Kilbourn, M.R. (1985): Brain glucose metabolism measured with PET and U-$^{11}$C-glucose. *J. Cereb. Blood Flow Metab.* **5**, (Suppl. 1) S623–S624.

Murase, K., Tanada, S., Inoue, T., Ochi, K., Fujita, H., Sakaki, S., Kimura, Y., Hamamoto, K. (1991): Measurement of the blood–brain barrier permeability of $^{123}$I-IMP, $^{99m}$Tc-HMPAO and $^{99m}$Tc-ECD in the human brain using compartmental model analysis and dynamic SPECT. *J. Nucl. Med.* **32**, 911.

Nakamura, K., Tukatani, Y., Kubo, A., Hashimoto, S., Terayama, Y., Amato, T. & Goto, F. (1989): The behaviour of $^{99m}$Tc-hexamethylpropyleneaminoxine ($^{99m}$Tc-HMPAO) in blood and brain. *Eur. J. Nucl. Med.* **15**, 100–107.

Neirinckx, R.D., Canning, L.R., Piper, I.M., Nowotnik, D.P., Pickett, R.D., Holmes, R.A., Volkert, W.A., Forster, A.M., Weisner, P.S., Marriott, J.A. & Chaplin, S.B. (1987): [$^{99m}$Tc]d,l-HM-PAO: A new radiopharmaceutical for SPECT imaging of regional cerebral blood perfusion. *J. Nucl. Med.* **28**, 191–202.

Nishizawa, S., Tanada, S., Yonekura, Y., Fujita, T., Mukai, T., Sagi, H., Fukuyama, H., Miyoshi, T., Harada, K., Ishikawa, M., Kikuchi, H. & Konishi, J. (1989): Regional dynamics of *N*-isopropyl-[$^{123}$I]-*p*-iodoamphetamine in human brain. *J. Nucl. Med.* **30**, 150–156.

Obrist, W.D., Thompson, H.K. & King, C.H. (1967): Determination of regional cerebral blood flow by inhalation of $^{133}$Xenon. *Circ. Res.* **20**, 124–135.

Obrist, W.D., Thompson, H.K., Wong, H.S. & Wilkinson, W.E. (1975): Regional cerebral blood flow estimated by $^{133}$Xenon inhalation. *Stroke* **6**, 245–256.

Obrist, W.D., Gennarelli, T.A., Segawa, H., Dolinshas, C.A. & Langfitt, T.W. (1979): Relation of cerebral blood flow to neurological status and outcome in head injured patients. *J. Neurosurg.* **51**, 292–300.

Phelps, M.E., Huang, S.C., Hoffmann, E.J., Selin, C., Sokoloff, L. & Kuhl, D.E. (1979): Tomographic measurement of local cerebral glucose metabolic rate in humans with (F-18) 2-fluoro-2-deoxy-D-glucose: validation of method. *Ann. Neurol.* **6**, 371–388.

Posner, M.I., Petersen, S.E., Fox, P.T. & Raichle, M.E. (1988): Localization of cognitive operations in the human brain. *Science* **240**, 1627–1631.

Raichle, M.E., Larson, K.B., Phelps, M.E., Grubb, R.L. Jr., Welch, M.J. & Ter-Pogossian, M.M. (1975): *In vivo* measurement of brain glucose transport and metabolism employing glucose-$^{11}$C. *Am. J. Physiol.* **228**, 1936–1948.

Raichle, M.E., Eichling, J.O., Straatman, M.G., Welch, M.J., Larson, K.B. & Ter-Pogossian, M.M. (1976): Blood–brain barrier permeability of $^{11}$C-labelled alcohols and $^{15}$O-labelled water. *Am. J. Physiol.* **230**, 543–552.

Raichle, M.E., Martin, W.R.W., Herscovitch, P., Mintun & M.A. & Markham, J. (1983): Brain blood flow measured with intravenous [$^{15}$O]H$_2$O. II. Implementation and validation. *J. Nucl. Med.* **24**, 790–798.

Redies, C., Diksic, M., Evans, A.C., Gjedde, A. & Yamamoto, Y.L. (1987): Double-label autoradiographic deoxyglucose method for sequential measurement of regional cerebral glucose utilization. *Neuroscience* **22**, 601–609.

Redies, C. & Diksic, M. (1989a): Tracer kinetic constants and simplified versions of the double-label deoxyglucose method. (Matters arising). *Neuroscience* **30**, 558–561.

Redies, C. & Diksic, M. (1989b): The deoxyglucose method in the ferret brain. I. Methodological consideration. *J. Cereb. Blood Flow Metab.* **9**, 35–42.

Reivich, M., Alavi, A., Wolf, A., Fowler, J., Russell, J., Arnett, C., MacGregor, R.R., Shiue, C.Y., Atkins, H., Anaud, A., Dann, R. & Greenberg, J.H. (1985) Glucose metabolic rate kinetic model parameter determination in humans: the lumped constant and rate constants for [$^{18}$F]fluorodeoxyglucose and [$^{11}$C]deoxyglucose. *J. Cereb. Blood Flow Metab.* **5**, 179–192.

Schmidt, K., Lucignani, G., Mori, K., Jay, T., Palombo, E., Nelson, T., Pettigrew, K., Holden, J.E. & Sokoloff, L. (1989): Refinement of the kinetic model of 2-[$^{14}$C]deoxyglucose method to incorporate effects of intracellular compartmentation in brain. *J. Cereb. Blood Flow Metab.* **9**, 290–303.

Schmidt, K., Lucignani, G., Moresco, R.M., Rizzo, G., Gillardi, M.C., Messa, C., Colombo, F., Fazio, F. & Sokoloff, L. (1992): Errors introduced by tissue heterogeneity in estimation of local cerebral glucose utilization with current kinetic models of the [18F]fluorodeoxyglucose method. *J. Cereb. Blood Flow Metab.* **12**, 823–834.

Sergent, J., Zuck, E., Terriah, S. & MacDonald, B. (1992): Distributed neural network underlying musical sight-reading and keyboard performance. *Science* **257**, 106–109.

Spence, A.M., Graham, M.M., Muzi, M., Abbott, G.L., Krohn, K.A., Kapoor, R., Woods, S.D. (1990): Deoxyglucose lumped constant estimated in a transplanted rat astrocytic glioma by the hexose utilization index. *J. Cereb Blood Flow Metab.* **10**, 190–198.

Talbot, J.D., Marrett, S., Evans, A.C., Meyer, E., Bushnell, C. & Duncan, G.H. (1991): Multiple representations of pain in human cerebral cortex. *Science* **251**, 1355–1358.

Thompson, C.J., Dagher, A., Meyer, E. & Evans, A.C. (1986): Imaging performance of a dynamic positron emission tomograph: Positome IIIp. *IEEE Trans. Med. Imaging* **MI-5**, 193–198.

Troutner, D.E., Volkert, W.A., Hoffmann, T.J. & Holmes, R.A. (1984): A neutral lipophilic complex of $^{99m}$Tc with a multidentute amino oxime. *Int. J. Appl. Radiat. Isot.* **35**, 467–470.

Winchell, H.S., Baldwin, R.M. & Liu, T.H. (1980): Development of $^{123}$I-labelled amines for brain studies; localization of $^{123}$I iodophenylalkylamines in rat brain. *J. Nucl. Med.* **21**, 940–946.

# Chapter 9

# Regulation of cerebral blood flow: an overview

## Wolfgang Kuschinsky

*Department of Physiology, University of Heidelberg, Heidelberg, Germany*

## Regulation by blood gases 'Chemical' regulation

The role of blood gases in the regulation of cerebral blood flow (CBF) has been documented in a large number of studies and reviews (e.g. Betz, 1972, 1975, 1981; Busija and Heistad, 1984; Kontos, 1981; Kuschinsky and Wahl, 1978). It is generally accepted that CBF is increased by a raised arterial $pCO_2$ and by a lowered arterial $pO_2$. In contrast, CBF is decreased by a lowered arterial $pCO_2$. In the present context only more recent aspects of mechanisms involved and of pathophysiology will be discussed.

## Arterial $pCO_2$, tissue pH: direct actions on brain vessels

The changes of CBF with variations of the arterial $pCO_2$ are basically mediated by concomitant changes in the pH of the brain tissue. A more extensive discussion can be found elsewhere (Kuschinsky, 1982) The essential findings concern the vasoactive action of $H^+$ and the release of $H^+$ during functional activity and hypercapnia. The vasodilative effect of increased $H^+$ concentration and the vasoconstrictor effect of decreased $H^+$ concentration has been shown in pial arteries using the technique of perivascular microapplication (Haller and Kuschinsky, 1981; Kuschinsky *et al.*, 1972). Figure 1 shows the typical reactions of pial arteries to changes in perivascular pH. A few microlitres of artificial cerebrospinal fluid are injected, under microscopic control, around the outside of single pial arteries. The vascular diameter is measured at the injection site. These reactions of pial arteries to changes in perivascular pH are comparable to reactions of intraparenchymal vessels which shows that there does not exist any systematic difference in reactivity of intra– and extraparenchymal vessels to pH changes although the quantitative congruence has not been proven. All these results show that the effect of $CO_2$ on CBF is mediated at least partly by concomitant changes in pH. This does not exclude an additional direct effect of $CO_2$ on brain vasculature. However, a direct effect of $CO_2$ apart from pH does not exist. This could be demonstrated by manipulating $pCO_2$ and pH independent from each other: alterations of the $HCO_3^-$ concentration and of the $pCO_2$ in artificial cerebrospinal fluid irrigating the brain surface only affect the vascular diameter of the underlying pial arteries when the pH of the cerebrospinal fluid is changed whereas an increased $pCO_2$ in the cerebrospinal fluid is not vasoactive as long as the $HCO_3^-$ concentration is increased in such a way that the pH of the cerebrospinal fluid remains constant.

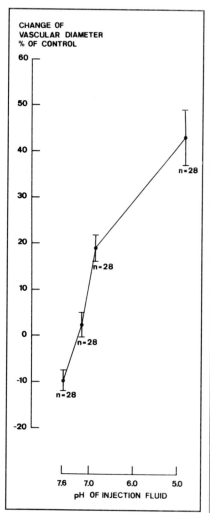

*Fig. 1. Dependency of pial arterial diameters on perivascular pH. Local changes in perivascular pH were induced by changes in the $HCO_3^-$ concentration. From Haller and Kuschinsky (1981).*

The conclusion that changes in brain tissue pH have a regulatory effect on cerebrovascular resistance depends on the demonstration of a vasoactive effect of such changes. In addition, the release of $H^+$ in situations of brain hyperaemia has to be demonstrated to reach such a conclusion. Whereas an increased brain $H^+$ concentration is undoubtedly found during systemic hypercapnia, its possible role is less clear in functional hyperaemia, i.e. during the hyperaemia induced by increased neuronal activity. Experiments investigating this question are difficult to perform: pH microelectrodes have been placed on the brain surface before and during extreme brain activation (seizure activity). The experiments using pH microelectrodes showed congruently decreases in extravascular pH and increases in local blood flow (Kuschinsky and Wahl, 1979; Leniger-Follert, 1984, Fig. 2). These experiments show the immediate interrelationship between decreases in brain pH and increases in CBF during neuronal activation.

## Potential indirect actions of pH changes on brain vessels

Apart from this direct effect of pH on cerebral vessels which has been established as a basic mechanism of regulation of CBF, secondary mechanisms are being discussed which may mediate vasoactive effects by a pH dependent alteration of the release of other vasoactive factors. Although such mechanisms are sometimes claimed to completely explain the vasoactive effects of pH they should only be regarded as supportive to the basic pH mechanism.

### Neural mechanisms

Neurogenic adrenergic influences can modify cerebrovascular pH reactivity. Pharmacological blockade of adrenergic mechanisms has been shown to increase the $CO_2$ sensitivity of CBF. Similar results have been obtained during denervation of the norepinephrine-containing fibres of the locus coeruleus which innervate brain vessels. On the other hand, transection of trigeminal fibres which innervate brain vessels by peptidergic, i.e. non-adrenergic mechanisms, had no influence on $CO_2$ reactivity (Moskowitz et al., 1988) which excludes nonspecific mechanisms induced by nerve lesions. Interactions between effects of pH and norepinephrine have also been shown in isolated pial and intraparenchymal brain vessels. This indicates an action which takes place, at least partly, at the neuromuscular junction or at the receptor site. The significance of cholinergic mechanisms is less well established. Systemic atropine, given at a dose which blocked muscarinic receptors, did not influence the cerebral vasodilatation induced by hypercapnia. Similarly, bilateral section of the VIIth to XIth cranial nerves did not effect the reactivity of CBF to $CO_2$. On the other hand, an efflux of acetylcholine from brain cortex has been measured during hypercapnia. This efflux could be blocked by transection of the midbrain. Together with the previously published findings of a decreased $CO_2$-reactivity after atropine and an enhanced $CO_2$-reactivity after physostigmine this was taken as evidence that the effect of $CO_2$ on brain vessels is mediated, at least partly, by cholinergic mechanisms.

### Prostaglandin mechanisms

Prostaglandins have been suggested as mediators of $CO_2$-dependent changes in cerebral blood flow. Such suggestions were based on the effects of the inhibitor of cyclooxygenase, indomethacin. Several authors could demonstrate, in different species, that CBF is decreased by indomethacin during normocapnia and that the increase in CBF during hypercapnia is strongly attenuated by indo-

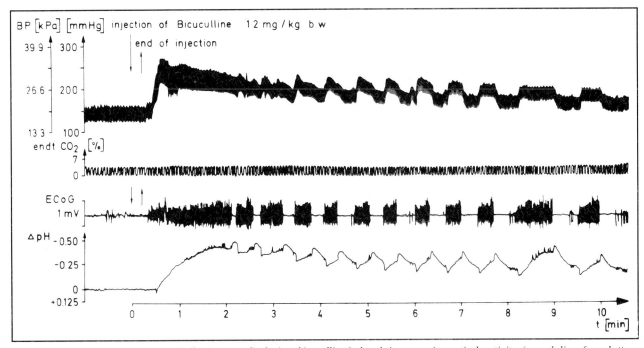

*Fig. 2. Changes of extracellular pH (bottom panel) during bicuculline-induced increases in cortical activity (second line from bottom: electrocardiogram). The acidotic shift parallel to the increased cortical activity is evident. Note also the parallel changes in the ECoG and pH during the later phases of the recordings. From Leniger-Follert (1984) with permission.*

methacin. These studies can be criticized mainly because no efforts were made to show an inhibition of prostaglandin synthesis parallel to the decreased $CO_2$ reactivity. More detailed studies showed that an alteration of prostaglandin synthesis in brain vessels and brain tissue does not necessarily lead to an alteration of the $CO_2$ response of CBF. The effects of indomethacin on cerebrovascular $CO_2$ reactions can therefore be ascribed to its nonspecific effects. The absence of effects of other cyclooxygenase inhibitors supports this conclusion. A further step in this kind of analysis of prostaglandin influences on CBF is the measurement of prostaglandin concentrations in the brain or in cerebrovenous blood during changes in arterial $pCO_2$. The results of such studies do not support the concept of a direct link between $pCO_2$, prostaglandin concentration and cerebrovascular resistance. Whereas all these data have been obtained from adult animals, the situation apparently is different in newborns: in newborn pigs, a prostaglandin product of the cyclo-oxygenase pathway appears to mediate the cerebrovascular response to $CO_2$ (Wagerle and Mishra, 1988). Thus, the role of prostaglandins in cerebrovascular $CO_2$ reactions appears to be limited, at least in adult animals.

## $CO_2$ reactivity, carbonic anhydrase

Tests of the $CO_2$ reactivity of CBF give information on the functional state of brain vessels. Such tests are therefore clinically applied to patients to evaluate the vasodilatory reserve which is decreased in areas downstream of stenosing lesions. In spite of their numerous applications it is still an open question whether the dependency of CBF on arterial $pCO_2$ is a linear or exponential one (for references see von Kummer, 1984). In general, an increase in arterial $pCO_2$ of 1 mmHg induces an increase in cortical CBF by 2–4 per cent in man. More local analysis of different brain structures using autoradiographic methods in animals (Smith *et al.*, 1983; Göbel *et al.*, 1989) have demonstrated the dependency of the hypercapnic increase in CBF on the resting blood flow of the respective brain structure: brain structures which have a high CBF under normocapnic control conditions show a higher absolute increase in CBF with hypercapnia than structures with a low resting flow. The low values of resting CBF in white matter and the low absolute changes during hypercapnia make it difficult to observe significant changes of CBF in white matter of patients, although they apparently exist (Reich and Rusinek, 1989). When averaged together with grey matter values, this may even mask the existing grey matter reactivity. An exception with regard to $CO_2$ reactivity is the neurohypophysis (median eminence and neural lobe) which does not show a change in blood flow with hypercapnia (Bryan *et al.*, 1988). Clinical tests of the $CO_2$ reactivity have been modified in several ways: instead of CBF, relative changes in middle cerebral artery blood flow velocity using transcranial Doppler ultrasonography have been measured for clinical purposes (Ringelstein *et al.*, 1988) and have been found to yield relevant results. Another modification of the test of $CO_2$ reactivity is the acetazolamide test. I.v. injection of acetazolamide induces an increase in CBF in patients as assessed by numerous studies of different groups (e.g. Vorstrup, 1988). Like tests of $CO_2$ reactivity, the acetazolamide test allows evaluation of the vasodilatory capacity, but the exact mechanism of action of acetazolamide is not known. What has been known for several decades is the fact that acetazolamide is an effective blocker of carbonic anhydrase. Although it is tempting to assume acidosis to be the cause of the cerebral hyperaemia induced by acetazolamide, an intracellular acidosis could not be detected in the brain following acetazolamide using [31]P nuclear magnetic resonance spectroscopy (Vorstrup *et al.*, 1989). This finding may be due to the low spatial resolution of the nuclear magnetic resonance spectroscopy method.

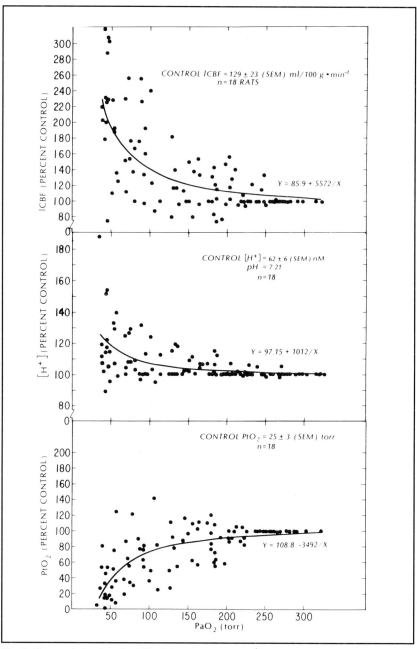

Fig. 3. *Changes in local cerebral blood flow, brain tissue $[H^+]_t$ and $P_tO_2$ in percentage of controls with arterial $pO_2$ ($p_aO_2$) during 40 per cent $N_2O$ analgesia in rats. Controls were obtained at $p_aO_2$ of 250 to 350 torr. From Shinozuka et al. (1989).*

Compartmentalization of tissue decarboxylations (which are the primary metabolic events) and inorganic phosphate stores (which are the signals measured) may favour a spatial separation and disequilibrium between compartments to occur shortly after inhibition of the carbonic anhydrase (Grieb, 1990) thus preventing the immediate effects of pH on the phosphate ratio from which intracellular pH is calculated.

## Arterial $pO_2$, chemoreceptor influences

### Effects of altering $pO_2$

The arterial $pO_2$ is a determinant of cerebrovascular resistance and hence of CBF. The general opinion is that a decrease in arterial $pO_2$ leads to an increase in CBF whereas an increase of arterial $pO_2$ beyond the normal values has no effect. This opinion is still grossly correct, although a more detailed analysis has shown that the reaction of CBF to changes in arterial $pO_2$ is a continuous and not a threshold phenomenon (Shinozuka et al., 1989) as believed previously (Betz, 1972; Kuschinsky and Wahl, 1978). In earlier studies, either hypoxia was tested resulting in an increased CBF or hyperoxia was investigated which did not induce significant decreases in CBF. Careful analysis of CBF over a range of arterial $pO_2$ of 40–300 mmHg showed an exponential increase in CBF with decreasing arterial $pO_2$ (Shinozuka et al., 1989, Fig. 3). At the normal arterial $pO_2$, the CBF was already elevated when compared to the hyperoxic blood flow indicating a sensitivity of cerebral resistance vessels to both increases and decreases in arterial $pO_2$. Local analysis of the hypoxic increase in CBF in anaesthetized rats showed close analogies between hypoxic and hypercapnic responses. Differences in absolute increases of CBF in different brain structures were found which depended on the basal flow rates. Therefore, on a relative basis, increases were more homogeneous than at first apparent (Smith et al., 1983). On the other hand, a greater increase in CBF during hypoxia has been measured in unanaesthetized cats in the caudal brain stem compared with cortical blood flow; the smaller effect on cortical flow was ascribed to an activation of the sympathetic nervous system which may affect cortical structures more than brain stem. A specific and extraordinary reaction of CBF to hypoxia has been reported for the neurohypophysis: neurohypophyseal blood flow is increased with hypoxic hypoxia, which is similar to other brain structures, but not with carbon monoxide hypoxia, which is a unique behaviour (Hanley et al., 1986). Concerning the flow in the cortical microcirculation, moderate hypoxaemia does not alter the mean microflow, whereas the scatter of microflows in different locations is increased. In contrast, strong hypoxaemia also increases mean microflow (Kozniewska et al., 1987).

## Mechanisms

Three mechanisms which may mediate this oxygen sensitivity of CBF are under discussion: 1. action of peripheral chemoreceptors, 2. direct effects of oxygen on resistance vessels, 3. indirect effects of oxygen mediated by tissue factors.

### Peripheral chemoreceptors

The old hypothesis that hypoxia increases CBF by activation of reflex pathways via peripheral chemoreceptors has now been abandoned. Carefully performed studies have shown that, at least in anaesthetized animals, chemoreceptor denervation does not influence the hypoxic response of CBF (e.g. Miyabe et al., 1989). Likewise, stimulation of vascularly isolated chemoreceptors by nicotine, hypoxia or hypocapnia did not influence CBF. Therefore, more local mechanisms are important for the oxygen sensitivity of CBF. One established neural effect which is mediated via peripheral chemoreceptors during hypoxia is an activation of the sympathetic nervous system which induces, if anything, a sympathetic vasoconstriction of cerebral vessels which would counteract the direct hypoxic vasodilation. Another chemoreceptor-mediated reflex concerns neurohypophyseal blood flow. Hypoxic hypoxia induces an increase in blood flow in this brain structure, as in others. However, in contrast to all other brain structures, this increase in neurohypophyseal blood flow can be blocked by combined denervations of carotid sinus and aortic arch chemoreceptors (Hanley et al., 1988). This suggests a specific influence of chemoreceptor activity on neurohypophyseal blood flow which seems likely to be linked with the simultaneous increase in vasopressin release (Hanley et al., 1988).

### Direct vascular effects of oxygen

Hypoxia exerts its effects on cerebrovascular resistance principally by two mechanisms: a direct effect on cerebral resistance vessels and an indirect effect which includes the release of vasoactive factors from brain tissue. The direct effect on brain resistance vessels can be deduced from the action of hypoxia on isolated brain vessels. Hypoxia induces a dilation of isolated larger conductance vessels (e.g. Pearce et al., 1989). This effect results from both a direct action on vascular smooth muscle cells which has been verified after endothelium removal and an action mediated by the endothelium. Endothelium-derived relaxing factor and endothelium-derived contracting factor are involved in the endothelial response to hypoxia (Pearce *et al.*, 1989). Concerning the contractile mechanisms, a direct inhibitory effect of hypoxia on sarcoplasmic $Ca^{2+}$ uptake has been postulated as a mechanism of action. Supportive of a local action of hypoxia on brain vessels are two further studies. It could be shown that the hypoxic vasodilatation of pial arteries can be eliminated by superfusion of the vessels with an oxygen-rich solution. In addition, the basic mechanism which mediates the increase in cerebral blood flow as measured in animals and in man does not adapt to sustained hypoxia of weeks and even years (Krasney *et al.*, 1990).

*Indirect vascular effects of oxygen (via tissue factors)*

The slow time course of the hypoxic increase in CBF after an acute decrease in arterial $pO_2$ which reaches a steady state only after minutes can be taken as an argument for the participation of tissue factors in the hypoxic vasodilatation. Such tissue factors are linked with the altered metabolism of the brain tissue during hypoxia. Local changes in the metabolic rate for glucose have been demonstrated both during hypoxia (e.g. Lockwood *et al.*, 1989) and during hyperbaric hyperoxia (Torbati & Lambertsen, 1985). Although these findings clearly indicate local heterogeneities in the metabolic reactions to hypoxia, studies of the importance of tissue factors for the hypoxic vasodilatation are only feasible when more global measurements of these factors are being performed. This may explain some of the

discrepancies in results and interpretations. The tissue factors discussed in this context are $H^+$ and adenosine. $K^+$ ions are also involved in hypoxic vasodilatation. Aspects of regulation of CBF by $K^+$ will be discussed in the context of coupling between function, metabolism and blood flow since this appears to be the most important role of $K^+$ in the regulation of cerebrovascular resistance.

The question as to whether pH-dependent mechanisms (mainly production of lactate) participate in the hypoxic vasodilatation has been a matter of controversy which has been discussed in several review articles (Betz, 1972, 1975, 1981; Busija and Heistad, 1984; Kuschinsky and Wahl, 1978). The problem of earlier approaches is that a direct local correlation between $pO_2$, pH and blood flow or metabolism could not be obtained. Therefore, negative results refuting the concept of a relationship between tissue $pO_2$ and tissue acidosis were not completely convincing. Two recent studies using more direct local approaches shed some more light on this issue. Lockwood *et al.* (1989) have related local changes in glucose utilization during moderate hypoxia (arterial $pO_2$ 40 mmHg) with local pH changes, both parameters being measured autoradiographically. They found that areas of increased glucose utilization were surrounded by an extensive zone of acidosis. Other areas showed normal glucose metabolism with acidosis which was explained by a direct effect of hypoxia on the sodium–hydrogen antiporter system. With a decreased ability of this antiporter to extrude $H^+$ from the cells, intracellular acidosis might develop although glucose metabolism is

not visibly changed. The most straightforward approach has been chosen by Shinozuka *et al.* (1989). These authors were able to measure simultaneously local $pO_2$, pH and blood flow in the brain cortex by means of inserted electrodes. In addition, they varied the values of arterial $pO_2$ between 40 and 300 mmHg. Therefore they could observe continuous changes in the measured parameters over a wide range of arterial $pO_2$ values. The results of this study appear therefore to be more meaningful than those of the usual two-point measurements, one taken at normoxia and one at hypoxia. Their results clearly showed that the decreasing tissue $pO_2$ was paralleled by an increasing blood flow and $H^+$ concentration. The curves showed that differences in all three parameters occurred already between arterial $pO_2$ values of 300 mmHg and the normal arterial $pO_2$ of close to 100 mmHg. These results show a close direct association between tissue $pO_2$, pH and local blood flow. If tissue acidosis exists during hypoxia, this may influence brain metabolism. The local acidosis may inhibit glucose utilization of the affected brain area (Kuschinsky *et al.*, 1981; van Nimmen *et al.*, 1986).

Another factor mediating vasodilatation during hypoxia is adenosine. Adenosine is released from the hypoxic brain (e.g. Park *et al.*, 1987a; Phillis *et al.*, 1987). It exerts a vasodilative effect on pial vessels (Wahl and Kuschinsky, 1976, Fig. 4) and increases local blood flow in the brain when locally applied (van Wylen *et al.*, 1989). The concentrations measured in the interstitial space of the brain during hypoxia are in a vasodilating range (Park *et al.*, 1987; Phillis *et al.*, 1987). Several studies have been performed using adenosine antagonists to assess the contribution of adenosine to hypoxic vasodilatation. The results of these studies are more controversial than one might expect. Some studies performed in different species and using different methods of determination of blood flow have shown an attenuation of hypoxic increases in cerebral blood flow by adenosine blockers, mostly theophylline, but also by other adenosine blockers such as caffeine. Conversely, negative results have also been reported from some studies performed under comparable conditions. Also studies using adenosine deaminase, and its inhibitors have yielded controversial results arguing in favour

and against adenosine as a mediator of hypoxic vasodilatation. It appears reasonable to conclude from these results that an effective action of the different drugs at the adenosine receptor site cannot be taken as certain under all experimental conditions. Therefore, the negative results should not be taken as an argument to refute any role of adenosine in hypoxic vasodilatation. They rather may show that adenosine may not be of equal value in mediating the vasodilatation of cerebral vessels under all of the varying conditions of hypoxia.

## Other aspects of regulation
## Relation between blood flow and metabolism

It is widely accepted that metabolism and blood flow in the brain are normally tightly coupled to the functional activity. This coupling is a strictly local phenomenon and can be demonstrated consistently only if local methods are applied. Coupling between function, metabolism and blood flow can be described in two different ways: one way is to quantify the steady state relationship between these parameters which exists at constant experimental conditions over a long time; the other way is to quantify the amount

of change of each of these parameters during altered experimental conditions. These two components of coupling may be termed the static and dynamic component.

### Static coupling

The availability of local methods to quantify blood flow and metabolism in the brain has allowed the definition of the relationship between local metabolism and blood flow in the brains of experimental animals as well as of man. Tight correlations have been found between local oxygen consumption and blood flow in man (e.g. Baron *et al.*, 1984; Fox and Raichle, 1986) as well as local glucose utilization and blood flow during normal experimental control conditions in experimental animals (e.g. Kuschinsky *et al.*, 1981, Fig. 5). These correlations are linear, as is evident from the example given in Fig. 5. The ratio of blood flow to glucose utilization has been shown to vary in normal conscious rats between about 1 and 2. Different experimental manipulations either did not change the correlation or induced an altered slope of the correlation. Uncoupling of this relationship was observed only during ischaemic conditions (Baron *et al.*, 1984) although uncoupling has been claimed for physiological conditions (Fox and Raichle, 1986; Lou *et al.*, 1987), as will be discussed later. There exists a long-term relationship between metabolism and blood flow in the brain, which is rather stable and is not, or only moderately, modified during different experimental conditions. Why does such a rather fixed relationship exist? Recently, the role of the local capillary pattern in the static coupling has been clarified.

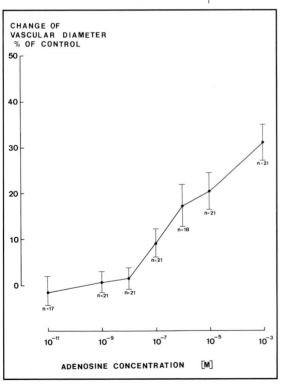

*Fig. 4. Dependency of pial arterial diameter on perivascular adenosine. From microapplication studies of Haller et al. (1986).*

Using morphometric methods, a tight correlation has been found between the capillary length per volume of brain and the local cerebral blood flow (Gjedde and Diemer, 1985). The perfused capillaries have been depicted with fluorescent dyes which are coupled to globulins and injected i.v.; brain slices have then been analysed and the number of capillary sections quantified for a given section area. These studies showed a strong heterogeneity of the perfused capillaries; moreover, the number of perfused capillaries in a specific brain area correlated highly with the blood flow and, even better, with the glucose utilization of the same brain structure (Klein *et al.*, 1986). By analogy, local heterogeneities have been also found for the capillary bed as verified by morphological methods (Göbel *et al.*, 1990). These studies have led to the conclusion that the varying metabolic rate in the different brain structures has triggered a varying capillary density; the capillary density then determines a basal level of blood flow. Since this component of the coupling mechanism is rather fixed by the structure, i.e. the capillaries, and since it persists over long periods, it can be termed the structural or static component of the coupling mechanism.

## Dynamic coupling

Although the static component is an important element of the coupling mechanism, it does not completely describe the phenomenon of coupling: coupling also includes the fact that the blood flow in the brain varies acutely and locally according to the demands of the tissue. Since this is a moment-to-moment adjustment of

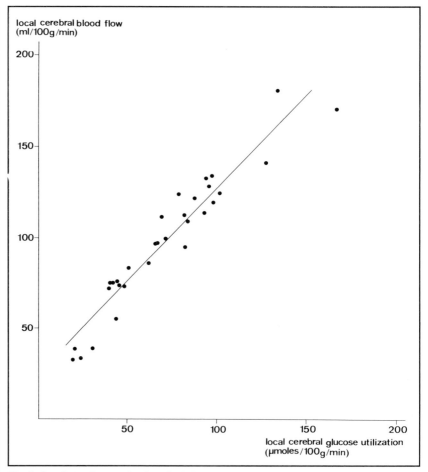

*Fig. 5. Correlation between local cerebral glucose utilization and local cerebral blood flow in the brains of conscious rats. Each point represents a brain structure. Glucose utilization and blood flow were measured using autoradiographic techniques. The tight correlation (r = 0.96) is the expression of the static coupling mechanism.*

the blood flow, it can be termed the dynamic or acute or short-term component of the coupling mechanism. This component depends on the regulatory action of the cerebral arteries and arterioles. Dynamic coupling takes place at a very local level. Thus, the cerebral resistance vessels are continuously influenced and controlled by the composition of the fluid which surrounds them. The basic mechanisms which are believed to mediate this dynamic coupling are

shown schematically in Fig. 6. Increased neuronal activity is manifested as an increased frequency of action potentials. This leads to a release of $K^+$ from the nerve cells to the extracellular space of the brain. Since this space is small in the brain (less than 15 per cent) and the intra/extracellular gradient is high, an increase of the extracellular $K^+$ concentration ensues. A raised extracellular $K^+$ concentration can dilate the cerebral resistance vessels (Kuschinsky *et*

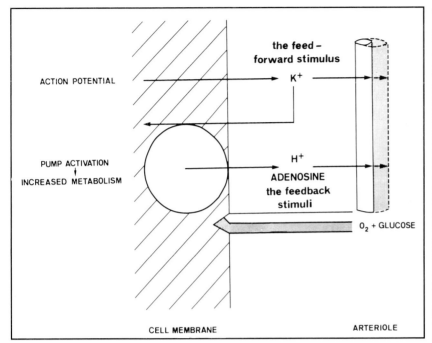

*Fig. 6. Mechanisms of dynamic coupling. Dynamic coupling consits of two elements: the function-dependent element ($K^+$) and the metabolic element ($H^+$, adenosine). Both elements are able to dilate the cerebral resistance vessels, if their concentration in the extravascular fluid is*

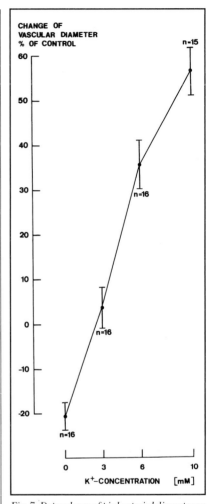

*Fig. 7. Dependency of pial arterial diameter on perivascular $K^+$. From microapplication studies of Haller and Kuschinsky (1981).*

*al.*, 1972; Kuschinsky and Wahl, 1978, Fig. 7): thus, the stimulus for the vasodilatation is released together with the information forwarded by the action potential. This means a very quick adjustment of blood flow to the increased functional activity. On a long-term basis, the original ionic gradients have to be reestablished. Pumps are activated to carry the $K^+$ back into (and the $Na^+$ out of) the cells. Increased pump activity means an increased metabolism: this is the basis of neuroimaging techniques which use metabolism as an indicator of function. The triggers for a metabolic vasodilatation, $H^+$ and adenosine, are released as long as a mismatch exists between the oxygen/glucose demand and delivery. The vasodilation induced by these factors counteracts this mismatch and thus permits a fine tuning of the regula-

tory mechanism to take place. Thus dynamic coupling occurs according to two principles: the feedforward principle, verified by $K^+$ because of its quick, instantaneous release, and the feedback principle utilized by $H^+$ and adenosine which permits fine tuning of the vascular diameter. Both principles are common in circulatory and respiratory physiology. Whereas $K^+$ is now generally accepted as a quick mediator of coupling, the role of $H^+$ and adenosine is more controversial. Although the dilating action of $H^+$ (Fig. 1) and adenosine (Fig. 2) is well established (Kuschinsky and Wahl, 1978), the conditions of their release are not completely defined. In some studies extravascular pH decreased with the start of the neuronal activation (e.g. Kuschinsky and Wahl, 1979); in other studies the decrease in pH lagged some seconds behind the

start of the neuronal activation (e.g. Leniger-Follert, 1984). For adenosine, the situation is even more complex; an exact time course is hard to define for the first seconds of functional hyperaemia, since a continuous on-line measurement of the adenosine concentration is not possible. However, analysis of brain intestitial fluid adenosine concentration using the brain dialysis technique (Benveniste, 1989; van Wylen *et al.*,

1988) has yielded some progress in this respect. The brain dialysis technique has shown increases in interstitial adenosine concentration under different conditions, such as systemic hypotension (Park *et al.*, 1988) and seizures. A detailed description of all aspects of adenosine in the relation of cerebral blood flow has been given by Phillis (1989).

The above-mentioned controversies about the role of $H^+$ and adenosine in functional hyperemia may be reconciled when one takes into account the conditions under which these local factors are emitted: whereas the release of $K^+$ depends on the neuronal firing rate, $H^+$ and adenosine accumulate in the tissue if there is a mismatch between oxygen demand and delivery as discussed recently for adenosine (Newby *et al.*, 1990). Therefore, the fine tuning of the $H^+$/adenosine metabolic control mechanism is likely to act after a time lag in most cases and will become more and more relevant in those cases in which the quick $K^+$ mechanism alone is not sufficient to yield an adequate $O_2$ supply of the tissue. The $H^+$/adenosine mechanism would then be of varying relevance under different conditions. Its impact would be small in such cases in which the $pO_2$ of the tissue increases as has been found for the starting phase (first seconds) of neuronal activation (Leniger-Follert, 1984). It would gain importance in all cases of oxygen lack, the extremes being hypoxia and ischaemia. The varying contribution of this metabolic control mechanism to the vascular adjustment would mean a varying ratio of metabolism to blood flow depending on the prevailing impact of the metabolic control mechanism and of the

$K^+$ mechanism, respectively. With less metabolic control, blood flow becomes less dependent on the actual metabolic rate but rather on the $K^+$ release. Under such conditions, an altered relationship between metabolism and blood flow is to be expected. It has been measured during somatosensory stimulation in man and has been interpreted as an indication of a physiological uncoupling (Fox and Raichle, 1986). This may be a premature interpretation since knowledge of the characteristics of static coupling is not sufficient to draw such a conclusion of physiological uncoupling, until the characteristics of dynamic coupling can be better defined. The fact that dynamic coupling consists of several components, as has been outlined, and that these components may change their impact, makes it unlikely that the characteristics of dynamic coupling can be defined by a constant ratio of metabolism to blood flow.

Regarding the metabolism in more detail, dissociations have been observed between the increase in glucose metabolism and oxygen consumption. During somatosensory stimulation in animals (Ueki *et al.*, 1988) and man (Fox *et al.*, 1988) local increases in glucose consumption have been measured which clearly exceeded the increases in local oxygen consumption. This indicates the stimulation of non oxidative glucose utilization in spite of a sufficient oxygen supply, during physiological conditions in the brain cortex. Such a phenomenon which is based on a delayed increase of oxygen utilization by the enzymes of the respiratory chain appears to represent an analogy to the oxygen debt in skeletal muscle.

## Capillary perfusion in the brain

Whereas dynamic coupling is mediated by local vasoactive factors which act on the resistance vessels of the brain according to the functional and metabolic needs, the mechanisms of the static coupling are less well elucidated. An important role can be ascribed to the capillary morphology since local differences in capillary density and a correlation between capillary density and local blood flow have been demonstrated in the brain (Fig. 8). Capillary density was also found to correlate with the local glucose utilization (Klein *et al.*, 1986, Fig. 9). It can be concluded from these data that the local metabolic rate in each brain structure determines its capillary density as already discussed. The local capillary density is the decisive parameter for the local blood flow. Local blood flow can then be varied by dynamic factors which are released depending on function and metabolism. The local enzymatic pattern may also vary in parallel to the capillary density: a negative correlation was found between capillary density and lactatedehydrogenase, whereas capillary density and cytochrome oxidase were positively correlated (Borowsky and Collins, 1989).

The question as to what fraction of the brain capillaries is perfused at any given moment under normal conditions is highly controversial. Perfusion fractions ranging from half to all of the existing cerebral capillaries have been reported. In numerous studies, Weiss (1988) has claimed that about half of the capillary bed is perfused at any moment. Capillary cycling has also been claimed by Jones *et al.* (1989), but later been revoked

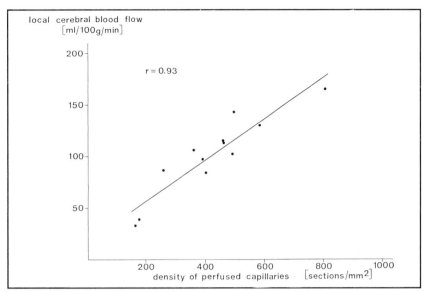

*Fig. 8. Correlation between the density of perfused capillaries in the brain and local blood flow. Each point represents a brain structure in which capillary density or blood flow have been measured. From Klein et al. (1986).*

(Williams *et al.*, 1993). A perfusion fraction of 70–80 per cent has been observed by Kikano *et al.* (1989). Collins *et al.* (1987) have concluded that more than 84 per cent of all capillaries

are perfused in the brain structure investigated, which was a part of the superior colliculus. Pawlik *et al.* (1981) found more than 90 per cent of all capillaries being perfused. It has also

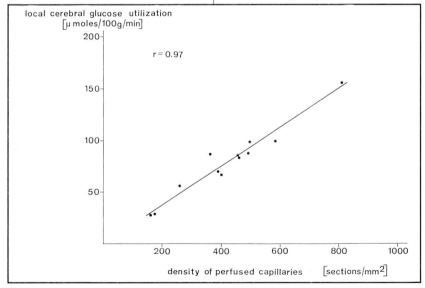

*Fig. 9. Correlation between the density of perfused capillaries in the brain and local glucose utilization. Each point represents a brain structure in which capillary density or glucose utilization have been measured. From Klein et al. (1986).*

been claimed that all capillaries in the brain are perfused at any given moment (Göbel *et al.*, 1990) which also excludes capillary recruitment during hyperaemic conditions (Göbel *et al.*, 1989). The finding of non-perfused capillaries under normal conditions may be due to methodological flaws of the staining methods employed (Kuschinsky & Paulson, 1992). The number of perfused and morphologically existing capillaries found by different groups appears to be a function of the sensitivity of the methods used.

## The nitric oxide hypothesis

An integrative model has been proposed which combines different aspects of coupling and neuronal function. The nitric oxide hypothesis of Gally *et al.* (1990) ascribes to the NO radical such an integrative function on different systems in the brain. One aspect of the action of NO is the control of blood flow by production and release from the endothelium. This aspect is discussed elsewhere in this book (Chapter 6). Other aspects concern a control of synaptic plasticity and long-term potentiation as well as the establishment of axonal projections. The validity of this attractive hypothesis has to be verified.

## Autoregulation
## Blood flow and perfusion pressure

## Description

Autoregulation is a basic vascular phenomenon which exists in most organs, including the brain. Paulson *et*

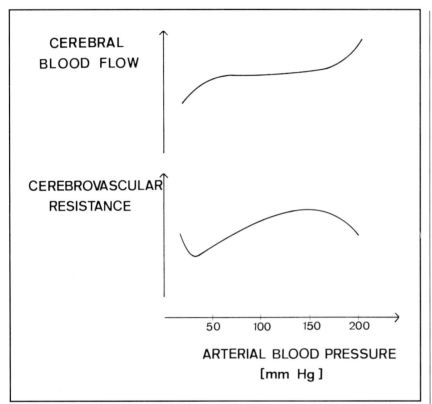

*Fig. 10. Schematic representation of an autoregulation curve. The plateau of the autoregulation curve (upper part) results from changes in cerebrovascular resistance (lower part) which parallel the changes in perfusion pressure. At the upper and lower limit of the autoregulation curve the capacity of the vessels to counteract the changes in perfusion pressure is exhausted or is insufficient.*

al. (1990) reviewed it. Autoregulation can be defined as the intrinsic ability of an organ to keep its blood flow constant during changing perfusion pressures. The perfusion pressure is the difference between arterial pressure and either cerebrospinal fluid pressure or cerebrovenous pressure, whichever of these two is larger. During physiological conditions, only arterial pressure is relevant. The brain, like other organs, maintains its blood flow largely, but not completely, constant by vasodilatation during decreases and vasoconstriction during increases in perfusion pressure. These vascular reactions have a time constant of many seconds, e.g. during an acute moderate hypertension, blood flow increases first and then slowly comes back to nearly its normotensive value (e.g. Aaslid *et al.*, 1989). The resulting autoregulation curve, as depicted schematically in Fig. 10, thus represents a steady state relationship. During the first few seconds after an acute change in perfusion pressure, the plateau part does not exist and a passive dependence of blood flow on perfusion pressure is found. Such passive behaviour of the vessels exists during pathophysiological conditions, such as ischaemia, trauma or inflammation. A gradual decline of autoregulatory reactions has been shown with increasing degrees of ischaemia (Dirnagl and Pulsinelli, 1990). More or less passive behaviour of the cerebral resistance vessels is seen during very high and very low perfusion pressures as is evident from the non-plateau part of the autoregulation curve. In these cases, the autoregulatory capacity is exhausted. At high perfusion pressures, vasoconstriction may alternate with passive distension of vessel segments thus resulting in a 'sausage string' appearance of cerebral vessel, which does not represent vasospasm. It should be mentioned that the plateau part is never complete, i.e. even in the plateau part there is some higher flow with higher perfusion pressure.

## Sympathetic influences

The inflection points where the autoregulation curve departs from its plateau part represent the upper and lower limit of the autoregulation curve. These limits can be shifted acutely and chronically. An acute shift to the right is found during an electrical stimulation of the cervical sympathetics. Conversely, sympathetic denervation induces a shift to the left. An acute increase in blood flow during the pressure rise is therefore blunted by sympathetic stimulation. From this, a protective effect of the sympathetic system has been deduced which, during acute increases in blood pressure, diminishs increases in cerebral blood flow. Severe hypertension also induces an increased permeability of the blood–brain barrier. This can be also attenuated by sympathetic stimulation thus yielding an additional protective effect of the cervical sympathetics. This protective effect at high perfusion pressures may change into a detrimental action at low perfusion pressures: a high sympathetic discharge during hypotension (e.g. during shock) may compromise the diminished blood flow by a constrictory action of the sympathetics on the cerebal vessels. A chronic shift of the upper and lower limits of the autoregulation curve to the right is observed during chonic hypertension. Whereas the shift of the upper limit makes higher perfusion pressures tolerable and induces vascular hypertrophy thus protecting cerebral vessels during hypertension (Mayhan *et al.*, 1987), the simultaneous shift of the lower limit can be detrimental during hypotensive periods, e.g. during forced antihypertensive therapy. A moderate therapy is therefore necessary.

## Renin–angiotensin system

Another system besides the sympathetic nervous system which can modulate autoregulatory reactions of cerebral vessels is the renin-angiotensin system. This has been concluded from studies using the angiotensin converting enzyme inhibitor captopril. Captopril has been shown to shift the limits of the autoregulation curve to lower blood pressure values in normotensive and spontaneously hypertensive rats (Paulson *et al.*, 1988; Waldemar *et al.*, 1989a). The effects appear to be less pronounced in man (Waldemar *et al.*, 1989b). They cannot be abolished by acute sympathetic denervation which points to the independence of mechanisms involving the sympathetic nervous system and the renin–angiotensin system in autoregulation (Waldemar, 1990).

## Possible mechanisms

The basic mechanism which mediates autoregulatory adjustments of cerebral vessels had not yet been completely clarified. Myogenic and metabolic mechanisms are under discussion; neurogenic mechanisms can probably be excluded since they only modulate but do not cause autoregulation.

## Metabolic mechanisms

The possibilities of directly verifying a metabolic mechanism of autoregulation are limited because changes in metabolic factors at the resistance vessels are hard to detect during autoregulatory adjustments: perivascular $H^+$ and $K^+$ activities have been measured close to pial arteries during their autoregulatory adjustments: the changes in these activities were small and can hardly explain the complete pial vascular reactions observed (Wahl and Kuschinsky, 1979). Adenosine concentrations in brain tissue change inversely with blood pressure; however, a quantitative analysis of the interstitial concentrations of adenosine using the brain dialysis technique has shown that adenosine could only be relevant at low blood pressures of less than 50 mmHg thus excluding all except an accessory role of adenosine in autoregulation (van Wylen *et al.*, 1988). Thus, although a metabolic component of the autoregulatory mechanism cannot be completely excluded, convincing evidence is scanty.

## Myogenic mechanisms

Myogenic mechanisms of autoregulation, i.e. stretch-dependent vasoconstriction and vice versa, have been demonstrated more convincingly, with the exception of one experimen-

tal model. This model is the increase in cerebral venous pressure. If a myogenic component of autoregulation were effective, a vasoconstriction should occur due to a raised intravascular pressure which could stimulate the myogenic response. The fact that an increased venous pressure does not change cerebral blood flow (e.g. McPherson *et al.*, 1988) and induces dilation of pial arteries instead of constriction (e.g. Wei and Kontos, 1984) has been interpreted by some to indicate the predominance of metabolic mechanisms in autoregulation. This indirect conclusion is opposed by the direct demonstration of myogenic mechanisms in the autoregulatory cerebral vascular adjustment. Myogenic mechanisms are reactions to changes in the transmural pressure of the vessel which oppose the passive behaviour. In isolated middle cerebral arteries, autoregulatory reactions could be clearly demonstrated (Vinall & Simeone, 1981; Katusic *et al.*, 1987; Rubanyi, 1988; Harder *et al.*, 1989; McCarron *et al.*, 1989). Most of these studies, with one exception (McCarron *et al.*, 1989), have found an endothelial component in this autoregulatory adjustment (Harder *et al.* 1989, Katusic *et al.*, 1987, Rubanyi, 1988). A calcium influx into the endothelial cells during stretch appears to be the primary event (Lansman, 1988). Congruent with direct evidence for a myogenic mechanism of autoregulation in isolated cerebral vessels are *in vivo* experiments, in which the transmural pressure of pial arteries was altered at an unchanged perfusion pressure. This was achieved by putting the animal into a sealed box with the cranial cavity open to the atmosphere. With increased pressure in the box – which means a vessel distension – pial arterioles constricted and vice versa (Bohlen and Harper, 1984). Altogether, most of the recent experimental evidence, with the exception of the experiments using increased venous pressure, argues for a myogenic component of autoregulation; moderate indications exist for an additional metabolic component. The experiments employing increases in cerebrovenous pressure should not be taken as firm evidence against myogenic mechanisms; they may show that, under these special conditions, a metabolic component or the reaction to cerebral hypoxia overrides the myogenic mechanism.

## Neurogenic regulation

Clearly, an overview of CBF regulation would be misleading if the major elements of neurogenic regulation were not introduced here. The facts and hypotheses concerning neurogenic regulation are dealt with systematically in the next three chapters which set out to describe the actions of the transmitters and related substances (Chapter 10), and the various influences and possible roles of both extrinsic (Chapter 11) and intrinsic (Chapter 12) neural elements. The extrinsic nervous system comprises the sympathetic, parasympathetic, and peptidergic (sensorial) innervations, and the intrinsic nervous system, the various central structures which appear to have a specific influence on local CBF (not solely through simple metabolic/flow coupling). Establishing the proof of such a non-metabolic relationship, i.e. modifications of CBF which cannot be explained solely by decreased or increased neuronal activity, represents one of the major challenges for the proponents of the hypothesis of intrinsic neurogenic regulation.

Lastly, it must be emphasized that regulation should not be conveniently subdivided into independent factors such as metabolic and neurogenic regulation, as used to be the tendency. Regulation is inevitably multifactorial as readily illustrated, for example, by the phenomenon of autoregulation, which involves intrinsic vascular reactions (endothelium and smooth muscle), the sympathetic nervous system, and perhaps under certain circumstances chemical (metabolic) factors. The following chapters should be read with this concept in mind.

## Summary

Cerebral blood flow (CBF) is influenced by the blood gases $CO_2$ and $O_2$. Whereas reactions of cerebral vessels to $CO_2$ are more or less exclusively due to a direct action of pH, oxygen reactivity appears to be mediated by both direct action and indirect mechanisms, like release of vasodilating adenosine and $H^+$. Clinical tests of the functional state of cerebral vessels are based on the $CO_2$ reactivity. Other tests examine autoregulation. Under physiological and pathophysiological conditions, autoregulation can be modified by the sympathetic and renin–angiotensin systems. Autoregulation appears to be based mainly on myogenic and endothelium-dependent mechanisms, whereas a metabolic component may become relevant under special experimental conditions. Under normal condtions, a tight coupling exists between the functional activity, metabolism and blood flow in the brain. This coupling consists of two components: a static long-term element appears to be based on local variations in the capillary pattern which are the consequence of local heterogeneity in the metabolic rate of different brain structures and which are responsible for different regional resting flows; a dynamic element mediates the fast adjustment of local blood flow to the functional and metabolic demands. The dynamic component of the coupling mechanism is based mainly on $K^+$, although metabolic factors, like $H^+$ and adenosine, are also involved. Nitric oxide is a further candidate.

## References: major reviews

Benveniste, H. (1989): Brain microdialysis. *J. Neurochem.* **52**, 1667–1679.

Betz, E. (1972): Cerebral blood flow: its measurement and regulation. *Physiol. Rev.* **52**, 595–630.

Betz, E. (1975): Experimental production of cerebral vascular disorders. In: *Handbook of experimental pharmacology*, eds. J. Schmier & O. Eichler, pp. 183–232. Berlin: Springer.

Betz, E. (1981): Physiologie und Pathophysiologie der Gehirndurchblutung. In: *Handbuch der medizinischen Radiologie*, eds. L. Diethelm & S. Wende, pp. 193–294. Berlin: Springer.

Busija, D.W. & Heistad, D.D. (1984): Factors involved in the physiological regulation of the cerebral circulation. *Rev. Physiol. Biochem.*

*Pharmacol.* **101,** 161–211.

Kontos, H.A. (1981): Regulation of the cerebral circulation. *Ann. Rev. Physiol.* **43,** 397–407.

Kuschinsky, W. (1982): Role of hydrogen ions in regulation of cerebral blood flow and other regional flows. In: *Ionic regulation of the microcirculation,* ed. B.M. Altura, pp. 1–19. Basel: Karger.

Kuschinsky, W. & Paulson, O. (1992): Capillary circulation in the brain. *Cerebrovasc. Brain Metab. Rev.* **4,** 261–286.

Kuschinsky, W. & Wahl, M. (1978): Local chemical and neurogenic regulation of cerebral vascular resistance. *Physiol. Rev.* **58,** 656–689.

Lou, H.C., Edvinsson, L. & MacKenzie, E.T. (1987): The concept of coupling blood flow to brain function: revision required? *Ann. Neurol.* **22,** 289–297.

Paulson, O.B., Strandgaard, S. & Edvinsson, L. (1990): Cerebral auto-regulation. *Cerebrovasc. Brain Metab. Rev.* **2,** 161–192.

Phillis, J.W. (1989): Adenosine in the control of the cerebral circulation. *Cerebrovasc. Brain Metab. Rev.* **1,** 26–54.

## Other references

Aaslid, R., Lindegaard, K.-F., Sorteberg, W. & Nornes, H.(1989): Cerebral autoregulation dynamics in humans. *Stroke* **20,** 45–52.

Baron, J.C., Rougemont, D., Soussaline, F., Bustany, P., Crouzel, C., Bousser, M.G. & Comar, D. (1984): Local interrelationships of cerebral oxygen consumption and glucose utilization in normal subjects and in ischaemic stroke patients: a positron tomography study. *J. Cereb. Blood Flow Metab.* **4,** 140–149.

Bohlen, H.G. & Harper, S.L. (1984): Evidence of myogenic vascular control in the rat cerebral cortex. *Circ. Res.* **55,** 554–559.

Borowsky, I.W. & Collins, R.C. (1989): Metabolic anatomy of brain: a comparison of regional capillary density, glucose metabolism, and enzyme activities. *J. Comp. Neurol.* **288,** 401–413.

Bryan, R.M., Myers, C.L. & Page, R.B. (1988): Regional neurohypophysial and hypothalamic blood flow in rats during hypercapnia. *Am. J. Physiol.* **255,**, R295–302.

Collins, R.C., Wagmann, I.L., Lymer, L. & Matter, J.M.(1987): Distribution and recruitment of capillaries in rat brain (abstract). *J. Cereb. Blood Flow Metab.* **7,** (Suppl 1) S336.

Dirnagl, U. & Pulsinelli, W. (1990): Autoregulation of cerebral blood flow in experimental focal brain ischaemia. *J. Cereb. Blood Flow Metab.* **10,** 327–336.

Fox, P.T. & Raichle, M.E. (1986): focal physiological uncoupling of cerebral blood flow and oxidative metabolism during somatosensory stimulation in human subjects. *Proc. Natl. Acad. Sci. USA* **83,** 1140–1144.

Fox, P.T., Raichle, M.E., Mintun, M.A. & Dence, C. (1988): Nonoxidative glucose consumption during focal physiologic neural activity. *Science* **241,** 462–464.

Gally, J.A., Montague, P.R., Reeke, G.N. & Edelman, G.M. (1990): The NO hypothesis: possible effects of a short-lived, rapidly diffusible signal in the development and function of the nervous system. *Proc. Natl. Acad. Sci. USA* **87,** 3547–3551.

Gjedde, A. & Diemer, N.H. (1985): Double-tracer study of the fine regional blood–brain barrier glucose transfer in the rat by computer-assisted autoradiography. *J. Cereb. Blood Flow Metab.* **5,** 282–289.

Göbel, U., Klein, B., Schröck, H. & Kuschinsky, W. (1989): Lack of capillary recruitment in the brain of awake rats during hypercapnia. *J. Cereb. Blood Flow Metab.* **66,** 491–499.

Göbel, U., Theilen, H. & Kuschinsky, W. (1990): Congruence of total and perfused capillary network in rat brains. *Circ. Res.* **66,** 271–281.

Grieb, P. (1990): Effect of acetazolamide on brain intracellular pH. Letter to the Editor. *J. Cereb. Blood Flow Metab.* **10,** 588–589.

Haller, C. & Kuschinsky, W. (1981): Reactivity of pial arteries to $K^+$ and $H^+$ before and after ischaemia induced by air embolism. *Microcirculation* **1,** 141–159.

Haller, C., Kuschinsky, W. & Reimnitz, P. (1986): The effect of gamma-hydroxybutyrate on the reactivity of pial arteries before and after ischaemia. *J. Cereb. Blood Flow Metab.* **6,** 658–666.

Hanley, D.F., Wilson, D.A., Feldman, M.A. & Traystman, R.J. (1988): Peripheral chemoreceptor control of neurohypophysial blood flow. *Am. J. Physiol.* **254,** H742–750.

Hanley, D.F., Wilson, D.A. & Traystman, R.J. (1986): Effect of hypoxia and hypercapnia on neurohypophyseal blood flow. *Am. J. Physiol.* **250**, H7–15.

Harder, D.R., Sanchez-Ferrer, C., Kauser, K., Stekiel, W.J. & Rubanyi, G.M. (1989): Pressure releases a transferable endothelial contractile factor on cat cerebral arteries. *Circ. Res.* **65**, 193–198.

Jones, S.C., Bose, B., Furlan, A.J., Friel, H.T., Easley, K.A., Meredith, M.P. & Little, J.R. (1989): $CO_2$ reactivity and heterogeneity of cerebral blood flow in ischaemic, border zone, and normal cortex. *Am. J. Physiol.* **257**, H473–482.

Katusic, Z.S., Shepherd, J.T. & Vanhoutte, P.M. (1987): Endothelium-dependent contraction to stretch in canine basilar arteries. *Am. J. Physiol.* **252**, 671–673.

Kikano, G.E., LaManna, J.C. & Harik, S.I. (1989): Brain perfusion in acute and chronic hyperglycemia in rats. *Stroke* **20**, 1027–1031.

Klein, B., Kuschinsky, W., Schröck, H. & Vetterlein F. (1986): Inter-dependency of local capillary density, blood flow and metabolism in rat brains. *Am. J. Physiol.* **251**, H1333–1340.

Kozniewska, E., Weller, L., Höper, J., Harrison, D.K. & Kessler, M. (1987): Cerebrocortical microcirculation in different stages of hypoxic hypoxia. *J. Cereb. Blood Flow Metab.* **7**, 464–470.

Krasney, J.A., Jensen, J.B. & Lassen N.A. (1990): Cerebral blood flow does not adapt to sustained hypoxia. *J. Cereb. Blood Flow Metab.* **10**, 759–764.

Kummer R. v.(1984): Local vascular response to change in carbon dioxide tension. Long term observation in the cat's brain by means of the hydrogen clearance technique. *Stroke* **15**, 108–114.

Kuschinsky, W., Suda, S. & Sokoloff, L. (1981): Local cerebral glucose utilization and blood flow during metabolic acidosis. *Am. J. Physiol.* **241**, H772–777.

Kuschinsky, W. & Wahl, M. (1979): Perivascular pH and pial arterial diameter during bicuculline induced seizures in cats. *Pflügers Arch.* **382**, 81–85.

Kuschinsky, W., Wahl, M., Bosse, O. & Thurau, K. (1972): Perivascular potassium and pH as determinants of local pial arterial diameters in cats. *Circ. Res.* **31**, 240–247.

Lansman, J.B. (1988): Going with the flow. *Nature* **331**, 481–82.

Leniger-Follert, E. (1984): Mechanisms of regulation of cerebral microflow during bicuculline-induced seizures in anaesthetized cats. *J. Cereb. Blood Flow Metab.* **4**, 150–165.

Lockwood, A.H., Peek, K.E., Izumiyama, M., Yap, E.W.H. & Labove, J. (1989): Effects of moderate hypoxemia and unilateral carotid ligation on cerebral glucose metabolism and acid-base balance in the rat. *J. Cereb. Blood Flow Metab.* **9**, 342–349.

Mayhan, W.G., Werber, A.H. & Heistad, D.D. (1987): Protection of cerebral vessels by sympathetic nerves and vascular hypertrophy. *Circulation* **75**, I107–I112.

McCarron, J.G., Osol, G. & Halpern W. (1989): Myogenic responses are independent of the endothelium in rat pressurized posterior cerebral arteries. *Blood Vessels* **26**, 315–319.

McPherson, R.W., Koehler, R.C. & Traystman, R.J. (1988): Effect of jugular venous pressure on cerebral autoregulation in dogs. *Am. J. Physiol.* **255**, H1516–1524.

Miyabe, M., Jones, M.D., Koehler, R.C. & Traystman, R.J. (1989): Chemodenervation does not alter cerebrovascular response to hypoxic hypoxia. *Am. J. Physiol.* **257**, H1413–1418.

Moskowitz, M.A., Wei, E.P., Saito, K. & Kontos, H.A. (1988): Trigeminalectomy modifies pial arteriolar responses to hypertension. *Am. J. Physiol.* **255**, H1–6.

Newby, A.C., Worku, Y., Meghji, P., Nakazawa, M. & Skladanowski, A.C. (1990): Adenosine: a retaliatory metabolite or not? *News Physiol. Sci.* **5**, 67–70.

Nimmen van, D., Weyne, J., Demeester, G. & Leusen, I. (1986): Local cerebral glucose utilization during intracerebral pH changes. *J. Cereb. Blood Flow Metab.* **6**, 584–589.

Park, T.S., van Wylen, D.G.L., Rubio, R. & Berne, R.M. (1987a): Increased brain interstitial fluid adenosine concentration during hypoxia in newborn piglet. *J. Cereb. Blood Flow Metab.* **7**, 178–183.

Park, T.S., Wylen van, D.G.L., Rubio, R. & Berne, R.M. (1987): Interstitial fluid adenosine and sagittal sinus blood flow during bicuculline-seizures in newborn piglets. *J. Cereb. Blood Flow Metab.* **7**, 633–639.

Park, T.S., Wylen van, D.G.L., Rubio, R. & Berne, R.M. (1988): Brain interstitial adenosine and sagittal sinus blood flow during systemic hypotension in piglet. *J. Cereb. Blood Flow Metab.* **8**, 822–828.

Paulson, O.B., Waldemar, G., Andersen, A.R., Barry, D.I., Pedersen, E.V., Schmidt, J.F. & Vorstrup S.(1988): Role of angiotensin in autoregulation of cerebral blood flow. *Circulation* **77**, I55–I58.

Pawlik, G., Rackl, A. & Bing, R.J. (1981): Quantitative capillary topography and blood flow in the cerebral cortex of cats: an *in vivo* microscopic study. *Brain Res.* **208**, 35–58.

Pearce, W.J., Ashwal, S. & Cuevas, J. (1989): Direct effects of graded hypoxia on intact and denuded rabbit cranial arteries. *Am. J. Physiol.* **257**, H824–833.

Phillis, J.W., Walter, G.A., O'Regan, M.H. & Stair, R.E. (1987): Increases in cerebral cortical perfusate adenosine and inosine concentrations during hypoxia and ischaemia. *J. Cereb. Blood Flow Metab.* **7**, 679–686.

Reich, Th. & Rusinek, H. (1989): Cerebral cortical and white matter reactivity to carbon dioxide. *Stroke* **20**, 453–457.

Ringelstein, E.B., Sievers, C., Ecker, S., Schneider, P.A. & Otis, S.M. (1988): Noninvasive assessment of $CO_2$-induced cerebral vasomotor response in normal individuals and patients with internal carotid artery occlusion. *Stroke* **19**, 963–969.

Rubanyi, G.M. (1988): Endothelium-dependent pressure-induced contraction of isolated canine carotid arteries. *Am. J. Physiol.* **255**, H783–788.

Shinozuka, T., Nemoto, E.M. & Winter, P.M. (1989): Mechanisms of cerebrovascular $O_2$ sensitivity from hyperoxia in the rat. *J. Cereb. Blood Flow Metab.* **9**, 187–195.

Smith, M.-L., Kagström, E. & Siesjö, B.K. (1983): Local cerebral blood flow in the rat brain during hypercapnia and hypoxia. *Acta Physiol. Scand.* **118**, 439–440.

Torbati, D. & Lambertsen, C.J. (1985): The relationship between cortical electrical activity and regional cerebral glucose metabolic rate in rats exposed to 3 atmospheres absolute oxygen. *Brain Res.* **344**, 186–190.

Ueki, M., Linn, F. & Hossmann, K.-A. (1988): Functional activation of cerebral blood flow and metabolism before and after global ischaemia of rat brain. *J. Cereb. Blood Flow Metab.* **8**, 486–494.

Vinall, P.E. & Simeone, F.A. (1981): Cerebral autoregulation: an *in vitro* study. *Stroke* **12**, 640–642.

Vorstrup, S. (1988): Tomographic cerebral blood flow measurements in patients with ischaemic cerebrovascular disease and evaluation of the vasodilatory capacity by the acetazolamide test. *Acta Neurol. Scand.* **77**, (Suppl. 114) 5–48.

Vorstrup, S., Jensen, K.E., Thomsen, C., Henriksen, O., Lassen, N.A. & Paulson, O.B. (1989): Neuronal pH regulation: constant normal intracellular pH is maintained in brain during low extracellular pH induced by acetazolamide – [31]P NMR study. *J. Cereb. Blood Flow Metab.* **9**, 417–421.

Wagerle, L.C. & Mishra, O.P. (1988): Mechanism of $CO_2$ reponse in cerebral arteries of the newborn pig: role of phospholipase, cylooxygenase, and lipoxygenase pathways. *Circ Res.* **62**, 1019–1026.

Wahl, M. & Kuschinsky, W. (1976): The dilatatory action of adenosine on pial arteries of cats and its inhibition by theophylline. *Pflügers Arch.* **362**, 55–59.

Wahl, M. & Kuschinsky, W. (1979): Unimportance of perivascular $H^+$ and $K^+$ activities for the adjustment of pial arterial diameter during changes of arterial blood pressure in cats. *Pflügers Arch.* **382**, 203–209.

Waldemar, G. (1990): Acute sympathetic denervation does not eliminate the effect of angiotensin converting enzyme inhibition on CBF autoregulation in spontaneously hypertensive rats. *J. Cereb. Blood Flow Metab.* **10**, 43–47.

Waldemar, G., Paulson, O.B., Barry, D.I. & Knudsen, G.M. (1989a): Angiotensin converting enzyme inhibition and the upper limit of cerebral blood flow autoregulation: effect of sympathetic stimulation. *Circ. Res.* **64**, 1197–1204.

Waldemar, G., Schmidt, J.F., Andersen, A.R., Vorstrup, S., Ibsen, H. & Paulson, O.B. (1989b): Angiotensin converting enzyme inhibition and cerebral blood flow autoregulation in normotensive and hypertensive man. *J. Hypertension* **7**, 229–235.

Wei, E.P. & Kontos, H.A. (1984): Increased venous pressure causes myogenic constriction of cerebral arterioles during local hyperoxia. *Circ. Res.* **55**, 249–252.

Weiss, H.R. (1988): Measurement of cerebral capillary perfusion with a fluorescent label. *Microvasc Res.* **36**, 172–180.

Williams, J.L., Shea, M. & Jones, S.C. (1993): Evidence that heterogeneity of cerebral blood flow does not involve vascular recruitment. *Am. J. Physiol.* **264**, 1740–1743.

Wylen van, D.G.L., Park, T.S., Rubio, R. & Berne, R.M. (1988): Cerebral blood flow and interstitial fluid adenosine during hemorrhagic hypotension. *Am. J. Physiol.* **255,** H1211–1218.

Wylen van, D.G.L., Park, T.S., Rubio, R. & Berne, R.M. (1989): The effect of local infusion of adenosine and adenosine analogues on local cerebral blood flow. *J. Cereb. Blood Flow Metab.* **9,** 556–562.

# Chapter 10

# Effects of neurotransmitters on cerebral blood vessels and cerebral circulation

## Jan Erik Hardebo and Christer Owman

*Department of Medical Cell Research and Neurology, University of Lund, Lund, Sweden*

## Introduction

The concept of a neurogenic control of the brain circulation did not start to take form until it was proved in 1967 that the brain vessels are supplied by a dense plexus of nerves innervating all parts of the vascular circuit. It has so far been demonstrated that some 15 transmitters or transmitter candidates are stored in various systems of nerve fibres associated with the cerebrovascular bed.

Most of the initial attention was focused on adrenergic neurotrans-

mission mechanisms in the cerebral circulation. A vast arsenal of drugs and experimental models has allowed pharmacological predictions of some important physiological aspects of the cerebrovascular system, such as influence on arterial resistance, blood volume regulation and importance for vascular growth, in the course of three decades of intensive research.

The latest period of research development in this area began with the discovery that not only 'classical' transmitter substances like noradrenaline (NA) and acetylcholine (ACh) but also certain polypeptides are present

in the cerebrovascular nerves, the first to be demonstrated in a long series being vasoactive intestinal polypeptide (VIP).

*In vitro* studies of isometric changes in tone on isolated segments of vessels have been widely used to define their receptor equipment. In such studies the neurotransmitter is administered to the organ bath, or endogenous transmitter is released by electrical or pharmacological stimulation of the perivascular nerves. The contractions elicited are observed at stretch-induced tone and dilatations at a pre-contractile tone induced by pharmacological agents or a depolarizing potassium solution. For practical reasons, vessels with diameters of 200 μm or more have been used in the majority of such studies. The importance of the findings for understanding how CBF is regulated is limited, because precapillary arterioles with diameters of about 50 μm or below are responsible for the greatest part of the cerebrovascular resistance. However, pial and parenchymal arteries also contribute to some extent to this regulation.

Below are summarized the effects on tone in the various segments along the vascular tree, i.e. pial arteries and arterioles, penetrating arteries and arterioles, microvessels and, finally, venules and veins, followed by the effect on cerebral blood flow (CBF) after administration of transmitters and mediators into the circulation or brain parenchyma. Recent reviews are referred to when appropriate.

## Catecholamines

### Noradrenaline

Pial and brain parenchymal arteries down to the arteriolar level are richly innervated by sympathetic nerves, located in the adventitia and adventitial–medial border, and in most species exclusively originating in the superior cervical ganglion (for references see Chapters 5 and 11). The main transmitter in these nerves appears to be NA (see also Chapter 11). The pial and parenchymal vessels are equipped with α- and β-adrenergic receptors, also at the microvascular level. The effect on cerebrovascular tone of NA depends upon which seg-

ment along the vascular tree is studied, and whether the monoamine reaches the vessel wall from the adventitial side (due to release from the sympathetic nerves, and during pathological conditions possibly also from NA neurons in the brain parenchyma) or from the intimal side (i.e. via the blood stream). During physiological conditions the morphological and enzymatic blood–brain barrier (BBB) effectively prevents circulating catecholamines from reaching beyond the endothelial lining and thus they do not reach the outer layer of the smooth musculature, where the

majority of receptors for catecholamines are present.

Various studies have revealed that the smooth muscle cell membranes of most *large pial arteries* from various species including man are equipped with vasocontractile α-adrenergic receptors, with little or no contribution from β-adrenergic receptors, except in rat basilar and pig large pial arteries where α-receptors predominate. The sensitivity of α-receptors in cerebral arteries is up to 10 times lower than that found in non-cerebral arteries, and in the rabbit the sensitivity is even

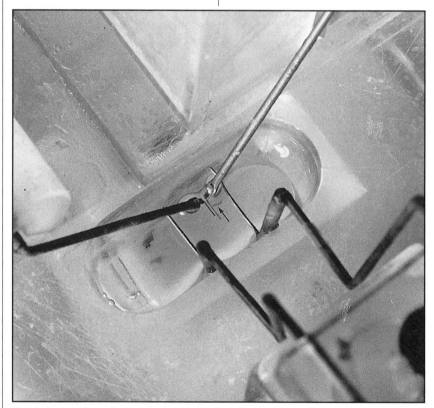

*Fig. 1. Commonly used set-up for in vitro studies of isometric changes in tone of isolated ring segments of pial and cortical arteries and veins. The vessel is mounted in the organ bath in the direction of the arrow on two L-shaped metal holders, one fixed to an adjustable unit for applying stretch-induced tension to the vessel, and the other fixed to a force-displacement transducer for continuous registration of changes in vascular tone upon administration of a vasoactive substance to the bath. Electrodes can be held in position on each side of the vessel, to study changes in tone following electrical field-induced transmitter release from the perivascular nerves.*

lower. The contractile response to NA is fairly homogeneous in the various main pial arteries, as studied in cat. The contractile reactivity to NA is similar, but the dilatory response to the β-adrenergic agent isoprenaline is greater in the distal compared to the proximal end of the middle cerebral artery, as studied in dog. High NA concentrations (above $10^{-4}$ M) produce further contraction, not involving α-receptor activation.

As studied in large pial arteries by both functional and ligand-binding studies the $\alpha_1$-receptor subtype predominates over the $\alpha_2$-subtype in man, monkeys and rats, whereas a low affinity $\alpha_2$-type seems to dominate in cats, and especially dogs, rabbits and cattle. In humans the $\alpha_1$ predominance is based on both ligand binding and autoradiographic studies. Ligand binding studies hve also revealed that both $\beta_1$ and $\beta_2$-receptors are present in human large and small pial vessels, the latter being more numerous. The effect on tone caused by exogenous NA in isolated pial arterial segments, although clearly indicative of the presence of adrenergic receptors, provides little information to elucidate the importance of their effect in the intact vasculature under physiological conditions. To meet these objections, the vasomotor response of pial arteries has been examined following perivascular microapplication or superfusion of the brain surface *in situ* with exogenous NA. Such studies in the mouse, rat and cat have revealed α-receptor-induced constrictions by NA.

As elsewhere in the body, $\alpha_2$-receptors are also located *prejunctionally* to control the neuronal release of NA.

The contraction upon exposure to α-receptor stimulating substances such as NA is more dependent upon influx of extracellular calcium in cerebral arteries (but not veins) than in non-cerebral arteries. Such dependence is higher in pial arteries from dog and man than in rabbit and monkey. It is possible, however, that the response to *neurally* released NA is less dependent upon extraceullar calcium, as demonstrated in non-cerebral arteries, on postjunctional $\alpha_1$-receptors.

Isolated segments of *penetrating intracerebral arteries* (with a diameter of around 100 μm ) *and arterioles* (diameter about 20–40 μm), branching off from the middle cerebral artery, have been studied in rats and rabbits *in vitro* and expressed as changes in isometric tone and perfusion pressure upon exposure to NA. Owing to the BBB, isolated vessel segments cannulated at both ends can be used for differentiation between an effect of NA applied extraluminally onto smooth muscle receptors and intraluminally onto endothelial receptors. A primarily extraluminal effect can also be determined through perivascular microapplication to pial vessels *in situ* or by stimulation of their sympathetic nerves, and the intraluminal effect can be demonstrated in vessels by removal of endothelium. The studies have revealed that small pial arteries are weakly contractile via α-receptors or unresponsive, and arterioles are either unresponsive or dilate via β-receptors to extraluminally applied NA. Further, the tonic response to NA in isolated rabbit and cat middle cerebral artery seems to involve an endothelium-dependent dilatory component. This

may comprise activation of endothelial α-receptors, leading to release of an endothelial factor attenuating NA-induced α-adrenergic smooth muscle contraction.

*Cerebral microvessels* (primarily capillaries) are also innervated by NA-containing nerves. These originate in the locus coeruleus (see Chapter 12). β-receptors, in the proportion $\beta_1/\beta_2 = 2/3$, and in some species also α-receptors (especially $\alpha_2$), are found in cerebral microvessels. Myofilaments are found in the endothelium of brain capillaries. These may subserve a contractile and relaxatory function of these vessels upon proper activation of endothelial receptors by mediators, such as NA, either from the circulation if the receptor is located on the luminal endothelial membrane, or from brain neurons if located on the abluminal membrane. Adrenergic receptors in cerebral microvessels may not only play a role in blood flow regulation (see Chapter 12), but also contribute to the control of vascular permeability via activation of β-receptor-linked adenylate cyclase. β-receptors have been shown to be expressed in cultured cerebral capillary endothelial cells, but not in cultured pericytes.

*Cerebral veins* are equipped with more sensitive contractile α-receptors than the cerebral arteries. Sympathetic activation induces a proportionally stronger constriction in small than in large pial veins, implying that venous blood will be drained out of the brain from these capacitance vessels. In this way perivascular sympathetic nerves may contribute to regulation of cerebral blood volume and hence intracranial pressure.

Due to the BBB and the presence of adrenergic receptors on the endothelium down to the microvascular level, the *effect on CBF by circulating NA* may be fundamentally different from that of sympathetically released NA. Systemic or intracarotid administration of NA, in experiments in which blood pressure and blood gas levels are controlled, results in a minor decrease in cerebral blood flow or, particularly in primates, in no response. Upon critical examination of the various studies it appears that only when the BBB has been intentionally or unknowingly opened beforehand will any substantial *increase* in CBF be observed. The same flow change can be obtained upon intraventricular or intraparenchymal injection of NA, and when NA is acutely released from intracerebral stores by intracarotid injection of reserpine. Most if not all of this flow increase occurs secondarily to stimulation by NA of the glucose metabolism in brain.

*At raised perfusion pressure and alkaline pH* the sensitivity of the contractile α-receptor is enhanced, as studied in large pial arteries and penetrating arterioles. At pH 7.80 a contractile α-response to NA predominates over the dilatory β-response in rat penetrating arterioles, starting at very low NA concentrations. There is reason to believe that the tonic effect of NA is stronger in vessels with only one smooth muscle cell layer, as in penetrating cerebral arterioles, provided a sensitive α/β-receptor is present.

The predominance at the adventitial-medial border of contractile α-receptors in the pial arteries, and of β-receptors in small parenchymal arteries and arterioles, is in keeping with the view that the perivascular sympath-etic nerves, activated during acute stress conditions, constrict only proximal cerebral arteries (with little flow regulating capacity) while dilating or leaving small distal vessels unaffected. This prevents a harmful over-distension of the small vessels by the wave of acutely raised blood pressure, while still ensuring an adequate blood supply to the brain in this acute situation. There is little sure indication of a tonic influence of the sympathetic nerves on CBF during resting conditions (see Chapter 11).

NA, released from perivascular sympathetic nerves, may not only influence tone in cerebral arteries, arterioles and veins. Other functions are possible, such as interaction with co-released neuropeptide Y (NPY), interaction with the release from adjacent parasympathetic or sensory nerves, and a trophic influence on the vascular bed (see Chapter 11).

## Adrenaline

Whereas circulating NA is mainly derived from perivascular sympathetic nerves, circulating adrenaline (A) originates from the adrenal medulla. There is no evidence for the presence of A in perivascular sympathetic, parasympathetic or sensory nerves. Whether intracerebral A neurons come into close contact with parenchymal microvessels is not known with certainty. Some release of A from brain neurons to the perivascular space may occur during pathological conditions.

If A gains access to brain perivascular α- or β-adrenergic receptors in sufficiently high concentrations to activate them (which is unlikely during physiological conditions) the local ac-tion on tone will be similar to that of NA. In comparison to NA, A is a slightly better stimulator of the $\alpha_2$- than $\alpha_1$-receptor. External application of A to the brain surface of cat constricts large pial arteries and veins, and perfusion with A of vessels in isolated cat and dog brains affects resistance vessels in a similar manner as NA.

Through an unknown mechanism adrenal catecholamines contribute, despite their inability to cross the BBB, to the cerebral flow increase seen upon stimulation of the dorsal medullary reticular formation in rat (see Chapter 12).

## Dopamine

The levels of dopamine (DA) in the circulation and cerebrospinal fluid are low. DA is present in perivascular sympathetic nerves, but merely as a precursor to NA. In some species, DA has been found in mast cells, in close contact with pial and brain parenchymal vessels. DA is a transmitter in several neuronal systems in the brain. Studies have made it likely that some of these DA neurons make contacts with cerebral arterioles and capillaries.

In studies on vascular tone where DA is applied to isolated unconstricted pial arteries *in vitro*, or to pial arterioles *in situ*, a constriction is obtained, which is mediated through stimulation of α-adrenergic and serotonergic receptors. When these receptors are blocked, DA or selective stimulators of $D_1$ (but not $D_2$) receptors cause relaxation of these vessels. Such relaxation is seen without prior α-blockade in human, monkey and rabbit pial arteries. If ever the perivascular concentration of DA rises to levels that can activate pial $D_1$ receptors, the primary source for such DA may be perivascular mast cells.

The effect of DA on cerebral venous tone has not been investigated.

Intravenous infusion of DA, which does not pass the BBB, leaves CBF unaffected, whereas BBB-permeable dopamine receptor stimulants, such as apomorphine and amphetamine, increase flow. The flow increase may involve activation of microvascular $D_1$ receptors, but the predominant mechanism responsible is an enhanced metabolic activity in the brain tissue.

## Serotonin

Serotonin (5-HT) may reach the cerebral vessel wall from several possible sources. First, it is likely that parenchymal microvessels come into close contact with raphe-derived central serotonergic neurons. Such fibres may contribute to the regulation of local CBF (see Chapters 5 and 12). 5-HT, present in the cerebrospinal fluid, particularly derived from ruptured platelets in conjunction with brain haemorrhage, is efficiently taken up by the perivascular sympathetic nerve fibres, and can be subsequently released upon sympathetic activation. The possibility even exists that pial vessels are innervated by authentic serotonergic fibres; this point has been the subject of a lively controversy for several years (see Chapter 5). Further, 5-HT is stored in brain perivascular mast cells of several species, in particular rodents, and can be released therefrom by a variety of stimuli.

5-HT is a moderate to strong constrictor when applied *in vitro* to isolated *large pial arteries* from all species including man, through activation of specific 5-HT-receptors. 5-HT-induced contractions are smaller in the distal than in the proximal middle cerebral artery, as studied in dog, and absent in some surface branches. In the open skull preparation perivascular pial application of 5-HT induces contraction in large pial arteries.

It appears that multiple subtypes of contractile $5\text{-HT}_1$ and $5\text{-HT}_2$ receptors exist in pial arteries of the same species, although one may predominate over the other, as particularly studied in dog and cat. In rats and monkeys the receptors are of the $5\text{-HT}_2$ type only, and in man of the $5\text{-HT}_1$ type. Further studies in the human pial artery have revealed that the $5\text{-HT}_1$ receptor cannot be defined as any of the known subtypes (A, B, C or D). The contraction upon exposure to 5-HT is more dependent upon influx of extracellular calcium in cerebral than in non-cerebral arteries. At higher concentrations of 5-HT, perivascular α-adrenergic receptors also become activated. Upon exposure to the endothelium of isolated pial arteries a dilatory component of 5-HT can be revealed.

A low dose of 5-HT markedly amplifies the contraction of pial arteries induced by NA; such interaction has also been demonstrated in non-cerebral vessels. Possibly, such an effect may be of significant clinical importance for the early vasospasm in subarachnoid haemorrhage, where 5-HT from disrupted platelets also reaches pial 5-HT receptors from the adventitial side.

A small contraction, or no response at all, is obtained when 5-HT is applied to isolated segments of *penetrating* cortical *arteries* of rabbit, and high 5-HT concentrations induces small contractions in isolated *penetrating arterioles* of rat with diameters of 30–70 μm. Also dilatation can be induced by 5-HT in rat cerebral arterioles. Perivascular application of 5-HT to the cat pia dilates arterioles with diameters below 70 μm, through activation of β-adrenergic-like or $5\text{-HT}_1$-like receptors. Dilatation is changed into a minor contraction in aged rats.

There are no positive reports on 5-HT receptors or binding sites in *brain microvessels*. Small amounts of 5-HT are actively taken up by cerebrovascular endothelial cells; such uptake may work as an inactivation mechanism for circulating 5-HT, and as the first step in an enzymatic BBB for the monoamine.

Large *pial veins* of various species constrict upon exposure to 5-HT *in vitro* or *in situ*, although at higher 5-HT concentrations than in the arteries, whereas a dilatory capacity is present in small veins (and in the pig also in large veins also).

Circulating 5-HT is prevented from gaining access to the receptors on the adventitial–medial border and in the brain parenchyma mainly by the enzymatic BBB. There is some evidence that 5-HT, administered to the luminal or abluminal side, is able to impair the morphological BBB capacity, but the question is far from resolved. Against this background it is not surprising that intracarotid injection of 5-HT in rat, baboon and man has no *effect on CBF* or causes only a weak reduction. A less effective endothelial barrier in the extracranial part of the vascular tree may explain why such injection in the baboon and man causes constriction of the internal carotid artery despite leaving CBF unchanged.

After experimental opening of the morphological or enzymatic BBB in primates, and in brain areas with a damaged morphological BBB in man, intravascular injection of 5-HT causes a reduction in CBF. There is evidence that the flow change primarily reflects a depression of metabolic and electrocortical activity. Similarly, intraventricular injection of 5-HT in rabbit causes a reduced flow locally. On the other hand, intrahypothalamic injection of 5-HT in rabbit increases local blood flow, possibly by a predominant direct activation of arteriolar 5-HT-receptors rather than action on the cerebral metabolism.

In summary, 5-HT from the circulation appears to have little influence on the cerebral vascular bed under physiological conditions. Instead, 5-HT most likely exerts its major function on the adventitial side, through the influence of the central serotonergic innervation on microvessels during physiological conditions, and through release from perivascular mast cells during inflammation and from ruptured platelets at cerebral haemorrhage.

## Acetylcholine

There is no good histochemical marker for the presence of ACh in nerves outside the central nervous system. This is the main reason why the question about the presence of ACh in postganglionic parasympathetic nerves in the cerebral vasculature is not yet definitely settled (see Chapters 5 and 11). There is no demonstration of an ACh-induced change in cerebrovascular tone upon stimulation of perivascular parasympathetic nerves *in vitro* or *in vivo*, although a neuronal release of radioactive ACh from pial vessels has been reported. Cholinergic nerve terminals have been observed in close proximity with cortical microvessels. Due to the high activity of cholinesterase in blood, circulating levels of ACh are very low. Local synthesis of ACh in the endothelium, as demonstrated by the presence of choline acetyltransferase in isolated cerebral vessels, particularly at the microvascular level, may represent one major source in the cerebrovascular wall.

Pharmacological studies on tonic changes upon ACh application *in vitro* and *in situ* have revealed the presence of muscarinic receptors inducing relaxation in *pial arteries and arterioles and penetrating* cortical *arterioles* (30–70 μm). Radioligand binding studies in various species have confirmed the presence of muscarinic receptors in pial arteries and arterioles. In some species a contractile muscarinic receptor in pial arteries is also revealed at high ACh concentrations, or after removal of the endothelium.

Interaction with the endothelial cells is essential for inducing the dilatation, as is also the case in non-cerebral vessels. The interaction elicits the release of endothelium-dependent relaxing factor. It is likely that most, if not all, of the cholinergic receptors are located on the endothelial cell membrane, and are reached by ACh after prior synthesis within the endothelium or following diffusion through the musculature from nerves located in the adventitia. In the endothelium of human cortical arterial branches, cholinergic receptors have been found to be most abundant on the luminal membrane. In human pial vessels these endothelial relaxatory receptors have been demonstrated to be of the M₃ subtype, whereas contractile receptors on the smooth muscle cells are of the M₁ subtype. The ACh-induced dilatation is less pronounced in pial arterioles from aged rats, and the response is more pronounced at alkaline pH.

*Isolated cerebral microvessels* also contain muscarinic receptors, as demonstrated by radioligand binding in various species and ACh-dependent formation of cyclic AMP in man. It seems likely that a large fraction of these receptors are not on endothelial cells but on astrocyte processes associated with the isolated microvessels.

In *pial veins* ACh only causes contraction, and at fairly high concentrations.

ACh is poorly penetrable across the luminal endothelial membrane. Despite this, ACh and muscarinic cholinomimetic agents increase *CBF* upon intravenous, intracarotid or intravertebral administration. This may indicate that muscarinic receptors are present also on the luminal endothelial membrane. Although one study has indicated a concomitant increase in cerebral oxygen consumption it is likely that the flow effect of ACh is primarily caused by direct activation of endothelial cholinergic receptors.

Evidence has been presented for a central origin in the cerebellar fastigial nucleus for cholinergic fibres to cortical microvessels. Stimulation of this nucleus causes an increase in CBF, not linked to changes in cerebral metabolism, which may indicate a direct action of such fibres on cerebral microvessels. Similar evidence has been put forward for the basal forebrain, which also projects cholinergic fibres to the cortex (see Chapter 12).

The physiological significance of CBF changes induced by ACh is questionable, since cholinergic blockade does not affect resting CBF. Further, the flow increase upon stimulation of postganglionic parasympathetic fibres to the brain circulation in rat is not affected by muscarinic blockade, and thus is probably attributable to release of VIP or some other cotransmitter alone.

Thus, in the brain circulation the functional role of ACh is still obscure. The endothelium-dependent relaxation to ACh may be a pharmacological curiosity; the cholinergic receptor may be present for other purposes than vasomotor activation. Despite the predominant dilator action of exogenous ACh, endogenous ACh may, alternatively, be primarily a vasoconstrictor substance, as evidenced from its direct effect on the smooth musculature in some species. Further, ACh may be involved in prejunctional blockade of NA release from cerebrovascular sympathetic nerves by stimulating muscarinic receptors on these nerves, it may interact with co-released VIP, it may release vasoactive substances from perivascular type II mast cells, and it may have trophic actions on the vascular bed.

## Histamine

There is no indication of the presence of histamine in pial perivascular nerves. Some central histaminergic nerve fibres have been found in close contact with parenchymal microvessels. The main perivascular source for histamine in the cerebral circulation is mast cells, where the amine is synthesized and stored. These cells are mobile along the vascular adventitia, and are found down to the level of small pial and parenchymal arteries, arterioles and venules. They are not located around capillaries, and only in the pia mater, choroid plexus and olfactory bulb are mast cells also found, at some distance from a blood vessel. Perivascular nerves make close contact with mast cells in the cerebral vessel wall. The number of mast cells is enhanced around pial arteries in patients with subarachnoid haemorrhage.

The net effect of histamine on the tone of isolated large *pial arteries* is constriction in rabbits and cats and dilatation in rats and man. As studied in dogs, contraction predominates over relaxation in the proximal middle cerebral artery, whereas the reverse is valid in the distal part. A great heterogeneity among the various large pial arteries is found for the histamine-induced maximal relaxation in cat: it is about 6 times larger in the posterior cerebellar artery than in the anterior communicating artery. However, all pial vessels from the various species studied appear to have both contractile and dilatory histamine receptors (the rat is an exception with no contractile receptors). The presence of histamine receptors in pial arteries has been confirmed by

ligand binding studies. The contractile receptor is of the $H_1$ type, whereas the dilatory receptor has been found to be of both $H_1$ and $H_2$ types, with the latter predominating in most species.

As studied in, for example, rat and rabbit pial arteries, many, though far from all, of the dilatory receptors are located on the endothelium or inner layer of the smooth musculature. The luminally oriented receptors are predominantly of the $H_1$ type (except in rabbits where they may be of the $H_2$ and $H_3$ types), whereas $H_2$ receptors are usually located in the outer layers near the adventitia. The contractile $H_1$ – receptor is highly dependent upon influx of extracellular calcium.

Isometric studies on isolated segments of *small pial* surface *arteries* (100–200 μm) from rabbits reveal only a contractile response to histamine, and the sensitivity is slightly lower than in the parent middle cerebral artery. In *penetrating* cortical *arteries* of rabbits (100 μm) studied by isometric tension and perfusion pressure techniques, a dilatory response is revealed when the contractile $H_1$ receptor is blocked. In perfusion pressure experiments with isolated rat *penetrating arterioles* (about 40–70 μm) extraluminally administered histamine induces only relaxation, mediated by $H_2$ receptors. Topical application of histamine in open-skull or cranial window preparations causes dilatation of cat pial arteries and arterioles (about 35–300 μm) through activation of $H_1$ receptors, whereas the pial arterioles constrict in rabbits.

Endothelial $H_1$- and $H_2$-receptors are found also in *brain capillaries*. Small amounts of histamine accumulate in the cerebrovascular endothelium, particularly at the microvascular level. This may be one source for the histamine acting on endothelial receptors. The major source is probably mast cells, from which histamine diffuses through the smooth muscle layer.

*Pial veins* from several species do not respond with dilatation or contraction *in vitro* or *in situ*; in rabbits and man a minor contraction and dilatation, respectively, have been observed.

Histamine penetrates poorly across the luminal endothelial membrane of the BBB. Intravenous or intracarotid injection of histamine causes no change or only a minor increase in *CBF*, as studied in various species including man. Only when the BBB has been opened does intracarotid histamine injection in rat result in a flow increase, through activation of both $H_1$ and $H_2$ receptors. The increase is not secondary to altered brain metabolism. In line with this finding, bypassing of the BBB through intraventricular injection or application to the brain surface enhances local CBF in cat.

A minor increase in BBB permeability occurs during intracarotid administration of histamine. Thus, upon release from mast cells the amine may participate in the development of vasogenic brain oedema. One further action of histamine has been demonstrated in cerebral vessels: histamine releases NA from cerebrovascular sympathetic nerves. Again, the local source for histamine may be type-II mast cells, which are often located in close contact with the nerve fibres.

## Neuropeptides

All of the peptides with which we are concerned have at least two functionally different amino acid sequences; an active site that stimulates the receptor and the post-receptor events, and an auxiliary amino acid sequence which may comprise elements of affinity, specificity, and protection from degradation. The tertiary structure and resulting flexibility or rigidity of the peptide provide additional means of achieving a multitude of actions and optimal adaptation during evolution. Peptide effects on the cardiovascular system can be caused through perivascular peptidergic innervation of pial vessels or brain parenchymal microvessels, by local generation of peptides in the vessel wall, and by hormonal effects of circulating peptides.

## Neuropeptide Y

NPY has an amino acid sequence that displays a marked degree of homology with bovine pancreatic polypeptide, avian pancreatic polypeptide and peptide YY. NPY is co-stored with NA in the majority of cerebrovascular sympathetic nerves and with VIP and probably ACh in many cerebrovascular parasympathetic nerve fibres (see Chapters 5 and 11).

Isolated segments of pial arteries from various species including man constrict slowly and to a modest degree upon exposure to NPY, probably through an action on specific NPY receptors. Micro application of NPY *in situ* constricts pial arterioles. The sensitivity to NPY is high, and similar in pial arteries, penetrating cortical arteries and arterioles, and pial veins.

Pial arterioles constrict to NPY as strongly as to NA, whereas penetrating arterioles constrict to NPY without constricting to NA. The contractile response to NPY does not involve endothelial mechanisms. Findings diverge about the dependency upon extracellular calcium for the response. Avian pancreatic polypeptide does not affect the tone of pial arteries.

NPY potentiates NA-induced contractions in peripheral arteries, but not to any substantial extent in pial arteries. There is some evidence for a presynaptic effect of NPY on cerebrovascular sympathetic nerves, by which the NA release is modulated, but others find no such effect. Possible interactions with VIP and ACh on cerebrovascular tone have not been studied.

NPY does not affect CBF in several brain regions upon intracarotid injection in rat, which is not surprising considering the efficient BBB to most circulating peptides. A slowly developing, long-lasting reduction in flow in certain regions has, however, been reported by some investigators upon systemic administration in rat and dogs. When the BBB is circumvented by administration of NPY into the striatum of rats a marked local reduction in flow occurs independent of any corresponding reduction in glucose utilization. This again indicates a direct constrictory action of NPY on intracerebral arterioles.

The presence of NPY in both sympathetic and parasympathetic nerves may indicate that the peptide has other functions than regulation of tone. It may be present as a trophic factor for the nerve or the vessel wall.

## Vasoactive intestinal polypeptide and related peptides

VIP is found in parasympathetic nerves in pial vessels of various species including man. They are located on the adventitial–medial border down to the level of penetrating arterioles and also in pial veins. The two major origins for these fibres are the sphenopalatine and otic ganglia, with a varying contribution from accessory ganglia at the base of the skull in the various species (see Chapters 5 and 11). Peptide histidine isoleucine (PHI) and its human counterpart, peptide histidine methionine (PHM-27), peptides coded by the same messenger RNA as VIP, are co-localized with VIP in these fibres. ACh is probably also present in some of these fibres.

VIP is the most potent vasodilator in the brain presently known. Even at very low (nanomolar) concentrations VIP causes a marked relaxation in isometric tension and perfusion pressure studies of isolated segments of pial arteries and penetrating cortical arteries and arterioles of all species studied including man. The relaxation is probably the result of action on specific VIP-receptors in the smooth musculature, and involves activation of adenylate cyclase. Ligand binding studies have confirmed the presence of specific sites, and one autoradiographic study locates them in the smooth muscle layer. Among large pial arteries in cats the maximum relaxation is least prominent in the basilar artery. Pial arterioles dilate more readily than arteries. Extraluminal application of VIP on pial arteries and arterioles *in situ* dilates the vessels. The VIP relaxation does not

involve endothelial factors, but may involve local formation of prostanoids.

VIP appears to interact with its co-transmitter ACh in various ways: they inhibit further release of each other, and VIP seems to stimulate the synthesis of ACh and its binding to the postsynaptic receptor. VIP does not influence the NA-release from the sympathetic nerves in pial arteries.

PHI is a less potent arterial dilator than VIP, both *in vitro* and *in situ*.

Pial veins appear to be less sensitive to VIP than the arteries, as studied *in vitro* and *in situ*.

Despite its independence of the endothelium for the vasodilatory effect, and its poor penetrability across the BBB, intracarotid injection of VIP increases CBF to an almost equally large magnitude as when given intraventricularly to circumvent the barrier. Intravenous injection in rabbits leaves CBF unaffected. Intraparenchymal injection of VIP increases both blood flow and local glucose consumption. Thus, when VIP gains access to the brain parenchyma the flow increase appears to be at least partly secondary to a VIP-induced activation of brain metabolism.

Nerve fibres containing VIP, together with ACh, have been demonstrated in close proximity to the walls of cerebral microvessels, apparently innervating these vessels. VIP-sensitive adenylate cyclase has been demonstrated in cerebral microvessels of rat, further suggesting a role for this neuropeptide in capillary reactivity. The microvascular nerves probably have a central origin, and are thought to be one of the mediators, besides local

metabolic factors, for cerebral vasodilatation in conjunction with enhanced brain neuronal activity. Evidence has been presented that VIP nerves originating or passing through the mesencephalic reticular formation exert such a function.

Besides a direct effect on cerebrovascular tone by VIP in parasympathetic nerves, and a mediation of neuronal metabolism-linked flow changes by central VIP nerves, a trophic function of the peptide in the parasympathetic nerve or pial vessel wall is possible.

## Substance P

There is strong evidence that substance P (SP) is a transmitter in trigeminal pain fibres (C-fibres), running in the adventitia along pial arteries, arterioles and veins. The majority of pial SP fibres originate in the ophthalmic division of the trigeminal ganglia (see Chapters 5 and 11). It is known from studies in man that acute head pain is felt when large proximal pial arteries and some veins at the base of the brain are electrically or mechanically stimulated. C-fibres in the vessel wall are sensitive to specific pain stimuli, and will also release SP and other transmitters locally (or in the vicinity through an axon reflex) when the free nerve ending is stimulated. The possibility of such an arrangement in pial arteries has been shown by electron microscopy. Capsaicin and potassium depolarization release SP from pial arteries and affect the vascular tone, which supports a motor function of SP-containing perivascular nerves besides transmission of pain.

In several species, including man, SP is a strong dilator of large pial arteries, whereas only a weak dilatory effect is seen in some species such as rat and rabbit, when the peptide is applied *in vitro* to isolated vessel segments. Studies with SP receptor antagonists have shown that the relaxation results from an activation of specific SP-receptors. The basilar artery is least sensitive to SP among feline pial arteries. The SP relaxation is entirely, or almost entirely, dependent upon an intact endothelium.

After removal of the endothelium a contractile response is obtained with SP in pial arteries. However, this direct effect on the smooth muscle cells, which predominates in other smooth muscle target organs like the iris, gastrointestinal and urogenital tract, is weak in the pial vessels and occurs only at high concentrations.

SP induces less pronounced relaxation in precontracted penetrating cortical arteries and arterioles *in vitro*, as well as *in situ* in small pial arteries and arterioles with diameters of 50–230 µm.

In pial veins from several species only a weak dilatory capacity of SP is revealed *in vitro*, whereas a substantial response is seen in cat pial veins *in situ*.

Some degree of tachyphylaxis to SP occurs in cerebral as well as non-cerebral vessels. This indicates that an SP release from perivascular nerves does not contribute to a continuous control of blood vessel tone. NA release from pial arteries is not influenced by SP.

SP poorly penetrates the BBB. Intravenous infusion of SP in man does not substantially affect internal carotid blood flow. Intrathecal injection of an SP agonist is without effect on brain and spinal cord blood flow, and intrahypothalamic injection of SP does not induce an effect on local blood flow on its own. This indicates that SP has little effect on flow in cerebral vessels, both with a luminal and abluminal vascular approach. The findings also indicate that no SP receptors are located on the luminal endothelial membrane.

Stimulation of the trigeminal ganglion in cats, or the pathway along the nasociliary nerve to the cerebral vessels in rats (with the central connection cut to avoid CBF changes due to pain perception) causes a minor increase in CBF. The mechanism for this small action may be a dilatation of pial arteries, since no primary sensory nerves have been demonstrated at the level of the brain parenchymal arterioles (see Chapter 11).

Central SP-containing nerves may come in close contact with cerebral microvessels. One action of this locally released SP on the microvessels may be a regulation of transport processes across the endothelium.

The presence of SP endings only in the outer adventitia, and the long diffusion distance from the adventitial nerve ending to the endothelium, questions the functional importance of a simple nerve–endothelial interaction of neural SP in these large arteries, and thus a role for SP in vasomotor control in these vessels despite the presence of SP receptors. The main *motor* function of SP in cerebrovascular nerve endings may rather be exerted via a release of histamine from closely located mast cells or by modulation of the activity of colocalized transmitters that more directly affect the smooth muscle cells.

A trophic function in the nerve or vessel wall is also possible. However, the major physiological function of cerebrovascular C-fibres is, in all probability, to transfer pain impulses from the vascular wall to the brain.

## Neurokinin A

Messenger RNA encoding for β and γ-preprotachykinin has been found in trigeminal ganglia. After cleavage of these preprotachykinins SP and a second tachykinin, neurokinin A (NKA) is formed. SP and NKA are found to coexist in trigeminal fibres running in the wall of large and small pial arteries. Higher concentrations of NKA than SP are needed to induce equally marked dilatations in large pial arteries of various species including man. Its action on small pial arteries, arterioles and veins has not yet been studied. The relaxation involves an endothelium-dependent mechanism.

## Calcitonin gene-related peptide

Calcitonin gene-related peptide (CGRP) co-exists with SP and NKA in the majority of cerebrovascular trigeminal SP/NKA-fibres. However, CGRP is also found in larger trigeminal neurons with no demonstrable SP/NKA activity, which also project to cerebral vessels (see Chapters 5 and 11). CGRP is found in fibres down to the level of small pial arteries and arterioles and sparsely also in veins.

Even at low (nanomolar) concentrations CGRP is a strong dilator *in vitro* of large pial arteries in all species including man, and in pial surface and

Table 1. Predominant direct effect on cerebrovascular tone of transmitters in the perivascular nerves after application to isolated vessel segments *in vitro* or to vessels on the brain surface *in situ*, and the effect on CBF after intravenous or intra-arterial administration (strong effect is indicated by bold type)

|  | Pial artery | Pial arteriole | Cortical artery–arteriole | Pial vein | CBF change at syst. inj. | Endothelium-linked dilat. shown |
|---|---|---|---|---|---|---|
| NA | **contr** | contr | 0–dilat | **contr** | 0 (–contr) | yes |
| DA | contr–dilat | contr–dilat |  |  | 0 |  |
| 5-HT | **contr** | 0–dilat | 0–contr | **contr** | 0 (–contr) | yes |
| ACh | **dilat** | **dilat** | **dilat** | contr | **dilat** | yes |
| NPY | contr | contr | contr | contr | 0–contr |  |
| VIP | **dilat** | **dilat** | **dilat** | 0–dilat | **dilat** |  |
| SP | dilat | 0–dilat | 0–dilat | 0–dilat | 0 | yes |
| CGRP | **dilat** | **dilat** | **dilat** | **dilat** | dilat |  |

contr = contraction; dilat = dilatation; 0 = no effect; syst. inj. = systemic injection.

penetrating branches, as demonstrated in rabbit (100 µm). Topical application of CGRP in an open-skull preparation in cat induces marked relaxations in pial arteries and arterioles (about 35–240 µm). Recent work with such preparations suggests that CGRP contributes to the vasodilation of pial vessels occurring during cortical spreading depression. The relaxation is probably caused by activation of specific CGRP-receptors, and involves mediation by adenylate cyclase. Generally, CGRP appears to be a more potent dilator in absolute terms than SP and NKA. The response is not dependent upon an intact cerebrovascular endothelium.

Human but not feline pial veins dilate upon exposure to CGRP.

CGRP probably penetrates poorly across the BBB. Nevertheless, intracarotid injection of CGRP in rat causes a long lasting increase in CBF, and intravertebral injection decreases vertebrobasilar resistance in dogs, whereas intravenous infusion reduces internal carotid resistance in rats, but has no effect on CBF in rabbits and man. Resting CBF remains normal after sectioning of the tri-

geminal nerves to the cerebral circulation in rats. There are indications that these nerves are involved in the restoration of normal CBF in conditions of threatened blood flow (see Chapter 11).

One function of CGRP in cerebrovascular nerves, besides regulation of vessel tone, may be to release histamine from perivascular mast cells; perivascular sensory fibres may come in close contact with such cells. Further, CGRP may modulate the activity of the co-transmitter SP: it prolongs the effect of SP by reducing its breakdown by endopeptidase, whereas SP shortens the action of CGRP through mast cell release of proteases which degrade CGRP. CGRP may also have trophic effects in the nerve or vessel wall.

## Opiates

*Met– and leu-enkephalin*, but not β-endorphin, have been demonstrated in adventitial cerebrovascular nerve fibres of pigs. It is not clear in which type of nerves the enkephalins are stored.

Enkephalins and morphine dilate pial

arteries *in vivo* and *in situ* through an action on opiate receptors, but only at high concentrations.

Circulating enkephalins are able to reach the cerebrovascular adventitia and brain parenchyma via a transport system across the BBB. Intravenous administration of enkephalins dilates pial arteries and veins, as measured through a cranial window in cats, and enhances CBF in mini-pigs, whereas intracarotid injection decreases CBF in cat. Intracerebroventricular injection of enkephalin and β-endorphin in rats, and intraparenchymal injection of enkephalin in cats, decreases CBF; at least part of the effect on flow is secondary to changes in cerebral metabolism.

Part of the pial arterial dilatation may be secondary to a reduced release of NA from perivascular sympathetic nerves, an effect of enkephalinergic agents that is also seen in non-cerebral vessels.

*Dynorphin B* is also found in adventitial cerebrovascular nerve fibres of several species including primates. In rats and guinea-pigs its presence in trigeminal SP/CGRP-containing nerves has been shown, whereas in guinea-pigs the opiate is also present in parasympathetic VIP-containing nerves.

This substance has no effect on the tone in pial arteries of rats and dogs, and no effect on NA release from perivascular sympathetic nerves.

Thus, an easily recognizable direct pharmacological action on cerebrovascular tone is not available for the opiates. The opiates may interact with co-stored transmitters in these nerves, e.g. to inhibit the release of SP from C-fibres, or may have trophic effects in the nerve or vessel wall.

## Galanin

Galanin is found in some perivascular nerve fibres in pial vessels from rat, guinea-pigs and cats. In rats the peptide is present together with SP and CGRP in trigeminal fibres.

It is without effect on the tone of isolated pial arteries, and does not affect NA release from cerebrovascular sympathetic nerves in rats. Galanin may interact with the co-existing transmitters and have trophic effects.

## Vasopressin

One study has demonstrated the presence of arginine-vasopressin (AVP) in guinea-pig pial arteries down to a diameter of about 100 μm. The peptide is probably located in trigeminal fibres.

In several species, including man, AVP and lysine-vasopressin induce marked contractions in isolated segments of pial and penetrating cortical arteries *in vitro* and of pial arteries and arterioles, but not veins, *in situ*. The response is probably caused by

activation of specific vasopressin $V_1$- and/or $V_2$-receptors. The contraction is greatest in the basilar territory, as studied in rats. In contrast, an endothelium-dependent relaxation has been demonstrated in the various pial arteries of dogs, with a contribution of a contractile response only in the middle cerebral artery.

Central AVP neurons have been found in close association with cerebral microvessels in rat. Possibly this finding is linked to the specific AVP receptors which are present on brain microvessels. Intraventricular AVP causes an increase in water permeability across BBB. Permeability to macromolecules is, however, not enhanced by intravenous AVP.

Vasopressin penetrates the BBB with the aid of a specific transport system. Therefore it is not surprising that AVP affects CBF upon systemic administration. Bolus injection into the internal carotid artery in rat increases CBF (via an action on $V_2$-receptors), at least partly due to enhanced oxygen consumption in the brain, whereas infusion into the internal maxillary artery in goats reduces CBF, and intravenous infusion in cats leaves CBF unaffected despite dilating large pial arteries. No effect on CBF is seen after intracerebroventricular administration of AVP, whereas microvascular spasms are seen when AVP is added to hippocampal slices in rats. The discrepancies in flow changes observed may be due to species differences and to the extent to which changes in brain metabolism contribute.

Thus, constriction of large cerebral arteries of abluminally acting vasopressin appears to be a major action,

but several other functions of the cerebral vessels may also be influenced by the peptide.

## Cholecystokinin

Central cholecystokinin (CCK)-containing nerve fibres have been found in close association with cerebral microvessels. Further, cerebrovascular trigeminal nerve fibres, immunoreactive for CCK in the cell body or the axon, have been found in rats, guinea-pigs and cats. It has been suggested, though, that the CCK antibodies used may have cross-reacted with CGRP.

CCK has no effect on tone in feline isolated pial arteries *in vitro* or pial arterioles *in situ*.

## Atrial natriuretic peptides

*Atrial natriuretic peptide* (ANP)-immunoreactive central nerve fibres have been found in close contact with parenchymal as well as with penetrating pial microvessels. ANP released from such fibres may affect these microvessels via specific receptors found on their surface. This could represent the morphological substrate for the prevention of fluid accumulation and the decrease in sodium content in brain seen after central administration of ANP.

No ANP-positive nerve fibres have been found in pial vessels. Although no natural source for ANP is known in these vessels, application of ANP and the closely related *atropeptin I* and *atropeptin II* on the pial surface in cats induces marked dilatation of small pial arteries and arterioles, the arteries being most sensitive. Intravenous infusion of atropeptin in rabbits does not alter CBF, presumably because the BBB prevents access to the cerebrovascular smooth muscle.

## Other putative cerebrovascular neuropeptides

*Oxytocin, bombesin, gastrin-releasing peptide, neurotensin and somatostatin* are all peptides that, based on single reports, have been suggested as transmitter candidates in pial vessels or in cerebral microvessels. Of these, only somatostatin (contraction/relaxation) and oxytocin (contraction/relaxation) affect pial arterial tone.

## Other peptides with effects on cerebrovascular tone

### Bradykinin

Bradykinin (BK) produces marked dilatation *in vitro* in pial and cerebral arteries and arterioles from several species including man, and *in situ* in rabbit and feline pial arteries and arterioles via an action on specific $B_2$-receptors. Pial arteries from rabbits and cats markedly dilate also upon local microapplication or superfusion with BK. Also endogenous BK, released from brain kininogen by cortical superfusion with kallikrein in rabbits, elicits a dilatation. The effect, which displays tachyphylaxis, is dependent upon an intact endothelium, and involves local formation of prostanoids, oxygen radicals or nitric oxide. A contraction via $B_2$-receptors is observed at fairly high concentrations (above $10^{-7}$ molar), particularly evident after removal of the endothelium. In dogs only a contraction is obtained. The

BK-induced relaxation is less pronounced in pial arterioles from aged rats. BK has also been shown to contract dog and rabbit basilar arteries, probably via muscular $\beta_1$ receptors.

Isolated human pial veins dilate upon exposure to BK; at high concentrations a contraction is seen. Superfusion with high concentrations of BK constricts feline pial veins, whereas lower concentrations have no effect on tone.

BK is poorly penetrable across the BBB. Intracarotid infusion of BK does not affect pial vascular diameter, which indicates that the $B_2$-receptors involved in regulation of vascular tone are not located on the luminal membrane of the endothelium. Cerebroventricular perfusion with BK enhances CBF, which can be explained as a direct vascular effect combined with an indirect, metabolism-linked response.

When administered either systemically or intracerebroventricularly BK opens the BBB to macromolecules, via an action on endothelial $B_2$-receptors. Such receptors must be located on the luminal surface of microvascular endothelium.

BK stimulates histamine secretion from perivascular mast cells. In this process interaction with neuronal SP and CGRP as well as locally synthesized prostanoids and leukotrienes may occur to produce a neurogenic inflammation in the microvessel wall. Whether such inflammation can take place also in the cerebral vessels is not known with certainty.

The various cerebrovascular effects of BK are involved in the formation of

brain oedema: uptake of intravascular kininogen and parenchymal release of BK and other kinins participate in the development and spread of brain oedema and secondary brain damage.

### Thyrotropin-releasing hormone

Thyrotropin-releasing hormone increases CBF upon intravenous injection. The peptide does not affect the tone of isolated pial arteries. It has been suggested that a direct effect on parenchymal vessels, in addition to activation of an intrinsic cerebral vasodilator pathway, is responsible for the flow effect.

### Angiotensin

Angiotensin (angiotensin I and II: AI, AII) markedly constricts isolated pial arteries from several species including man. Likewise, microapplication of AI and AII to feline pial arteries *in situ* causes constriction, with AII being considerably more potent than AI. Most of the contractile action of AI occurs after its conversion to AII by the locally present angiotensin converting enzyme in the vessel wall (endothelium) and CSF. In the isolated middle cerebral artery of dog a relaxation is obtained by AII, more pronounced distally than proximally. Such relaxation, dependent upon an intact endothelium, is also seen in pial arterioles upon topical application in rats. The response is generated through local formation of prostanoids.

The angiotensin converting enzyme forms part of a vascular renin–angiotensin system throughout the body, and also contributes to maintaining the lower limit of CBF autoregulation, probably through a continuous formation of AII within the wall of large cerebral arteries.

Owing to its poor penetrability across the BBB, intracarotid AII does not affect pial arterial diameter or CBF in various species including man.

## Endothelin

Endothelin (endothelin-1, –2, and –3) is synthesized in vascular endothelial cells. It is the most potent vasoconstrictor known to date in pial arteries and arterioles of several species including man, both when studied *in vitro* and *in situ* (AVP may be equally potent in cerebral arteries from some species). Endothelin-1 is the most potent subtype, the contractile response being highly dependent upon extracellular calcium. The receptors are not located on the endothelial cells.

Upon microapplication of endothelin to pial arterioles in rat and piglet a relaxation is seen at low concentrations, followed by contraction at higher concentrations. Evidence has been presented for an endothelium-linked dilatory component of the peptide in cerebral vessels, with the formation of a cyclooxygenase product.

The low degree of dependency of extracellular calcium for contractions to various agents in pial veins may explain why isolated human pial veins are less sensitive to endothelin than human pial arteries. On the other hand, upon perivascular microapplication, feline pial veins and arterioles are equally sensitive to endothelin.

Endothelin does not cross the BBB or affect BBB permeability. Luminal administration of endothelin to the cerebral vasculature does not affect pial arterial diameter, whereas intracisternal injection causes a long-lasting constriction and, in addition, a decreased CBF.

As evidenced by the high sensitivity of pial arteries it is plausible that the endothelial production of endothelin is an important mechanism whereby cerebral vessels can regulate local tone. The physiological and pathophysiological stimuli eliciting such a response are probably to be found on the circulatory side of the endothelial layer.

## Prostanoids and leukotrienes

Arachidonic acid, which is a component in phospholipids of all cell membranes, can be oxidized enzymatically to a large number of products. In the cyclooxygenase pathway, via the endoperoxides $PGG_2$ and $PGH_2$, the various *prostaglandins* ($PGF_{2\alpha}$, $PGD_2$, $PGE_2$ and further $PGA_2$ and $PGB_2$) are formed, as well as *prostacyclin* ($PGI_2$) and *thromboxanes* ($TXA_2$ and further $TXB_2$). In the lipoxygenase pathway mainly *leukotrienes* ($LTA_4$ and further $LTB_4$, $LTC_4$ and $LTE_4$) are synthesized, and via other lipoxygenases a variety of hydroperoxy– and hydroxy-derivatives are formed.

## Prostanoids

The main prostanoid (i.e. cyclooxygenase pathway product) formed in pial arteries and cerebral microvessels appears to be $PGI_2$, probably followed by $PGF_{2\alpha}$. Prostanoids are synthesized in all cell types of the vessel wall, but most $PGI_2$ is formed in the endothelium whereas most of the other prostaglandins appear in the smooth muscle cells. Fish eaters receive proportionally more eicosapentaenoic acid at the expense of arachidonic acid, which results in the formation of $PGF_{1\alpha}$, $PGE_1$, $PGA_1$, $PGD_1$ and $PGB_1$. Stable analogues for endoperoxides (U 44069) and $TXA_2$ (U 46619 and carboxylic $TXA_2$) are also used in the pharmacological characterization of vascular effects of prostanoids.

The prostanoids are strong local modulators of vascular tone. Most prostanoids contract isolated pial arteries *in vivo* and *in situ* and pial arterioles *in situ* in various species including man. Most potent are $TXA_2$, its stable analogues and $PGB_2$. Species differences exist: $PGF_{2\alpha}$ is a poor constrictor in rabbit pial arteries, and is without effect on tone in pial arterioles *in situ*; in small penetrating arteries of rabbits even a dilatation is seen. As studied in rats, pial arteries constrict to $PGF_{2\alpha}$ at lower concentrations than mesenteric arteries. Based on the order of agonist potency, cerebrovascular contractions appear to be the result of activation of TP-receptors (with high sensitivity to $TXA_2$) and possibly also of FP-receptors (with high sensitivity to $PGF_{2\alpha}$), as primarily studied in the feline basilar artery. Most of the contraction is due to release of calcium bound to the outside of the smooth muscle cell membrane.

Low concentrations of $PGA_1$, $PGA_2$, $PGD_2$, $PGE_1$, $PGE_2$ and in particular $PGI_2$ induce dilatation, followed by contraction at increasing concentrations in all species studied, except in dogs where the opposite pattern occurs for $PGE_1$. The relaxation to $PGE_1$ is more marked in small penetrating

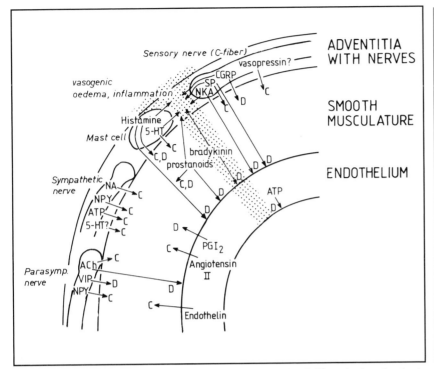

*Fig. 2. Main sources in the pial (and cortical) arterial wall and probable main sites of action in the vessel wall for vasoactive monoamines, peptides, prostanoids and purines. Changes in vascular tone are indicated by C = contraction, D = dilatation.*

arteries. $PGI_2$ and $TXA_2$ are chemically unstable and the tonic response is therefore transient. The hydrolysis products 6-keto-$PGF_{1\alpha}$ and $TXB_2$, respectively, are vasoactive only at high concentrations, and $PGI_2$ enzymatically transformed into 6-keto-$PGE_1$ is less potent as a vasodilator than $PGI_2$. Relaxant responses in cerebral vessels seem to be mediated by a $PGI_2$-sensitive receptor on smooth muscle cell membranes.

Cerebral microvessels are under continuous tonic influence of locally synthesized prostanoids, since systemic administration of indomethacin affects CBF: a rapid reduction in flow is seen in the majority of studies in all species including man. This indicates the presence of prostanoid receptors

also at the level of microvessels.

Human pial veins constrict upon exposure to $PGF_{2\alpha}$.

Intravascular administration of $PGF_{2\alpha}$ and $PGE_2$ in primates reduces CBF but in addition cerebral oxygen consumption. Intracarotid infusion of $TXA_2$ in rabbits markedly reduces CBF. Intracarotid infusion of $PGI_2$ in the baboon strongly enhances CBF. $TXA_2$ and $PGI_2$ are lipid-soluble and may reach receptors on the smooth muscle cells to exert this effect.

Hypercapnic vasodilatation may represent a situation in which the cerebrovascular synthesis of prostanoids is enhanced (see Chapter 9). Further, local synthesis is stimulated by many vasoactive agents, whereby the activ-

ity of these agents may be modulated. Cerebrovascular dilatation produced by VIP, endothelin, BK and AII (both endothelium-linked) is at least partly caused by formation of relaxatory prostanoids, or by oxygen radicals generated in association with the accelerated arachidonic acid metabolism. There is some evidence that such formation also attenuates the constriction caused by NA, ACh, SP, $PGE_2$ and $PGF_{2\alpha}$. NA and ATP may also cause release of a contractile prostanoid from pial arteries, enhancing the contraction. When released from aggregating platelets, the vasocontractile action of $TXA_2$ is modulated by local endothelial formation of the vasodilator $PGI_2$.

## Leukotrienes

Leukotrienes, and also prostanoids, may be formed in or around the vessel wall during inflammation. In some non-cerebral vessels leukotrienes induce potent contractions. However, leukotrienes have been found to be practically without effect on tone in all studies except one on isolated pial arteries *in vitro* from various species including man, and in rabbit pial arterioles *in situ*. A slight constriction of large intracranial arteries has been demonstrated upon intravascular administration in rats.

## Purinergic agents

*Adenosine 5'-triphosphate (ATP)* is present in perivascular sympathetic nerves not only in the periphery but also in cerebral vessels, at least in some species. Upon release it may act as a modulator of the tonic NA-response postsynaptically by a direct contractile effect on the smooth muscle cell, through stimulation of $P_{2x}$-receptors. ATP appears to be responsible for the fast depolarization recorded during stimulation of perivascular sympathetic nerves, whereas NA is responsible for the slow, maintained depolarization seen in vessels equipped with $\alpha$-adrenergic receptors. ATP is rapidly metabolized to *adenosine 5'-monophosphate* (AMP) and adenosine which mediate smooth muscle relaxation through stimulation of postsynaptic $P_1$-receptors, and also inhibit the further release of NA through stimulation of presynaptic $P_1$-receptors.

ATP is synthesized and contained within all cells. It is released in high amounts into the systemic circulation from skeletal muscle during exercise. Locally, high amounts of ATP can be released from endothelial and smooth muscle cells, and from aggregating platelets. This holds true also for brain microvascular endothelium. Such ATP and *adenosine 5'-diphosphate* can elicit vascular dilatation via an action on $P_{2y}$-receptors on the abluminal endothelial membrane, or, after rapid local breakdown to AMP and adenosine, on adventitial smooth muscle $P_1$-receptors.

Adenosine is also taken up, stored in and released (partly as ATP after phosphorylation) from endothelial cells in, e.g., brain vessels. Even if higher concentrations of adenosine are usually needed to cause a vascular dilatation similar to ATP, the former may be responsible for maintaining the dilatation. High amounts of purines are also released from brain neurons into the extracellular fluid compartment during hypoxia, ischaemia and seizures. These local sources

for purinergic agents are believed to play a role in the vasodilatation initiated by hypoxia and ischaemia (see Chapter 9).

Luminal relaxatory $P_{2y}$-receptors, abluminal constrictory $P_{2x}$- receptors and abluminal relaxatory $P_1$-receptors (of the $A_2$ subtype) have been demonstrated in isolated pial arteries *in vitro* from several species including man. Regional differences exist between the various pial arteries, and greater relaxations to adenosine are seen in the distal than in the proximal middle cerebral artery of dogs. Microapplication or superfusion on pial arteries and arterioles and cortical arterioles of ATP in cats and adenosine in rats, cats and dogs causes marked relaxation even at fairly low concentrations ($10^{-7}$ molar). Contraction via $P_{2x}$-receptor stimulation may involve synthesis of contractile prostanoids, and $P_{2y}$-receptor stimulation causes release of endothelium-derived relaxing factor. The endothelium-dependent relaxation to ATP appears at lower concentrations in cerebral than in non-cerebral arteries.

Adenosine receptors ($A_1$ and $A_2$ subtype) are also present on brain microvessels.

Whether ATP or adenosine affects tone of isolated pial veins *in vitro* has not been studied. Rat pial venules do not react to adenosine *in situ*.

ATP is poorly permeable across the luminal endothelial membrane. Nonetheless, intracarotid injection of ATP increases CBF in all species studied, in addition to dilating pial arteries and arterioles in cats, which indicates that relaxatory $P_{2y}$-receptors are present on the luminal endothelial membrane. The flow increase in

baboons is coupled to increased oxygen consumption in the brain. The changes in flow and metabolism are not potentiated after opening of the BBB. This indicates that ATP, perhaps after partial conversion to adenosine, is able to cross the intact BBB in sufficient amounts to activate brain metabolism.

Adenosine penetrates the BBB more readily than ATP. An increase in CBF is seen upon administration of adenosine or a stable AMP analogue into the internal carotid artery in rabbits, goats and baboons. In contrast, intracarotid infusion of potent adenosine analogues in rat reduces CBF, without affecting brain metabolism. Upon intravenous administration of adenosine in rabbits, and intra-arterial administration in cats and dogs, no or only a local increase in CBF is observed. The negative findings in cat and dog may be explained on an anatomical basis, because the presence of a carotid rete allows only small amounts of the injected agent to reach the brain. In man intravenous adenosine increases CBF. Superfusion of the brain cortex with adenosine or a potent analogue increases local cortical blood flow, at least in part secondary to an enhanced neuronal metabolism. Ventriculocisternal perfusion with adenosine in piglets increases CBF without altering oxygen consumption in the brain.

It has been suggested that circulating purines mediate a form of metabolic communication in the body: purines released from, for example, working skeletal musculature may reach arterial blood levels high enough to affect blood flow and metabolism in the brain. Adenosine has been suggested as one of the factors coupling local

**Table 2. Probable main sources and functions on the cerebral vasculature of the various monoamines, peptides, prostanoids, purines and amino acids**

| Substance | Source | Function |
|---|---|---|
| NA | Perivascular sympathetic nerves | Constrict pial arteries and veins<br>Dilate intracerebral arterioles? |
| A | Circulation (adrenal medulla) | |
| DA | Brain neurons | Enhance local CBF |
| 5-HT | Perivascular mast cells | Constrict pial arteries<br>Modulate vasogenic oedema |
| ACh | Periascular parasympathetic nerves<br>Brain neurons | Modulate the activity of parasympathetic transmitters?<br>Induce mast cell degranulation<br>Modulate local CBF |
| Histamine | Perivascular mast cells | Vasogenic oedema<br>Constrict/dilate pial arteries |
| NPY | Perivascular sympathetic and para-sympathetic nerves | Modulate the activity of sympathetic and parasympathetic transmitters? |
| VIP | Perivascular parasympathetic nerves<br>Brain neurons | Dilate pial arteries<br>Enhance local CBF |
| SP, NKA, CGRP | Perivascular sensory nerves | Sensory transmission<br>Dilate pial arteries<br>Induce mast cell degranulation |
| Opiates | Perivascular sensory and parasympathetic nerves? | Modulate the activity of sensory and parasympathetic transmitters? |
| Galanin | Perivascular sensory nerves | Modulate the activity of sensory transmitters? |
| Vasopressin | Perivascular sensory nerves? | Constrict pial arteries |
| Atrial natriuretic peptides | Brain neurons | |
| Bradykinin | Local synthesis in vessel wall | Vasogenic oedema |
| Angiotensin | Local synthesis in endothelium | Modulate local CBF? |
| Endothelin | Local synthesis in endothelium | Modulate local CBF |
| Prostanoids | Local synthesis in vessel wall | Modulate local CBF |
| Leukotrienes | Local synthesis in vessel wall | Vasogenic oedema |
| Purinergic agents | Local synthesis in vessel wall<br>Brain neurons | Modulate local CBF?<br>Enhance local CBF during hypoxia–ischaemia |
| Amino acids | Perivascular nerves (GABA)?<br>Brain neurons | Modulate the activity of sensory and parasympathetic transmitters and the sympathetic transmitter release (GABA) |

cerebral flow to the momentary metabolic demand, not only during hypoxia and ischaemia but also during physiological conditions (see Chapter 9).

Apart from a direct effect on vascular tone, ATP can also induce histamine release from perivascular mast cells.

## Amino acids

### Glutamate and aspartate

Excitatory amino acids, particularly glutamate and aspartate, are neurotransmitters in several neuronal pathways in the brain. Levels in the extracellular fluid compartment become elevated during ischaemia and hypoglycaemia. Glutamate may also be found in cranial primary sensory neurons. The amino acids penetrate the BBB through carrier-mediated facilitated diffusion. Circulating levels of glutamate and aspartate are elevated during migraine attacks. Thus, there are several sources from which brain vessels may become exposed to excitatory amino acids.

Pial arteries of rats, cats and man and penetrating pial arterioles of rats do not constrict or dilate *in vitro* upon exposure to glutamate, aspartate or other stimulators of excitatory receptors (kainate, *N*-methyl-D-aspartate (NMDA), quisqualate and quinolinate). In the piglet high doses of glutamate, aspartate and NMDA dilate pial arterioles *in situ*. It is likely that such concentrations are in excess of those achieved even in cerebral ischaemia.

No NMDA receptors are found in ovine brain microvessels.

Intrahypothalamic injection of glutamate increases local blood flow secondary to activation of brain metabolism.

Thus, it is likely that any change in CBF during release of neuronal, or exposure to circulating, excitatory amino acids is due to altered neuronal metabolism alone.

## GABA and glycine

Gamma-aminobutyric acid (GABA), formed from glutamate, is an inhibitory neurotransmitter in several neuronal pathways in brain. Further, high activity of the GABA-synthesizing enzyme, glutamic acid decarboxylase, and the catabolizing enzyme, GABA transaminase, are found in brain vessels, particularly at the level of microvessels but also in the endothelium of pial arteries. The two enzymes have also been shown to be present in non-sympathetic nerves at the adventitial–medial border in pial arteries of several species, and these vessels contain high levels of GABA. This opens the possibility for an influence on cerebral tissue perfusion by GABA.

GABA causes a minor relaxation of pial arteries *in vitro* and *in situ* (in dogs also contraction), blockable by GABA antagonists. GABA and the inhibitory amino acid glycine does not affect tone in penetrating cerebral arterioles *in vitro* or pial arterioles *in situ*.

The potent GABA agonist muscimol binds to cerebral microvessels.

GABA is unable to cross the BBB. Despite this, GABA administered into the cerebral circulation in conscious goats increases CBF. Intravenous administration of muscimol, which crosses the BBB more

freely, causes an increase in CBF in anesthetized rats. In the conscious rat, however, a minor decrease in CBF is seen, which is fully attributable to reduction in brain metabolic activity.

Despite evidence for a GABA-synthesizing capacity in pial perivascular nerves and cerebrovascular endothelium, and the presence of GABA receptors in brain vessels, no clear evidence has emerged for a role of GABA (or glycine) in CBF regulation under physiological conditions. GABA may be present in pial perivascular nerves to modulate the activity of cotransmitters (see Chapter 11). The presence of GABA-synthesizing and γ-catabolizing enzymes in the endothelium may represent an enzymatic BBB for glutamate and GABA.

## Summary

This chapter summarizes the effects on vascular tone by monoamines and peptides found in perivascular nerves along pial vessels and their cortical branches (noradrenaline, serotonin, acetylcholine, neuropeptide Y, vasoactive intestinal polypeptide, substance P, neurokinin A, and calcitonin gene-related peptide). Such effects by other monoamines and peptides, as well as prostanoids, purines, and amino acids, present or synthesized perivascularly (mast cells, smooth muscle cells, endothelial cells, central nerve fibres) or in the circulation are also reviewed. Details are given about dependency on endothelium and extracellular calcium for the tonic response, regional, segmental and species differences as well as receptor types involved. The effects are integrated with data on changes in brain metabolism and blood–brain barrier permeability. Interactions and trophic effects are briefly discussed. An attempt is made to synthesize the major functions of the various substances during physiological and pathological conditions.

## References (for further reading)

### Monoamines (catecholamines, serotonin, acetylcholine and histamine)

#### Major reviews

Auer, L.M. (1986): Sympathetic control of pial vessels under *in vivo* conditions: In: *Neural regulation of brain circulation*, eds. C. Owman & J.E. Hardebo, pp. 497–513. Amsterdam: Elsevier.

Duckles, S.P. (1986): Cholinergic innervation of cerebral arteries. In: *Neural regulation of brain circulation*, eds. C. Owman & J.E. Hardebo, pp. 235–243. Amsterdam: Elsevier.

Gross, P.M. (1982): Cerebral histamine: indications for neuronal and vascular regulation. *J. Cereb. Blood Flow Metab.* **2**, 3–23.

Hamel, E. & Estrada, C. (1989): Cholinergic innervation of pial and intracerebral blood vessels: evidence, possible origins and sites of action. In: *Neurotransmission and cerebrovascular function II*, eds. J. Seylaz & R. Sercombe, pp. 151–173. Amsterdam: Excerpta Medica.

Hardebo, J.E. & Owman, C. (1980): Barrier mechanisms for neurotransmitter monoamines and their precursors at the blood–brain interface. *Ann. Neurol.* **8**, 1–11.

Kobayashi, H., Magnoni, M.S., Govoni, S., Izumi, F., Wada, A. & Trabucci, M. (1985): Neuronal control of brain microvessel function. *Experienta* **41**, 427–458.

Lee, T.J.F. (1989): Recent studies on serotonin-containing fibers in cerebral circulation. In: *Neurotransmission and cerebrovascular function II*, eds. J. Seylaz & R. Sercombe, pp. 133–149. Amsterdam: Excerpta Medica.

MacKenzie, E.T., Scatton, B. (1987): Cerebral circulatory and metabolic effects of perivascular transmitters. *CRC Crit. Rev. Clin. Neurobiol.* **2**, 357–419.

Moro,

Owman, C. (1986): Neurogenic control of the vascular system: focus on cerebral circulation. In: *Handbook of physiology. The nervous system. intrinsic regulatory systems of the brain*. Vol. IV, Section 1, pp. 525–580. Baltimore: Waverly Press.

Owman, C., Chang, J.Y. & Hardebo, J.E. (1990): Presence of 5-hydroxytryptamine in adrenergic nerves of the brain circulation: its role in sympathetic neurotransmission and regulation of the cerebral vessel wall. In: *Cardiovascular Pharmacology of 5-hydroxytryptamine: prospective therapeutic applications*, eds. P.R. Saxena, D.I. Wallis, W. Wouters & P. Bevan, pp. 211–230. Dordrecht: Kluwer Academic.

Palmer, G.C. (1986): Neurochemical coupled actions of transmitters in the microvasculature of the brain. *Neurosci. Behav. Rev.* **10**, 79–101.

Saxena, P.R., Bom, A.H. & Verdouw, P.D. (1989): Characterization of 5-hydroxytryptamine receptors in the cranial vasculature. *Cephalalgia* **9**, (Suppl. 9) 15–22.

Sercombe, R. & Verrecchia, C. (1986): Role of vascular endothelium in modulating cerebrovascular effects of neurotransmitters. In: *Neural regulation of brain circulation*, eds. C. Owman & J.E. Hardebo, pp. 27–41. Amsterdam: Elsevier.

Sercombe, R., Lacombe, P. & Seylaz, J. (1984): Functional significance of the cerebrovascular reactivity to autonomic neurotransmitters. In: *Neurotransmission and the cerebral circulation*, eds. E.T. MacKenzie, J. Seylaz & A. Bès, pp. 65–89. New York: Raven Press.

Sharkey, J. & McCulloch, J. (1986): Dopaminergic mechanisms in the regulation of cerebral blood flow and metabolism: role of different receptor subtypes. In: *Neural regulation of brain circulation*, eds. C. Owman & J.E. Hardebo, pp. 111–127. Elsevier: Amsterdam.

Toda, N. (1985): Reactivity of human cerebral artery: species variation. *Fed. Proc.* **44**, 326–330.

Vanhoutte, P.M. (1989): Endothelium and control of vascular function. *Hypertension* **13**, 658–667.

Wahl, M. (1985): Local chemical, neural and humoral regulation of cerebrovascuolar resistance vessels. *J. Cardiovasc. Pharmacol.* **7**, (Suppl. 3) S36–S46.

#### Recent references of particular importance

Ayajiki, K., Okamura, T. & Toda, N. (1992): Involvement of nitric oxide in endothelium-dependent phasic relaxation caused by histamine in monkey cerebral arteries. *Japan. J. Pharmacol.* **60**, 357–362.

Dimitriadou, V., Aubineau, P., Taxi, J. & Seylaz, J. (1987): Ultrastructural evidence for a functional unit between nerve fibers and type

II cerebral mast cells in the cerebral vascular wall. *Neuroscience* **22**, 621-630.

Hamel, E., Edvinsson, L. & MacKenzie, E.T. (1988): Heterogenous vasomotor responses of anatomically distinct feline cerebral arteries. *Br. J. Pharmacol.* **94**, 423-436.

Hardebo, J.E., Kåhrström, J., Owman, C. & Salford, L.G. (1987a): Vasomotor effects of neurotransmitters and modulators on isolated human pial veins. *J. Cereb. Blood Flow Metab.* **7**, 612-618.

Mayhan, W.G., Faraci, F.M., Baumbach, G.L. & Heistad, D.D. (1990): Effects of aging on responses of cerebral arterioles. *Am. J. Physiol.* **258**, H1138-H1143.

Nosko, M., Krueger, C.A., Weir, B.K.A. & Cook, D.A. (1986): Effects of nimodipin on *in vitro* contractility of cerebral arteries of dog, monkey and man. *J. Neurosurg.* **65**, 376-381.

Ottosson, A., Jansen, I. & Edvinsson, L. (1988): Characterization of histamine receptors in isolated human cerebral arteries. *Br. J. Pharmacol.* **94**, 901-907.

Parsons, A.A. & Whalley, E.T. (1989): Further characterization of the 5-HT$_1$-like receptor present on human isolated basilar artery. In: *Neurotransmission and cerebrovascular function I*, eds. J. Seylaz & E.T. MacKenzie, pp. 229-232. Amsterdam: Excerpta Medica.

Sercombe, R., Hardebo, J.E., Kåhrström, J. & Seylaz, J. (1990): Amine-induced responses of pial and penetrating cerebral arteries: evidence for heterogenous responses. *J. Cereb. Blood Flow Metab.* **10**, 808-818.

Steinbusch, H.W.M. & Verhofstad, A.A.J. (1986): Immunocytochemical demonstration of noradrenaline, serotonin and histamine and some observations on the innervation of the intracerebral blood vessels. In: *Neural regulation of brain circulation*, eds. C. Owman & J.E. Hardebo, pp. 181-194. Amsterdam: Elsevier.

Toda, N. & Miyazaki, M. (1984): Heterogenous responses to vasolidators of dog proximal and distal, middle cerebral arteries. *J. Cereb. Blood Flow Metab.* **6**, 1230-1237.

Toda, N., Okamura, T. & Miyazaki, M. (1985): Heterogeneity in the response to vasoconstrictors of isolated dog proximal and distal middle cerebral arteries. *Eur. J. Pharmacol.* **106**, 291-299.

## Neuropeptides and other peptides with effects on cerebrovascular tone

## Major reviews

Banks, W.A. & Kastin, A.J. (19876): Saturable transport of peptides across the blood-brain barrier. *Life Sci.* **41**, 1319-1338.

Brayden, J.E. & Bevan, J.A. (1986): Acetylcholine and vasoactive intestinal polypeptide in the cerebral circulation: histochemical and biochemical indices of innervation. In: *Neural regulation of brain circulation*, eds. C. Owman & J.E. Hardebo, pp. 371-381. Amsterdam: Elsevier.

Kelly, P.A.T. (1987): Vasoactive intestinal polypeptide: functional significance in cerebral organization and in the control of cerebral blood flow. In: *Peptidergic mechanisms in the cerebral circulation*, eds. L. Edvinsson & J. McCulloch, pp. 100-116. Chichester/Veinheim: Ellis Horwood/VCH.

McCulloch, J. (1984): Perivascular nerve fibers and the cerebral circulation. *Trends Neurosci.* **7**, 135-138.

Shigeno, T. & Mima, T. (1990): A new vasoconstrictor peptide, endothelin: profiles as vasoconstrictor and neuropeptide. *Cerebrovasc. Brain Metab. Rev.* **2**, 227-239.

Tuor, U.I., Kelly, P.A.T., Tatemoto, K., Edvinsson, L. & McCulloch, J. (1986): Neuropeptide Y and the cerebral circulation. In: *Neural regulation of brain circulation*, eds. C. Owman & J.E. Hardebo, pp. 333-354. Amsterdam: Elsevier.

Uddman, R. & Edvinsson, L. (1989): Neuropeptides in the cerebral circulation. *Cerebrovasc. Brain Metab. Rev.* **1**, 230-252.

Wahl, M. (1985): The effect of peptides on cerebrovascular resistance. In: *Oxygen transport to tissue VII*, eds. F. Kreuzer, S.M. Cain, Z. Turek & T.K. Goldsvich, pp. 121-130. New York: Plenum Press.

Wahl, M. & Schilling, L. (1993): Effects of bradykinin in the cerebral microcirculation. In: *The regulation of cerebral blood flow*, ed. J.W. Phillis, pp. 315-328. Boca Raton: CRC Press.

Waldemar, G. & Paulson, O.B. (1989): Angiotensin converting enzyme inhibition and cerebral circulation - a review. *Br. J. Clin. Pharamacol.* **28**, 177S-182S.

## Recent references of particular importance

Dacey, Jr, R.G., Bassett, J.E. & Takayasu, M. (1988): Vasomotor responses of rat intracerebral arterioles to vasoactive intestinal peptide, substance P, neuropeptide Y, and bradykinin. *J. Cereb. Blood Flow Metab.* **8,** 254–261.

Faraci, F.M., Mayhan, W.G., Schmid, P.G. & Heistad, D.D. (1988): Effects of arginine vasopressin on cerebral microvascular pressure. *Am. J. Physiol.* **255,** H70–H76.

Hardebo, J.E., Kåhrström, J., Owman, C. & Salford, L.G. (1989a): Endothelin is a potent constrictor of human intracranial arteries and veins. *Blood Vessels* **26,** 249–253.

Hardebo, J.E., Suzuki, N. & Owman, C. (1991): Dynorphin B is present in sensory and parasympathetic nerves innervating pial arteries. *Neurosci. Letters* (submitted).

Itakura, T., Okuno, T., Ueno, M., Nakakita, K., Nakai, K., Naka, Y., Imai, H., Kamei, I. & Komai, N. (1988): Immunohistochemical demonstration of vasopressin nerve fibers in the cerebral artery. *J. Cereb. Blood Flow Metab.* **8,** 606–608.

Koskinen, L.O.D. (1989): Cerebral and peripheral blood flow effects of TRH in the rat – a role of vagal nerves. *Peptides* **10,** 933–938.

Macrae, I.M., Graham, D.I. & McCulloch, J. (1987): Vasomotor effects of atrial natriuretic peptides on feline pial arterioles. *Brain Res.* **435,** 195–201.

Markowitz, S., Saito, K., Buzzi, M.G. & Moskowitz, M.A. (1989): The development of neurogenic plasma extravasation in the rat dura mater does not depend upon the degranulation of mast cells. *Brain Res.* **477,** 157–165.

Moskowitz, M.A., Saito, K., Brenza, L. & Dickson, J. (1987): Nerve fibers surrounding intracranial and extracranial vessels from human and other species contain dynorphin-like immunoreactivity. *Neuroscience* **23,** 731–737.

Suzuki, N., Hardebo, J.E., Kåhrström, J. & Owman, C. (1989): Galanin-positive nerves of trigeminal origin innervate rat cerebral vessels. *Neurosci. Lett.* **100,** 123–129.

Suzuki, N., Hardebo, J.E., Kåhrström, J. & Owman, C. (1990): Neuropeptide Y: coexistence with vasoactive intestinal polypeptide and acetylcholine in parasympathetic cerebrovascular nerves originating in the sphenopalatine, otic, and internal carotid ganglia of the rat. *Neuroscience* **36,** 507–519.

Suzuki, Y. Satoh, S.-I., Kimura, M., Oyama, H., Asano, T., Shibuya, M. & Sugita, K. (1992): Effects of vasopressin and oxytocin on canine cerebral circulation *in vivo. J. Neurosurg.* **77,** 424–431.

Thureson-Klein, Å., Kong, J.Y. & Klein, R.L. (1989): Enkephalin and neuropeptide Y in large cerebral arteries of the pig after ischaemia and reserpine. *Blood Vessels* **26,** 177–184.

# Prostanoids and leukotrienes

## Major reviews

Uski, T.K. (1986): Cerebrovascular smooth muscle effects of prostanoids. In: *Neural regulation of brain circulation*, eds. C. Owman & J.E. Hardebo, pp. 245–260. Amsterdam: Elsevier.

Walker, V. & Pickard, J.D. (1985): Prostaglandins, thromboxane, leukotrienes and the cerebral circulation in health and disease. *Adv. Tech. Stand. Neurosurg.* **12,** 3–90.

## Recent references of particular importance

Högestätt, E.D. & Uski, T.K. (1987): Actions of some prostaglandins and leukotrienes on rat cerebral and mesenteric arteries. *Gen. Pharmacol.* **18,** 111–117.

Sercombe, R., Hardebo, J.E., Kåhrström, J. & Seylaz, J. (1989): Penetrating cortical arterioles in rabbit dilate strongly to noradrenaline, $PGE_1$, $PGF_{2\alpha}$ and $TXA_2$ mimetic. *J. Cereb. Blood Flow Metab.* **9,** (Suppl. 1) S686.

Uski, T. (1989): Pharmacological characterization of contraction-mediating cerebrovascular prostanoid receptors. In: *Neurotransmission and cerebrovascular function I*, eds. J. Seylaz & E.T. MacKenzie, pp. 141–144. Amsterdam: Excerpta Medica.

Wahl, M., Schilling, L. & Whalley, E.T. (1989): Cerebrovascular effects of prostanoids. *In-situ* studies in pial arteries of the cat. *Naunyn Schmied. Arch. Pharmacol.* **340,** 314–320.

## Purinergic agents

### Major reviews

Burnstock, G. & Kennedy, C. (1986): A dual function for adenosine 5′-triphosphate in the regulation of vascular tone. Excitatory cotransmitter with noradrenaline from perivascular nerves and locally released inhibitory intravascular agent. *Circ. Res.* **58**, 319–330.

Phillis, J.W. (1989): Adenosine in the control of the cerebral circulation. *Cerebrovasc. Brain Metab. Rev.* **1**, 26–54.

### Recent references of particular importance

Hardebo, J.E., Kåhrström, J. & Owman, C. (1987b): $P_1$- and $P_2$- purine receptors in brain circulation. *Eur. J. Pharmacol.* **144**, 343–352.

Ibayashi, S., Ngai, A.S., Meno, J.R. & Winn, H.R. (1991): Effects of topical adenosine analogs and forskolin on rat pial arterioles *in vivo. J. Cereb. Blood Flow Metab.* **11**, 72–76.

Torregrosa, G., Miranda, F.J., Salmon, J.B., Alabadi, J.A., Alvarez, C. & Alborch, E. (1990): Heterogeneity of $P_2$-purinoceptors in brain circulation. *J. Cereb. Blood Flow Metab.* **10**, 572–579.

## Amino acids

### Major reviews

Kelly, P.A.T. (1984): GABAergic influences on cerebral tissue perfusion. In: *Neurotransmitters and cerebral circulation*, eds. E.T. MacKenzie, J. Seylaz & A. Bès, pp. 175–190. New York: Raven Press.

### Recent references of particular importance

Hardebo, J.E., Wieloch, T. & Kåhrström, J. (1989): Excitatory amino acids and cerebrovascular tone. *Acta Physiol. Scand.* **136**, 483–485.

Imai, H., Okuno, T., Wu, J.Y. & Lee, T.J.F. (1991): GABAergic innervation in cerebral blood vesels: an immunohistochemical demonstration of L-glutamic acid decarboxylase and GABA transaminase. *J. Cereb. Blood Flow Metab.* **11**, 129–134.

# Chapter 11

# Neurogenic regulation of cerebral blood flow: extrinsic neural control

**Peter J. Goadsby*** and **Richard Sercombe[†]**

*Institute of Neurology, The National Hospital for Neurology and Neurosurgery, Queen Square, London WC1, UK; [†]Laboratoire de Recherches Cerebrovasculaires; CNRS UA 641, Université Paris VII, 10 Ave de Verdun, 75010 Paris, France*

## Introduction

Until comparatively recently the issue of whether or not nerves influence the cerebral circulation was open to some debate, despite the description by Thomas Willis of the presence of nerve fibres in the walls of the cerebral vessels. In recent years it has been established by precise anatomical means that such nerves exist, so that experimental attention has turned to their role in normal physiology and possible role in the pathophysiology of such diverse conditions as migraine, dementia and stroke. In this chapter we shall examine the major divisions of the extrinsic innervation of the cer,ebral circulation. To define this system it shall be taken to mean: those nerve pathways that course from the central nervous system through the peripheral nervous system and innervate the cerebral vessels, that is extramedullary in respect of the brain and spinal cord. These may be divided further into three well defined groups, viz., the sympathetic nerves, parasympathetic nerves and a group of sensory nerves, the trigeminovascular system. The chapter will describe the basis for this classification, the transmitters they employ and the effects of either activation or blockade upon cerebral blood flow (CBF). A more detailed account of the pharmacology of the various transmitters has been presented in Chapter 10 and for a systematic bibliography readers are directed to several recent monographs and major reviews (Seylaz and Sercombe, 1989; Olesen and Edvinsson, 1988; MacKenzie *et al.*, 1984; Owman and Hardebo, 1986; Busija and Heistad, 1984; MacKenzie and Scatton, 1987).

## A. The sympathetic nervous system

### Anatomy

#### Generalities

The sympathetic nervous system has been characterized as consisting of preganglionic neurons originating in the central nervous system, and postganglionic neurons with their cell bodies in one of the sympathetic ganglia. The preganglionic fibres arise from the thoracic and upper lumbar cord, to enter the sympathetic chain of ganglia located adjacent to the vertebral bodies. Only the upper two ganglia, supplied by efferent fibres from the thoracic spinal cord, give rise to postganglionic fibres innervating cerebral vessels and other vessels of the head (Fig. 1). The lower of these ganglia, the stellate ganglion, is formed by the fusion of the inferior cervical ganglion and the first two or three thoracic ganglia. Sometimes a middle cervical ganglion is present (Fig. 1), but functionally it is undoubtedly a subsection of the stellate ganglion. The upper ganglion, called the superior cervical ganglion (SCG), receives its major imput from neurons in segments C8 to T5 (90 per cent in T1–T3). In the cervical sympathetic trunk all the preganglionic fibres are directed cranially, but there is also a contingent of postganglion fibres originating in the SGC, and contingents of postganglionic fibres directed cranially and originating in the middle and stellate ganglia.

#### Postganglionic pathways

##### The ganglion cells

The superior cervical ganglion has

been extensively studied, and although certain differences with respect to, for example, the coeliac and mesenteric ganglia are becoming evident, its general features are also shown by the stellate ganglion. In addition to the principal ganglion cells (giving rise to postganglionic fibres), there are several other cell types, including small intensely fluorescent cells (probably secretory in nature), glial type cells, mast cells, fibroblasts and vascular cells (mainly endothelial cells). The vascularization consists essentially of capillaries, some of which are fenestrated, allowing proteins to diffuse rapidly into the intercellular space of the ganglion.

The principal ganglion cells in the SCG vary in number between species; for example estimates for the rat SCG are in the range 15 000 to 45 000, and in man nearly one million. Transmission across the ganglia is by cholinergic synapses operating mainly through nicotinic receptors. Only a limited topographic subdivision of ganglionic neurons is apparent, although neurons projecting to the internal carotid artery are located mainly in the cranial part of the SGC and those projecting to the external carotid artery are mainly located in the caudal portion.

The great majority of neurons in sympathetic ganglia are adrenergic as detected histochemically by fluorescence microscopy. A small proportion of ganglionic neurons – about 4 per cent in the rat SCG – are cholinergic, and supply cholinergic dilator fibres to some blood vessels and secretomotor fibres to eccrine sweat

glands. However, in the cerebral circulation no evidence of cholinergic sympathetic fibres has been found. In fact, the presence of noradrenaline-induced fluorescence appears to be a universal marker for the sympathetic perivascular fibres: all noradrenaline-containing fibres disappear from cerebral vessels after sympathectomy, whereas cholinergic markers in perivascular fibres (especially choline acetyl transferase, ChAT) are unaffected.

### The distribution of sympathetic fibres

#### General features

The reference method for identifying sympathetic fibres around the cerebral vessels remains therefore the detection of this specificity, either by fluorescence histochemistry of catecholamines (based on the formaldehyde method of Falck and Hillarp, or the glyoxylic acid method) or the immunohistochemical methods for detecting dopamine β-hydroxylase (DBH), tyrosine hydroxylase (TH) or monoamine oxidase (MAO), enzymes necessarily involved in the metabolism of adrenergic neurons. Although sympathetic (adrenergic) fibres have been identified on all pial arteries and the internal carotid and vertebral arteries, the density of the plexus appears greatest on major rostral arteries and the circle of Willis (see Chapter 5). Both the SCG and the stellate ganglion contribute fibres, those of the former following the internal carotid artery and those of the latter accompanying the vertebral arteries (see Fig. 1). Considerable overlap of fibres of these two ganglia occurs in

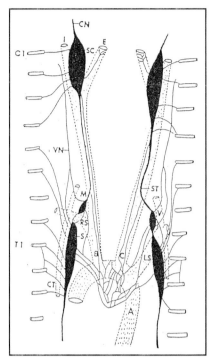

Fig. 1. Semidiagrammatic composite drawing of the cervical sympathetic trunk and its branches of the rat. A = Aorta; B = brachiocephalic trunk; C = left common carotid artery; CN = carotid nerve; CT = costocervical trunk; C1 = first cervical spinal nerve; I = internal carotid artery; LS = left subclavian artery; S = stellate ganglion; SC = superior cervical ganglion; ST = sympathetic trunk; V = vertebral artery; VN = vertebral nerve; E = external carotid artery; M = middle (intermediate) cervical ganglion; RS = right subclavian artery; T1 = first thoracic spinal nerve. From Hedger and Webber (1976), with permission.

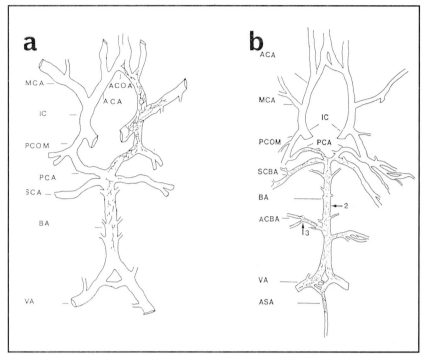

Fig. 2. Schematic representations of the distribution of anterogradely labelled nerve fibres on whole mounts of the arterial tree on the ventral surface of the rat brain. Injection of wheat germ agglutinin-horseradish peroxidase into (a) the superior cervical ganglion and (b) the stellate ganglion. ACA, anterior cerebral artery; ACBA, anterior inferior cerebellar artery; ACOA, anterior communicating artery; ASA, anterior spinal artery; BA, basilar artery; IC, internal carotid artery; MCA, middle cerebral artery; PCA, posterior cerebral artery; PCOM, posterior communicating artery; SCA/SCBA, superior cerebellar artery; VA, vertebral artery. From Arbab et al. (1986) and Arbab et al. (1988) with permission.

the basilar artery and to a lesser degree in the posterior cerebral arteries, as has been demonstrated by elegant sympathectomy and tracing experiments (both retro- and anterograde) in the rat by Arbab and co-workers (1986, 1988). According to these authors, in this species the SCG supplies the anterior cerebral, the middle cerebral, the anterior communicating, the internal carotid and the posterior cerebral arteries ipsilat-

erally, in confirmation of a number of previous fluorescence histochemical studies (Fig. 2a). Furthermore, they also claim that the basilar artery and some side branches are innervated (bilaterally) from the SCG, this innervation extending to the distal vertebral arteries, whereas the stellate ganglion fibres innervate the basilar artery and branches, and the vertebral and posterior cerebral arteries mostly ipsilaterally (Fig. 2b). Catecholamine fluorescence studies suggested a relatively reduced or absent contribution by the SCG to the vertebro-basilar system; the greater sensitivity of the WGA-HRP anterograde tracing technique may explain the

comparatively greater density of SCG-derived fibres observed in these recent studies.

### Species variations
There is little direct evidence of wide species variation in the supply of sympathetic fibres to the cerebral vessels, although no other species has received as much attention as the rat. The strictly ipsilateral pattern of innervation from the SCG may not hold for larger species: for example, in the rabbit, a small number of fibres of contralateral origin have been observed in the proximal middle cerebral artery and especially the anterior cerebral artery. Obviously, in man

tracing experiments cannot be performed, but extrapolation from careful dissection studies and fluorescence histochemistry of whole mounts of rapidly obtained cerebral arteries suggest that the pattern of sympathetic innervation is essentially identical to that of common laboratory species.

### Arterioles and veins

Fluorescence histochemistry, and especially the glyoxylic acid technique, has demonstrated the presence of (noradrenaline containing) fibres on intraparenchymal arteries and arterioles in several laboratory animals and primates (baboons and man), although the innervation of small species such as the rat may not extend as far as the penetrating arterioles. Cerebral veins are also innervated by sympathetic fibres: in general the plexuses appear less dense than in the major arteries (see Chapter 5). Fluorescent fibres can be seen on deep cerebral and collecting veins on the brain surface and also the dural sinuses (with greatest density), but their presence on intraparenchymal venules seems more doubtful. The fibres on the superior saggital sinus have been shown to originate in the SCG.

### Meningeal arteries

Like the cerebral arteries, they also are endowed with a multiple innervation, including sympathetic fibres originating in the SCG. The plexus is dense on the rat middle meningeal artery, and 'free' sympathetic fibres are also scattered over the dura, and extend along the branches of the meningeal arteries.

## Non-adrenergic markers

At least two other putative transmitters have been described in the cerebrovascular sympathetic fibres, the peptide neuropeptide Y (NPY) and the indoleamine derived from tryptophan, serotonin or 5-hydroxytryptamine (5-HT).

## Neuropeptide Y

Initially, due to cross-recognition of a similar molecule, it was thought that the cerebral arteries were endowed with fibres containing avian polypeptide (APP), but this is now known to be due to NPY which is also found in non-cerebrovascular sympathetic fibres (postganglionic). It was at first believed that a complete sympathectomy eliminated all NPY-reactive fibres, but the question as to the specificity of NPY as a marker of sympathetic fibres has recently received careful attention. Since 1985, five groups have found that a proportion of NPY-reactive fibres on cerebral arteries (in all cases a substantial minority of the total number) did not degenerate after sympathectomy, whether surgical or chemical. Suzuki *et al.* (1990) were able to totally eliminate NPY-reactive fibres in the anterior cerebral, the internal ethmoidal and the distal middle cerebral arteries of the rat after removal of the SCG and the sphenopalatine ganglion. This accords with Gibbins and Morris (1988) who showed NPY to be localized in 86 per cent of noradrenergic fibres and 18 per cent of VIP-containing fibres (many of which originate in the sphenopalatine ganglion; see part B). Interestingly, according to these authors, after sympathectomy the proportion of VIP-containing fibres reacting to NPY rose to 70 per cent in

the rostral circle of Willis, indicating that an interaction occurs between the sympathetic and the parasympathetic fibres regarding storage and/or synthesis of NPY. In contrast to the cerebral arteries, NPY-reactive fibres in the dura and on the superficial temporal artery all appear to derive from the sympathetic ganglia.

## Serotonin

The serotonin (5-HT) identified in the major pial arteries and originally believed to represent serotonergic fibres originating in the raphe nuclei, now seems to represent, at least partially, 5-HT captured from the perivascular environment by specific carrier mechanisms in sympathetic fibres. There is no biochemical evidence of 5-HT synthesis in the sympathetic neurons under normal conditions. Whether recaptured 5-HT can play any role *in vivo* remains to be determined (see Chapter 10). Its presence or absence in sympathetic fibres apparently depends on the experimental conditions during the artery preparation. The question as to whether neuronal 5-HT in the cortex may include fibres of central origin innervating arterioles remains to be clarified. Evidence for and against the existence of a true serotonergic innervation originating in the raphe has been summarized in Chapter 5.

## Effects of ablation, section or blockade

This section deals with *acute* effects of interruption of the sympathetic nervous output to the cerebral vessels on flow and metabolism in the brain, and also on phenomena such as cerebral autoregulation (see Chapter 9), hypertensive insult, etc. Long-term actions of sympathectomy (2 weeks) are incorporated in the section on the trophic influence of the sympathetic nervous system (see 'Trophic influences' below).

## Effects on 'resting' CBF

It is a fairly widely held opinion that sympathetic denervation of the cerebral vasculature under 'resting' conditions has no immediate influence on blood flow or other vascular parameters. While many investigations have indeed provided evidence of this nature, one should perhaps bear in mind that *this result has not been universally found*, and that consideration should be given to factors such as: (1) the meaning of 'resting' conditions, (2) the species, (3) the anaesthetic conditions, (4) the techniques used, and (5) the latency of the measurement after denervation.

By 'resting' conditions most authors mean that no experimental manoeuvre is undertaken which might modify blood flow/vessel diameter (such as systemic hypercapnia) or alter the basal level of sympathetic activation. It is the latter factor which is difficult to be sure of, since most anaesthetics will have directly or indirectly an inhibitory influence on the level of activation, and certain non-anaesthetized preparations may on the other hand be subject to some degree of exacerbation of this activation (pain, discomfort, restraint/immobilization).

It has been well established that species differences exist with regard to the sympathetic innervation and, perhaps more importantly, to the vascular reactivity to the sympathetic transmitter including the relative importance of α- and β-adrenergic receptors (see Chapter 10). For example, few significant changes have been observed in rats, which possess no α-adrenergic receptors on the basilar artery (only β-receptors), and in which intracortical arterioles have been shown to react more readily to β-adrenergic receptors (see Chapter 10). One study has shown a significant increase in *overall* CBF after α-adrenergic blockade by phentolamine (+ 17 per cent) or phenoxybenzamine (+ 28 per cent), but, notably, no regional changes in the cerebral cortex. In the absence of indices of cerebral metabolism, it cannot be excluded in such studies that the α-blockers, by increasing the local production of vasodilator metabolites, increased cerebral blood flow indirectly. In studies using radioactive microspheres, the inherently high scatter of the estimated flow values may explain why small effects (usually increases in CBF or decreases in cerebrovascular resistance) may have escaped detection. Insufficient spatial resolution in the measurements may have also contributed. Results of several such investigations on dogs have been negative: apart from methodological reasons, one can evoke the relatively poor development of the sympathetic innervation in the cerebral vasculature of this species, with correspondingly low levels of noradrenaline in the pial arteries. In the conscious goat, using a technique of input flow monitoring after ligation/thrombosis of the extracerebral circulation, it has been shown that SCG removal increased CBF by 66 per cent immediately, and infusion of phentolamine increased it by 30 per cent. Species differences, the method

of measurement, or the conscious state of the animal, may all contribute to explain this relatively large flow change. For instance, the cerebral influx measured must traverse the rete mirabilis, which may be under greater sympathetic influence than the cerebral circulation as generally represented in other species (internal carotid artery, circle of Willis, pial arteries and intracerebral arterioles). One might also suspect that some contamination of extracerebral circulation persists despite the surgical procedures employed to eliminate it.

When sympathectomy is obtained by post-ganglionic section (or SCG removal), the post ganglionic fibres degenerate, leading in the first 1–2 days to a period of increased transmitter release, followed after a week or more by the development of supersensitivity of the vascular receptors. Various other compensatory mechanisms may come into action, and the net result seems to be, ultimately, the recovery of blood flow levels similar to those preceding the sympathectomy. For example, 2 weeks after the SCG removal in the goat, CBF had returned to control levels.

Work on primates has not been notably prolific. The available data on anaesthetized monkeys have ranged from no significant change to about 30 per cent increase in CBF (global or large regional flow measurements), whereas in one study on unanaesthetized man the use of intravenous phentolamine to induce α-adrenergic blockade did not significantly affect regional CBF. It may be that the dose of the α-adrenergic blocking agent used was insufficient to gain access to the smooth muscle receptors across the blood–brain barrier; whether or not this was the case, the question of the tonic influence of the sympathetic innervation on the cerebral circulation under 'resting' conditions cannot be considered as settled.

Taken overall, the question of the 'resting' tonic action of the sympathetic innervation on cerebral blood flow would appear to be an elusive problem, inadequately addressed to date, but this influence can be reasonably estimated to be of fairly minor importance. However, as the following section will show, there is good evidence that the sympathetic nervous system comes into play under various conditions of modified cardiovascular function, both physiological and pathological.

## Effects during hypertension or on CBF autoregulation

### Experimental conditions

The term 'autoregulation' as applied to the cerebral circulation has been defined in Chapter 9, and the general characteristics described there. The overall body of evidence on the question supports the notion of multiple underlying mechanisms, probably not all intervening at any one time, one of which could be the activity of the sympathetic nervous system (SNS). In point of fact, the available evidence rather suggests that the SNS plays a most significant role under rather extreme conditions, i.e. when perfusion pressure is reduced to levels where flow begins to diminish, or when a hypertension of a high level is attained (close to, or above the point at which flow tends to follow blood pressure changes passively). The most convincing demonstration requires that the experimental condi-

tions be chosen such that a normal reflex physiological reaction of the SNS is obtained: first, under haemorrhagic hypotension, the baroreceptor reflexes activate the cardiovascular sympathetic pathways; second, if hypertension is induced by 'physiological' reflex excitation, again the cardiovascular sympathetic pathways are activated. Under these conditions, sympathectomy will inform us whether the sympathetics play any role in adjusting flow. If, however, hypertension is induced pharmacologically, or if anaesthesia is too sympathetically depressant, there will be little spontaneous sympathetic tone and cervical sympathectomy will have no effect on CBF.

### Haemorrhagic hypotension

Controlled haemorrhage-induced hypotension, by decreasing arterial blood pressure progressively, stimu-

lates baroreceptors and progressively activates cardiovascular hypertensive reflexes. The tonic cerebrovascular sympathetic activity thus induced has been shown, in baboons, dogs, goats and cats, to significantly influence cerebral blood flow, although in one cat study, with microspheres for CBF measurement, the effects were apparently small (< 10 per cent), perhaps because the hypotension was not severe enough to reduce CBF. At pressures below the autoregulated range, where CBF normally begins to fall with arterial pressure, acute section of the sympathetics or α-adrenergic blockade reduces cerebrovascular resistance and restores to some extent the level of CBF at a given pressure towards the normal level (Fig. 3). In a rabbit study with microspheres, no increase in flow was seen in the denervated hemisphere, but since the hypotension did not in any case re-

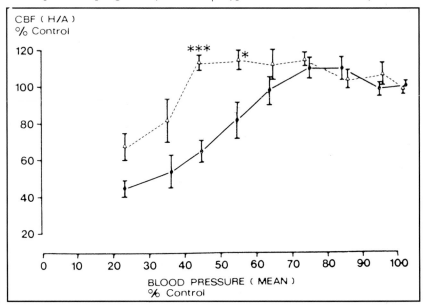

Fig. 3. Effect of decreasing mean arterial blood pressure on mean CBF in baboons subjected to haemorrhagic hypotension alone (●) and those subjected to haemorrhagic hypotension plus acute cervical sympathectomy (Δ). Cerebral blood flow at 55 and 45 per cent of the initial value was significantly greater in the baboons subjected to haemorrhage plus an acute surgical sympathectomy (* P < 0.05, *** P < 0.001). Values are means ± SE. From Fitch et al. (1975) with permission.

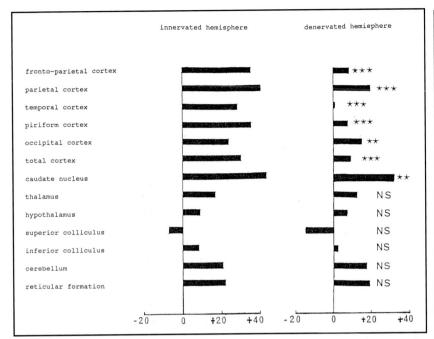

*Fig. 4. Percent changes in local cerebrovascular resistance compared with control during a reflex hypertension induced by bilateral section of the sinus nerves in the rabbit. 'Innervated hemisphere', side with intact sympathetic nerves; 'denervated hemisphere', side with superior cervical ganglion removed. \*\* significant difference, P < 0.01; \*\*\* significant difference, P < 0.001; NS, difference not significant. Modified from Lacombe et al. (1985) with permission.*

duce CBF, it seems likely that the hypotension was not severe enough to activate cardiovascular reflexes greatly. In a cat study using $H_2$ clearance, use of specific α-adrenergic blockade showed that the receptors involved, in this species, in the sympathetically-induced increase in cerebrovascular resistance during haemorrhagic hypotension were $α_2$-receptors, corresponding to *in vitro* studies on the pharmacology of cerebral vessels (Chapter 10). Overall, it seems reasonable to accept that, in primates as in lower species, the sympathetic activity induced by haemorrhagic hypotension has a distinct though moderate effect on CBF.

## Reflex-induced hypertension

Denervation of the baroreceptors has been performed in the cat and rabbit to induce reflex hypertension: in the rabbit the moderate increase in arterial pressure (30–35 mmHg) remained within the autoregulated range, whereas in the cat the increase was much greater, to beyond the autoregulated range (226 mmHg initially). In both cases, acute unilateral sympathetic denervation led to increased ipsilateral CBF and relatively decreased ipsilateral cerebrovascular resistance (Fig. 4), the effects being more widespread in the rabbit than in the cat. Parallel results were obtained on cortical blood flow by others in the dog, in which a reflex moderate hypertension (to about 165 mmHg) was induced by stimulation of chemoreceptors with nicotine. These studies again indicate that reflex acti-

vation of the SNS has moderate but definite effects on CBF and cerebrovascular resistance, both within and above the autoregulated range of arterial pressure.

## Non reflex-induced hypertension

When arterial pressure has been increased in the rat by infusion of vasoconstrictor drugs, the interruption of the theoretically less active sympathetic pathways (due to reflex inhibition) has nonetheless resulted in a shift of the upper limit of the autoregulation plateau to a slightly lower level (in cortex and thalamus). This result is difficult to understand unless we assume that the reduced sympathetic activity had a substantially amplified influence on flow at the high pressures investigated, as has been suggested on the basis of stimulation experiments (see 'Stimulation during hypertension' below). It has also been suggested that, in the piglet, sudden arterial pressure increases to above the autoregulated range (induced by adrenaline infusion) may result in higher flow-levels in the sympathetically denervated animals, but this result is tarnished by the existence of a higher arterial pressure in the latter animals.

## Effects during hypercapnia, hypocapnia or hypoxia

During systemic hypercapnia, sympathetic section or blockade has proportionately larger effects than during normocapnia: this has been demonstrated in unanaesthetized rabbits, by microsphere measurement of CBF, and in anaesthetized cats, by both microsphere measurement of CBF and *in situ* measurement of pial artery diameter. In both cats and rabbits a decrease of about 20 per cent in cerebrovascular resistance was found after bilateral sympathectomy under a $PaCO_2$ of 62/67 mmHg and a relatively high CBF. The diameter of pial arteries initially exceeding 100 μm increased less than those initially less than 100 μm during hypercapnia, but after ipsilateral sympathectomy or topical phentolamine, their diameter increased as much as that of the smaller vessels. These results have been interpreted as evidence of reflex activation of sympathetic influence on the cerebral circulation, although it is to be wondered whether the amount of activation was not limited by the anaesthesia (in cats). These experiments confirmed earlier work on the anaesthetized baboon in which a 10 per cent decrease in cerebrovascular resistance (and a similar increase in CBF) were observed at a $PaCO_2$ of 59 mmHg.

Hypocapnia has been little studied and the available data suggest that, during it, sympathectomy has rather little influence on CBF.

In contrast, two studies in which sympathectomy or blockade was performed to test the influence of the SNS on the hypoxic cerebral vasodilatation, both revealed a significant, even substantial, further increase in CBF during hypoxia after sympathetic ganglionectomy. The evidence is compatible with a reflex activation of the SNS during hypoxia which causes a sizeable decrease in flow, from about 20 per cent in the medulla to about 50 per cent in the cortex.

## Effects of sympathetic stimulation

### Stimulation under resting conditions

Although, for several decades, the effectiveness of sympathetic stimulation in influencing cerebral blood flow remained controversial, a certain consensus emerged in the 1980s. It is now widely agreed that small to moderate decreases in ipsilateral flow (i.e. 10–40 per cent) can be induced by pre– (caudal to the SCG) or post-ganglionic electrical excitation, although the actual amplitude of the response observed is dependent on both methodological and anatomo-physiological factors (anaesthesia, brain region, species, side and duration of stimulation). For reasons explained elsewhere (Chapter 8), measurements made with microspheres appear to indicate smaller changes than other techniques, whereas brain inflow and outflow techniques have frequently suggested relatively high levels of flow reduction. When tested, the flow reductions have been found sensitive to α-adrenergic blockers. What is probably important is that effects vary with the experimental conditions (blood pressure, levels of $PaO_2$, $PaCO_2$) and that a physiological sympathetic discharge does not reduce flow to ischaemic levels. Effects on cerebral metabolism seem practically absent, although it has been suggested that the escape phenomenon is dependent on an increase in tissue $pCO_2$ (see 'The escape phenomenon' below).

### Regional and segmental variation

The regional variation of flow reduction to sympathetic stimulation shows an evident correlation with the density of the sympathetic innervation: flow changes are greatest in structures nourished via the internal carotid arteries, the plexus of adrenergic fibres being most dense on the interior carotid, anterior cerebral and middle cerebral arteries (Chapter 5). Effects of stimulation have been shown by some authors to be not purely unilateral, since bilateral stimulation had greater effects than ipsilateral stimulation alone (40 per cent compared to 21 per cent at 16 Hz) and contralateral stimulation reduced the diameter of small pial arteries. This can be explained by some degree of 'cross-over' by sympathetic fibres via interhemispheric anastomoses.

Several lines of evidence suggest that sympathetic stimulation constricts large cerebral arteries more than small arteries:

(1) direct observation of pial arteries via a cranial opening, whether closed or paraffin-protected, has shown that vessels below about 80–100 μm diameter constrict relatively less than larger vessels which narrow by some 10–15 per cent;

(2) measurement of pressure changes in rabbit pial arteries of mean diameter 128 μm and cat arteries of mean diameter 221 μm showed that the constriction of larger arteries reduced pressure at the point measured, whereas flow was not significantly changed (microsphere measurement);

(3) there is evidence that rabbit small penetrating arteries (< 100 μm) react to noradrenaline by a

dilatation and that small surface arteries may be non-reactive;

(4) a laser-Doppler study by Saeki *et al.* (1990) in the rat demonstrated increased local cortical flow at low frequency stimulation, sensitive to β-adrenergic blockade, and decreased flow at high frequency stimulation, sensitive to α-adrenergic blockade.

Small and penetrating arteries may thus not only constrict less but possibly even dilate in response to sympathetic stimulation. Such a concept certainly helps explain the moderacy of the sympathetic influence compared to other vascular beds, and may be involved in the escape phenomenon described below.

## The escape phenomenon

The existence, under certain conditions, of escape from sympathetic-induced reduction in CBF was first analysed in some detail by Sercombe *et al.* (1979), although evidence of this phenomenon could be found in figures of other authors in several earlier studies. It may be defined as the non-stabilization of blood flow at a reduced level during continued sympathetic stimulation, the level in fact returning progressively towards the control (non-stimulated) level (Fig. 5). It is clear that techniques of CBF measurement which are not continuous and which require even quite short periods of stable flow for an estimate to be made cannot be used to observe maximal effects with any reliability. This said, the escape phenomenon has been shown to be attenuated by increasing barbiturate anaesthesia, and to be only moderately dependent on stimulation frequency. Although some workers using microsphere measurements and barbiturate + chloralose anesthesia were not able to demonstrate an escape phenomenon, this phenomenon has been clearly demonstrated in more recent cat studies in which cortical blood flow and blood volumes, and pial arterial and venous diameters

were measured. Smaller arteries (< 100 μm or 50 μm) showed distinctly more tendency to escape, whereas arteries greater than 100 μm in diameter escaped little or not at all. Veins showed a weak trend towards escape. Both cerebral blood volume and intracranial pressure (initially reduced) also escaped, reaching a suprabaseline level before the end of the stimulation.

The analyses performed to date do not allow definitive conclusions to be drawn regarding the mechanisms of escape. Evidence has been adduced for and against a metabolic mechanism, i.e. the accumulation of a metabolite – notably $CO_2$ – in the parenchyma which then redilates intraparenchymal arteries and arterioles (and small pial arteries?). There is evidence that myogenic mechanisms could cause dilatation in small arteries because of the lowered intraluminal pressure. There might well be some β-adrenergic-induced ('active') vasodilatation in penetrating vessels as such reactivity has been shown to predominate in intraparenchymal arterioles. Any combination of these mechanisms, or none, may intervene according to the state of the preparation (especially anaesthesia) and the cerebral region or type of vessels considered (vascular reactivity may change according to depth within the brain tissue). Finally, as a possible contributing factor, albeit not directly demonstrated so far, prejunctional inhibition of sympathetic transmitter release may be born in mind : inhibitor candidates are GABA, noradrenaline, acetylcholine and adenosine (see 'Prejunctional inhibition' below).

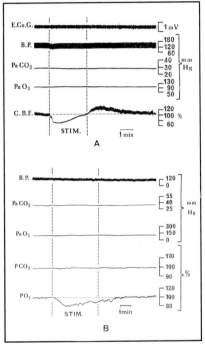

*Fig. 5. Continuous recordings in the rabbit showing the time course of effects of cervical sympathetic nerve stimulation on (A) local blood flow measured by thermal clearance in the caudate nucleus (C.B.F.) and (B) partial pressures of $O_2$ ($PO_2$) and $CO_2$ ($PCO_2$) measured by mass spectrometry in the tissue of the caudate nucleus. Other variables measured continuously were the electrocorticogram (E.Co.G.), arterial blood pressure (B.P.), and partial pressures of $O_2$ and $CO_2$ in arterial blood ($PaO_2$, $PaCO_2$). Notice the initial fall of blood flow and local $PO_2$, followed by a progressive return towards the baseline. No modification of the other variables occurred, except for a slight, transient fall in BP in recording A. Stimulation (STIM.) parameters were (A) 10 Hz, 2 ms, 4 V; (B) 15 Hz, 2 ms, 6 V. From Sercombe et al. (1979) with permission.*

## Stimulation during hypertension

During both moderate hypertension, where autoregulation maintains CBF approximately constant, and severe hypertension to above the usual upper limit of autoregulation, sympathetic stimulation has been shown to have a significant protective action on CBF and blood–brain barrier (BBB) permeability in several species. Thus in cats it was found that a moderate, sudden increase in arterial pressure was accompanied by an abrupt increase in CBF, with progressive return to baseline within 60 s, and that the size and duration of the flow increase could be reduced by sympathetic stimulation. This result was not confirmed in dogs, but this may well be in relation to the generally less developed sympathetic innervation in this species, and a highly efficient autoregulatory reaction.

Perhaps more important physiologically, and of greater amplitude, are the effects of sympathetic stimulation during severe hypertension, where CBF is prevented from rising above the plateau level (within certain limits of arterial pressure). First demonstrated by Bill and Linder (1976), this phenomenon has been confirmed in rats, baboons, and other cat studies, and it has been confirmed in many cases that gross extravasation of Evans blue (complexed with albumin) into the brain is significantly reduced, notably in the cerebral cortex. It has also been found that bilateral stimulation in the baboon, in comparison to unilateral stimulation, further prolonged the autoregulatory plateau, as measured in one hemisphere, indicating an appreciable functional crossover of sympathetic fibres. Data from

pial artery pressure measurements in cats suggest that the greatest influence of sympathetic stimulation occurs during severe, as compared to modest, hypertension. However, both 'large' (> 200 μm) and 'small' (< 200 μm) pial arteries showed increased resistance by sympathetic stimulation under these conditions, indicating that the protection of the BBB involves decreased pressure in more distal segments (capillaries, venules).

## Stimulation during hypoxia or hypercapnia

Corroborating denervation and blockade experiments, sympathetic stimulation during hypoxia (with basal flow substantially increased) has been found to induce significant decreases in total CBF (–25 per cent) in newborn lambs, with a maximum change of –37 per cent in cerebral cortex and a minimum of zero change in the medulla (measurements by microspheres). In another study with microspheres, no significant change in CBF was found under hypoxia in the anaesthetized rabbit. However, percent flow reductions under sympathetic stimulation have been generally lower when measured by microspheres compared with other techniques, and one may speculate that greater effects might have been obtained with other techniques under the same conditions (see also Chapter 8).

The effects of stimulation during systemic hypercapnia have been more widely studied, and it is clear that responses are more elevated than during normocapnia including in newborn animals (Fig. 6). For example, in

a microsphere study on anaesthetized cats it was found that unilateral stimulation increased cerebrovascular resistance by 15 per cent, and bilateral stimulation increased it by 66 per cent (corresponding to a 30 per cent decrease in cortical flow), whereas the same workers had not found significant effects in normocapnia with this species. Previous studies with baboons had also demonstrated increased effects of sympathetic stimulation under hypercapnia and on this basis Harper *et al.* (1972) proposed a hypothesis of differential reactivity of intra- and extracerebral resistance vessels, according to which intracerebral arterioles dilated during extracerebral arterial constriction to sympathetic stimulation. Subsequent studies, including those on the escape phenomenon, have tended to lend support to this notion.

## Stimulation in foetal or neonatal subjects

Foetal and newborn lambs and newborn piglets have been the subject of a certain number of investigations of the capacity of action of the sympathetics in immature subjects. CBF measurements have been performed with microspheres. Evidence from lambs shows clearly that foetal reactivity to both noradrenaline and sympathetic stimulation (pial arteriolar diameter) was strong (21 per cent constriction for stimulation), but tended to diminish with age. Newborns were less reactive to noradrenaline and adults not reactive at all, and most likely reactivity to sympathetic stimulation would be correlated. The significant fall in the CBF of hypoxic newborn lambs during sympathetic stimulation has been mentioned. In piglets, a signifi-

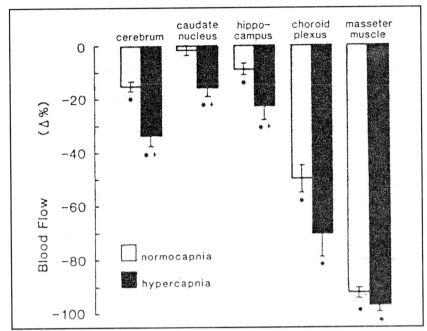

*Fig. 6. Effect in the piglet of electrical stimulation (15 Hz, 15 V, 3 ms) of the right cervical sympathetic trunk on regional cerebral blood flow in the ipsilateral hemibrain during normocapnia (PaCO_2 = 38 mm Hg, n = 41) and hypercapnia (PaCO_2 = 64 mm Hg, n = 5). % is the per cent change in flow = (right–left)/left × 100. Values are mean ± SE. \* significantly different from zero. ⁺significantly different from normocapnia. P < 0.05. From Wagerle et al., 1986, with permission.*

cant modest fall (15–16 per cent) in cerebrum blood flow has been shown in several studies and a 15 per cent decrease in pial arteriolar diameter reported elsewhere. CBF responses were increased in hypercapnia (to 34 per cent in cerebrum) (Fig. 6). The $\alpha_1$-adrenergic receptor appears to be responsible for constriction in both species. In lambs with induced seizures (bicuculline), sympathetic denervation significantly increased CBF (+14 to +38 per cent), and stimulation reduced it by –17 to –27 per cent. Overall, the evidence strongly supports a significant role for sympathetic regulation of CBF during the frequent, life-endangering situations occurring in foetal and newborn animals.

## *In vitro* studies

### Technical difficulties

A large number of *in vitro* studies on the effects of electrical field stimulation on isolated cerebral arteries have contributed a wealth of information on the modalities of the influence of perivascular nerves on the smooth muscle. The method comprises a number of technical difficulties and interpretation must be made in the light of strict control experimentation. As emphasized by systematic studies by Hardebo and colleagues, cerebral arteries are especially delicate material in that it is extremely difficult to obtain purely neurogenic responses, i.e. responses elicited via the nerves without any directly in-

duced contraction or relaxation. Stimulation parameters, especially pulse duration, must be appropriate, and proof of nerve conduction obtained by abolition of responses with tetrotoxin (TTX) at concentrations up to about $10^{-6}$ M. The pattern of stimulation pulses, i.e. in regular trains or in brief bursts at higher frequency, appears to be important too.

A second difficulty, variable according to species, is that responses obtained via non-sympathetic nerves may confound the interpretation: it is thus essential to confirm the implication of sympathetic fibres by use of specific pharmacological tools such as guanethidine or bretylium (inhibitors of noradrenaline release), or by denervation. It has been suggested on the basis of such experiments that some resistance to TTX may occur, despite evidence of sympathetic terminal involvement, because the varicosities are depolarized directly by the field stimulation (no nerve conduction involved). Furthermore, non-TTX sensitive responses are subject to fading. Thus, without evidence of TTX sensitivity of the responses (preferably leading to a *large* reduction in the response), and evidence of *sympathetic* fibre involvement, no definite conclusions can be drawn.

### Response to sympathetic stimulation

On the above bases, it is generally agreed that electrical field stimulation may cause contraction of smooth muscle in cerebral arteries of rabbit, primates, rats, guinea-pigs and goats via activation of sympathetic fibres. In cats, some workers have found only relaxation, whereas others have described a partially TTX- and bretylium-sensitive contraction, together with evidence of noradrenaline release. In dog and sheep arteries, neurogenic contraction and relaxation can both be obtained according to the conditions and antagonists present, although dilator effects (non-sympathetic) seem more prominent in sheep. In pig arteries, only relaxations have been obtained, but this is not surprising since the overall response to noradrenaline is known to be predominantly via beta-adrenergic receptors.

Despite the unanimity on the type of response obtained, and for several species the evidence of release of neuronal noradrenaline (cattle, rabbits, cats, dogs), responses to electrical stimulation were not blocked by α-adrenergic antagonists in rabbit, rat and dog. Indeed, in rabbit and dog cerebral arteries phentolamine and/or phenoxybenzamine even potentiate both contractions and noradrenaline release.

These results may be interpreted as indicating the absence of significant post-junctional α-adrenergic activation and the presence of inhibition of transmitter release by activation of α-adrenergic autoreceptors on the prejunctional membrane. They were mostly obtained on the basilar arteries of rabbit, rat and dog, with extremely poor sensitivity to noradrenaline (Chapter 10), so that other neurotransmitters must be responsible for the contractions obtained in this artery and in these species (e.g. NPY). In goat middle cerebral arteries (MCA) phentolamine blocked the tonic phase, but not the brief early phasic contraction. In dog basilar artery the early contractile phase was enhanced by phentolamine, as was the release of tritiated noradrenaline and purines. Correspondingly, purine release was inhibited by noradrenaline and noradrenaline release inhibited by ATP. In sheep MCA the contractions were blocked by phentolamine and prazosine, but also partially by ketanserin (5-HT2 antagonist) and atropine (muscarinic anta-

**Table 1.** *In vitro* field electric stimulation of cerebral arteries: attempts to elicit responses to sympathetic nerve excitation

| Species | Major response to stimulation | TTX and guanethidine sensitivity of constriction | Evidence of NA release | Effect of α-blocker | Observations |
|---|---|---|---|---|---|
| Rabbit | C | TTX[†], G | + | Potentiation | Non-adrenergic transmitter? (NPY?) |
| Sheep | D > C | TTX, G | 0 | Blockade | Transmitter is NA (possible release of 5-HT also) |
| Cat | D (C*) | (TTX, G[‡]) (partial block) | + | (Inhibited* by 20%) | Probably NA is transmitter |
| Guinea-pig | C/D | – # | 0 | No change | Main transmitter neither NA nor ATP |
| Rat | C | – | + | No effect on e.j.p.s. (basilar artery) | NA not transmitter (basilar artery) |
| Goat | C | TTX, G | + | Blockade of tonic phase | Main transmitter is NA |
| Dog | C/D | TTX, G | + | Potentiation | Main transmitter not NA? |
| Pig | D | TTX, G | 0 | No significant change (α- or β-blocker) | Transmitter not NA |
| Monkey | C | TTX, G | + | Blockade | Main transmitter is NA |
| Human | C | TTX, G | + | Partial blockade | One transmitter may be NA |

[†]Certain authors have not been able to obtain TTX-sensitive responses.
* = Results of 1 report (Salaices *et al.*, 1983).
[‡] = Sympathetic denervation or reserpine treatment prevented response.
e.j.p. = excitatory junction potential.
C = constriction; D = dilatation; TTX = tetrodotoxin; G = guanethidine; NA = noradrenaline; ATP = adenosine triphosphate; 5-HT = 5-hydroxytryptamine.

gonist). 5-HT is known to be present in sympathetic fibres under certain conditions, so it can be assumed that it was also released by the stimulation. The muscarinic effect presumably reflected release of acetylcholine from parasympathetic fibres. In monkey MCA the contractions were inhibited by phentolamine and prazosine, but not yohimbine (below $10^{-7}$ M), or blockers of 5-HT$_2$, H$_1$, muscarinic, or P$_2$ receptor-mediated effects. In contrast, in guinea-pig basilar artery there were no effects of phentolamine or guanethidine, despite the abolition of responses of vessels from reserpine-treated or denervated animals. In cat posterior communicating artery (with about 50 per cent inhibition by TTX) a 20 per cent inhibition was obtained with phentolamine ($10^{-6}$ M). Excitatory junction potentials recorded in cat basilar artery after excitation with scorpion venom were also blocked by phentolamine. These results are summarized in Table 1.

It thus appears that the primary sympathetic transmitter in primate, goat and cat cerebral arteries is probably noradrenaline. Candidates for part or all of the response in the large arteries of other species include ATP, serotonin, neuropeptide Y, but in some species other candidates should be sought. It should also be born in mind that differences may exist between different cerebral arteries of the same species. What might be suspected is that a wide variety of transmitters may potentially be released from all sympathetic varicosities, but that the exact proportions vary according to species, vessel and experimental conditions. One major variant in the conditions is the *in vivo/in vitro* alternative, and, so far, *in vivo* work has sug-

gested that noradrenaline is the major transmitter. This fact might be related to the observations in a variety of vessels, cerebral and especially non-cerebral, suggesting that the co-transmission by noradrenaline and ATP occurs in variable proportions, but that very short stimulation trains (more frequently used *in vitro*) tend to favour the relative contribution of ATP to the contraction. Finally, the basic tone of the vascular preparation can profoundly affect the response to ATP: for example, it has been shown that at low perfusion pressure ATP constricts the mesenteric arterial bed, whereas at high perfusion pressure it dilates it.

## Prejunctional inhibition

The phenomenon of prejunctional inhibition (and enhancement) of sympathetic transmitter release in blood vessels has been recognized for 15–20 years. In the peripheral (non-cerebral) circulation such inhibition is known to occur via α$_2$-adrenergic, prostanoid, cholinergic, serotonergic, histaminergic, peptidergic (e.g. NPY, vasopressin), purinergic, and GABAergic receptors situated presumably on the sympathetic nerve terminals. β$_2$-Adrenergic receptors mediate enhancement of transmitter release.

In the case of cerebral vessels, evidence exists for most of these prejunctional effects, but apparently no work has yet been performed on prostanoid or β$_2$-adrenergic actions. α-Adrenergic inhibition has been amply demonstrated *in vitro* in rabbit, dog and cat arteries, but was not detected in one study on monkey arteries. Both contraction and transmitter release were modified in these ex-

periments, even though in dog arteries the *effective* transmitter is not noradrenaline (see above), but the data shows that ATP release can be inhibited by noradrenaline. The use of yohimbine and clonidine by Sakakibara *et al.* (1982) indicated that the α$_2$-receptor subtype was responsible. Some *in vivo* experiments on piglets and rabbits (microsphere measurements of CBF) tend to confirm the notion that α$_2$-adrenergic inhibition can occur.

Following the original work by Edvinsson *et al.* (1977) on cat MCA, suggesting that cholinergic inhibition of ³H-NA release could occur, apparently mediated by nicotinic receptors, confirmation of this hypothesis was provided by *in vivo* studies in the rabbit and cat, and although formal identification of the receptor subtype was not possible, it was shown not to be muscarinic. In the goat, both *in vivo* and *in vitro* experiments have suggested that the muscarinic receptor was involved, although, surprisingly, only the phasic contractile component was inhibited, the tonic component being potentiated (*in vitro* experiments). Recent *in vitro* work on cats suggests that M$_2$ muscarinic receptors are responsible for carbachol-induced inhibition of ³H-NA release, as also seems to be the case in non-cerebral arteries. Further investigation seems necessary to clarify this question, especially since a muscarinic blocking agent had no effect in monkey arteries, and the *in vivo* goat preparation might not represent purely cerebrovascular responses as discussed above (see 'Effects on "resting" CBF').

Concordant results have been obtained in both rabbit basilar and goat

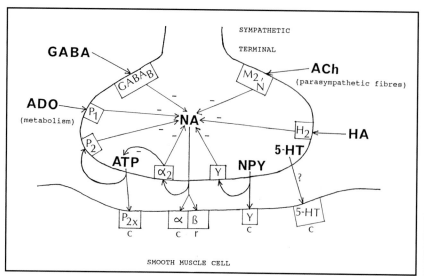

*Fig. 7. Hypothetical scheme of the possible mechanisms controlling the release of noradrenaline and other substances from the sympathetic varicosity/terminal in cerebral blood vessels (based on published data). Neurotransmitters/neuromodulators in heavy type: ACh, acetylcholine; ADO, adenosine; ATP, adenosine 5'-triphosphate; GABA, gamma-amino butyric acid; HA, histamine, 5-HT, serotonin; NA, noradrenaline; NPY, neuropeptide Y. In boxes, in fine type, are the proposed receptors involved (N = nicotinic). The sign – indicates the inhibitory influence on release. c = contraction induced; r = relaxation induced.*

middle cerebral arteries, indicating that gamma-amino butyric acid (GABA) can inhibit sympathetic-induced contraction *in vitro* probably via activation of GABA_B receptors. Similar results have been found in the rabbit central ear artery, but not the mesenteric or internal carotid arteries. In goats, confirmation was obtained by *in vitro* $^3$H-NA release studies and *in vivo* CBF measurement. Evidence of prejunctional inhibition by histamine via H$_2$ receptors has also been found in cat and goat preparations.

A limited amount of evidence for purinergic inhibition (by adenosine and ATP) has been published, and NA release from sympathetic fibres on the rat basilar artery can be inhibited by NPY (although NA is certainly not responsible for contraction in this vessel). Clearly, intensive investiga-tion is required in order to elaborate comprehensive hypotheses of the functioning of the sympathetic neuromuscular junction in cerebral vessels (see Fig. 7).

## Other influences of sympathetic nerves

### Cerebrospinal fluid production

#### Morphological bases

About 80 per cent of the cerebrospinal fluid (CSF) is secreted by the choroid plexuses which are highly irrigated, specialized organs in contact with the cerebral ventricles. The general relationship of the CSF with the other liquid compartments of the brain is shown in Chapter 5. After its production, the CSF flows through the ventricles posteriorly; via aper-tures in the IVth ventricle it joins the cisternal magna and the space surrounding the central nervous system between the arachnoid and the pial membranes. From there it is reabsorbed into the venous blood via the arachnoid villae in the dural sinuses. In most laboratory mammals the CSF secretion rate is between 0.4 and 0.6 per cent of the total volume per min.

Sympathetic and parasympathetic fibres innervate both the vascular smooth muscle and the secretory epithelium of the choroid plexus. The secretion of CSF is clearly under inhibitory sympathetic tone under steady state conditions. The general pattern of distribution of adrenergic fibres is very similar in all species: the nerve terminals form networks around small vessels and in the plexus tufts, and isolated fibres and terminals run in close relation to the epithelium. In most species the density of innervation is in the order IIIrd ventricle > lateral ventricle > IVth ventricle. This innervation is abolished in all plexuses of the cat and rat after bilateral excision of the SCG, but in the rabbit a small proportion of fibres supplying the IVth ventricle plexus remains and presumably originates in the stellate ganglion.

### The sympathetic influence

On the basis of *in vitro* studies on choroid artery reactivity and plexus epithelial functions, and *in vivo* studies on CSF production, it seems likely that the sympathetic influence is mediated by effects on both the vascular bed and the secretory cells. Sympathetic stimulation causes a 20–40 per cent decrease in CSF production rate and a corresponding decrease in

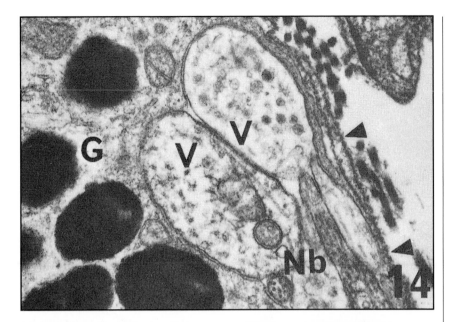

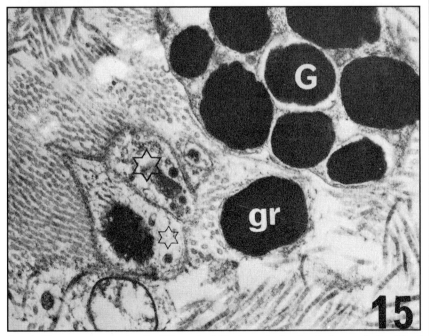

*Fig. 8. Electron micrographs of rabbit cerebral blood vessels showing intimate relationship of adventitial granular cells (G) (mast cells) and nerve varicosities (V). Notice in (a, top), that the mast cell and the varicosities are surrounded by the same basement membrane (arrowheads), and that the varicosities are situated within an invagination of the mast cell membrane. In (b) the denervated vessel has been treated by Tranzer and Richards's method, and reveals that one of the varicosities is sympathetic (small asterisk) since it shows signs of degeneration. The other varicosity contains small clear and large dense-cored vesicles (large asterisk). Magnification: (a) × 26 768; (b, bottom) × 19 684. From Dimitriadou et al., 1987, with permission.*

intracranial pressure. Parallel effects can be obtained by intravenous infusion of noradrenaline, and the decrease in CSF production is counteracted by α-adrenergic and potentiated by β-adrenergic blockers. In contrast, the effects of intraventricular noradrenaline (essentially similar) can be inhibited by β-blockade. These results can probably best be interpreted by assuming there is a stimulation of both α-(dominant) and β-receptors on the choroidal arteries, and stimulation via β-receptors of the epithelial secretory cells. Sympathetic activation can thus be assumed to act not only via vascular receptors but also via β (β1 or β2 according to species)-adrenergic receptors activating adenylate cyclase in the epithelial cells, the end-product of which (cAMP) must inhibit CSF production. Choroid plexus choline transport and sodium-potassium ATPase activity are increased 1 week after sympathectomy, as is carbonic anhydrase activity, considered essential for CSF production. Finally, the parasympathetic nervous system has been proposed to intervene both by a direct action of vasoactive polypeptide (VIP) and acetylcholine on the secretory cells, and by a VIP-induced increase in noradrenaline release from sympathetic neurons.

## Possible action on perivascular mast cells

Mast cells of at least two types exist in the central nervous system, a considerable proportion of which are located around the blood vessels. The dura is also the seat of a large number, some of them situated perivascularly (around the meningeal arteries), but it has been suggested they are of a

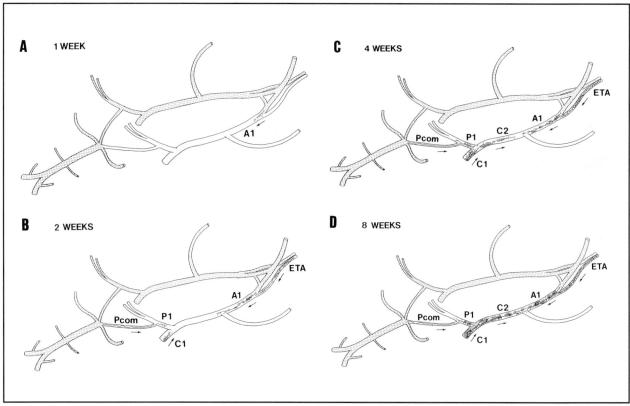

Fig. 9. Distribution pattern of the maximal sprouting of the reinnervating nerve fibres originating in the intact (left) superior cervical ganglion (SCG) on the denervated cerebral arteries 1 week (A), 2 weeks (B), 4 weeks (C) and 8 weeks (D) after removal of the right SCG. The area of screen tone indicates the area normally innervated by the SCG. C1, proximal portion of the internal carotid artery (ICA) to the bifurcation of the posterior cerebral artery (PCA) and ICA; C2, distal part of the ICA from the bifurcation of the PCA to the bifurcation of the middle cerebral artery (MCA); A1, proximal portion of the anterior cerebral artery (ACA) from the bifurcation of the MCA to the bifurcation of the anterior communicating artery (Acom); P1, proximal portion of the PCA from the bifurcation of the ICA to the posterior communicating artery (Pcom); P2, distal portion of the PCA. From Handa et al. (1991) with permission.

different type ('connective tissue' type).

According to Ibrahim (1974), the two cerebral types (type I and type II or neurolipomastocytes) vary in number and proportion according to species and brain area. Type II mast cells exclusively, and type I frequently, lie in the advential layer of arteries and arterioles, especially at branch points. Serotonin assay and fluorescence are useful methods for studying both dural and cerebral mast cells in rodents, this amine being released dur-

ing the degranulation induced by compound 48/80 and cholino-mimetic agents.

A possible role of the sympathetic nerves in mast cell functioning is suggested by (1) morphological evidence demonstrating an intimate relation between type II mast cells and nerve varicosities some of which appear to be sympathetic as shown by ganglionectomy experiments (Fig. 8, Dimitriadou et al., 1987); (2) evidence of degranulation and loss of 5-HT fluorescence of dural mast cells after sym-

pathetic ganglionectomy, and conversely, increased fluorescence following stimulation (Ferrante et al., 1990). Interactions between mast cells and peptidergic (substance P and CGRP-containing) fibres seem probable in the brain, as elsewhere, so that the three extrinsic nervous systems of the cerebral circulation may have complex interrelations through these representatives of the immune system, some of which may be concerned in migraine for example (see section C and Chapter 14).

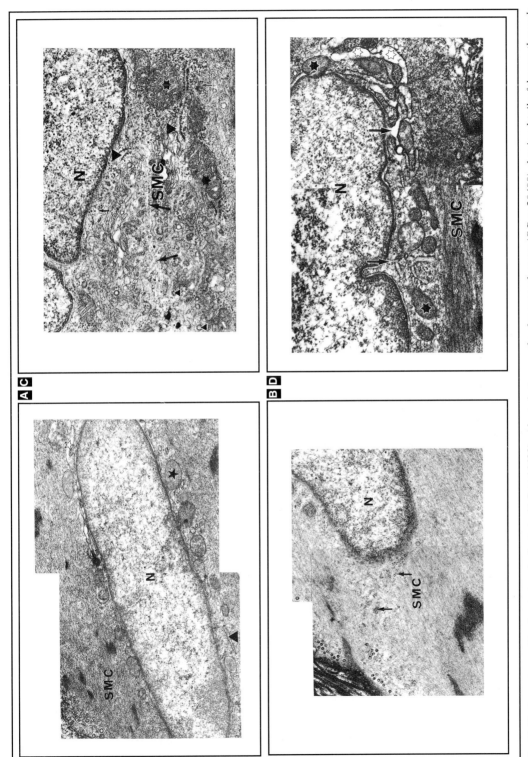

Fig. 10. Electron micrographs of control cerebral arteries (A,B; × 14688) and arteries two months after sympathectomy (C,D; × 20400), showing details of the smooth muscle cytoplasm. A,B: Notice the paucity in cytoplasmic organelles. SMC = Smooth muscle cell; N = nucleus; filled star = mitochondria; open star = ergastoplasmic cisternae; arrowhead = RER (rough endoplasmic reticulum); arrow = ribosomes. C: Notice the abundance of the cytoplasmic perinuclear organelles, especially RER (large arrowhead), ribosomes (small arrow), Golgi apparatus, dictyosomes (empty asterisk), transition or maturation cytoplasmic vesicles and coated vesicles (small arrowhead) and microtubules (large arrows). D: Notice the well-developed RER cisternae (large empty asterisk) arranged close to the nucleus and their contiguity with the nuclear envelope (large arrow). Filled asterisks = mitochondria.

Fom Dimitriadou et al. (1988) with permission.

## Trophic influences

It has been known for many years that the sympathetic nervous system not only acts on effector systems on a moment-to-moment basis, but exerts a long-term influence on their growth, development and even reactivity (Azevado and Osswald, 1986). A multitude of experiments have confirmed this general observation in the cerebral circulation even though controversy persists, notably over the role of sympathetic nerves in protecting the brain from infarction and other dangers during chronic hypertension. Denervation experiments, by both chemical and surgical intervention, have been the usual means of approaching the problem, and the results may for convenience be divided into two parts, observations at the cellular level and morphometric observations on vessel segments, and functional effects.

It is important to emphasize here that, although unilateral denervation experiments seem ideal, enabling side-to-side comparisons to be made, one should be aware than reinnervation from the contralateral SCG can occur in certain segments of cerebral arteries. A study by Handa *et al.* (1991) on rats denervated at 3–4 weeks showed that sprouting from the intact innervation begins at only 1 week after sympathectomy along the anterior cerebral artery, and at 2 weeks is also visible on the ethmoidal, posterior communicating, and internal carotid arteries of the denervated side (Fig. 9). The middle cerebral and posterior cerebral arteries seem to be relatively spared, but considerable reinnervation of the circle of Willis occurs by 8 weeks. Similar observations have been made in young rabbits, but this may not occur in rabbits denervated when adult (2.3–3 kg).

## Cellular and morphometric observations

In large pial arteries (rabbit middle cerebral artery), the effect of denervation in adult animals presents 6 weeks later as a profound modification of the internal structure of the smooth muscle cells which tend to take on the aspect of young (embryonic) cells with essentially a synthetic phenotype. In particular, a large increase in the number of organelles, especially the Golgi apparatus, rough endoplasmic reticulum, free ribosomes, coated vesicles and microtubules can be observed (Fig. 10). Frequently, the nucleus of smooth muscle cells became polylobular and indented in the denervated artery. Ultrastructural modifications also occur in the adventitia, notably increased rough endoplasmic reticulum in fibroblasts, and more extracellular collagen is present in both the adventitia and the media. According to Dimitriadou *et al.* (1988), these changes are accompanied by increased wall and media thickness, increased smooth muscle cell section, and wall to lumen ratio. Similar intracellular changes, and modest increases in the morphometric parameters, have also been reported for animals sympathectomized at an early age (7–10 days). However, no hypertrophy of the artery wall was found by another group using 'young' rabbits (9–11 weeks); on the contrary, there was a loss of wall mass in both cerebral and non-cerebral arteries by up to 20 per cent.

Studies in normotensive rats, involving sympathectomy in young animals (3–4 weeks), suggest that long-term sympathectomy leads to a decreased smooth muscle mass in equivalent artery/arteriolar segments. Baumbach and Heistad (1989) have pointed out the problems raised by the measurements made *post-mortem* on cerebral vessel dimensions, in particular suggesting that the use of the internal elastic lamina as a measure of the vessel circumference may not be totally reliable under all circumstances, and that point counting stereology in completely relaxed vessels is probably the most reliable method. With this problem in mind, one can tentatively conclude that if the decrease in wall mass and wall/lumen ratio is real for normotensive rats, it is far more marked in the case of spontaneously hypertensive rats (SHRs). Furthermore, denervation apparently leads to increased stiffness of the arterioles studied which may be related to a relative increase in the amount of non-distensible material such as collagen. Related to these findings is the reproducible observation that the sympathetic innervation of cerebral arteries, both in absolute terms and relative to other types of nerves, is significantly denser in SHRs than in normotensive rats. Indeed, it has been emphasized that this hyper-innervation precedes the development of hypertension and the medial hypertrophy.

It is important to note here that the delay between sympathectomy and the moment of measurement has been variable, and may be important in determining the precise changes reported in the vascular morphology. The age at which sympathectomy is performed also seems determinant, more profound and lasting effects being observed if the animals are denervated when still immature.

## Functional changes

One of the major hypotheses proposed by a number of groups working on rats is that the sympathetic-induced hypertrophy protects SHRs from haemorrhagic accidents, stroke, and impairment of the blood–brain barrier. This is correlated with findings by the same groups of higher wall/lumen ratio and wall mass in innervated hemispheres, but contrasts with others' results showing no significant increase in risk for the denervated hemispheres of SHRs (and, correspondingly, no increases in wall mass and media/lumen ratio – see Johansson & Nordborg, 1989). Whether these discrepancies can be explained by differences in the diets and the methods of morphometric analysis or reinnervation problems may require further investigation.

In adult rabbits, *in vivo* experiments 8–10 weeks after sympathectomy have revealed significantly decreased cortical CBF (–17 per cent), but no significant effect of infused noradrenaline which was taken to indicate absence of supersensitivity of the cerebrovascular α-adrenergic receptors. However, *in vitro*, increased responses to hypercapnia and serotonin were observed by the same workers, who had previously noted increased thickness of the media and adventitia in this model.

Thus, while it seems certain that the sympathetic nerves have a trophic influence on cerebral blood vessels, as elsewhere, some confusion exists about the exact nature of the changes occurring in the vessel wall. Perhaps an important point in this debate is the fact that by far the most substantial changes, when observed, occurred in SHRs, in which the density of sympathetic fibres is known to be significantly increased.

## Overview: role of the sympathetic nervous system

We have seen that the sympathetic innervation is widely distributed throughout the cerebral vascular system (with some predominance in rostral vessels) and the dura. Although the main transmitter appears to be noradrenaline, with NPY as a co-transmitter, it seems likely that other transmitter substances may be released under specific conditions, especially ATP, and perhaps serotonin, although the latter does not seem to be synthesized in sympathetic nerves. There seems to be rather little tonic constrictory action on cerebral vessels under 'resting' conditions, but such influence may be increased notably during hypoxic or hypercapnic dilatation, and during reflex hypertension (i.e. involving sympathetic activation). The sympathetic nervous system undoubtedly then plays a homoeostatic role, limiting flow increases and contributing importantly to the autoregulation phenomenon. However, the effects of electrical stimulation under 'resting' conditions are very moderate in amplitude, the constrictions being limited by an escape phenomenon; there may even be a predominant dilatory phase in the cerebral cortex. The sympathetic nervous system acts independently of the metabolic driving forces acting on CBF, and may be seen as a kind of negative feedback system attempting to limit potentially exaggerated changes in cerebral vascular resistance. The protection of blood–brain barrier integrity is an important aspect of this role. During haemorrhagic hypotension, however, its activity may be detrimental to cerebral blood flow maintenance and hence brain function.

There is clear evidence that sympathetic activity is likely to be subject to modulation prejunctionally by a number of neuromodulators released from brain parenchyma and other perivascular nerves (parasympathetic). It undoubtedly exerts a very profound controlling influence on CSF production by acting on both the choroid plexus blood flow and the secretory cells. It may also play a regulatory role in mast cell activity (especially those in the vascular system), but this requires working out by further investigation. Finally, but not the least important, it has far-reaching long-term influences on structure and function of the vascular wall. In its absence the phenotype of smooth muscle cells (and fibroblasts) tends to retrograde to that of synthetic (secretory) cells, with less contractile machinery. One may surmise that these effects are more pronounced and more permanent if denervation is performed in immature subjects. Although some controversy exists, there is considerable evidence that the vascular wall changes induced by the presence of the sympathetic nerves (with increased density) in hypertensive rats protect the brain tissue from the dangers of hypertension.

## B. The parasympathetic nervous system

### Anatomy

The argument concerning the existence of a parasympathetic innervation of the cerebral vessels has been confused in the past by the lack of a clear anatomical definition for this system. On the basis of the accumulated data that will be presented here the most correct view is to regard the primary autonomic innervation of the cerebral vessels to be divided into two, a classical sympathetic system (discussed in section A) and a parasympathetic that is fundamentally vasodilator. This latter system can be characterized most simply as arising from the facial nerve and distributing fibres through the sphenopalatine and otic ganglia to dilate vessels almost certainly by way of a peptidergic transmitter. In some species, including man, there may be additional variably sized microganglia located in the wall of the internal carotid artery that contribute to this system. It is defined further by the presence of one or more substances, such as acetycholine, vasoactive intestinal polypeptide (VIP) and peptide histidine (methionine in man; PHI(M)). This system is the subject of this section of the chapter.

### Origin

Nearly a century and a half has passed since it was first proposed that the greater superficial petrosal (GSP) branch of the facial or seventh cranial nerve innervated the cranial vessels. Little was done to refute or clarify this concept until Chorobski and Penfield (1932) described the anatomy of the facial nerve dilator pathway that runs from the medulla via the GSP to the pial vessels of the cat and monkey.

For some years thereafter studies examining the parasympathetic innervation of the cerebral vessels did so by defining them as cholinergic. Although this has emerged to be incomplete as a method of characterization, it has served to illustrate several important features of this system. Careful studies have shown acetychlolinesterase by histochemical means on the large cerebral vessels. Biochemical and histochemical evidence of choline acetyltransferase activity and the presence of a high-affinity choline uptake system on the vessels has also

**Table 2. Biochemical markers of cholinergic innervation[†]**

| Species | Origin | Value |
|---|---|---|
| **Choline uptake** | | |
| Cat | Middle cerebral a. | 1.32 nmol/g/min |
| | Circle of Willis | 1.38 nmol/g/min |
| | Basilar a. | 1.55 nmol/g/min |
| Rabbit | Circle of Willis | 1.4 nmol/g/min |
| | Basilar a. | 1.68 nmol/g/min |
| Rat | Major arteries | 0.17 pmol/mg protein/min |
| **ChAT activity** | | |
| Cat | Middle cerebral a. | 170 nmol/g/h |
| | Circle of Willis | 211 nmol/g/h |
| | Basilar a. | 270 nmol/g/h |
| Rabbit | Circle of Willis | 199 nmol/g/h |
| | Basilar a. | 145 nmol/g/h |
| | Major arteries | 280 nmol/g/h |
| | | 85 pmol/mg protein/min |
| Rat | Major arteries | 325 pmol/mg protein/min |
| **ACh release** | | |
| Cat | Middle cerebral a. | 9 nmol/g |
| | Circle of Willis | 19 nmol/g |
| | Basilar a. | 35 nmol/g |
| Rabbit | Circle of Willis | 8 nmol/g |
| | Basilar a. | 14 nmol/g |
| **ACh release** | | (% increase from control) |
| Cat | Middle cerebral a. | 376% |
| Rabbit | Basilar a. | 528% |

ACh, acetylcholine; ChAT, choline acetyltransferase.
[†]Modified with permission from Hamel, E. and Estrada, C. In: *Neurotransmission and cerebrovascular function II*, J. Seylaz and R. Sercombe (eds).

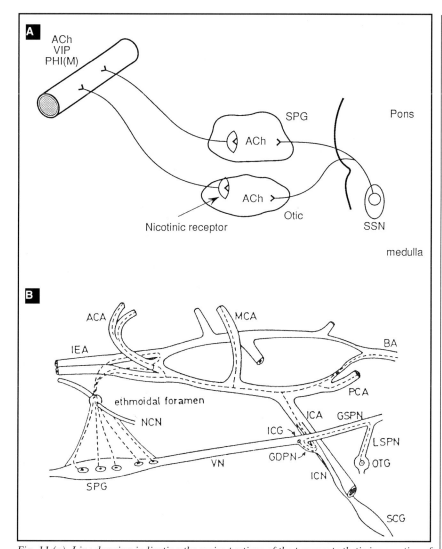

*Fig. 11 (a). Line drawing indicating the major portions of the parasympathetic innervation of the cranial vessels. The cell bodies of origin centrally are in the superior salivatory nucleus (SSN) in the pons and pass via the facial (VIIth cranial) nerve to the sphenopalatine (SPG, pterygopalatine in man) and otic ganglia. Here they synapse to release acetylcholine (ACh) or to classical nicotinic autonomic receptors. The post-ganglionic nerves contain ACh, vasoactive intestinal polypeptide (VIP) or peptide histidine isoleucine [methionine in man, PHI(M)] or various combinations of each of these. The action of each of these peptides is vasodilator.*
*(b) This line drawing taken from Suzuki et al. (J. Cereb Blood Flow Metab, 10, 399–408, 1990, by permission) is a schematic representation of the peripheral anatomical distribution of the cerebral parasympathetic (ChAT positive) innervation in the rat. ACA, anterior cerebral artery; BA, basilar artery; GDPN, greater deep petrosal nerve; ICN, internal carotid nerve; IEA, internal ethmoidal artery; MCA, middle cerebral artery; NCN, nasociliary nerve; PCA, posterior cerebral artery; SCG, superior cervical ganglion; VN, vidian nerve.*

nonadrenergic transmitters. The innervation with cholinergic nerves is regionally variable anterior to posterior and the concentration of nerves along any vessels varies, with the densest innervation being at branching points, particularly in the pial vessels. In the rat where detailed studies are available, the densest innervation is that of the anterior vessels when acetylcholine esterase is used as the marker. Marked variation in choline-acetyltransferase activity is seen between species, with the middle cerebral artery density being less than the basilar in the cat and other variables such as age being recognized.

The facial/greater superficial petrosal nerve pathway courses in the cat to the ipsilateral sphenopalatine (called pterygopalatine in man) and otic ganglia (Fig. 11a). The course that this innervation takes thereafter is via the ethmoidal nerve in the rat (Fig. 11b) but is not fully characterized in many species. Ablation of the sphenopalatine ganglion leads to a marked reduction in cholinesterase fibre density in the anterior vessels of the circle of Willis and it has been shown by direct means that nerves in the posterior circulation innervating the basilar artery also course to the sphenopalatine ganglion. Furthermore using the transganglionic tracer pseudorabies virus it has been established that the central nucleus for these fibres is in the superior salivatory nucleus of the pons.

## Transmitters and neuromodulators

Although each of the identified transmitters in this system will be dealt with separately, there is considerable evi-

provided a marker for the parasympathetic system (Table 2). Ultrastructural studies have noted a small number of agranular vesicles that store

dence for co-localization. Whether these anatomical subgroups with one, two, three or even four co-localized transmitters have functional correlates is not clear. It does seem likely, however, that this issue will emerge as crucial in coming years. There are sub-classes of cell bodies with most combinations of ACh, VIP and PHI(M).

## Acetycholine

The cerebral vessels in isolation bind [$^3$H]-QNB and thus various muscarinic receptors have been shown to exist on the cerebral vessels. Acetylcholine dilates most of the cerebral vessels either *in vitro* or *in vivo* by an atropine-sensitive receptor, an effect that is mediated via the release of an endothelium-derived relaxing factor. The extent of the atropine sensitivity has been questioned but the variations are likely to represent only species and minor methodological differences. It is of interest that acetylcholine-induced relaxation of isolated cerebral vessels is not as potent as that seen in the periphery. Local application of a cholinomimetic (carbechol) into the cerebospinal fluid causes pial vessel dilatation which is again antagonized by atropine and parenteral administration of acetylcholine increases cerebral blood flow, an effect again inhibited by atropine (see Chapter 10).

## Vasoactive intestinal polypeptide

The characterization of neuropeptides particularly over the last two decades has led to a substantial re-evaluation of virtually all neuronal systems with respect to transmitter content. The parasympathetic system is no exception. Potent vasodilator peptides have been characterized and identi-fied in the parasympathetic nerves of the cranium, particularly vasoactive intestinal polypeptide (VIP) and co-existent with it, peptide histidine iso-leucine (methionine in man; PHI(M)). VIP is a 28 amino acid basic polypeptide that was first isolated from the porcine duodenum. It belongs to a structural super-family of peptides along with glucagon, secretin and gastrin inhibitory peptide. Human and porcine VIP are identical. The family is characterized by he-lodermin/helospectin-like peptides that are distributed in the central nervous system and in endocrine cells, such as the C cells of the thyroid.

It has been clearly established that the large cerebral vessels and cortical pial vessels have a rich VIPergic innervation that can be immunohistochemically identified and measured by radioimmunoassay. Indeed using ultrastructural techniques it can be seen that vasoactive intestinal polypeptide is found in large dense core neuronal vesicles in perivascular nerve terminals on the vessels. Two important features characterize this innervation, first, it is predominantly in the anterior segments of the circle of Willis and secondly, the fibres may be seen to follow the vessels and penetrate into the parenchyma. However, it has been shown that VIP-immunoreactive nerves that innervate the pial vessels may arise at least in part from intra-cortical neurons. The pattern of innervation of the vessels has further been characterized as having a spiral distribution with respect to the lumen and importantly this innervation is seen in human vessels.

The origin of the nerves is essentially as it is for the cholinergic system. In the guinea pig, the cerebrovascular VIP-containing nerves originate in the microganglia of the internal carotid while those of the extracerebral circulation are located in the spheno-palatine and otic ganglia. In the rat, the major origin of the fibres is from the sphenopalatine ganglion itself while it has been shown in the dog that there may be an additional contralateral contribution from the sphenopalatine ganglion.

## Peptide histidine isoleucine (methionine)

The third major marker for the parasympathetic system is peptide histidine isoleucine (methinonine in man; PHI(M)). This is cleaved from the same prepropeptide as VIP and there is at least a 50 per cent sequence homology. PHI and PHM are almost identical with a two amino acid residue difference (92: lysine for arginine; 107: methionine for isoleucine). Immunohistochemical studies have confirmed the existence of PHI(M)-like immunoreactivity on cerebral vessels and the distribution is essentially parallel to that of VIP. PHI(M) elicits a less potent dose-dependent vasodilatation than VIP *in vitro* and *in vivo*. Microapplication of PHI(M) dilates both arteries and veins *in situ*.

## Effect of parasympathetic blockade

Given that the cranial parasympathetic outflow to the cerebral vessels is via the facial nerve and is marked by three neurotransmitters or neuromodulators (acetylcholine, VIP and PHI(M)), what is the effect of blocking this outflow? The responses that characterize normal cerebral blood flow

**Table 3. Influence of extrinsic neural systems on basic cerebrovascular physiology**

| System | Resting | | Autoregulation | | Hypercapnia | Hypoxia | Other |
|---|---|---|---|---|---|---|---|
| | CBF[†] | CGU[‡] | Normal limits | Hypertension | | | |
| Sympathetic | Nil or small reduction | Nil | Extends upper limits | Reduces flow increase | Nil | Nil or reduction | Trophic influences |
| Parasympathetic | Nil | Nil | Nil | – | Nil | Nil | ? threat |
| Trigeminal | Nil | Nil | Nil | Promotes vasodilation | Nil | Nil | Protective role |

†CBF, cerebral blood flow; ‡CGU, cerebral glucose utilization.

are the hypercapnic vasodilator response, the autoregulatory response to changes in blood pressure and the hypoxic vasodilator response (see Chapter 9).

## Resting CBF and autoregulation

Few experiments have addressed the question of the facial nerve and its role in autoregulation. Although there is some indirect evidence using cholinesterases in the baboon that such a mechanism may play a role in autoregulation, this has not been confirmed in the cat by direct observations. Sectioning the facial nerve does not alter autoregulation in this latter setting. In addition no data suggest that in carefully prepared animals resting cerebral blood flow or glucose utilization are affected by section of the facial nerve.

## Hypercapnia and hypoxia

There is limited evidence in the baboon that after seventh cranial nerve section the vasodilator responses to hypercapnia and hypoxia are blocked. Such a role has not been confirmed by others in the baboon, cat, dog, rat or rabbit and now seems to be unlikely role. In contrast, by pharmacological techniques it has been suggested that a cholinergic mechanism may have a role in hypercapnic va-

sodilatation although this point is controversial, or even that a cholinomimetic may increase cerebral blood flow without altering cerebral glucose utilization. Importantly, Seylaz and colleagues (1988) have demonstrated that the main parasympathetic outflow ganglia, the sphenopalatine ganglion, may be ablated without any alteration of hypercapnic vasodilatation (Table 3).

# Stimulation of the parasympathetic nerves

## In vitro stimulation

To further characterize the relationship between anatomically defined parasympathetic nerves and their transmitters, studies have carefully examined the release of the various marker substances of this system in vitro. Incubation of cerebral vessels from either cats or rabbits with labelled choline chloride, a precursor in the synthesis pathway for acetylcholine, permits measurement of labelled acetylcholine when nerves surrounding the vessels are stimulated. This response may be blocked if calcium is removed from the buffer or tetrodotoxin added suggesting an active neural process. The method of

transmural nerve stimulation has been employed in isolated vessel preparations to examine the possible role of identified putative transmitters. Except in porcine vessels, relaxation is only seen if the vessels are precontracted. Transmural nerve stimulation leads to contraction of rabbit, goat and human pial vessels in contrast to relaxation in the cat, dog and pig (Table 1). For the cat and pig it is clear that a non-adrenergic non-cholinergic dilator mechanism is operating and available data suggest that it is mediated by VIP. Similar studies have also been used to determine what substances are released when cerebral nerves are stimulated directly in vitro. VIP is released when cerebral arterial nerves are stimulated. Indeed, although no antagonist to VIP is available, specific VIP antiserum has been used to inhibit non-cholinergic/non-adrenergic dilator responses in vivo study has demonstrated local cortical release of VIP with facial nerve stimulation. Finally, a recent in vivo study has demonstrated local cortical release of VIP with facial nerve stimulation. Finally, a recent in vivo study has demonstrated local cortical release of VIP with facial nerve stimulation. This release is blocked by hexamethonium demonstrating release of VIP in the context of a classical nicotinic autonomic ganglion in the cerebral circulation.

## Effect of direct parasympathetic stimulation on cerebral blood flow

### Facial nerve stimulation

Direct stimulation of the facial nerve in humans leads to an increase in total cranial blood flow as does facial nerve nucleus stimulation in the monkey or an area just dorsal to it in the cat. Stimulation of the nerve in the primate and cat with its proximal end intact, that is attached to the brainstem, reduces cerebral blood flow while stimulation of the sectioned distal segment will increase it. In the baboon using the $^{133}$Xe clearance method it has been shown that facial nerve stimulation increases cerebral blood flow and a similar effect is seen in the dog, rat and rabbit. In the cat there are two conflicting studies: whereas it was demonstrated in the α-chloralose anaesthetized cat using radiolabelled microspheres that greater superifical petrosal nerve stimulation did not alter cerebral blood flow, it has been demonstrated in similarly anaesthetized animals with iodoantipyrine or laser Doppler flowmetry that such stimulation can increase cerebral blood flow. The major difference in these studies was in the dissection performed to stimulate the nerve. The dissection for the greater superficial petrosal nerve is much more extensive with a significant removal of extracranial bone and therefore blood loss compared to the simple posterior fossa approach used by Goadsby. Consequently the published control blood flow values in the

microsphere study were much lower than those previously reported from that laboratory or noted elsewhere for similarly anaesthetized animals. The lack of a response was then most likely due to the experimental preparation consequent upon the extensive dissection. On balance it is clear that the facial nerve system can increase cerebral blood flow when activated. In addition it does so without altering metabolic activity as reflected by stable glucose utilization and sagittal sinus oxygen content. These responses are mediated through a classical parasympathetic ganglion as they can be blocked by hexamethonium and probably use vasoactive intestinal polypeptide as the major transmitter of the system.

Finally, it has also been suggested that the greater superficial petrosal nerve dilator fibres distribute with the trigeminal system peripherally. There is no anatomical connection to support this notion in the periphery, although a reflex connection through the brainstem does exist. This latter system can alter cerebral blood flow and in part mediates both the cerebral and extra-cerebral vasodilatation seen when the trigeminal ganglion is stimulated, although at least some 20 per cent of the extracerebral response may be due to antidromic activation of the trigeminal system (see Section C, this chapter). Moreover, at least two other major brainstem structures entrain the facial nerve dilator system to increase extracerebral blood flow when stimulated, the nucleus locus coeruleus and the dorsal raphe nucleus.

## Sphenopalatine (pterygopalatine) ganglion

Studies of the peripheral ganglion mediating facial nerve vasodilatation have included peripheral reflex (trigeminal ganglion) and central structure (locus coeruleus) stimulation. It is clear that facial nerve mediated effects can be blocked by ganglion removal and that this same response is again VIP mediated. Indeed, stimulation of the sphenopalatine ganglion in cats and rats have been shown to produce strong frequency-dependent cerebral vasodilator responses using at least three different methods of measurement of cerebral blood flow. Importantly, cerebral glucose utilization and tissue pCO$_2$ were not increased and the response is thus truly neurogenic.

*In summary*, the cranial parasympathetic pathway to the cerebral vessels arises in the superior salivatory nucleus in the pons, it traverses the facial nerve joining the greater superficial petrosal nerve to be distributed to the vessels after synapsing chiefly in the sphenopalatine or otic ganglia (Fig. 11a and b). A variable small number of fibres in different species (including man) have this peripheral synapse located in microganglia on the wall of the internal carotid artery particularly near the carotid siphon. The transmitters contained in this system are acetylcholine, VIP and PHI(M). The ganglionic transmission in the periphery is mediated by a classical parasympathetic nicotinic ganglion while current data would suggest that VIP is the major neuroeffector substance at the nerve/smooth muscle junction. The pathway does not play a role in either hypercapnic or hypoxic vasodilator responses or autoregulatory responses to changes in arterial perfusion pressure. The system can be activated by either direct stimulation or via connections with other important central neural vasoactive nuclei to increase cerebral blood flow independent of cerebral metabolic needs.

## Physiological significance

Although there have been considerable advances in understanding the basic capabilities and connectivity of the parasympathetic innervation of the cerebral circulation, these data have not indicated a clear physiological role for the system. The nerves are not directly involved in the most basic cerebrovascular responses, such as hypoxic or hypercapnic vasodilatation nor do they appear to play a role in autoregulation (Table 3). Their effects are, however, independent of direct metabolic intervention. This latter fact suggests a role in times of threat such as during ischaemia or in vasospastic conditions such as subrachnoid haemorrhage. The parasympathetic system is ideally placed to be engaged to increase cerebral blood flow when ordinary metabolic driving factors are impaired. This protection may, however, be regionally variable since the posterior circulation innervation with VIP is much less than that seen anteriorly. This finding may have implications in situations where predominantly posterior changes are reported such as migraine (see Chapter 14). As yet this question has not been adequately addressed. It may be too narrow a perspective to consider only a vasomotor function for these nerves as there is evidence that cholinergic mechanisms can alter capillary permeability including the movement of amino acids. Whatever their function is ultimately revealed to be it is now clear that the evidence for the existence of the parasympathetic innervation of the cerebral vessels is beyond question.

## C. Trigeminal innervation of the cerebral vessels

The trigeminocerebrovascular system is in a unique, indeed pivotal, position in the cerebral vasculature. It is the sole sensory (afferent) innervation of the cerebral vessels and is thus crucially placed to affect all the systems that are discussed in this monograph. In this section of the chapter the very considerable and recent explosion of knowledge in this system will be analysed and inferences will be made as to its function. A discussion of its possible dysfunction and role in diseases, specifically migraine, is taken up in Chapter 14.

### Anatomy of the trigeminocerebrovascular system

#### Structures

Of the systems discussed in this chapter the trigeminal system is certainly the simplest to define. It consists of those neurons innervating the cerebral vessels whose cell bodies are located in the trigeminal ganglion. The ganglion contains bipolar cells with the peripheral fibre making synaptic connection with the vessels, and other cranial structures, and the centrally projecting fibre synapsing in the low brainstem or high cervical cord (Fig. 12a). Initial tracing studies identified the trigeminal nerve as the major afferent pathway for pain from the vessels and dura mater. Some years later with a resurgence of interest in this system and its relationship to headache the innervation of the pial vessels from the trigeminal system was confirmed with sensitive tracing techniques. These fibres are predominantly found in the first (ophthalmic) division of the trigeminal nerve and have widely ramifying axons that may innervate several vessels ipsilaterally. These systems are, in evolutionary terms, well conserved and may be found in monkeys, cats, guinea-pigs, gerbils and rats. Some of the projections have been noted to involve both cerebral (middle cerebral artery) and extra-

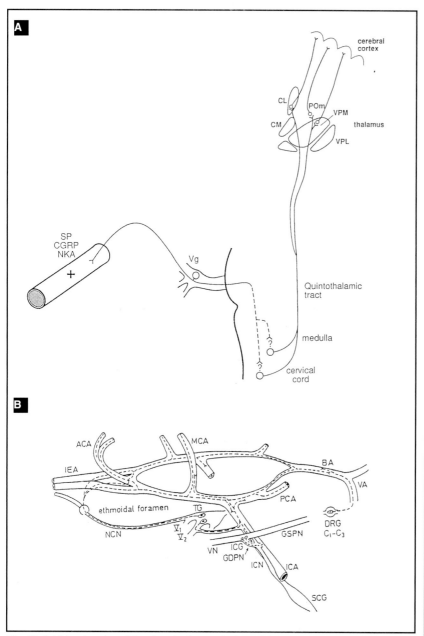

*Fig. 12. (a) Line drawing of the trigeminal system in its peripheral extent. The cell bodies for this system are found in the trigeminal ganglion (Vg) and are bipolar. Substance P (SP), calcitonin gene-related peptide (CGRP) and neurokinin A (NKA), all vasodilator peptides, are released peripherally. Central connections employ as yet undetermined transmitters in the high cervical cord and low medulla, synapsing to relay input to the thalamus. Nuclei of the thalamus: CL, centrolateral; CM, centromedian; POm, medial nucleus of the posterior complex; VPL, ventroposterolateral; VPM, ventroposteromedial. (b) Schematic drawing of the peripheral distribution of the trigeminal cerebrovascular nerves. The abbreviations are as for Fig. 1b with V1, ophthalmic division of the trigeminal nerve; V2, maxillary division of the trigeminal nerve. From Suzuki et al. (1989) by permission.*

cerebral (middle meningeal artery) vessels. The trigeminal innervation is predominantly to the forebrain but extends posteriorly to the rostral basilar artery whereas the more caudal vessels are innervated by $C_2$ and $C_3$ dorsal roots that also synapse with the central trigeminal neurons (Fig. 12b).

## Transmitters

The description and characterization of neuropeptide transmitters in the trigeminal neurons has greatly aided the study of this system. Four substances, substance P (SP), calcitonin gene-related peptide (CGRP), neurokinin A (NKA) and cholecystokinin (CCK), have been identified in trigeminal neurons (Table 3). These are all powerful vasodilator peptides and their characterization has not only provided an anatomical tool but also more recently a marker of functional activity for the trigeminal system.

### Substance P

The initial characterization of SP-containing fibres from the trigeminal nerve involved an elegant immunocytochemical demonstration that sectioning of the nerve eliminated, at least ipsilaterally, SP-like immunoreactivity. These fibres are unmyelinated with neuropeptide-containing axonal swellings and an abundance of mitochondria. Interestingly, SP-like immunoreactivity has also been demonstrated in fibres in the dura mater with a particularly prominent density around the superior sagittal sinus. SP is an undecapeptide first described some 60 years ago and initially suggested to be involved in nociceptive transmission. SP is a potent vasodilator in many vascular beds including the cerebral circulation and its effect is endothelium-dependent.

*In vitro* studies with cow and guinea-pig cerebral vessels have shown that SP can be released by potassium, capsaicin or electrical stimulation and that this is abolished if calcium ions are removed from the superfusion buffer. Isolated cerebral vessels from man, cats and guinea-pigs relax in a concentration-dependent manner with SP application. SP also dilates both arteries and veins *in situ*. Interestingly, SP can induce protein extravasation in the periphery and intradermal injection produces wheal, flare and itching. A similar response is seen in the isolated dura mater with protein extravasation and mast cell degranulation *in vitro*.

## Neurokinin A (NKA)

The β-preprotachykinin that produces substance P may be differentially cleaved to yield a decapeptide with a similar profile of action and localization in the trigeminal system, NKA. SP and NKA coexist in perivascular nerve fibres in both peripheral and cerebral vessels. NKA can relax cerebral vessels both *in vitro* and *in vivo* although it is only one-tenth as potent as SP. The receptor that is activated by NKA has been termed neurokinin-2 (NK-2; substance K in the old tachykinin terminology) and has been sequenced and cloned. In this new nomenclature the NK-1 receptor is preferentially activated by SP, and the NK-3 receptor by a further product of the preprotachykinins, neurokinin B (NKB). This latter peptide does not seem to be located in the trigeminal nerves.

## Calcitonin gene-related peptide (CGRP)

CGRP is the third, most potent and perhaps the most interesting of the neuropeptides that have been mapped to the trigeminal system. It is derived by alternative processing of the calcitonin gene messenger RNA. Biochemical determinations of CGRP by HPLC and radioimmunoassay have demonstrated the substance in cat cerebral vessels. This has been confirmed by immunocytochemical characterization of fine varicose CGRP-containing fibres on cerebral vessesl that are not found if the animal has had a trigeminal nerve section. The trigeminal ganglion contains numerous CGRP immunoreactive cells and CGRP can be released by these cells in culture by potassium. Some of the trigeminal ganglion CGRP-containing cell bodies also contain SP and NKA while there is a distinct subpopulation that projects to the cerebral vessels that is enriched with CGRP. CGRP is the most potent vasodilator transmitter thus far identified in the cerebral circulation and its action is endothelium-independent (see Chapter 10). This dilatation is paralled by an increase in vessel wall cyclic-AMP. However, in contrast to SP, CGRP does not dilate cerebral veins although CGRP-containing fibres are found on the veins.

## Cholecystokinin (CCK)

CCK-8-containing fibres have been identified on the cerebral vessels and are reduced by either capsaicin treatment or by trigeminal ganglionectomy. There is also co-localization of SP and CCK-8 in the trigeminal ganglion cells that project to the cerebral vessels. There is no known function for these fibres.

---

## Effect of blocking the trigeminal system

### Resting cerebral blood flow

Cerebral blood flow measured with iodoantipyrine and tissue autoradiography is not altered in the cat after trigeminal ganglion section. Indeed, after unilateral section, flow is identical to homologous contralateral cortex. Furthermore glucose utilization is not affected by trigeminal ganglion section and thus the usual close relationship between flow and metabolism in not disturbed.

### Hypercapnia and hypoxia

In a series of well conceived studies in the rat it has been shown using the radiolabelled microsphere technique that trigeminal ganglionectomy does not alter cerebrovascular responsiveness to hypercapnia. However, in the cat the hyperaemia following temporary occlusion of the brachiocephalic and left subclavian vessels is mediated in part (some 60 per cent) by a trigeminal mechanism. The nature of the transmitters and mechanisms for the this response remains to be determined.

### Autoregulation

Induction of hypertension outside the autoregulatory range or induction of a tonic/clonic seizure with bicuculline in trigeminal ganglionectomized animals leads to a reduced vasodilator response when compared with the non-ganglionectomized side.

The effect is seen in the cortical grey matter not in the diencephalon or white matter. These experiments cleverly delineated that the responses observed were due to the axon-reflex part of the trigeminovascular system since trigeminal root section prior to entry of the nerve to the brainstem does not reduce the effect whereas ganglionectomy does. Similarly, trigeminal fibres appear to limit noradrenaline-induced constrictor responses in the pia and play a role in restoring constricted vessels to their normal calibre. Normal perfusion range autoregulation is unimpaired by trigeminal root section. These data suggest that the trigeminal system by its sensory role detects changes in perfusion and has, in the setting of the of raised pressure, a facilitatory role. It is tempting to speculate that this vasodilator role balances the well-described augmentation of vasoconstriction mediated by the sympathetic nerves, an imbalance of either mechanism having implications for pathophysiology in humans in such conditions as stroke and prolonged seizures.

## Trigeminal system stimulation

### In vitro

There is limited but convincing evidence that stimulation of pial perivascular trigeminal fibres *in vitro* can result in the release of substance P. This is likely to represent depolarization-dependent activation of trigeminal neurons and thus satisfy one of the criteria for substance P being a neurotransmitter in this system. Such data are not directly available for CGRP or NKA.

### In vivo

#### Craniovascular circulation

It has been known for some years that stimulation of the trigeminal ganglion in man by either thermocoagulation or injection of alcohol can cause facial flushing usually in the division or divisions appropriate to the manipulation. In addition it has been shown that this flushing is accompanied by an increase in facial temperature of 1–2 °C. Corresponding with this flush there is an increase in the dilator peptides SP and CGRP in the

external jugular but not in the peripheral circulation. Such changes are also seen in the cat. In addition, trigeminal ganglion stimulation in the cat leads to a diminution of carotid arterial resistance, with increases in both blood flow and facial temperature through a predominantly reflex mechanism. The afferent limb of this arc is the trigeminal nerve and the efferent the facial/greater superficial petrosal nerve (parasympathetic) dilator pathway. About 20 per cent of the dilatation remains after facial nerve section and is probably mediated by antidromic activation of the trigeminal system directly. The portion running through the parasympathetic outflow traverses the sphenopalatine (ptery-gopalatine) and otic ganglia and employs vasoactive intestinal polypeptide as its transmitter. Essentially compatible results were obtained in a recent human study in which the trigeminal ganglion was stimulated by thermocoagulation (treatment for trigeminal neuralgia).

Using the isolated cross-perfused dog brain it has been shown that stimulation of the trigeminal ganglion can increase cerebral blood flow by 19 per

**Table 4. Extrinsic neural systems: anatomy, transmitters and effect of activation**

|  | Effect of stimulation on CBF[†] | Ganglia | Transmitters or neuromodulators |
|---|---|---|---|
| Sympathetic NS | ↓ | Superior cervical (± stellate) | Noradrenaline<br>Neuropeptide Y (NPY)<br>Peptide YY (PYY)[§] |
| Parasympathetic NS | ↑ | Sphenoplatine<br>Otic<br>IC ganglia[‡] | Acetylcholine<br>Vasoactive intestinal polypeptide (VIP)<br>Peptide histidine isoleucine (methionine; PHI(M)) |
| Trigeminal nerve | ↑ | Trigeminal | Substance P<br>Calcitonin gene-related peptide (CGRP)<br>Neurokinin-A (NKA)<br>Cholecystokinin-8 (CCK-8) |

†CBF, cerebral blood flow; ‡IC ganglia, internal carotid ganglia; §May be cross-reactivity with NPY.

cent, a response not al-
tered by distal nerve section. Measured as internal
carotid blood flow in the
monkey, trigeminal gan-
glion stimulation does not
alter total flow, whereas in
another study looking at
regional cerebral blood
flow certain areas have increased flow. Unilateral
ganglion stimulation in-
creases frontal and parietal
cortical blood flows bilaterally without affecting
other cortical areas, white
matter or brainstem flow.
These flow changes in the
cerebral circulation can be
blocked by facial nerve section. Stimulation of the superior sagittal sinus, a pain-
sensitive centrally-placed
intracranial structure (see
Chapter 14), leads to a
quantitatively greater increase in cerebral blood
flow, again measured with
iodoantipyrine and brain
dissection in the cat.

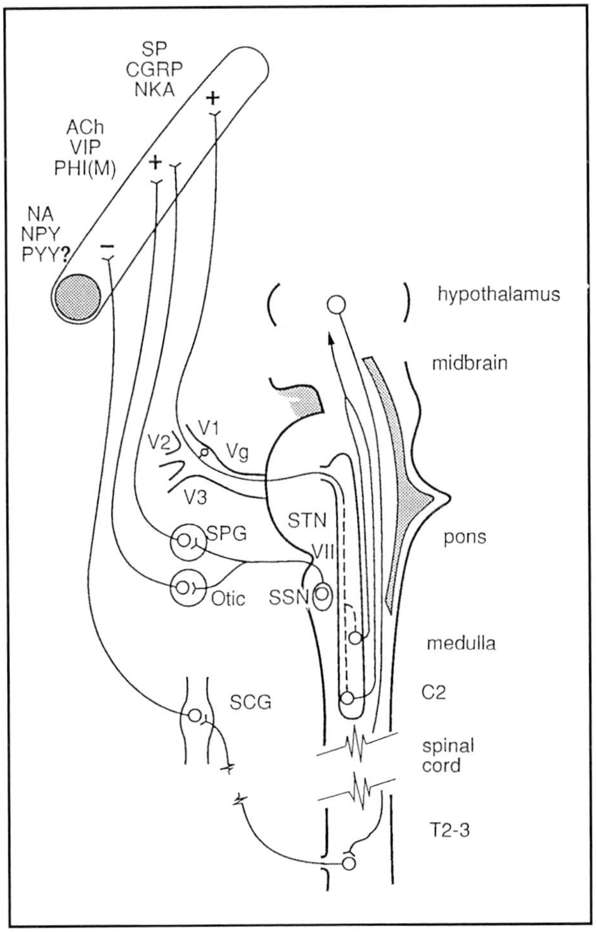

*Fig. 13. A schematic summary of the extrinsic innervation of the cranial circulation. The sympathetic system has its cell bodies of origin in the hypothalamus passing to the thoracic cord and synapsing in the intermediolateral cell column. The output is then to the stellate and superior cervical ganglion (SCG) and thence to the vessels releasing noradrenaline (NA), neuropeptide Y (NPY) and possibly peptide YY (PYY). The action is vasoconstrictor (−). The trigeminal system has its cell bodies of origin in the trigeminal ganglion (Vg) and these are bipolar. The three divisions are marked (V1, ophthalmic; V2 maxillary; V3, mandibular) and it is V1 that is largely responsible for craniovascular innervation. The vasodilator (+) transmitters substance P (SP), calcitonin gene-related peptide (CGRP) and neurokinin A (NKA) are released peripherally. The central termination of these neurons is within the spinal tract of the trigeminal nerve (STN) and is on both medullary and high cervical neurons (C2). The parasympathetic system arises from cell bodies in the superior salivatory nucleus (SSN) passing out to the sphenopalatine (SPG) and otic ganglia via the facial (VIIth cranial) nerve. Acetylcholine (ACh), vasoactive intestinal polypeptide (VIP) and peptide histidine isoleucine (methionine, PHI(M)), are all vasodilator (+) peptides.*

## Dura mater and mast cells

The dura and vessels of the cerebral
circulation are the protection and
first portal of entry of any compounds
exogenous to the central nervous system. The dura mater is an important
pain-sensitive intracranial structure
that has recently been the subject of
detailed study. The dura mater is innervated by branches of the trigeminal nerve whose cell bodies are to be
found in the trigeminal ganglion. The
innervation is somatotopic with the
first division giving rise to fibres that
innervate the anterior fossa and and

tentorium cerebelli, the second division innervating the orbital roof and
the third structures in the middle
fossa. The tentorial nerve arises from
the ophthalamic nerve proximal to
the superior orbital fissure and runs
along the edge of the tentorium. The
supply from the mandibular nerve
enters the cranial vault by way of the
foramen spinosum arising from an
extracranial portion of the nerve. The
supratentorial dural sympathetic innervation arises from the ipsilateral
superior cervical ganglion as does
that innervating the dura of the posterior fossa. In the posterior fossa the
dural innervation has important con-

tributions from the trigeminal ganglion and from the upper two dorsal
root ganglia. The dural innervation
has been further characterized by
examining the distribution of neuropeptide immunoreactivity around
the superior sagittal sinus. There is a
dense plexus of SP and CGRP-containing nerves around the sinus that is
equally distributed antero-posteriorly
whereas the distribution of VIP varies
from dense anteriorly to sparse posteriorly, as it does for the vessels. Moreover, there is evidence that the VIP-
ergic nerves arise from the same site
as those for the cerebral circulation,
the sphenopalatine ganglion.

Recent exciting data have linked the trigeminal system to a neuroexocrine function. Ultrastructural evidence has suggested that the sympathetic nerves arising from the superior cervical ganglion may make functional connections with the mast cells in the walls of large cerebral vessels (see section A, this chapter). It has been shown in the rat that unilateral trigeminal ganglion stimulation can cause mast cell degranulation in the dura mater and tongue. The possible interaction between these highly humorally active cells and the protective covering of the brain is of great interest. The cells are located randomly over the rat dura with some concentration at the vessels but in monkeys, man and cats they are only located at the branching of vessels and are thus well placed for some regulatory role. The localization of various mediators of inflammation and the stabilization and protection of this degranulation by antimigraine drugs forms a leading edge of current basic research in the trigeminal system (see Chapter 14).

## Possible role

The trigeminocerebrovascular system defined as the innervation of the cerebral circulation by trigeminal neurons is well-placed for an important role in the physiology of the brain

circulation (Fig. 12). The trigeminal system is placed as a watchdog and receives early warning of impending changes in cerebral haemodynamics, particularly threats to the circulation. Its function does not appear to be in the maintenance of resting cerebral flow or in the ordinary operation of flow/metabolism coupling (Table 3). Autoregulation and hypercapnic vasodilatation also proceed without apparent trigeminal interaction. What then are its roles? The trigeminal system responds to threat to the integrity of the cerebral blood supply. In situations such as hypoxia and post-seizure, where some insult has occurred, it ensures that an adequate hyperaemia will redress the imbalance in metabolic fuels by mediating a large neurogenic vasodilatation. This would appear to be exclusively a function of the axon-reflex portion of the trigeminal connections although its powerful central synapse with parasympathetic neurons, capable also of vasodilator responses, may play an as yet undetermined role in these settings (Fig. 13).

In terms of pathophysiology and disease processes, the trigeminal system is certainly involved in migraine as it is clear that it is activated both in terms of the pain that patients report and the recent description of neuropeptides from the trigeminal nerve being

released locally in the cranial circulation. The issue will be more fully addressed in Chapter 14. It has also become clear that in the subarachnoid haemorrhage, where blood released in the cerebrospinal fluid from a burst aneurysm or arteriovenous malformation causes spasm of the cerebral vessels and thus compromises the cerebral circulation, again the trigeminal system is activated. Markers of trigeminal activation, such as CGRP, can again be found in the cranial circulation and CGRP is discharged from trigeminal neurons as evidenced by reduced immunohistochemical marking in patients who have died (Chapter 14). Indeed the trigeminovascular system and its central connections are most likely to be a sensor/effector system that protects the brain when its circulation is threatened. It will be the further characterization of this system (Fig. 13) that will enable its manipulation to augment its role by therapeutic intervention which in many situations will be life saving.

**Acknowledgements:** The work of the author (P.J.G.) included in this chapter has been supported by the National Health and Medical Research Council of Australia and by grants from Warren and Cheryl Anderson, The J.A. Perini Family Trust, the Basser Trust and Australian Brain Foundation. P.J.G. is a Wellcome Senior Research Fellow.

## Summary

The extrinsic nervous control of the cerebral circulation is described in terms of the three types of innervation, i.e. the sympathetic (motor), the parasympathetic (motor) and the trigeminovascular (sensory-motor) systems. The sympathetic nerves originate mostly in the superior cervical ganglion, and to a minor extent in the stellate ganglion, and are distributed to cerebral arteries, arterioles and veins with somewhat greater density in rostral vessels. Although there appears to be a small or negligible resting sympathetic tone, and the stimulation of sympathetic nerves leads to only modest decreases in cerebral blood flow, these nerves play a certain role in protecting the cerebral circulation from reflex hypertensive insult and contribute to the autoregulation phenomenon. They are also involved in the control of cerebrospinal fluid production and they probably, in conjunction with parasympathetic and trigeminal nerves, influence the dural and cerebrovascular mast cells. The sympathetic nervous system also exerts important trophic effects on the vascular smooth muscle cells, notably by modifying their phenotype.

The parasympathetic nerves originate in the sphenopalatine and otic ganglia, as well as micro-ganglia on the internal carotid artery. Their influence is vasodilatory, but they do not apparently exert any basal tonic effect. Like the sympathetics, their distribution (both arteries and veins) is denser on rostral vessels. Their role is largely unknown, but they may well be involved in reflex mechanisms augmenting cerebral blood flow without metabolic changes.

The trigeminovascular system, a purely peptidergic innervation, consists of bipolar sensory neurons whose peripheral fibres are distributed on cerebral vessels (arteries and veins) and whose central fibres are projected to and synapse in the low brainstem or high cervical cord. Most of the neurons are situated in the trigeminal ganglion which innervates rostral vessels, but vessels caudal to the rostral basilar artery are innervated by cells in the $C_2$ and $C_3$ dorsal roots. Although again there appears to be no tonic action of this system, its excitation leads to increased cerebral blood flow (probably partly through a reflex mechanism) and its destruction seems detrimental to flow regulation in situations involving vasoconstriction. Excitation of the trigeminal ganglion also causes inflammatory changes in the vessels due to degranulation of the mast cells. The role of this system is as yet largely conjectural: it is suggested to respond to threats to the integrity of the blood supply (via axon reflexes?) and it apparently participates in pathological processes involving pain, such as migraine. The probable neurotransmitters and neuromodulators of these three systems of innervation are described.

# References for further reading (including references cited)

## General reviews

Busija, D.W. & Heistad, D.D. (1984): Factors involved in the physiological regulation of the cerebral circulation. *Rev. Physiol. Biochem. Pharmacol.* **101,** 161–211.

Edvinsson, L., Mackenzie, E.T. & McCulloch, J. (1993): *Cerebral blood flow and metabolism.* New York: Raven Press.

Mackenzie, E.T., Seylaz, T. & Bes, A. (eds.) (974): *Neurotransmitters and the cerebral circulation.* New York: Raven Press.

Mackenzie, E.T. & Scatton, B. (1987): Cerebral circulatory and metabolic effects of perivascular neurotransmitters. *CRC Crit. Rev. Clin. Neurobiol.* **2,** 357–419.

Olesen, J. & Edvinsson, L., eds. (1988): *Basic mechanisms of headache.* Amsterdam: Elsevier.

Owman, Ch. & Hardebo, J.E., eds. (1986): *Neural regulation of brain circulation.* Amsterdam: Elsevier.

Seylaz, J. & Sercombe, R., eds. (1989): *Neurotransmission and cerebrovascular function*, Vol. II. Amsterdam: Elsevier.

## Part A. The sympathetic nervous system

### Anatomy

Arbab, M.A.R., Wiklund, L. & Svendgaard, N.A. (1986): Origin and distribution of cerebral vascular innervation from superior cervical, trigeminal and spinal ganglia investigated with retrograde and anterograde WGA-HRP tracing in the rat. *Neuroscience* **19,** 695–708.

Arbab, M.A.R., Wiklund, L., Delgado, T. & Svendgaard, N.A. (1988): Stellate ganglion innervation of the vertobrobasilar arterial system demonstrated in the rat with anterograde and retrograde WCrA-MRP tracing. *Brain Res.* **445,** 175–180.

Gibbins, I.L. & Morris, J.L. (1988): Coexistence of immunoreactivity to neuropeptide Y and vasoactive intestinal peptide in non-noradrenergic axons innervating guinea-pig cerebral arteries after sympathectomy. *Brain Res.* **444,** 402–406.

Hedger, J.H. & Webber, R.H. (1976): Anatomical study of the cervical sympathetic trunk and ganglia in the albino rat (*Mus norvegicas albinos*). *Acta Anat.* **96,** 206–217.

Keller, J.T., Marfurt, C.F., Dimlich, R. & Tierney, B.E. (1989): Sympathetic innervation of the supratentorial dura mater of the rat. *J. Comp. Neurol.* **290,** 310–321.

Suzuki, N., Hardebo, J.E., Kahrström, J. & Owman, Ch. (1990): Neuropeptide Y coexists with vasoactive intestinal polypeptide in parasympathetic cerebrovascular nerves originating in the sphenopalatine, otic, and internal carotid ganglia of the rat. *Neuroscience* **36,** 507–519.

## Sympathectomy

Busija, D.W. & Heistad, D.D. (1984): Effects of activation of sympathetic nerves on cerebral blood flow during hypercapnia in cats and rabbits. *J. Physiol.* **347,** 35–45.

Fitch, W., Mackenzie, E.T. & Harper, A.M. (1975): Effects of decreasing arterial blood pressure on cerebral blood flow in the baboon. Influence of the sympathetic nervous system. *Circ. Res.* **37,** 550–557.

Harper, A.M., Deshmukh, V.D., Rowan, J.O. & Jennett, W.B. (1972): The influence of sympathetic nervous activity on cerebral blood flow. *Arch. Neurol.* **27,** 1–6.

Kissen, I. & Weiss, H.R. (1989): Cervical sympathectomy and cerebral microvascular and blood flow responses to hypocapnic hypoxia. *Am. J. Physiol.* **256,** H460–H467.

Lacombe, P., Miller, M.C. & Seylaz, J. (1985): Sympathetic regulation of cerebral blood flow during reflex hypertension. *Am. J. Physiol.* **249,** H672–H680.

Szabo, L., Kovach, A.G.B. & Babosa, M. (1983): Effect of autonomic blocking agents on local cerebral blood flow in the rat. *Acta Physiol. Hung.* **62,** 191–204.

## Sympathetic stimulation

Alonso, M., Arribas, S., Marin, J. Balfagon, G. & Salaices, M. (1991): Presynaptic $M_2$-muscarinic receptors on noradrenaergic nerve endings and endothelium-derived $M_3$ receptors in cat cerebral arteries. *Brain Res.* **567,** 76–82.

Auer, L.M. & Ishiyama, N. (1986): Pial vascular behaviour during bilateral and contralateral cervical sympathetic stimulation. *J. Cereb. Blood Flow Metab.* **6,** 298–304.

Baumbach, G.L. & Heistad, D.D. (1983): Effects of sympathetic stimulation and changes in arterial pressure on segmental resistance of cerebral vessels in rabbits and cats. *Circ. Res.* **52,** 527–533.

Bill, A. & Linder, J. (1976): Sympathetic control of cerebral blood flow in acute arterial hypertension. *Acta Physiol. Scand.* **96,** 114–121.

Edvinsson, L., Falck, B. & Owman, C. (1977): Possibilities for a cholinergic action on smooth musculature and on sympathetic axons in brain vessels mediated by muscarinic and nicotinic receptors. *J. Pharmacol. Exp. Ther.* **200,** 117–126.

Gaw, A.J., Wadsworth, R.M. & Humphrey, P.P.A. (1990): Neurotransmission in the sheep middle cerebral artery: modulation of responses by 5-HT and haemolysate. *J. Cereb. Blood Flow Metab.* **10,** 409–416.

Hardebo, J.E., Hanko, J. Kahrström, J. & Owman, C. (1986): Electrical field stimulation in cerebral and peripheral arteries: a critical evaluation of the contractile response. *J. Auton. Pharmacol.* **6,** 85–96.

Hardebo, J.E. (1992): Influence of impulse pattern on noradrenaline release from sympathetic nerves in cerebral and some peripheral vessels. *Acta Physiol. Scand.* **114,** 333–339.

Laher, I., Germann, P. & Bevan, J.A. (1994): Neurogenically evoked cerebral artery constriction is mediated by neuropeptide Y. *Can. J. Physiol. Pharmacol.* **72,** 1086–1088.

Lee, T.J.F., Chiueh, C.C. & Adams, M. (1980): Synaptic transmission of vasoconstrictor nerves in rabbit basilar artery. *Eur. J. Pharmacol.* **61,** 55–70.

Mackenzie, E.T ., McGeorge, A.P., Graham, D.I., Fitch, W., Edvinsson, L. & Harper, A.M. (1979): Effects of increasing arterial pressure on cerebral blood flow in the baboon: influence of the sympathetic nervous system. *Pflügers Arch.* **378,** 189–195.

Miranda, F.J., Torregrosa, G., Salom, J.B., Campos, V., Alabadi, J.A. & Alborch, E. (1989): Inhibitory effect of GABA on cerebrovascular sympathetic neurotransmission. *Brain Res.* **492,** 45–52.

Miranda, F.J., Torregrosa, G., Salom, J.B., Alabadi, J.A. Alvarez, C. & Alborch, E. (1992): Direct and neuromodulatory effects of histamine on isolated goat cerebral arteries. *J. Auton. Pharmacol.* **12,** 25–36.

Saeki, Y., Sato, A., Sato, Y. & Trzebski, A. (1990): Effects of stimulation of cervical sympathetic trunks with various frequencies on the local cortical cerebral blood flow measured by laser-Doppler flowmetry in the rat. *Jpn. J. Physiol.* **40,** 15–32.

Sakakibara, Y., Fujiwara, M. & Muramatsu, I. (1982): Pharmacological characterization of the alpha-adrenoceptors of the dog basilar artery. *Naunyn-Schmiedebergs Arch. Pharmacol.* **319,** 1–7.

Salaices, M., Marin, J., Marco, E.J., Conde, M.V., Gomez, B. & Lluch, S. (1983): Neurogenic vasoconstriction of cat cerebral and femoral arteries. *Gen. Pharmacol.* **14,** 355–360.

Sercombe, R., Lacombe, P., Aubineau, P., Mamo, H., Pinard, E., Reynier-Rebuffel, A.M. & Seylaz, J. (1979): Is there an active mechanism limiting the influence of the sympathetic system on the cerebral vascular bed? Evidence for vasomotor escape from sympathetic stimulation in the rabbit. *Brain Res.* **164,** 81–102.

Tamaki, K. & Heistad, D.D. (1986): Response of cerebral arteries to sympathetic stimulation during acute hypertension. *Hypertension* **8,** 911–917.

Uematsu, D., Gotoh, F., Fukuuchi, Y. Amano, T., Suzuki, N., Kobari, M., Kawamura, J. & Ito, N. (1987): Comparison between pial and intraparenchymal vascular responses to sympathetic stimulation under hypercapnic conditions. With special reference to the mechanism for escape phenomenon. *J. Neurol. Sci.* **78,** 303–311.

Wagerle, L.C., Kumar, S.P. & Delivoria-Papadopoulos, M. (1986): Effect of sympathetic nerve stimulation on cerebral blood flow in newborn piglets. *Pediatr. Res.* **20,** 131–135.

Wagerle, L.C., Kurth, C.D. & Roth, R.A. (1990): Sympathetic reactivity of cerebral arteries in developing fetal lamb and adult sheep. *Am. J. Physiol.* **258,** H1432–H1438.

## Other influences

Aubineau, P., Reynier-Rebuffel, A.M., Bouchaud, C., Jousseaume, O. & Seylaz, J. (1985): Long term effects of superior cervical ganglionectomy on cortical blood flow of non-anaesthetized rabbits in resting and hypertensive conditions. *Brain Res.* **338,** 13–23.

Azevado, I. & Osswald, W. (1986): Trophic role of the sympathetic innervation. *J. Pharmacol.* (Paris) **17,** (Suppl. II), 30–43.

Baumbach, G.L. & Heistad, D.D. (1989): Trophic role of sympathetic nerves during hypertension. In: *Neurotransmission and cerebrovascular function,* ed. J. Seylaz & R. Sercombe, pp. 279–291. Amsterdam: Elsevier.

Baumbach, G.L., Heistad, D.D. & Siems, J.E. (1989): Effect of sympathetic nerves on composition and distensibility of cerebral arterioles in rats. *J. Physiol.* **416,** 123–140.

Dimitriadou, V., Aubineau, P., Taxi, J. & Seylaz, J. (1987): Ultrastructural evidence for a functional unit between nerve fibers and type II cerebral mast cells in the cerebrovascular wall. *Neuroscience* **22,** 621–630.

Dimitriadou, V., Aubineau, P., Taxi, J. & Seylaz, J. (1988): Ultrastructural changes in the cerebral artery wall induced by long-term sympathetic denervation. *Blood Vessels* **25,** 122–143.

Ferrante, F., Ricci, A., Felici, L. Cavallotti, C. & Amenta, F. (1990): Suggestive evidence for a functional association between mast cells and sympathetic nerves in meningeal membranes. *Acta Histochem. Cytochem.* **23,** 637–646.

Handa, Y., Nojyoy, Y. & Hayashi, M. (1991): Patterns of reinnervation of denervated cerebral arteries by sympathetic nerve fibers after unilateral ganglionectomy in rats. *Exp. Brain Res.* **86,** 82–89.

Ibrahim, M.Z.M. (1974): The mast cells of the mammalian central nervous system – I. Morphology, distribution and histochemistry. *J. Neurol. Sci.* **21,** 431–478.

Johansson, B.B. & Nordborg, C. (1989): Denervation related morphological alterations in cerebral vessels during hypertension. In: *Neurotransmission and cerebrovascular function,* eds. J. Seylaz & R. Sercombe, pp. 267–278. Amsterdam: Elsevier.

Kondo, M., Miyazaki, T., Fujiwara, T., Yano, A. & Tabei, R. (1991): Increased density of fluorescent fibers around the middle cerebral arteries of stroke-prone spontaneously hypertensive rats. *Virchows Archiv. B Cell. Pathol.* **61,** 117–122.

Lindvall, M. & Owman, C. (1981): Autonomic nerves in the mammalian choroid plexus and their influence on the formation of cerebrospinal fluid. *J. Cereb. Blood Flow Metab.* **1,** 245–266.

Lindvall, M., Gustavsson, A., Hedner, P. & Owman, C. (1985): Stimulation of cyclic adenosine 3′,5′-monophosphate formation in rabbit choroid plexus by β-receptor agonists and vasoactive intestinal polypeptide. *Neurosci. Lett.* **54,** 153–157.

Nilsson, C., Kannisto, P., Lindvall-Axelsson, M., Owman, C. & Rosengren, E. (1990): The neuropeptides vasoactive intestinal polypeptide, peptide histidine isoleucine and neuropeptide Y modulate [$^3$H]noradrenaline release from sympathetic nerves in the choroid plexus. *Eur. J. Pharmacol.* **181,** 247–252.

## Part B. The parasympathetic nervous system

### Anatomy

Baramidze, D.G., Reidler, R.M., Gadamski, R. & Mchedishvili, G.I. (1982): Pattern and innvervation of pial microvascular effectors which control blood supply to the cerebral cortex. *Blood Vessels* **19**, 284–291.

Chorobski, J. & Penfield, W. (1932): Cerebral vasodilator nerves and their pathway from the medulla oblongata. *Arch. Neurol. Psychiat.* **28**, 1257–1289.

Duckles, S.P. (1982): Choline acetyltransferase in cerebral arteries: modulator of amino acid uptake? *J. Pharmacol. Exp. Ther.* **223**, 716–720.

Eckenstein, F. & Baughman, R.W. (1984): Two types of cholinergic innervation in cortex, one co-localized with vasoactive intestinal polypeptide. *Nature* **309**, 153–155.

Edvinsson, L., Nielsen, K.C., Owman, C. & Sporong, B.L. (1972): Cholinergic mechanisms in pial vessels. Histochemistry, electron microscopy and pharmacology. *Z. Zellforsch Mikroskop. Anat.* **134**, 311–325.

Edvinsson, L., Fahrenkrug, J., Hanko, J., Owman, C., Sundler, F. & Uddman, R. (1980): VIP (vasoactive intestinal polypeptide)-containing nerves of intracranial arteries in mammals. *Cell Tissue Res.* **208**, 135–142.

Edvinsson, L. & McCulloch, J. (1985): Distribution and vasomotor effects of peptide HI (PHI) in feline cerebral blood vessels *in vitro* and *in situ*. *Regul. Pept.* **10**, 345–356.

Edvinsson, L., Ekman, R., Jansen, I., Ottosson, A. & Uddman, R. (1987): Peptide-containing nerve fibers in human cerebral arteries: immunohistochemistry, radioimmunoassay and *in vitro* pharmacology. *Ann. Neurol.* **21**, 431–437.

Florence, V.M. & Bevan, J.A. (1979): Biochemical determinations of cholinergic innervation in cerebral cortex. *Circ. Res.* **45**, 212–218.

Gibbins, I.L., Brayden, J.E. & Bevan, J.A. (1984): Perivascular nerves with immunoreactivity to vasoactive intestinal polypeptide in cephalic arteries of the cat: distribution, possible origins and functional implications. *Neuroscience* **13**, 1327–1346.

Hara, H., Hamill, G.S. & Jacobowitz, D.M. (1985): Origin of cholinergic nerves to the rat major cerebral arteries: coexistence with vasoactive intestinal polypeptide. *Brain Res. Bull.* **14**, 179–188.

Iwayama, T., Furness, J.B. & Burnstock, G. (1970): Dual adrenergic and cholinergic innervation of the cerebral arteries of the rat. An ultrastructural study. *Circ. Res.* **26**, 635–646.

Lee, T.J-F. (1981): Ultrastructural distribution of vasodilator and vasoconstrictor nerves in the cat cerebral arteries. *Circ. Res.* **49**, 971–979.

Matsuyama, T., Shiosaka, S., Matsumoto, M., Yoneda, S., Kimura, K., Abe, H., Hayakawa, T., Inoue, H. & Tohyama, M. (1983): Overall distribution of vasoactive intestinal polypeptide-containing nerves on the wall of the cerebral arteries: an immunohistochemical study using whole-mounts. *Neuroscience* **10**, 89–96.

Saito, A., Wu, J.Y. & Lee, T.J-F. (1985): Evidence for the presence of cholinergic nerves in cerebral arteries: an immunohistochemical demostration of choline acetyltransferase. *J. Cereb. Blood Flow Metab.* **5**, 327–334.

Spencer, S.E., Sawyer, W.B., Wada, H., Platt, K.B. & Loewy, A.D. (1990): CNS projections to the pterygopalatine parasympathetic preganglionic neurons in the rat: a retrograde transneuronal viral cell body labeling study. *Brain Res.* **534**, 149–169.

Suzuki, N. & Hardebo, J.E. (1993): The cerebrovascular parasympathetic innervation. *Cerebrovasc. Brain Metab. Rev.* **5**, 33–46.

Suzuki, N., Hardebo, J.E. & Owman, Ch. (1990): Origins and pathways of choline acetyltransferase-positive parasympathetic nerve fibres to cerebral vessels in rat. *J. Cereb. Blood Flow Metab.* **10**, 399–408.

Uemura, Y., Sugimoto, T., Kikuchi, H. & Mizuno, N. (1988): Possible origins of cerebrovascular nerve fibers showing vasoactive intestinal polypeptide-like immunoreactivity: an immunohistochemical study in the dog. *Brain Res.* **448**, 98–105.

Walters, D.W., Gillespie, S.A. & Moskowitz, M. (1986): Cerebrovascular projections from the sphenopalatine and otic ganglia to the middle cerebral artery of the cat. *Stroke* **17**, 488–494.

### Parasympathetic blockade

Busija, D.W. & Heistad, D.D. (1982): Atropine does not attentuate cerebral vasodilatation during hypercapnia. *Am. J. Physiol.* **242**, H683–H687.

Dauphin, F., Lacombe, P., Sercombe, R., Hamel, E. & Seylaz, J. (1991): Hypercapnia and stimulation of the substantia innominata increase rat frontal cortical blood flow by different cholinergic mechanisms. *Brain Res.* **553,** 75–83.

Goadsby, P.J. (1989): Effect of stimulation of the facial nerve on regional cerebral blood flow and glucose utiilization in cats. *Am. J. Physiol.* **257,** R517–R521.

Goadsby, P.J., (1991): Characteristics of facial nerve-elicited cerebral vasodilatation determined using laser Doppler flowmetry. *Am. J. Physiol.* **260,** R255–R262.

Linder, J. (1981): Effects of facial nerve section and stimulation on cerebral and occular blood flow in hemorrhagic hypotension. *Acta Physiol. Scand.* **112,** 185–193.

Scremin, O.U., Rubinstein, E.H. & Sonnenschein, R.R. (1978): Cerebrovascular $CO_2$ reactivity: role of a cholinergic mechanism modulated by anesthesia. *Stroke* **9,** 160–165.

Seylaz, J., Hara, H., Pinard, E., Mraovitch, S., MacKenzie, E.T. & Edvinsson, L. (1988): Effect of stimulation of the sphenopalatine ganglion on cortical blood flow in the rat. *J. Cereb. Blood Flow Metab.* **8,** 875–878.

## Parasympathetic stimulation

Bevan, J.A., Moskowitz, M.A., Said, S.I. & Buga, G. (1984): Evidence that vasoactive intestinal polypeptide is a dilator transmitter to some cerebral and extracerebral cranial arteries. *Neuroscience* **19,** 597–604.

Brayden, J.E. & Bevan, J.A. (1985): Vasoactive intestinal polypeptide (VIP) is a cerebral vasodilator transmitter. *Stroke* **16,** 136.

Busija, D.W. & Heistad, D.D. (1981): Effects of cholinergic nerves on cerebral blood flow in cats. *Circ. Res.* **48,** 62–69.

Duckles, S.P. (1981): Evidence for a functional cholinergic innervation of cerebral arteries. *J. Pharmacol. Exp. Ther.* **217,** 544–548.

Goadsby, P.J. & MacDonald, G.J. (1985): Extracranial vasodilatation mediated by VIP (vasoactive intestinal polypeptide) *Brain Res.* **329,** 285–288.

Goadsby, P.J. & Duckworth, J.W. (1987): Effect of stimulation of trigeminal ganglion on regional cerebral blood flow in cats. *Am. J. Physiol.* **253,** R270–R274.

Goadsby, P.J. (1989): Effect of stimulation of the facial nerve on regional cerebral blood flow and glucose utilization in cats. *Am. J. Physiol.* **257,** R517–R521.

Goadsby, P.J. (1990): Sphenopalatine ganglion stimulation increases regional cerebral blood flow independent of glucose utilization in the cat. *Brain Res.* **506,** 145–148.

Goadsby, P.J. & Shelley, S. (1990): High frequency stimulation of the facial nerve results in local cortical release of vasoactive intestinal polypeptide in the anesthetized cat. *Neurosci. Lett.* **112,** 282–289.

Goadsby, P.J. (1991): Characteristics of facial nerve elicited cerebral vasodilatation determined using laser Doppler flowmetry. *Am. J. Physiol.* **260,** R255–R262.

Pinard, E., Purves, M.J., Seylaz, J. & Vasquez, J.V. (1979): The cholinergic pathway to cerebral blood vessels. II. Physiological studies. *Pflügers Arch.* **379,** 165–172.

Seylaz, J., Hara, H., Pinard, E., Mraovitch, S., Mackenzie, E.T. & Edvinsson, L. (1988): Effects of stimulation of the sphenopalatine ganglion on cortical blood flow in the rat. *J. Cereb. Blood Flow Metab.* **8,**875–878.

Suzuki, N., Hardebo, J.E., Kahrström, J. & Owman, C. (1990): Selective stimulation of postganglionic cerebrovascular parasympathetic nerve fibres originating from the sphenopalatine ganglion enhances cortical blood flow in the rat. *J. Cereb. Blood Flow Metab.* **10,** 383–391.

## Part C. The trigeminovascular system

### Anatomy

Arbab, M.A-R., Wiklund, L. & Svendgaard, N.A. (1986): Origin and distribution of cerebral vascular innervation from superior cervical trigeminal and spinal ganglia investigated with retrograde and anterograde WGA-HRP tracing in the rat. *Neuroscience* **19,** 695–708.

Buzzi, M.G., Dimitriadou, V., Theoharides, T.C. & Moskowitz, M.A. (1992): 5-Hydroxytryptamine receptor agonists for the abortive treatment of vascular headaches block mast cell, endothelial and platelet activation within the rat dura mater after trigeminal stimulation. *Brain Res.* **583,** 137–149.

Edvinsson, L., Rosendahl-Helgesen, S. & Uddman, R. (1983): Substance P: localization, concentration and release in cerebral arteries, choroid plexus and dura mater. *Cell Tissue Res.h* **234,** 1–7.

Edvinsson, L., Brodin, E., Jansen, I. & Uddman, R. (1988): Neurokinin A in cerebral vessels: characterization, localization and effects *in vitro. Regul. Peptides* **20,** 181–197.

Lee, Y., Kawai, Y., Shiosaka, S., Takami, K., Kiyama, H., Hillyard, C.J., Girgis, S., MacIntyre, I., Emson, P.C. & Tohyama, M. (1985): Coexistence of calcitonin gene-related peptide and substance P-like peptide in single cellls of the trigeminal ganglion of the rat: immunohistochemical analysis. *Brain Res.* **330,** 194–196.

Liu-Chen, L.Y., Han, D.H. & Moskowitz, M.A. (1983): Pia-arachnoid contains substance P originating from trigeminal neurons. *Neuroscience* **9,** 803–808.

Liu-Chen, L.Y., Gillespie, S.A., Norregaard, T.V. & Moskowitz, M.A. (1984): Co-localization of retrogradely transported wheat germ agglutinin and the putative neurotransmitter substance P within trigeminal ganglion cells projecting to cat middle cerebral artery. *J. Comp. Neurol.* **225,** 187–192.

Liu-Chen, L.Y., Norregaard, T.V. & Moskowitz, M.A. (1985): Some cholecystokinin-8 immunoreactive fibers in large pial arteries originate from trigeminal ganglia. *Brain Res.* **359,** 166–176.

McCulloch, J., Uddman, R., Kingman, T.A. & Edvinsson, L. (1986): Calcitonin gene-related peptide: functional role in cerebrovascular regulation. *Proc. Natl. Acad. Sci. (USA)* **83,** 1–5.

Moskowitz, M.A., Brody, M. & Liu-Chen, L.Y (1983): *In vitro* release of immunoreactive substance P from putative afferent nerve endings in bovine pia-arachnoid. *Neuroscience* **9,** 809–814.

O'Connor, T.P. & Van Der Kooy, D. (1988): Enrichment of a vasoactive neuropeptide (calctonin gene-related peptide) in trigeminal sensory projection to the intracranial arteries. *J. Neurosci.* **8,** 2468–2476.

Ruskell, G.L. & Simons, T. (1987): Trigeminal nerve pathways to the cerebral arteries in monkeys. *J. Anat.* **155,** 23–37.

Suzuki, N., Hardebo, J.E. & Owman, Ch. (1989): Origins and pathways of cerebrovascular nerves storing substance P and calcitonin gene-related peptide in rat. *Neuroscience* **31,** 427–438.

Uddman, R., Edvinsson, L., Ekman, R., Kingman, T. & McCulloch, J. (1985): Innervation of the feline cerebral vasculature by nerve fibers containing calcitonin gene-related peptide: trigeminal origin and co-existence with substance P. *Neurosci. Lett.* **62,** 131–136.

Yamamoto, K., Matsuyama, T., Shiosaka, S., Inagaki, E., Senba, Y., Shimizu, I., Ishimoto, T., Hayakawa, M., Matsumoto, M. & Tohyama, M. (1983): Overall distribution of substance P containing nerves in the wall of the cerebral arteries of the guinea pig and its origins. *J. Comp. Neurol.* **215,** 421–426.

## Trigeminal blockade

Edvinsson, L., McCulloch, J., Kingman, T.A. & Uddman, R. (1986): On the functional role of the trigemino-cerebrovascular system in the regulation of cerebral circulation. In: *Neural regulation of brain circulation*, eds. Ch. Owman & J.E. Hardebo, pp. 407–418. Amsterdam: Elsevier.

McCulloch, J., Uddman, R., Kingman, T.A. & Edvinsson, L. (1986): Calcitonin gene-related peptide: functional role in cerebrovascular regulation. *Proc. Natl. Acad. Sci. (USA)* **83,** 1–5.

Moskowitz, M.A. Wei, E.P., Saito, K. & Kontos, H.A. (1988): Trigeminalectomy modifies pial arteriolar responses to hypertension or norepinephrine. *Am. J. Physiol.* **255,** H1–H6.

Moskowitz, M.A. (1989): Trigeminovascular system: form and function in the cephalic circulation. In: *Neurotransmission and cerebrovascular function*, Vol. III, eds. J. Seylaz & R. Sercombe, pp. 311–328. Amsterdam: Elsevier.

Sakas, D.E., Moskowitz, M.A., Wei, E.P., Kontos, H.A., Kano, M. & Ogilvy, C. (1989): Trigeminovascular fibers increase blood flow in cortical grey matter by axon-dependent mechanisms during severe hypertension or seizures. *Proc. Natl. Acad. Sci. (USA)* **86,** 1401–1405.

## Trigeminal stimulation

Dimitriadou, V., Buzzi, M.G., Moskowitz, M.A. & Theoharides, T.C. (1991): Ultrastructural evidence for neurogenically mediated changes in blood vessels of the rat dura mater and tongue following antidromic trigeminal stimulation. *Neuroscience* **48,** 187–203.

Goadsby, P.J., Lambert, G.A. & Lance, J.W. (1986): Stimulation of the trigeminal ganglion increases flow in the extracerebral but not the cerebral circulation of the monkey. *Brian Res.* **381,** 63–67.

Goadsby, P.J. & Duckworth, J.W. (1987): Effect of stimulation of trigeminal ganglion on regional cerebral blood flow in cats. *Am. J.*

*Physiol.* **253,** R270–R274.

Goadsby, P.J., Edvinsson, L. & Ekman, R. (1988): Release of vasoactive peptides in the extracerebral circulation of man and the cat during activation of the trigeminovascular system. *Ann. Neurol.* **23,** 193–196.

Keller, J.T., Saunders, M.C., Beduk, A. & Jollis, J.G. (1985): Innervation of the posterior fossa dura of the cat. *Brain Res. Bull.* **14,** 97–102.

Lambert, G.A., Bogduk, N., Goadsby, P.J., Duckworth, J.W. & Lance, J.W. (1984): Decreased carotid arterial resistance in cats in response to trigeminal stimulation. *J. Neurosurg.* **61,** 307–315.

Lambert, G.A., Goadsby, P.J. Zagami, A.S. & Duckworth, J.W. (1988): Comparative effects of stimulation of the trigeminal ganglion and the superior sagittal sinus on cerebral blood flow and evoked potentials in the cat. *Brain Res.* **453,** 143–149.

Moskowitz, M.A., Brody, M. & Liu-Chen, L.Y. (1983): *In vitro* release of immunoreactive substance P from putative afferent nerve endings in bovine pia arachnoid. *Neuroscience* **9,** 809–814.

Suzuki, N., Hardebo, J.E., Kahrstrom, J. & Owman, C. (1990): Selective stimulation of postganglionic cerebrovascular parasympathetic nerve fibres originating from the sphenopalatine ganglion enhances cortical blood flow in the rat. *J. Cereb. Blood Flow Metab.* **10,** 383–391.

Tran Dinh, Y.R., Thurel, C., Cunin, G., Serrie, A. & Seylaz, J. (1992): Cerebral vasodilation after the thermocoagulation of the trigeminal ganglion in humans. *Neurosurgery* **31,** 658–633.

# Chapter 12

# Neurogenic regulation of cerebral blood flow: intrinsic neural control

## Sima Mraovitch

*Laboratoire de Recherches Cerebrovasculaires; CNRS UA 641, Université Paris VII, 10 Ave de Verdun, 75010 Paris, France*

## Introduction

It is now widely accepted that the central nervous system (CNS) plays a critical role in regulating the systemic circulation. Many CNS nuclei are shown to be involved in the control of cardiovascular functions. The output from these nuclei reaches intermedio-lateral cell columns of the spinal cord via specific brainstem regions such as the C1 area of the rostral ventrolateral medulla (RVL) where they influence the innervation of the heart and the peripheral vascular beds.

The way in which the CNS regulates its own circulation has also been a topic of enquiry for almost a century. Indeed, for many years it has been recognized that brainstem regions can profoundly influence cerebral blood flow. Numerous experiments have shown that electrical or chemical stimulation of these regions elicit changes in cerebral blood flow (CBF) and/or cerebral metabolism. Evidence has accumulated in support of the view that the brain, through intrinsic neuronal pathways (i.e. pathways whose neurons and their processes are entirely contained within the CNS) can control its own blood flow and metabolism.

In recent years, two schools of thought have emerged in investigations of the brain mechanisms whereby CBF and/or cerebral metabolism are regulated. The first, and experimentally the most promising, has focused on studying cerebrovascular and metabolic changes elicited by stimulating or lesioning specific brain regions, such as the locus coeruleus and the dorsal raphe nucleus, believed to have anatomical connections with the intracerebral blood vessels. The second, less obvious and often contested on anatomical grounds, has been based on the assumption that the brain regions known to be involved in the control of the peripheral circulation might also participate in mechanisms controlling cerebral blood flow.

Advances in technical methods, such as those described in Chapter 8, whereby CBF and metabolism are measured regionally and locally during activation of anatomically specific

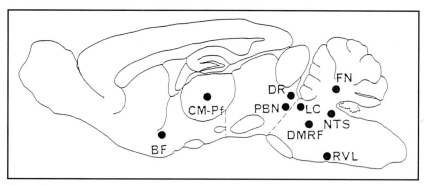

*Fig. 1. Scematic drawing of sagital section through the rat brain showing the sites from which electrical and or chemical stimulation elicets alteration in CBF and CGU.*

brain nuclei has provided convincing evidence of neurogenic regulation of CBF by several brain regions throughout the brainstem and forebrain (Fig. 1). Interestingly, changes in CBF and/or metabolism due to electrical or chemical stimulation of brain areas which directly project to intracerebral blood vessels have been rather small and to a certain degree controversial. Stimulation of brain regions implicated in cardiovascular regulation has yielded in the last decade the most exciting results.

The neurogenic control of cerebral blood flow has been the object of several recent reviews (Reis & Iadecola, 1989; Mraovitch, 1989). Therefore, only the most recent original papers will be cited below.

This chapter will focus first on experimental and conceptual considerations followed by the functional organization of central pathways implicated in CBF regulation and the efferent mechanisms, including intrinsic–extrinsic and neuronal–hormonal interaction upon CBF and metabolism. Finally, some data will be presented concerning a new and rapidly developing field: nitric oxide and CBF regulation.

## Experimental considerations

By definition, when a specific brain region system, subjected to an electrical or chemical stimulation, elicits changes in local or regional cerebral blood flow and/or metabolism it is considered to be implicated in cerebrovascular regulation.

## Chronic and acute experimental preparations

Chronic animal preparation, have been widely used in studying changes in CBF and/or cerebral glucose utilization (CGU) during stimulation of the brain. However, the introduction of quantitative autoradiography coupled to the stereotaxic exploration of the brain regions implicated in the control of CBF required, for obvious reasons, anaesthetized preparations. Alpha-chloralose is an anaesthetic agent that has been widely used for such cerebrovascular experimentation. The argument against the anaesthetized preparation is that intravenous as well as inhaled anaesthetics, including alpha-chloralose, result in the heterogeneous alterations of CBF and glucose utilization throughout the CNS (Dudley *et al.*,

1982). The magnitude of this change depends upon the anaesthetic employed and the depth of anaesthesia.

The effect of alpha-chloralose including some other anaesthetic agents such as halothane, pentobarbital and nitrous oxide has been recently studied on the functional-metabolic coupling (Ueki *et al.*, 1992). It was concluded that during somatosensory stimulation the pattern of metabolic activation of the primary somatosensory cortex of rats anaesthetized with alpha-chloralose was similar to that of the awake animal. Moreover, it appears that a-chlorolose plus halothiene induction remains the most suitable anaesthetic regimen for study of cerebrovascular reactivity (Bonvento *et al.*, 1994).

Despite the limitation of general anaesthetics, the experimental conditions of the anaesthetized and paralysed preparation provide important advantages for analysing and interpreting altered cerebrovascular as well as metabolic responses during activation of specific brain regions. In contrast to the awake unrestrained or lightly restrained animal, the anaesthetized experimental preparation limits activation of those pathways conveying information about position and movement, and hence the reflex activation of the thalamus and cortex that may be unrelated to the experimental question. For example, during electrical stimulation of the brain in a conscious animal, the values of local CBF and/or CGU will comprise the effects due to the specific activation of the stimulated region and 'non-specific' activation of reflex pathways due to limb movements of the behaving animals. Thus, the interpretation of the results and their statistical analysis become difficult and often meaningless. Acute animal preparations have been used to establish the appropriate experimental conditions for further investigations of anatomical and neurochemical mechanisms via which the altered CBF and metabolic changes occur during specific brain stimulation.

## Electrical and chemical stimulation

In the past decade, substantial advances have been made in identifying regions of the brain participating in the central neural control of CBF. Progress has been made by using classical neuroscience techniques, including the use of electrical and chemical stimulation coupled with autoradiographic and, most recently, laser Doppler flow methods for evaluating cerebral blood flow.

### Electrical stimulation

Electrical stimulation of the specific brain region has been the most commonly used technique for eliciting cerebrovascular and/or metabolic responses. The proper selection of stimulus temporal frequency and intensity provides a means of maximizing the response. The optimal stimulus parameters used during stimulation eliciting CBF responses were at first defined by observing peripheral vascular phenomena such as arterial blood pressure changes. The standard experimental procedure has been to identify the regions within the brain from which electrical stimulation will elicit a change in arterial pressure. This site is then used to determine the optimal frequency and the threshold of the response. The threshold is defined as the minimum

intensity that changes arterial pressure reliably from baseline. Using this procedure optimal stimulus parameters are established and used to study cerebrovascular responses. Following the introduction of mass spectroscopy and, more recently, laser-Doppler methods which allow continuous monitoring of cerebrovascular and cardiovascular responses, both variables may be monitored and their respective stimulus parameters determined. As can be seen in Fig. 2, the characteristics of the CBF responses are determined independently of the arterial pressure responses, and are comparable in the different techniques used (Nakai et al., 1982; Iadecola, 1990a). For example, the CBF increases during fastigial nucleus stimulation are dependent on both the stimulus frequency (optimal level 100–200 Hz) and intensity (threshold stimulus varies between 10 and 20 μA) (Mraovitch et al., 1986).

Electrical stimulation, however, may be a source of potential errors concerning the interpretation of results. It is well known that electrical stimulation of the brain depolarizes cell bodies, leading to the propagation of action potentials along their efferents to activate projection areas. In addition, the stimulation also excites dendritic arborizations, local neuronal circuits, multisynaptic pathways, glia and fibres of passage. Despite this handicap, electrical stimulation is still a valuable tool in studying the mechanisms of the cerebrovascular responses elicited by brain nuclei.

### Chemical stimulation

To determine whether it is intrinsic neurons or fibres of passage within

the brain regions which effectively participate in the control of cerebral blood flow and metabolism, some authors have recently used a chemical instead of electrical stimulus. Local microinjection of kainic acid has been used to stimulate neurons of FN (Chida et al., 1989) and RVL (Underwood et al., 1992a) regions. The cholinergic agonist carbachol has been microinjected in the CM-Pf (Mraovitch et al., 1989) of the thalamus. As one would expect, the results obtained with chemical stimulation within some regions (FN, CM-Pf) differed markedly from those observed during electrical stimulation. For example, while electrical stimulation of the FN, as will be described later in the chapter, elicited global vasodilatation not associated with an increase in glucose metabolism in several brain regions except the cerebral cortex, kainic acid stimulation caused significant and proportional reduction in CBF and CGU throughout the brain. Electrical and chemical stimulation of the CM-Pf provide differing differential patterns of lCBF and lCGU responses throughout the brain (Mraovitch et al., 1989, 1992). In contrast, the effects of electrical or chemical stimulation of the RVL on CBF are in many respects the same in distribution and magnitude. At present there are no available data for CGU during chemical stimulation of the FN and the RVL.

One of the problems with chemical stimulation is difficulty in estimating the diffusion of the microinjected substance. Typically, dyes such as eosin have been microinjected together with the stimulating agent in order to evaluate the stimulation site. However, the sphere of diffusion may not

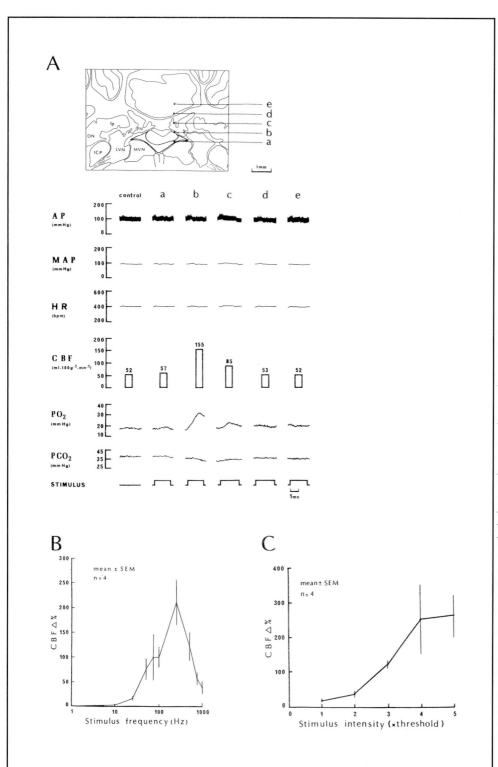

Fig. 2. A Changes of AP, HR, CBF, and intracerebral $PO_2$ and $PCO_2$ elicited by electrical stimulation of the caudal FN. Upper drawing represents a coronal section of rat cerebellum $4.0 \pm 0.25$ mm anterior to the calamus scriptorius from a representative experiment. An electrode tract, which penetrates the FN, is shown with stimulus sites 500 μm apart, illustrated by filled circles. At each site the cerebellum was stimulated with an intermittent stimulus train (1 s on/1 s off) of square-wave pulses (0.5 ms in duration, 50 Hz, and constant intensity of 50 μA). Stimulus sites with letters a–e correspond to polygraph tracings in lower panel. Note that the greatest rise in CBF (d and e) was not associated with significant change in AP.

B. Relationship between stimulus frequency and increase in local cerebral cortical blood flow (CBF) during fastigial nucleus stimulation at constant stimulus intensity (50 μA) from four representative experiments. The rise in local CBF appeared at 10 Hz, was increased by increasing the stimulus frequency, and reached a maximum between 100 and 500 Hz. Highest point was found at 250 Hz. Note that further increase in stimulus frequency up to 1000 Hz sharply decreased the local CBF response.

C. Relationship between stimulus intensity and magnitude of local cerebral blood flow (CBF) responses (as % changes from control) in four representative experiments. Local CBF response appeared at a stimulus current of 15 μA (1 × threshold current) and increased in proportion to the stimulus intensity until it reached a plateau at approximately 4 × threshold. (modified from Mraovitch et al., 1985).

be the same for the dye and the stimulating agent. In order to define the extent of the diffusion area one can use a stimulating agent labelled with either $^3$H or $^{14}$C. In a separate experiment, using autoradiography one can estimate the area of diffusion and the concentration gradient. For example, in order to define the extent of the diffusion sphere of 8-OH-DPAT (5-HT$_{1A}$ agonist), Bonvento *et al.* (1992) microinjected [$^3$H]8-OH-DPAT together with cold 8-OH-DPAT. The brain was then processed for autoradiography, and the size of the injection was directly visualized and the volume of diffusion quantified.

## What are the parameters of cerebrovascular function?

Classically, changes in CBF have been viewed as a consequence of change in brain neuronal activity and, as a corollary, metabolism. Thus, it has long been considered that the flow of blood to a given brain region is coupled to the level of neural activity and metabolism in the region.

The interrelation between functional activity, metabolism and blood flow was first suggested in 1880 by C. R. Roy and C. S. Sherrington, who observed that the onset of an epileptic seizure, reminiscent of an increase in neuronal activity, was followed by a swelling of the brain reflecting an increase in the blood supply. This and subsequent observations have led to the belief that brain function may be localized by studying variations in local blood flow. In addition to metabolic coupling, there is ever growing physiological and anatomical evidence showing that specific CNS regions, via intrinsic pathways, can markedly influence CBF independently of metabolism. In light of recent results, it should be recognized that, although changes in cerebral blood flow and metabolism may be interdependent variables, electrical or chemical intracerebral stimulation may elicit alterations of cerebral blood flow independent of metabolism.

## Neural control of CBF

As described in Chapter 5, there is abundant innervation of cerebral blood vessels, both pial and to a lesser extent intracerebral. While numerous studies have demonstrated that stimulation of the sympathetic and para-sympathetic (extrinsic) innervating systems, as shown in Chapter 11, can modulate CBF, the evidence for CBF regulation by intrinsic neural systems is yet to be shown convincingly on anatomical grounds. The main objection frequently raised is that, despite abundant physiological observations, there is a lack of anatomical evidence linking the neural systems to the intracerebral vascular beds. Despite this objection, several lines of evidence presented in this chapter have established the facts that:

(a) CBF responses (cerebrovasodilations or cerebrovasoconstrictions) are elicited from anatomically restricted brain regions. In general, these regions are closely related to brain nuclei involved in central autonomic regulation including cardiovascular regulation;

(b) CBF responses may be independent of changes in metabolism;

(c) CBF responses may be independent of changes in arterial pressure;

(d) CBF responses are dependent on the integrity of the pathways through which the effects are mediated.

## Neural control of enzymes involved in energy metabolic responses

As discussed in Chapter 7, in the brain, glucose is the primary substrate for energy metabolism via glycolysis and the citric acid cycle. The brain does not utilize glucose uniformly everywhere. Instead, histochemical staining and biochemical assays for energy-related enzymes have demonstrated (Borowsky & Collins, 1989a,b) that different regions of the brain express different enzymes. In recent years several lines of evidence have shown that the activity of mitochondrial enzymes of the tricarboxylic acid cycle in the brain are directly related to synaptic input and are regulated by the functional activity of the brain region. One of the key issues in understanding the mechanisms implicated in neural regulation of metabolism is to answer the following question: are there functional relationships between pathway-specific synaptic inputs within a brain region and their metabolic properties?

In mice, the neurons innervating facial whiskers project through several synapses to anatomically distinct contralateral somatosensory cortical regions called barrels. Clipping all of the whisker hair on one side of the face for 60 days leads to a significant ipsilateral (not sensory deprived) increase in the level of energy-related enzymes such as citrate dehydrogenase, malate

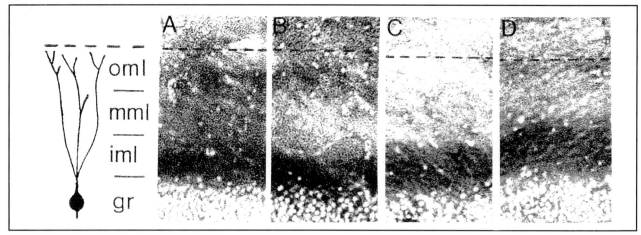

*Fig. 3. Dentate gyrus stained for LDH at various times following entorhinal lesions. (A) Contralateral to lesion. (B–D) Ipsilateral to lesion. (B) Twenty-four hours after entorhinal lesion. (C) Eight days post-lesion. (D) Ninety days post-lesion. Note the outward expansion of the intense band of staining for LDH in the inner molecular layer induced by the lesion. Dashed lines indicate the hippocampal fissure. oml, outer molecular layer; mml, middle molecular layer; iml, inner molecular layer; gr, granule cell layer. Magnification × 140 (from Borowsky and Collins, 1989b).*

dehydrogenase and glucogen phosphorylase (Dietrich *et al.*, 1982). The enzymatic level returns to normal values once the sensory deprivation is restored. Thus, the level of energy-related enzymes in the brain may increase due to increased utilization of the intact sensory input.

Sensorimotor impairment resulting from injury of the mesotelencephalic dopaminergic projection leads to a long-lasting differential pattern of changes in the enzyme succinate dehydrogenase (Marshall *et al.*, 1981). While the level of the enzyme decreased in the basal ganglion ipsilateral to the lesioned site, the globus pallidus, the entopendicular nucleus and the substantia nigra pars reticulata showed enhancement. Thus, the long-term effect of interrupting ascending dopaminergic axons are related to metabolic alteration within neurons of the basal ganglion.

The hippocampus and dentate gyrus of the rat are precisely laminated. Afferent fibres arising from the entorhinal cortex (perforant pathway) termi-

nate in the molecular layer (lacunosum-moleculare and dentate moleculare) while intrahippocampal commissural and associational fibres synapse within the striatum oriens and radium of the hippocampus and the inner one-third of the dentate molecular layer. In the normal animal, synaptic terminal fields of the perforant pathway show a high level of cytochrome oxidase (CO) activity whereas the terminal field of the commissural and associational fibres show a high level of lactate dehydrogenase (LDH) activity. Borowsky & Collins (1989a,b) have postulated that regional differences in the level of enzymes involved in oxidative and glycolytic metabolism reflect the specific afferent connection.

Unilateral lesion of the entorhinal cortex destroys, as expected, the perforant pathway, producing extensive terminal degeneration of the outer two-thirds of the dentate moleculare, in the striatum lacunosum-moleculare of the hippocampus, and elicits a marked sprouting of the remaining afferents (commissural and associ-

ational) to the dentate gyrus. On the side of the lesion, a significant reduction in staining density was found for CO and LDH in the deafferented outer and middle third of the molecular layer, while in the inner third of the molecular layer a relative increase in staining of both CO and LDH was observed. Moreover, entorhinal lesions produce a region of high reactivity for LDH in the inner molecular layer which expands into the adjacent middle layer to parallel the expansion of the commissural and associational pathways (ca) fibre sprouting after perforant pathway lesion. Thus, ca afferent fibre sprouting is accompanied by a corresponding change in the pattern of enzyme activity in the dentate gyrus. The increase in width of the LDH activity (Fig. 3) correlates well with the extent of ca afferent fibre sprouting. Reinnervation of a portion of the territory is accompanied by physiological changes and corresponding alterations in enzyme activity.

In addition to anatomical and histochemical observations leading to the conclusion that afferent pathways may influence the activity of the energy-related enzymes, a series of biochemical investigations has indicated that neurotransmitters may also regulate the enzymes within the specific brain region. The experimental evidence indicates that among the various putative neurotransmitters present in the cerebral cortex, such as glutamate, aspartate, GABA and acetylcholine, only norepinephrine (NE) and vasoactive intestinal peptide (VIP) induce a concentration-dependent glycogen hydrolysis (Magistretti *et al.*, 1981). NE-stimulated glycogenolysis is mediated by β-adrenergic receptors. VIP elicits glycogen hydrolysis rapidly and reversibly (50 per cent of the maximal effect being reached within 1 min). The glycogenolytic effect of VIP and NE is presumed to be mediated by cyclic AMP formation. Glycogen hydrolysis induced by VIP and NE will result in an increase in glucose availability for the generation of phosphate-bound energy in those cellular elements receiving terminals from VIP and NE projections. This further indicates that energy metabolism is regulated within the precise anatomical region of the CNS.

In view of these results, including those obtained from 2-DG measurement, it seems highly probable that differential metabolic requirements within various brain regions are regulated by their afferent inputs.

## Functional organization of central pathways implicated in the regulation of CBF and metabolism

The central nervous system, via specialized receptors, receives information from the external and/or internal (humoral or visceral) environment, which stimulates specific afferent pathways. This information is centrally processed by specific brain regions which, in turn, provide neural output to appropriately adjust functional activity, metabolism and blood flow. In so doing, the cerebrovascular control system is designed, therefore, to provide adequate blood flow to various brain regions under different physiological conditions.

Substantial advances have been made in the past several years in identifying the regions in the brain stem, cerebellum, and more recently in the thalamus and forebrain, which participate in the regulation of cerebral blood flow and glucose metabolism (Table 1). This section focuses on the data available on the neuroanatomical connections between regions implicated in cerebrovascular and metabolic regulation. While the regions discussed below participate in a number of physiological functions and have extensive afferent and efferent projections, we will discuss only those connections that appear at present to be relevant to cerebrovascular regulation (Fig. 4).

### Sensory inputs

From studies in man and animals it is evident that stimulation of somatic sensory, visual and auditory pathways can elicit changes in CBF and metabolism (see Reis and Iadecola, 1989).

Tactile stimulation elicits topographically selective increases in CBF in man and animals. In rats, stimulation of single vibrissae can increase rCBF within the territory of a single cortical barrel field, an effect which is paralleled by a comparably restricted change in rCGU. Thus, it has long been assumed that the increase in

Table 1. Cerebral blood flow (CBF) and cerebral glucose utilization (CGU) responses to electrical and/or chemical* stimulation of CNS regions

| CNS region | CBF | CGU |
|---|---|---|
| I. *Basal forebrain* | | |
|     Basal nucleus | ↑ | → |
| II. *Thalamus* | | |
|     Centromedian-parafascicular(*) | ↑ | ↑ or → |
| III. *Pons* | | |
|     Dorsal raphe nucleus | ↑ or ↓ | ↑ or → |
|     Locus coeruleus(*) | ↓ or → | — |
|     Nucleus parabrachial lateral | ↓ | → |
|     Nucleus parabrachial medial | ↓ | ↓ |
| IV. *Cerebellum* | | |
|     Fastigial nucleus | ↑ | ↑ or → |
| V. *Medulla* | | |
|     Dorsal medullary reticular formation | ↑ | ↑ |
|     C1 adrenergic neurons(*) | ↑ | ↑ |
|     Nucleus of the tractus solitari | ↑ | ↑ or → |

↑ increase; ↓ decrease; → no changes; — not measured.

CBF within the somatotopically corresponding cortical region results from metabolic activation due to increased local neuronal activity. However, there is now evidence in man to suggest that the initial event may be a primary vasodilation not coupled to metabolism, with the metabolic changes in response to sensory stimulation being a later contribution. Conceivably, the coupling of flow and rCGU recorded autoradiographically within barrel fields in experimental animals may be artifactual, reflecting differences in timing between measurements of rCBF and rCGU (see Chapter 8).

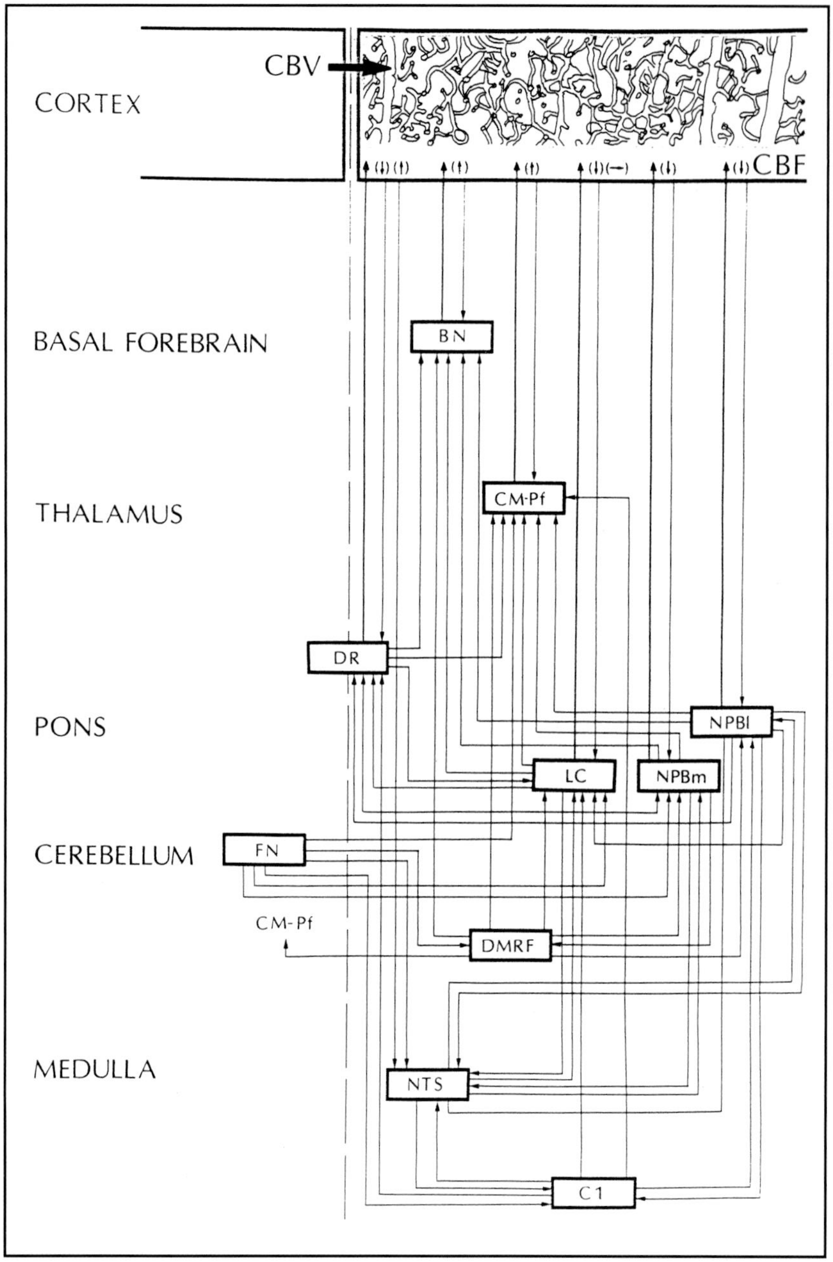

The increase in tissue perfusion in the somatosensory cortex elicited by electrical stimulation of the sciatic nerve is associated with an increase in the diameter of pial vessels on the cortical surface. The increase in vascular diameter is localized only to those arterial branches supplying the cortical representation of afferents from the sciatic nerve and is unaffected by sympathectomy. These observations are particularly intriguing since they raise the question of how restricted neural activity within brain might elicit compensatory and/or retrograde dilation in the afferent artery to cortex.

Less clearly established, surprisingly, is the pattern of activation of the brain in response to painful stimuli. According to Ngai *et al.* (1988), electrical stimulation of the sciatic nerve with stronger stimuli activating pain fibres increases the magnitude of vasodilation in cortical arterioles over that obtained with weaker stimuli. Studies by Tsubokawa *et al.* (1981) suggest that pain may elicit widespread activation of flow, although in that study the possibility that the associated elevations in arterial pressure exceeded the autoregulated range was not evaluated.

Fig. 4. Schematic diagram of connections of sites in the medulla, cerebellum, pons, thalamus and basal forebrain involved in the regulation of the cerebrovascular system. The cortical vascular network, at the level of the parietal cortex was made visible with Evans blue. Abbreviation: cortical blood flow: ↑ increase, ↓ decrease, → no change (from Mraovitch, 1989).

## Medulla

### Nucleus tractus solitarii (NTS)

The NTS is a column-shaped collection of neurons (see Chapter 1) extending from the level of the rostral medulla to the spinal-medullary junction. The NTS is the principal site of termination of afferent fibres from the viscera and arterial chemo- and baroreceptors. In addition, the NTS is innervated by neurons arising from the fastigial nucleus, the parabrachial nuclear complex, the C1 adrenergic neurons, the locus coeruleus, and the cerebral cortex. NTS neurons give rise to efferent projections to the parabrachial nuclear complex, the locus coeruleus, the C1 neurons and the dorsal medullary reticular formation.

Lesions of the NTS (Fig. 5) in anaesthetized rats elicit a potent elevation of AP, the so-called NTS hypertension. Associated is a substantial and global increase in rCBF. However, when the rise in AP is controlled, NTS lesions have no effect upon resting rCBF and rCGU, but the lesions will abolish cerebrovascular autoregulation globally and locally. The abolition of autoregulation is not replicated by transection of the IXth and Xth cranial nerves which interrupt all input from arterial baro- and or chemoreceptors as well as other cardiopulmonary afferent nerves. These studies suggest that neurons of the NTS may tonically regulate rCBF throughout the brain by some as yet unknown regulatory process influencing cerebrovascular autoregulation. Moreover, this indicates that cerebrovascular autoregulation is not only a process regulated at the local level of the microvasculature, as has been widely

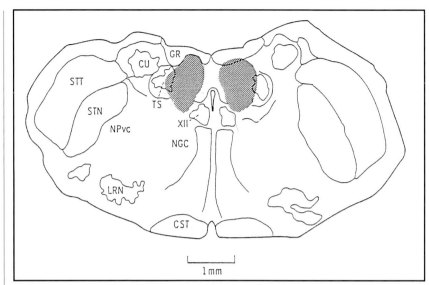

Fig. 5. Location and extent of typical bilateral electrolytic lesion of nucleus tractus solitarii (NTS) (shaded area) in a diagram of a representative section of rat brain 0.5 mm caudal to obex (from Ishitsuka et al., 1986).

thought to be the case, but may itself be regulated by neuronal networks represented at remote sites. Further support for the possible role of neurogenic mechanisms in cerebrovascular autoregulation comes from recent findings that autoregulation of CBF may be influenced by a number of neuropeptides including β-endorphin, α-MSH, and TRH.

### The C1 area of the rostral ventrolateral medulla (RVL)

Adrenaline synthesizing neurons corresponding to the C1 area described by Hökfelt et al. (1974) occupy the ventral surface of the rostral ventrolateral medulla (Fig. 6). The C1 area receives a major projection from baroreceptor innervated areas in the NTS. The C1 neurons also receive projections from the fastigial nucleus and the lateral parabrachial nucleus. The C1 neurons give rise to efferent fibres ascending to the parabrachial complex, dorsal raphe nucleus, NTS,

intralaminar thalamus and to the locus coeruleus (see Chapter 1 and Ruggiero et al., 1989).

### Dorsal medullary reticular formation (DMRF)

Since there is no specific data on the morphology and connections of the region located at the junction of the gigantocellular and parvocellular reticular nuclei of the dorsal medulla, called the dorsal medullary reticular formation (DMRF), the available anatomical information has been gathered considering its medial and lateral constituents: the gigantocellular and the parvocellular reticular nuclei (Fig. 7). The gigantocellular reticular nucleus (NGC) occupies the mid-portion of the medulla and is distinguished by its giant and medium-sized neurons. The parvocellular reticular nucleus (PCR) is composed primarily of small neurons.

Afferents to the DMRF originate from

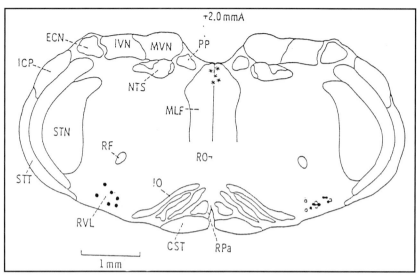

*Fig. 6. Diagram of a representative brain section 2.0 mm rostral to the obex through the medulla showing the sites of electrical and chemical stimulation (from Underwood et al., 1992a).*

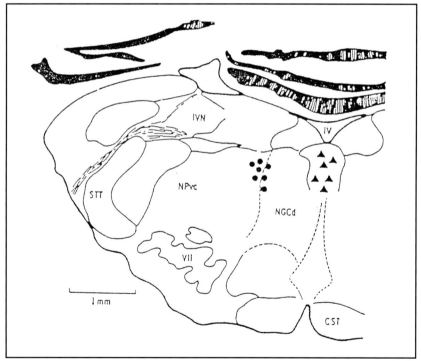

*Fig. 7. Diagram of representative brain section through the pontomedullary junction, showing the sites stimulated in the dorsal medulla. Filled circles represent sites from which electrical stimulation elicited an increase in arterial pressure, bradycardia, and an increase in rCBF; the area lies between the nucleus parvocellular and gigantocellularis dorsalis. Filled triangles represent sites in the midline at the level of the DMRF (interstitial nucleus of the medial longitudinal fasciculus) where electrical stimulation (100 μA) failed to produce any cardiovascular response and failed to increase rCBF (from Iadecola et al., 1983a).*

the fastigial nucleus and parabrachial nuclear complex. Fibres from the NGC innervate ipsilateral and both ipsilateral and contralateral locus coeruleus. Rostrally, the major long ascending fibres are most prominent on the contralateral side, reaching the centromedian-parafascicular complex of the intralaminar thalamus and the substantia innominata of the basal forebrain. The parvocellular projections ascend to the parabrachial nucleus and, in contrast to the NGC efferents, relatively few fibres from the PCR project beyond the rostral pons (for references see Mraovitch, 1989).

## Cerebellum

### Fastigial nucleus (FN)

The fastigial nucleus is the most medial of the deep cerebellar nuclei, and lies near the midline in the roof of the fourth ventricle (Fig. 8). The FN consists of a population of densely packed cells of varying size.

Afferent fibres to the FN arise primarily from cerebellar Purkinje cells, the inferior olivary nucleus and the vestibular complex. Fastigial efferent projections to the brainstem emerge from the cerebellum via the uncinate fasciculus and the juxtarestiform body. While descending fibres (crossed and uncrossed) are distributed differentially in parts of all of the vestibular nuclei including the DMRF, the C1 area and the NTS, ascending fibres decussate within the cerebellum and ascend contralaterally to reach the locus coeruleus, the parabrachial nucleus and the centromedian-parafascicular complex.

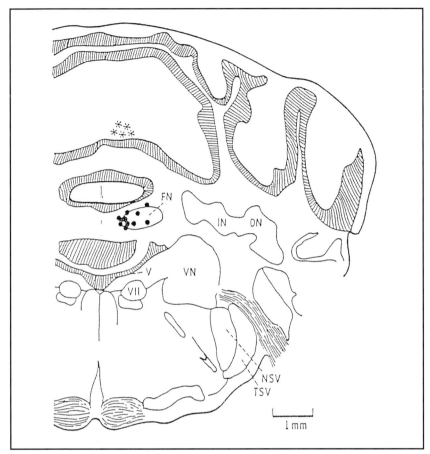

*Fig. 8. Representative brain section through rostral fastigial nucleus showing electrode localizations within cerebellum. Solid circles, distribution of sites in rostral fastigial nucleus (FN) from which electrical stimulation elicited an increase of arterial pressure, tachycardia, and regional cerebral blood flow (rCBF) in anaesthetized, paralysed rats. Asterisks, sites from which electrical stimulation failed to increase rCBF (from Nakai et al., 1983).*

## Pons

### Locus coeruleus (LC)

The locus coeruleus is located medial to the parabrachial nucleus, near the fourth ventricle (Fig. 9). The LC consists of a collection of medium-sized pigmented cells, and is also referred to as the nucleus pigmentosus pontis. In addition, the LC contains numerous small pigment-free cells.

The afferent projections to the LC are diverse. There is evidence for projections from the cortex, the dorsal raphe, the centromedian-parafascicular complex, the fastigial nucleus, the nucleus tractus solitar, and the parabrachial complex. Neurons of the LC give rise to extensive noradrenergic fibre projections, including fibres to the dorsal raphe, the NTS, the basal forebrain, the CM-Pf, the parabrachial nucleus, and the cortex (see Fig. 4).

Studies on the effects of LC stimulation upon CBF have not been consist-

ent. In part, the discrepancies may reflect the fact that the LC lies in a region of the dorsal pons close to other autonomically active centres (e.g. parabrachial nucleus) and near major ascending visceral autonomic bundles in the periventricular grey matter. Such proximity raises questions about the anatomical effects of lesions and/or stimulation of the nucleus. Nevertheless, the bulk of studies have suggested that electrical stimulation of the LC in monkey, rat or cat acts to increase cerebral vascular resistance and to reduce rCBF (for further references see Reis and Iadecola, 1989).

### Parabrachial nuclear complex (PBN)

The parabrachial nuclear complex consists of a collection of small neurons surrounding the brachium conjunctivum in the dorsolateral pons (Fig. 9). In the cat and rat, it is divided cytologically into two major subdivisions: a medial division and a lateral division. The ventrolaterally situated Kolliker–Fuse nucleus is also considered to be a part of the PBN. Recently (Bernard et al., 1993), the medial and lateral divisions were further subdivided according to the organization of the efferent projections to the amygdala.

Afferent and efferent connections of the parabrachial complex are extensive (for references see Mraovitch et al., 1985). Neurons projecting to the PBN are found in the NTS, the C1 adrenergic neurons, the regions comprising the DMRF and the FN. The parabrachial efferent projections innervate the C1 neurons, the NTS, the DMRF, the LC, the dorsal raphe, the CM-Pf and the cerebral cortex (Fig. 4).

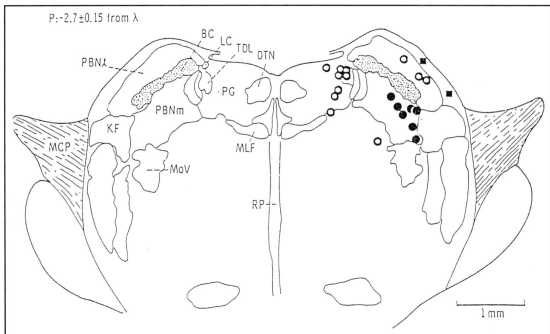

*Fig. 9. Diagram of a representative brainstem section through the rostral pons, illustrating the rCBF changes elicited by electrical stimulation of the area surrounding the brachium conjunctivum. Stimulation of adjacent sites, including the LC and pontine grey, failed to reproduce a similar pattern of rCBF changes (from Mraovitch et al., 1985).*

A number of studies have demonstrated that the PBN participates in neural mechanisms subserving a variety of visceral, neuroendocrine and behavioural functions, including respiration, locomotion, secretion of ACTH, sleep, grooming and attack behaviour as well as taste sensations and cardiovascular regulation (for references see Mraovitch *et al.*, 1982). That the PBN could modify the cerebral circulation was first noted by Mraovitch *et al.* (1985).

## Mesencephalon

### Dorsal raphe nucleus (DR)

The dorsal raphe nucleus is located in the caudal part of the central grey and the rostral part of the periventricular grey of the 4th ventricule (Fig. 10). Three types of neurons, according to the shape of the cell bodies and dendritic morphology, have been identified in the DR: type 1 neurons are polygonal in shape, type 2 have fusiform cell bodies, and type 3 have small, piriform cells.

Serotonergic neurons are the most characteristic cell type in the DR. The serotonin system has been shown to be involved in the regulation of sleep, temperature, pain, aggression, locomotor activity, sexual behaviour, avoidance learning and water consumption (for review see Azmitia, 1978). However, the presence of a large population of non-serotonergic neurons in the DR has to be considered in the interpretation of functional studies involving stimulation and/or lesion of the DR.

The afferent and efferent connections of the raphe nucleus are extensive (for review see Jacobs and Azmitia, 1992). The afferent fibres to the DR arise from the cingulate cortex, the LC, and the parabrachial nuclear complex. The efferent projections can be divided into the descending and the ascending pathways. The DR projects to the LC, the nucleus basalis, the CM-Pf and the cerebral cortex.

The effects of electrical stimulation of the most rostral nucleus in the raphe group, the DR, which provides the principal serotonergic innervation of telencephalon, has been studied by several investigators (see 'Primary cerebral vasoconstriction orginating in the DR'). A decrease of CBF in anaesthetized animals and increase in unanaesthetized animals have been observed.

## Thalamus

The thalamus plays a key role in relaying and transforming information that reaches the cerebral cortex. It is considered to be a processing station for input from widely different sources within the brain. As such, the thalamus has evolved in a close rela-

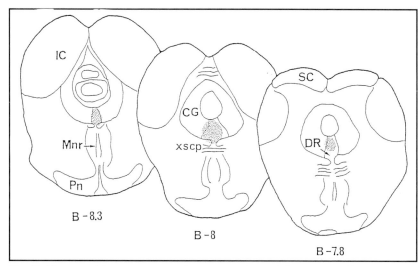

*Fig. 10. Diagram of three representative brainstem section at different levels of the rostral pons. The DR was visualized following injection of (3H) 8-OH-DPAT using the autoradiography and Cresyl violet staining (from Bonvento et al., 1992).*

tionship with the cortex.

With rare exceptions, all the specific sensory pathways, including some autonomic visceral fibres arising from the NTS and the PBN projecting to the cortex, do so through specific relay nuclei in the thalamus. In contrast, inputs from various nuclei of the reticular formation project to the cortex via relay stations within the intralaminar or non-specific thalamic nuclei.

Little is known about the effects of thalamic stimulation or lesion on rCBF using contempory techniques. The only region of the thalamus which has been systematically studied is the centromedian-parafascicular complex (CM-Pf). However, it is likely that primary sensory relay thalamic nuclei must also influence cortical blood flow in view of the evidence that, in man, somatosensory stimulation elicits both primary and secondary vasodilation in the cerebral cortex (Katayama *et al.*, 1986).

## Centromedian-parafascicular complex (CM-Pf)

The centromedian parafascicular complex is the most posterior group of intralaminar thalamic nuclei (Fig. 11). It consists of closely packed medium-sized cells which surround the fasciculus retroflexus. Numerous studies have demonstrated that the CM-Pf participates in the neural mechanisms subserving cortical desynchronization and recruitment phenomena, conditioning, pain-related neural responses and blood pressure regulation (for references see Mraovitch *et al.*, 1986).

The CM-Pf receives afferents from all cortical areas and from a number of brain-stem structures, including the parabrachial nuclei, a region of the DMRF, the FN, the DR, the LC, and the C1 area. Neurons of the CM-Pf have diffuse connections with the cortex.

## Forebrain

### Basal forebrain (BF)

The basal forebrain or substantia innominata/basal nucleus of Meynert is located on the ventral aspect of the forebrain. The main regions included in the basal forebrain structure are the bed nucleus of the stria terminalis, the substantia innominata, the nucleus accumbens, the diagonal band nuclei, and the ventral portion of the globus pallidus (Fig. 12). The basal forebrain is associated with nervous system disorders such as schizophrenia, Parkinson's disease, Alzheimer's disease as well as behaviour ranging from basic drives and emotions to higher cognitive functions (see Alheid and Heimer, 1988).

The afferent connections to the nucleus basalis in the BF include fibres from the LC, the DR, the PBN, and the DMRF.

The basal forebrain provides the major source of cholinergic efferent fibres to the cerebral cortex. The highest concentrations of cholinergic terminals are in layers V and VI, while layers I and III contain moderate concentrations.

## Cerebral cortex

On the basis of anatomical and physiological studies, the cerebral cortex is neither a structural nor a functional entity. In Golgi preparations, where axons and dendrites are made visible, the cortex appears to be divided into six separate horizontal layers. Electrophysiological studies have led to the concept that the elementary organization of the cortex is a 'column' of neurons oriented perpendicular to the cortical surface. Different cortical re-

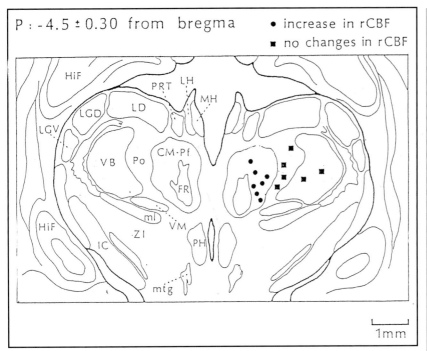

*Fig. 11. Composite drawing of a representative brain section 4.5 ± 0.3 mm posterior to the bregma, illustrating the rCBF changes evoked by electrical stimulation of the CM-Pf and the adjacent thalamic area. Stimulation of adjacent sites including posterior nucleus of the thalamus and the thalamic ventrobasal complex, failed to elicit rCBF changes (from Mraovitch et al., 1986).*

gions have different morphological patterns of cytoarchitecture.

Afferent information is relayed to the cortex by three major sensory ascending pathways: dorsal column-medial lemniscal, anterolateral (spinothalamic and spinotectal) and ascending activating pathways. With some exceptions, all of these afferents, including the visual pathway, converge in the thalamus where sensory and visceral information is coded and processed to the cortex.

In recent years, a number of non-thalamic cortical afferents have been described. There are direct ponto-cortical projections from the PBN, the LC and the raphe nuclei. The parabrachial-cortical fibres are a rostrolateral extension of the dorsal tegmental bundle ascending pathway. These fibres innervate layers I and V of the insular cortex and layer V of the frontal cortex. Neurons of layer V project back upon the parabrachial region. The majority of the serotonergic innervation reaches the cortex via the dorsal raphe-cortical tract, forming a diffuse pattern throughout the cortical layers. The efferent projection of the LC to the cortex ascends within the dorsal bundle joining the median forebrain bundle at the level of the fasciculus retroflexus. Axons from the LC enter the cortex at the frontal region and run rostro-caudally in layer VI. The LC also innervates layers I, IV and V.

The major cortical efferent pathways are the cortico-thalamic, cortico-tectal and cortico-spinal pathways which include those to the basal forebrain, the centromedian-parafascicular complex, the dorsal raphe, the locus coeruleus, the parabrachial nuclear complex and the nucleus tractus solitari.

Although, the cerebral cortex is the most important neural structure for perception, thought, and other higher functions, a number of studies have shown its role in autonomic functions. Alterations in autonomic activity, including changes in respiration, arterial pressure, and heart rate, can be elicited by electrical stimulation of a cerebral cortex (Ruggiero *et al.*, 1987).

Recent physiological and anatomical findings have suggested that afferent projections from the subcortical 'cerebrovascular' areas to cerebral cortex do not directly contact (except fibres of the locus coeruleus) cortical blood vessels, but do so through pathways involving local cortical neurons.

## Local cortical neurons

The term 'local cortical neurons', derived from 'local circuit neurons' (see Rakic, 1975), implies that the cells are connected to other cells within the vicinity of the local region of the cortex, rather than to cells in distant brain regions. The cortex has a large number of different types of neurons with a variety of dendritic processes.

VIP-containing neurons in the cerebral cortex are bipolar cells oriented in a plan perpendicular to the pial surface (Fig. 13) with dendritic arborizations in layers I and V. VIP neurons are distributed in a non-cluster manner so that there is, on average, one VIP-containing neuron for each 30 μm diameter of cortical column (Magistretti and Morrison, 1988). Ap-

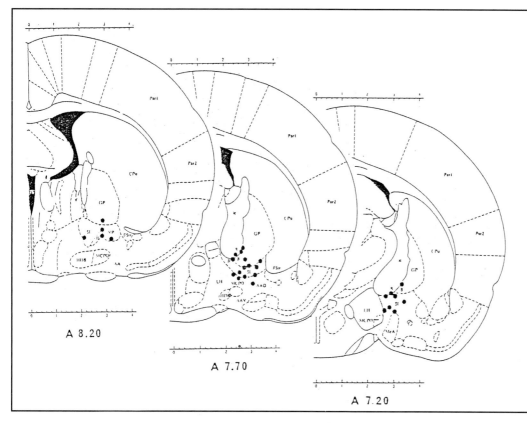

*Fig. 12. Stimulated sites (n = 28) in the SI region giving rise to flow responses without notable hypertension, agitation or generalized movement. The sites considered have been placed on three consecutive coronal planes of the atlas with simplified nomenclature at: 7.20, 7.70, and 8.20 mm anterior to the interaural line (from Lacombe et al., 1989).*

proximately 1 per cent of the neurons in the rat primary visual cortex contain VIP. Experimental data indicate that VIP and acetylcholine are colocalized in the same population of intracortical neurons (Eckenstein and Boughman, 1984).

The use of the term 'local cortical neurons' with respect to cortical vasodilation was first introduced by Iadecola *et al.* (1987) and Arneric *et al.* (1987). They observed that a neural system mediating the elevation in cortical blood flow elicited by FN stimulation involves a restricted population of local cortical neurons.

Indeed, within the cerebral cortex, the vasodilation appears to involve the action of local neurons in part associated with local release of ace-

tylcholine. Destruction of neurons, but not axons, within a restricted cortical area by the excitotoxin ibotenic acid abolishes the vasodilation elicited by FN stimulation without affecting resting rCBF or the vasodilation evoked by hypercarbia. Topical application of atropine reduced, by 60 per cent, the response locally. These findings suggest that the cortical vasomotor response to FN stimulation is not mediated by a direct neurovascular projection originating from or passing through basal forebrain but that this projection must engage a local interneuron. Moreover, since the amount of acetylcholine released by FN stimulation in cortex is undetectable in contrast to the substantial release evoked by basal forebrain stimulation, it seems probable that the

source of ACh involved in the vasodilation may be local cholinergic neurons or even the capillary endothelium. In agreement with these observations was an observation that a burst of electrocortical activity generated by the activity of local cortical neurones and probably driven from subcortical pacemakers (CM-Df and/or thalamic specific nuclei) is associated with spontaneous cerebrovascular waves (Golanov *et al.*, 1994).

## The cerebrovascular and metabolic response patterns elicited by specific neural systems

In the preceding section we reviewed studies which demonstrated that electrical or chemical stimulation of well-defined regions in brain may elicit vasodilatation or vasoconstriction. Cerebral vasodilatation is effected via two major regulatory mechanisms: metabolic or chemical (see Chapter 9) and neurogenic. Neurogenic influence on the cerebral blood flow is believed to be mediated through intracerebral innervation of the cerebrovascular smooth muscle.

The cerebrovascular responses were defined as *primary* when the effect of the stimulation upon the vascular bed was independent of changes in metabolism, and secondary, when the cerebral blood flow alterations were associated with proportionally similar changes in local metabolism.

## Central neurogenic vasodilation

### Primary neurogenic cerebral vasodilation

Cerebral vasodilatory responses unassociated with corresponding changes in cerebral glucose utilization has been elicited through neural pathways intrinsic to the brain originating within the several brain regions including the FN, BF, CM-Pf and RVL.

Except for the BF, all of these regions have been implicated in the control of cerebrovascular functions (see Mraovitch, 1989; Reis and Iadecola, 1989).

### Primary cerebral vasodilation originating in the FN

The possibility that the cerebellum could modify the rCBF is derived from studies of the potent vasopressor response elicited by electrical stimulation of the FN, the fastigial

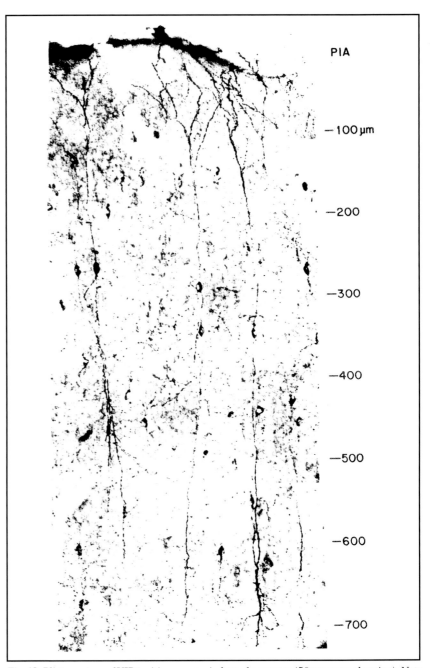

*Fig. 13. Photomontage of VIP-positive neurons in lateral neocortex (50 μm coronal section). Note the following features: (1) most cells bodies are ovoid, with major ascending and descending process. (2) These processes extend radially for several hundred within the plane of section and do not arborize extensively, except within layers I (pia-100 μm) and superficial V (600–700 μm). (3) The distal branches are more heavily stained and of larger diameter than the segments closer to the cell bodies. (4) Processes that arise from proximal segments may be axons, as they are very fine, highly varicose and more variable in orientation. Note example marked by arrow (from Morrison et al., 1984).*

pressor response (FPR) (for references, see Reis & Iadecola, 1989).

Using quantitative autoradiography to assay local cerebral blood flow and glucose metabolism, Nakai *et al.* (1983) were the first to obtain experimental evidence leading to the confirmation of the hypothesis that intracerebral pathways might act upon cortical blood flow independent of metabolism. Indeed, these authors demonstrated that electrical stimulation of the FN in chloralose-anaesthetized rats results in a marked increase in CBF in 24 of 28 brain areas examined. The largest increases in flow, up to two-fold, occurred in areas of the cerebral cortex; flow was significantly increased in frontal (215 per cent of control), parietal (153 per cent) and auditory (181 per cent) cortex. In contrast, glucose utilization following FN stimulation was altered in only 15 of 28 regions, the smallest changes being in the cerebral cortex. The relationship between the percent increase in

CBF and CGU in brain regions elicited by FN stimulation is shown in Fig. 14. This analysis demonstrates that in some brain areas such as in the cerebral cortex there were significant increases in CBF without changes in metabolism. These findings demonstrated that excitation of neurons originating in or passing through FN will increase CBF within the brain by at least two mechanisms, one in the cortex, independent of CGU (primary vasodilatation), and the other linked to metabolism (secondary vasodilatation). This conclusion was recently challenged by Talman *et al.* (1991), demonstrating that changes in cerebral blood flow elicited by the FN stimulation were not due to a direct neurogenic mechanism but secondary to a metabolic effect. In contrast to these results, it is of interest to note that Takahashi *et al.* (1995) fail to support a specific role for the FN in physiological regulation of cerebral blood flow in unanaesthetized rats.

The identity of neural elements in the

FN initiating the cardio- and cerebrovascular responses to electrical stimulation of FN was not known. It now seems likely (for references see Reis and Iadecola, 1989) that the FPR and the cerebrovascular vasodilation are a consequence of stimulation of local axons in FN and not FN neurons. Indeed, stimulation of local FN neurons with the excitatory amino acid kainic acid will elicit a dose-dependent *fall* of arterial pressure, heart rate, and lCBF, effects abolished by destruction of intrinsic FN neurons with ibotenic acid. Moreover, after destruction of local neurons with ibotenic acid the FPR and the cerebrovascular vasodilation persists. Such axons may arise from neurons originating in the PBN, the NTS, the periaqueducal grey, which also project to the C1 area. A likely hypothesis that antidromic activation of this network is responsible for the FPR. It is also possible that the representation in the cerebellum of the systemic and cerebrovascular components of the FPR may be anat-

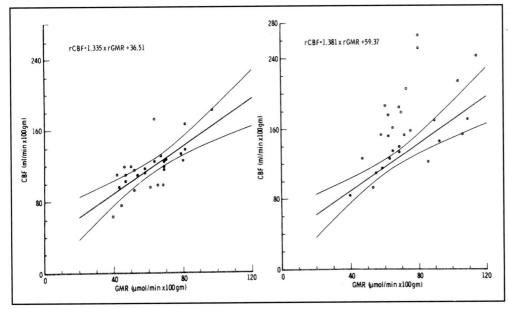

Fig. 14. Plotting of rCBF in function of rGMR in 28 brain regions in sham-operated unstimulated rats (left) and during FN stimulation (right). Rats were anaesthetized and paralysed. The straight lines represent linear regression lines and symmetrical curves, drawn above and below, represent the 95 per cent confidence interval. Note that in controls most of the points lie close to the regression line, with a correlation coefficient r = 0.733. During FN stimulation (right) the linear relationship observed in controls is deeply pertubed (r = 0.568), most of the points lying away from the regression line and outside of the 95 per cent confidence interval (from Nakai et al., 1983).

omically distinct (Mraovitch *et al.*, 1986).

Chemical stimulation of the FN elicits a fall of arterial pressure and heart rate, the fastigial depressor response (FDR). Associated with FDR is a widespread reduction in rCBF coupled to a proportional reduction in metabolism which also is dependent upon the integrity of the C1 area (Chida *et al.*, 1989).

### Primary cerebral vasodilation elicited by the BF

Several recent findings suggest a role for cholinergic BF neurons in the primary vasodilatation of the cortical microvasculature. Local blood flow (by helium clearance), combined with tissue gas partial pressures ($pO_2$ and $pCO_2$) as metabolic indices, were measured in the frontal and parietal cortices in unanaesthetized animals during electrical stimulation via a chronically implanted electrode (Lacombe *et al.*, 1989). During stimulation, both frontal (+ 114 per cent) and parietal (+ 28 per cent) cortical blood flow increased significantly. The stimulation also increased cortical $pO_2$ and decreased $pCO_2$. In view of these observations the authors called into question the existence of a pure metabolic coupling since the tissue gas changes accompanying flow increase during BF stimulation did not show interdependence. More conclusive evidence of primary vasodilatation has been given in recently published articles by Kimura *et al.* (1990). Using quantitative autoradiography techniques in adult Wistar rats artificially ventilated and under halothane anaesthesia, electrical stimulation of

the BF produced a significant increase in CBF in the frontal, parietal and occipital cortices on the ipsilateral side. In contrast, ipsilateral stimulation of the BF did not modify CGU in any brain region analysed including cortical areas. Thus, the authors concluded that the increase in cortical CBF following BF stimulation was not accompanied by an increase in cortical CGU, indicating that increase in cortical CBF was not a consequence of an increase in the metabolic rate.

### Primary cerebral vasodilation elicited by the CM-Pf

Since monosynaptic projections arising from the FN were found to terminate in the CM-Pf but not in the cortex (for references see Mraovitch *et al.*, 1986), it was suggested that the increase in cortical blood flow elicited by FN was, in part, mediated via diffuse CM-Pf cortical projections. In a series of studies (see Mraovitch, 1989; Mraovitch *et. al.*, 1992) it was, therefore, examined whether electrical or chemical stimulation of the CM-Pf would modify lCBF and, if so, whether the changes were associated with corresponding alterations in CGU.

Electrical stimulation restricted to the CM-Pf resulted in a widespread and bilateral rise in CBF (Fig. 15). The greatest increase in CBF (Fig. 16) was seen in mesencephalic reticular regions (415 ± 122 per cent), zona incerta (532 ± 99 per cent) including subcortical (substantia nigra compacta 212 ± 44 per cent, globus pallidus (118 ± 24 per cent) and the cortex (parietal cortex 211 ± 35 per cent). In contrast to the marked increase in CBF, electrical stimulation of the CM-

Pf had a rather small effect on the CGU except in the regions comprising the reticular formation (zona incerta (123 ± 28 per cent and anterior pretectal area (121 ± 13 per cent).

Thus, the CM-Pf stimulation elicited a potent and widespread, but heterogeneous, increase in lCBF. In contrast, lCGU remained unchanged in a number of brain regions including the cerebral cortex, indicating primary cerebrovascular dilation. Further analysis, in which global data analysis was extended to anatomically and functionally well-defined neural systems (such as the extrapyramidal motor system, the limbic system, and the reticular formation), led to the suggestion that vascular and metabolic changes elicited by CM-Pf stimulation within the individual brain regions and specific neural systems might involve different regulatory mechanisms (neurogenic, metabolic and hormonal).

The CM-Pf is a site of termination of cholinergic afferent fibres originating from neurons of the upper brainstem core including the PBN and a region comprising the DMRF, both of which participate in the regulation of CBF (for references see Mraovitch *et al.*, 1989). Cholinergic stimulation of neurons within the CM-Pf by unilateral microinjection of a cholinergic agonist (carbachol) resulted in a rise in lCBF (Mraovitch *et al.*, 1989). Under these conditions, however, the rise in lCBF associated with lCGU differed from the electrically elicited lCBF and lCGU alterations: the mechanism whereby the CM-Pf cholinergic activation may differentially alter lCBF and/or lCGU is not clear. One possibility is that CCh may activate specific, as yet unidentified CM-Pf cerebrovascular pathways.

*Fig. 15. See page 355 for colour reproduction.*

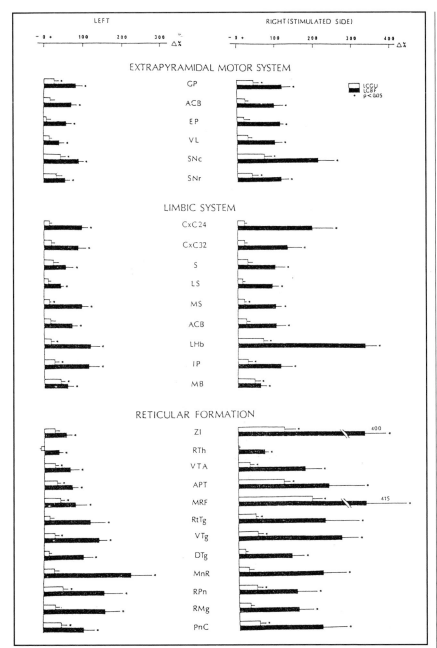

*Fig. 16. Effect of CM-Pf stimulation on lCBF and lCGU (percentage of unstimulated sham control) of brain regions comprising the EMS, the LS and the RF not affected by a cortical spreading depression in anaesthetized and paralysed rats. In the EMS note that the bilateral increase in lCBF (except in globus pallidum and substantia nigra) was not associated with a significant rise in metabolism. In the LS, significant increases in lCGU accord in four (medial septum, lateral habenular nucleus, interpeduncular nucleus and mammillary body) out of nine structures. In the RF, bilateral increases in lCBF (except in the reticular thalamic nucleus, dorsal tegmental nucleus, median raphe and raphe magnus nucleus) were associated with significant rises in metabolism (from Mraovitch et al., 1992).*

## Primary cerebral vasodilation elicited by the RVL

As shown in 'The C1 area of the rostral ventrolateral medulla (RVL)' above, the RVL has direct anatomical connection to other brain regions with demonstrated cerebrovascular action. In urethane-anaesthetized rats, electrical or chemical (l-glutamate) stimulation of the RVL produced an increase in cortical cerebral blood flow as measured by laser-Doppler flowmetry (Saeki et al., 1989).

The effects of electrical and chemical stimulation of the C1 area of the rostral ventrolateral medulla (RVL) on rCBF and regional cerebral glucose utilization (rCGU) in chloralose-anaesthetized, paralysed and ventilated rats were recently examined by Underwood et al. (1992a). lCBF and lCGU were measured using iodoantipyrine and deoxyglucose as tracers. Maintaining arterial pressure by controlled haemorrhage comparable to control animals, unilateral electrical stimulation of C1 area of the RVL markedly increased rCBF bilaterally and symmetrically in most regions measured (Fig. 17) including hypothalamus (141 per cent of control), frontal cortex (145 of control), parietal cortex (154 per cent) and thalamus (165 per cent of control). In contrast, electrical stimulation of the RVL had no effect upon rCGU.

Microinjection of kainic acid in the RVL increased rCBF significantly and bilaterally in most brain regions including frontal cortex (139 per cent of vehicle), parietal cortex (145 per cent), and thalamus (162 per cent).

It is interesting that the pattern and magnitude of the rCBF responses were not different between electrical and chemical stimulation of the RVL indicating that both stimuli have activated the same neural mechanisms.

Bilateral electrolytic lesions of the C1 area will also abolish both the cerebrovascular vasodilation associated with electrical stimulation of the cerebellar fastigial nucleus (FN) and the fall of rCBF accompanying stimulation of intrinsic neurons of the FN. Thus, neurons of the C1 area also appear essential for the expression of the cerebrovascular effects during the stimulation of the FN.

Apparently contradicting results have been obtained by Maeda *et al.* (1991). Using labelled microspheres in anaesthetized (alpha-chloralose and urethane) rats, they have stimulated caudal ventrolateral medulla by microinjection of l-glutamate which elicited a pronounced decrease in cerebral blood flow within the ipsilateral cortex (from 41 ml/100 g/min to 29 ml/100 g/min) and the upper brainstem.

The discrepancy between these two observations might be due to the different physiological role of the neurons in the rostral and caudal ventrolateral medulla on the cerebral circulation. The increase of CBF observed by Maeda *et al.* may be mediated via the RVL. Indeed, these authors proposed that chemical stimulation of the caudal ventrolateral medullary neurons is expected to inhibit the RVL neurons which, in turn, is expected to result in a decrease in CBF.

### Intracerebral pathways mediating primary cerebral vasodilatory responses

The fact that FN stimulation produces widespread and bilateral cortical vasodilation is intriguing since the projection pathways from the FN do not innervate the cerebral cortex. Following a unilateral BF lesion, FN stimulation no longer elicited vasodilatation in the cerebral cortex ipsilateral to the BF lesion. This finding provides strong evidence that FN-elicited cortical vasodilatation is mediated via neural pathways originating in, or projecting through, the BF. Since the BF is the site of origin of the ascending cholinergic system to the cortex, it was hypothesized that these cholinergic projections mediate the FN-elicited cortical vasodilatation. However, several other pathways can be considered through which the FN could act on BF neurons, including projections from the FN to the medullary reticular nucleus and the parabrachial complex (Fig. 4). In addition, one cannot exclude that FN-elicited cortical vasodilatation is mediated by a polysynaptic input to the cortex such as that which may be implicated in the medial thalamic projection system that includes the CM-Pf complex.

As indicated above, the BF is widely recognized as the predominant source of cholinergic efferent fibres innervating the cerebral cortex. However, the morphological link between BF neurons and cortical microvessels may implicate either cortical or subcortical synapses. Indeed, it has been recently demonstrated (Linville & Arneric, 1991) that BF stimulation elicits an ipsilateral increase in thalamic CBF which may be mediated by BF-thalamic projections (Levey *et al.*, 1987). It is thus conceivable that the CM-Pf may be activated during BF stimulation which will in turn elicits cortical vasodilation. Reciprocal interaction between the CM-Pf and the BF may also occur.

The relationship between the CM-Pf efferents and cerebral blood vessels has not yet been investigated. It is conceivable that CM-Pf stimulation could activate mono- or polysynaptically other neural systems known to participate in CBF regulation. Activation of the CM-Pf could also alter CBF by direct CM-Pf cortical projections or by implicating local cortical neurons.

The pathways mediating cerebrovasodilatation evoked by RVL stimulation are also undetermined. It is well known that neurons of the RVL do not innervate the cerebral cortex. Thus, as for to the FN, elevation of CBF in the cerebral cortex must implicate multisynaptic pathways. Due to its thalamic projections, one of the candidates are diffuse thalamocortical projections arising within the intralamina thalamus such as the CM-Pf.

### Cholinergic mechanisms and role of local cortical neurons in primary vasodilation

The demonstration that lesion of the BF blocks the increase in rCBF elicited by FN stimulation raised the question whether the primary cerebrovascular vasodilatation is mediated by a cholinergic mechanism. In their experiments, Iadecola *et al.* (1986) and Arneric *et al.* (1987) showed that blockade of the muscarinic cholinergic receptors by systemic administration or topical cortical application of atropine abolishes the increase in rCBF elicited by FN stimulation, and that the cholinergic link in the FN-elicited increase in cortical rCBF may reside locally in the cortex. The evidence for the second

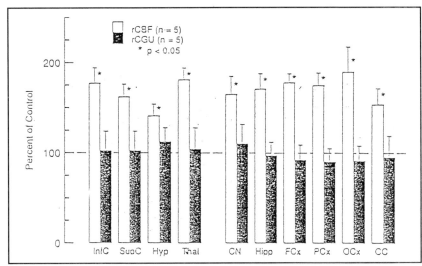

*Fig. 17. Effect of electrical stimulation of the rostral ventrolateral medulla (RVL) on regional cerebral blood flow (rCBF) and regional cerebral glucose utilization (rCGU). To control for the 45 min of stimulation required for measurement of rCGU, rCBF was measured in an additional group of animals following 45 min of stimulation. Note that the increase in rCBF to RVL stimulation was sustained and significant in all regions. Furthermore, the increase in rCBF was unassociated with any similar change in rCGU (from Underwood et al., 1992).*

conclusion was substantiated with several observations: first, the increases in cortical rCBF elicited by FN stimulation were in large part mediated by local muscarinic cholinergic receptors and second, the release of acetylcholine from the cortical surface was, in general, reduced during FN stimulation.

The source of ACh released within the cortex during FN stimulation and the neural pathway from the FN to the cortex is unknown. The most likely sources for the cholinergic link could be local cholinergic neurons of the cortex or nerve terminals of afferent cholinergic fibres arising from or passing through the BF as mentioned in the previous section.

Additional evidence that cholinergic mechanisms are implicated in cortical vasodilatation has been provided by studies on the BF elicited vasodilata-

tion. Using the mass spectroscopy technique that enables repeated measurement of local CBF it was shown that the vasodilatations elicited in the frontal cortex by electrical stimulation of the BF were modulated by cholinergic drugs (Dauphin *et al.*, 1991). Physostigmine (the acetylcholine esterase inhibitor) potentiated the stimulation-elicited increase in cortical blood flow by a factor of 2.2. In contrast, scopolamine (muscarinic receptor antagonist) administration resulted in a 65 per cent decrease in frontal cortex flow response to BF stimulation.

Thus, activation of cholinergic muscarinic receptors appears to occur when either the BF or the FN are activated, eliciting cortical vasodilatory responses.

Primary vasodilatation could be a consequence of two possible neurogenic

mechanisms acting upon the smooth muscle cells of the cortical vascular bed: direct innervation arising from the subcortical neural systems and/or a morphological link implicating local cortical neurons.

A monosynaptic cholinergic connection linking the subcortical neural system and cortical microvessels has not yet been described. In contrast, the primary vasodilatation appears to involve the action of local neurons in part associated with local release of acetylcholine: destruction of neurons, but not axons, within a restricted cortical area by the excitotoxin ibotenic acid abolishes the vasodilatation elicited by FN stimulation without affecting resting rCBF or the vasodilatation evoked by hypercarbia. These findings suggest that the cortical vasomotor response to FN stimulation is not mediated by a direct neurovascular projection originating from or passing through the BF, but that this projection must engage a local interneuron. Moreover, since the amount of acetylcholine released by FN stimulation in cortex is undetectable in contrast to the substantial release evoked by basal forebrain stimulation, it seems probable that the source of ACh involved in the vasodilatation may be local cholinergic neurons.

### Non-cholinergic mechanisms in primary cerebral vasodilation

Two neural systems which were shown to elicit a rise in cortical CBF independently of glucose metabolism, the CM-Pf and the RVL, do not implicate cholinergic mechanisms. Using laser Doppler flowmetry, Goadsby *et al.* (1993) have shown that cerebrovascular effects elicited by electrical stimulation of the CM-Pf do not involve cholinergic or adrenergic mechanisms. The cortical vasodilatory response to electrical stimulation of the CM-Pf was not blocked by the muscarinic antagonist scopalamine (1 mg/kg) or by the nicotinic antagonist mecamylamine (4 mg/kg). The response was also unaffected by the β-adrenoceptor antagonist propranolol (1.5 mg/kg). In this study the authors did not assess directly the prospect of a peptide transmitter but this could be a possible link from the CM-Pf to the cerebral circulation. A potential candidate as mediator for cortical vasodilation elicited by the CM-Pf stimulation is nitric oxide (NO). The role of nitric oxide in the cortical vasodilation during stimulation of intrinsic neural pathways will be discussed later in this chapter.

Although the non-cholinergic nature of the RVL-elicited cortical vasodilatory response has been recently demonstrated, no detailed pharmacological study has examined the question.

### Functional significance of primary cerebral vasodilation

The evidence indicates that anatomical connections arising within specific neural systems dedicated to cerebrovascular action reaches cortical blood vessels to modify cortical blood flow independently of metabolism. Functionally, it is conceivable that the vascular action may be closely related to the anticipating behaviour (FN), cognitive processes (BC), complex behaviour (CM-Pf) or adjustement to critical situations such as hypoxia and/or ischaemia (RVL).

FN stimulation, in addition to cerebrovascular changes, may elicit anticipatory changes in CBF to increase flow and substrate availability for cere-

bral metabolism. This would require anticipation of the possibility of a widespread increase in metabolism elicited by the actual performance of a given behaviour. As previously indicated, the BF is implicated in learning and memory processes. In addition, it is well known that this region undergoes marked degeneration in dementia of the Alzheimer type. It is, therefore, conceivable that the BF could also regulate cerebral cortical blood flow in order to ensure efficient and adequate blood flow and energy substrates needed for cognitive processes. The CM-Pf, having a close relationship with neural systems implicated in arousal, pain, motor control and cerebrovascular control, could conceivably play a pivotal role in the neural integration of CBF in a complex behaviour.

Concerning the functional significance of the primary vasodilatation elicited by the RVL, it is conceivable that the RVL may be of importance in mediating autonomic adjustments to cerebral hypoxia and/or ischaemia. Indeed, it has been argued that the RVL may serve as an oxygen sensor analogous in function to chemoceptors of the carotid body (Underwood et al., 1992b). The fact that cerebral hypoxia elicits a marked elevation of rCBF unassociated with changes in rCGU raises the possibility that the RVL may act not only to elicit the systemic circulation adjustment to hypoxia but also those of the cerebral circulation.

### Secondary neurogenic cerebral vasodilation

### Secondary cerebral vasodilation elicited by DMRF

It has been known for many years that the brainstem may regulate cerebral circulation (Reis, 1984). For example, it was demonstrated that electrical stimulation of an anatomically specific region within the medullary reticular formation can modify global cerebral metabolism and consequently cerebral blood flow (Iadecola et al., 1983b). Thus, in contrast to eliciting changes in CBF unassociated with changes in glucose metabolism, the increase in CBF elicited by stimulation of the DMRF is proportional to changes in CGU (Iadecola et al., 1983). rCBF increased significantly in 25 of 27 brain areas analysed. The largest increases were seen in the cerebral cortex (up to 163 per cent), globus pallidus (151 per cent), intralaminar thalamus (169 per cent), parabrachial complex (186 per cent) including regions in the lower brain stem such as the nucleus reticularis gigantocellularis (188 per cent). In comparison, rCGU significantly increased in 21 of the 27 brain structures analysed. The most pronounced changes occurred in the cerebral cortex (up to 168 per cent in the frontal cortex), the intralaminar thalamus (156 per cent), and the nucleus reticularis gigantocellularis (197 per cent).

That the widespread cerebral vasodilation evoked by DMRF stimulation is a consequence of a widespread increase in cerebral metabolism is well illustrated using linear correlation analysis. Indeed, the relationship between CBF and CGU is highly correlated in controls as well as during

DMRF stimulation, and the regression slopes under these two conditions are not significantly different.

### Functional significance of the secondary neurogenic cerebral vasodilation

The finding that a small region of the dorsal medullary reticular formation can increase global cerebral metabolism has been associated with the possibility that a small population of neurons may be pacemakers regulating cerebral metabolism. However, lesion studies have not been performed to prove this hypothesis.

## Central neurogenic vasoconstriction

Several previous observations have demonstrated that the brain contains networks with the capacity to reduce its own metabolism and flow. Cerebral vasoconstriction has mostly been elicited from the neural systems located within the rostral pons: the LC, the DR and the PBN. Potent cerebral vasoconstrictory responses have also been elicited from the FN. As primary vasoconstriction is defined as those cerebrovascular changes elicited by an intrinsic neural network when the decrease in CBF is independent of glucose metabolism, and as secondary vasoconstriction when the CBF decrease is associated with a parallel decrease in brain metabolism.

## Primary neurogenic vasoconstriction

Primary cerebral vasoconstriction elicited from a specific neural system was found to be regionally restricted to the lateral PBN and the DR.

### Primary cerebral vasoconstriction originating in the lateral PBN

Unilateral electrical stimulation of the lPBN, in alpha-chloralose anaesthetized rats, significantly decreases regional CBF in all areas of the ipsilateral cerebral cortex. The largest reduction (–36 per cent) was observed in the occipital and the smallest (–25 per cent) in the parietal cortex. In contrast to flow, stimulation of the PBN did not alter rCGU in the cortex.

In addition to primary cortical vasoconstriction, electrical stimulation of the PBN elicits a selective and regionally differentiated effect upon rCBF and metabolism. For example, during the same experiment, rCBF was significantly increased bilaterally in the hypothalamus and unilaterally in the medulla (+ 31 per cent), but was unchanged in the hippocampus, the caudate nucleus and interestingly in the contralateral cerebral cortex. PBN stimulation significantly increased rCGU bilaterally in the medulla, thalamus and hypothalamus. Thus, the PBN induces primary vasoconstriction restricted to the ipsilateral cerebral cortex.

### Primary cerebral vasoconstriction originating in SNc

Another example of a restricted primary cerebral vasoconstriction was recently observed within the caudate nucleus following unilateral lesion of the mesencephalic dopaminergic neurons in the rat (Mraovitch et al., 1993).

Changes in local CBF and CGU were assessed in dopaminergic primary target areas in the rat 6 weeks after unilateral lesion of the substantia nigra pars compacta (SNC) and the adjacent ventrotegmental area (VTA) using 6-hy-droxydopamine (6-OHDA). lCBF and lCGU were determined using autoradiographic methods. Dopaminergic deafferentation provoked a marked unilateral decrease in lCBF in the dorso-lateral portion of the rostral caudate-putamen. The decrease in lCBF was not associated with significant changes in glucose metabolism. It was concluded that lesion of dopaminergic afferents to the caudate-putamen appear to provoke a sustained decrease in the basal blood flow with unchanged local metabolic activity.

It is thus conceivable that the decrease in lCBF might be the consequence of withdrawal of a dopaminergic vasodilatory excitation creating a long-term imbalance between inhibitory and excitatory vasomotor activity.

### Primary cerebral vasoconstriction originating in the DR

The role of the dorsal raphe nucleus in cerebrovascular regulation has been studied by several investigators (for references see Underwood et al., 1992b).

Although, previous investigation of the effect of the DRN stimulation on CBF have yielded conflicting results with both cerebral vasodilation (Goadsby et al., 1985a,b) and cerebral vasoconstriction (Bonvento et al., 1989; Cao et al., 1992) being reported, the most recent experimental findings based on the precise anatomical localization of stimulating sites have demonstrated that either an increase or a decrease in cortical blood flow can be elicited depending on the area of stimulation within the DR (Fig. 18) (Underwood et al., 1992b). Sustained DR stimulation resulted in increased cortical blood flow from the caudal portion and decreased cortical blood flow from the rostral portion of the DR.

The effect of focal chemical (l-glutamate) stimulation of the DR on the regional blood flow in cerebral cortex was studied in urethane-anaesthetized rats using laser-Doppler flowmetry. Microinjection of l-glutamate into the DR produced a decrease in cortical blood flow (Cao *et al.*, 1992) The decrease was abolished by prior i.v. administration of either methylsergide (a nonselective 5-HT antagonist) or ketanserin (a selective 5-HT2 antagonist) demonstrating that the DR-elicited release of serotonin in the cortex constricting microvessels by activating 5-HT2 receptors.

Relatively few studies have examined alterations in lCGU during electrical and/or chemical stimulation of the DR. In unanaesthetized rats electrical stimulation of the DR increased lCGU as measured with 2-DG autoradiographic technique (Cudennec *et al.*, 1988).

Electrical stimulation of the DR (Bonvento *et al.*, 1989) elicited widespread decreases in lCBF unassociated with changes in cerebral glucose utilization. A weak but significant decrease in rCBF was found in 24 out of 33 regions investigated, for example, the frontoparietal cortex (–19 per cent), the accumbens (–16 per cent), the median raphe nucleus (–23 per cent) and the caudate nucleus (–21 per cent). In a subsequent study (Bonvento *et al.*, 1991), the same authors using a similar experimental preparation showed that DR stimulation induced a significant increase in rCGU in eight brain regions out of 33 investigated. The greatest increases were seen in the dorsal tegmental nucleus (+52 per cent), the red nucleus (+29 per cent) and the zona inserta (+24 per cent). These results indicate that the DR via ascending serotonergic system may regulate cerebral circulation by a primary vasoconstrictory effect on cerebral blood vessels.

## Secondary neurogenic vasoconstriction

The activation of two neural systems, the medial parabrachial nucleus and the fastigial nucleus, produced cerebral vasoconstriction secondary to the decrease in cerebral metabolism.

### Secondary vasoconstriction originating in the medial PB

Electrical stimulation of sites restricted to the medial portion of the PBN resulted in a widespread reduction of rCBF. The most pronounced changes were seen in the cerebral cortex where the decreases were bilateral and ranged from –35 per cent of control in the occipital to –31 per cent in the parietal cortex. A substantial decrease also occurred bilaterally in the caudate nucleus (–29 per cent). The reduction in rCBF was associated regionally with a decrease in rCGU. Significant hypometabolism was seen in the cerebral cortex (frontal and occipital, –24 per cent and parietal –28 per cent) as well as in the hippocampus (–33 per cent). Linear regression analysis showed that the reductions of rCBF were proportional to the reductions in rCGU.

This study demonstrates that the PBN may have a role in regulating the cerebral as well as the systemic circulation. Moreover, it confirms that resting rCBF and metabolism are not at the lowest possible level but are set near a mid-range so that reduction as well as enhancement can be produced by the action of intrinsic neural networks.

### Secondary vasoconstriction originating in the FN

Cerebral vasoconstriction secondary to a decrease in metabolism was recently observed during chemical stimulation of the FN in anaesthetized rats. Microinjection of kainic acid in the FN resulted in widespread and symmetrical reductions in rCBF. The reduction in rCBF was most pronounced in the cerebral cortex where flow in frontal, parietal and occipital cortices ranged from –44 per cent to –52 per cent of control. In addition to the cerebral cortex, kainic acid significantly decreased rCGU bilaterally in most of the brain regions studied.

## Efferent mechanisms of cerebrovascular responses elicited by intrinsic neural systems

The neural systems that regulate cerebral blood flow are anatomically complex and employ a number of neurotransmitters. Although direct anatomical evidence for local neurogenic control of CBF is still incomplete, our understanding of the mechanisms by which intrinsic neural systems influence and regulate cerebral vascular smooth muscle via neurotransmitters has received much attention in recent years.

In this section we will provide a comprehensive presentation of current understanding of the modulatory effect of the extrinsic neural system, the circulating catecholamines and nitric oxide on cerebral blood flow elicited

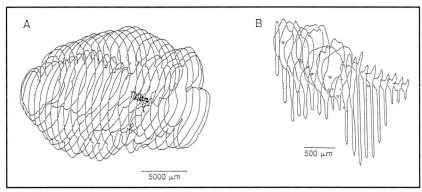

*Fig. 18. Three-dimensional reconstruction of the rat brain and dorsal raphe nucleus and functional organization of cerebrovascular responses elicited by electrical stimulation within different regions of the DRN. The left panel (A) illustrates a total reconstruction of the rat brain and the orientation of the DRN within it. The right panel (B) is a magnification of the DRN with a similar parasagittal rotation as panel A and illustrates the localization of stimulation sites eliciting either cerebrovasodilation (+: n = 5), cerebrovasoconstriction (Δ: n = 9), or no change in CoBF (×: n = 2). Note that stimulation within rostroventral portions of DRN elicited decreased CoBF, whereas stimulation in caudal portions of DRN resulted in increased CoBF (from Underwood et al., 1992b).*

by the intrinsic neural systems.

## Effects of neuronal intrinsic–extrinsic interactions upon cerebral blood flow

As seen in Chapters 5 and 11, it is now well established that most portions of the cerebral vascular tree, extraparenchymal and intraparenchymal, including arteries, arterioles, veins and capillaries are innervated or at least have close contacts with nerve varicosities. While pial arteries are densely innervated by sympathetic nerves arising in the superior cervical and stellate ganglia, innervation of intraparenchymal arteries and arterioles may not extend deeper than the penetrating arterials. In contrast to the sympathetic system, cholinergic fibres innervating extraparenchymal arterial force, do not appear to penetrate into the brain.

In this section, the functional implication of the sympathetic nervous system related to the intrinsic regulation of the CBF will be discussed.

### Effect of cervical sympathectomy on primary vasodilation

The effect of unilateral electrical stimulation of the FN, RVL, and CM-Pf was analysed in animals in which the ipsilateral preganglionic trunk was transected.

The increase in rCBF during stimulation of either the FN (Nakai *et al.*, 1982) or the RVL (Underwood *et al.*, 1992a) did not differ between the sympathetically innervated and denervated side. This indicates that the FN or the RVL elicited increase in CBF is independent of postganglionic sympathetic nerves arising from the superior cervical ganglion (SCG).

In contrast, the CM-Pf elicited cerebrovasodilation is shown to be, in part, dependent on the integrite of the sympathetic innervation arising from the SCG (Mraovitch *et al.*, 1986).

After ipsilateral sympathectomy, the CM-Pf-elicited cortical rCBF was significantly reduced on the denervated

side. For example, in the frontal cortex CM-Pf stimulation increased rCBF up to $130 \pm 11$ ml/100g/min. Following sympathectomy the CM-Pf elicited increase was only $102 \pm 4$ml/100g/min.

The mechanisms by which acute sympathectomy coupled to the CM-Pf stimulation reduces elevated cortical blood flow remains to be elucidated.

### Effect of cervical sympathectomy on secondary vasodilation

Following unilateral transection of the cervical sympathetic trunk, the DMRF stimulation increased CBF bilaterally in all brain regions. In the cortical regions (frontal and parietal cortices) and in the caudate nucleus, the increase in CBF was slightly but significantly higher in the denervated side. Thus, the DMRF elicited cerebrovascular dilation is in part dependent on sympathetic excitation. However, the finding that the DMRF elicited cortical and caudate cerebrovasodilation following blockade of nerve traffic through the SCG increases in magnitude is in opposition to the finding that sympathectomy coupled to the DMRF stimulation decreased CBF responses. The reasons for this discrepancy is presently unknown.

### Effect of cervical sympathectomy on secondary vasoconstriction

Since, at present, there are no available data on the effect of acute sympathectomy on the primary cerebral vasoconstriction elicited by a specific brain stimulation, we shall review data concerning unilateral transection of the cervical sympathetic trunk associated with electrical stimulation of the mNPB and the FN known to elicite secondary cerebral vasoconstriction.

After transection of the CST, the mNPB elicited decrease in rCBF in most brain regions did not differ between the innervated and denervated side. However, in hippocampus, caudate nucleus and frontal cortex, rCBF on the denervated side was slightly (less than 10 per cent) but significantly higher than that on the innervated side. Thus, transection of the CST did not substantially modify the decrease in rCBF elicited by mNPB stimulation.

The effect of unilateral transection of the CST also studied after microinjection of kainic acid into the FN (Chida *et al.*, 1989) which is known to produce a global reduction in CBF. Transection of the CST had no effect on the magnitude of the reduction of rCBF elicited by chemical stimulation of the FN.

## What are the neural pathways mediating neuronal intrinsic–extrinsic interaction upon cerebral blood flow?

The sympathetic adrenergic nerves innervating cerebral vessels arise from cell bodies contained in the superior cervical ganglion (SCG). Sympathetic fibres have been identified on all pial arteries and the internal carotid and vertebral arteries, thus innervating predominantly extracerebral vasculature (Fig. 19). As shown in Chapter 11, one of the important features of the sympathetic innervation of the cerebral circulation is the fact that it does not, at least in the rat, extend as far as the penetrating arterioles.

## Effect of hormonal–neuronal interaction upon cerebral blood flow and metabolism

Over the last few years evidence has accumulated and revealed a modulatory role of adrenal hormones on CBF increases during stimulation of intrinsic neural pathways. In addition to the well-demonstrated role of adrenal catecholamines in cerebrovasodilation evoked from DMRF stimulation, the CM-Pf-evoked rise in cerebral blood flow was also found to depend, in part, on the integrity of the adrenal glands.

### Effect of adrenalectomy on the primary cerebrovasodilation

The electrical stimulation of the intralaminar thalamus including the region of the CM-Pf has been shown to release plasma adrenal catecholamines (Gauthier, 1981).

The possible contribution of adrenal hormones to the cerebrovascular responses elicited by the CM-Pf stimulation has been compared in animals with and without acute bilateral adrenalectomy. Acute adrenalectomy decreased elevated rCBF during CM-Pf stimulation in all brain regions (Mraovitch *et. al.*, 1986). The most pronounced decrease was observed bilaterally in cortical region (frontal cortex, –29 per cent; parietal cortex, –29 per cent; and occipital cortex, –27 per cent).

### Effect of adrenalectomy on the secondary cerebrovasodilation

The observation that after transection of the spinal cord at the first cervical segment, the increase in rCBF elicited by DMRF stimulation is reduced or

abolished (Iadecola *et. al.*, 1983a,b) suggested that probably adrenal medullary catecholamines, presumably released into the systemic circulation during the DMRF stimulation, may modulate the cerebral blood flow increase.

In the subsequent study Iadecola *et al.* (1987) sought to answer two basic questions: are cathecholamines released from the adrenal glands during DMRF stimulation and, if so, do they participate in the increase in cerebral blood flow? Stimulation of DMRF markedly elevated epinephrine and norepinephrine. After adrenalectomy, DMRF stimulation failed to increase epinephrine significantly. The elevation of catecholamine was specific for the DMRF region since stimulation of the adjacent median logitudinal fasciculus failed to alter the level of catecholamine in plasma. Bilateral adrenalectomy markedly reduced or abolished the elevation in rCBF elicited by DMRF stimulation. The greatest reduction occurred in frontal (–32 per cent) and parietal (–30 per cent) cortex. In some brain regions the increase was abolished (hippocampus and colliculi); in other areas (thalamus, caudate nucleus) the increase did not reach statistical significance. Based on these results the authors distinguished adrenal-dependent and adrenal-independent components of the cerebral vasodilation elicited from the DMRF (Reis and Iadecola, 1986a,b).

These findings suggest that circulating catecholamines released from the adrenal medulla contribute to the flow increase mediated by beta-adrenergic receptors was the fact that propranolol significantly reduced the increase in rCBF elicited by DMRF stimulation.

## What are the neural pathways mediating adrenal-dependent CBF changes?

Adrenal-dependent modulation of cerebral blood flow suggests that DMRF (and possible CM-Pf) stimulation activates descending pathways to preganglionic sympathetic neurons in the intermediolateral columns of the spinal cord innervating the adrenal medulla. Since catecholamines do not cross the blood–brain barrier (BBB), it is possible that the increase in cerebral blood flow is then modulated via multisynaptic neural pathways originating in the regions lacking the BBB, such as the cirumventricular organs (CVO) area postrema and subfornical organs, for example (Fig. 20). Thus, DMRF or CM-Pf activation will stimulate release of catecholamines and 'sensitize' the specific brain region CVO, to circulating epinephrine which in turn elicits a metabolically dependent increase in cerebral blood flow.

## Effects of nitric oxide–intrinsic neural system interaction upon cerebral blood flow

The evidence presented recently suggests that nitric oxide may participate in the neurogenic vasodilatation (for review see Iadecola *et al.*, 1994). Several studies have indicated that nitric oxide participates in the increases in cerebral blood flow elicited by stimulation of intrinsic neural systems and their pathways.

As shown in the section 'Basal forebrain (BF)' the basal forebrain provides most of the cholinergic fibres innervating the cortex. There is increasing evidence which suggests that the cortical vasodilation elicited by acetylcholine is mediated by nitric oxide (NO) (Palmer *et al.*, 1987) and that NO release may have an important role in the regulation of CBF (Gally *et al.*, 1990). In the brain, NO is synthesized from L-arginine by the enzyme NO synthase (NOS). It has recently been demonstrated (Toda and Okamura, 1990) that electrical stimulation of cholinergic nerves innervating cerebral arteries produced a dilation which was abolished by L-NMMA (*N*-monomethyl-L-arginine), an NO synthesis inhibitor. The inhibition was reversed by L-arginine.

Following studies showing that NO participates in the increase in cerebral blood flow elicited by hypercapnia (Iadecola, 1992) the question has arisen: is NO involved in the increase in cortical blood flow following the stimulation of cholinergic fibres originating in the BF or the FN, polysynaptically activating the BF? Iadecola *et al.* (1993) have shown that the FN-elicited increase in CBF is mediated by NO via activation of intrinsic neural pathways.

Using laser Doppler flowmetry in urethane-anaesthetized rats Raszkiewicz *et al.* (1992) have assessed changes in cortical CBF elicited by electrical stimulation of the BF. They found that the stimulus-elicited increase in cortical CBF was significantly reduced following NOS inhibition, and reduction

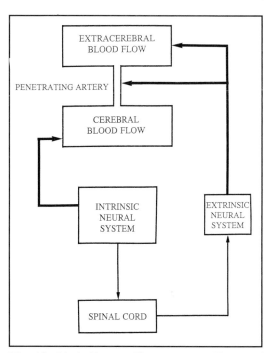

*Fig. 19. Block diagram illustrating possible neural intrinsic–extrinsic interaction upon CBF.*

was dose-related. In contrast, L-arginine significantly potentiated the BF-elicited cortical blood flow response up to 38 per cent. Adachi *et al.* (1992b) using a similar animal preparation have confirmed that NO is involved in the increase in cortical blood flow following the stimulation of the cholinergic fibres originating in the BF.

## Origin and distribution of nitric oxide synthase-containing nerve fibres

Although mapping of the distribution of neuronal NOS in rat brain (Schmidt *et al.*, 1992) has shown that NOS is, among other brain regions, located in the CM-Pf region of the thalamus, the site of NO production during FN stimulation is not known.

## Integration of efferent mechanisms

In the preceding sections we have described the fundamental ingredients of cerebrovascular reflex control: (1) specific brain regions, (2) afferent and efferent pathways controlling the flow of action potentials, and (3) effector mechanisms.

Thus heterogeneous changes in lCBF and/or lCGU during brain stimulation involve differential contributions of metabolic, hormonal (epinephrine) and neurogenic (extrinsic and intrinsic) mechanisms (Fig. 21). Each of these mechanisms contributes to the appropriate adjustment of the immediate and/or anticipated local energy and blood flow requirements within a given brain region or neural network, necessary for expression of a specific neurogenic function.

## Conceptual considerations

## Coupling and uncoupling of CBF and metabolism

The interrelation between brain functional activity, metabolism and cerebral blood flow, known as the metabolic homeostasis hypothesis, was first suggested in 1890 by Roy and Sherrington as we saw in 'What are the parameters of cerebrovascular function?' earlier. Numerous recent studies using brain stimulation coupled with quantitative autoradiography for measuring both lCBF and lCGU have shown parallel changes in blood flow and glucose utilization throughout the brain (for reviews see Reis & Iadecola, 1989; Mraovitch *et al.*, 1989). However, it has also become clear that the metabolic homoeostasis hypo-

thesis cannot explain observations made during intracerebral stimulation of some regions such as the FN and the CM-Pf which elicit two- or threefold increases in local CBF with small or insignificant changes in local CGU. Thus, it seems that under certain experimental conditions lCBF and/or lCGU responses can be independently altered, implying that independent pathways may regulate the expression of these variables at their respective target areas.

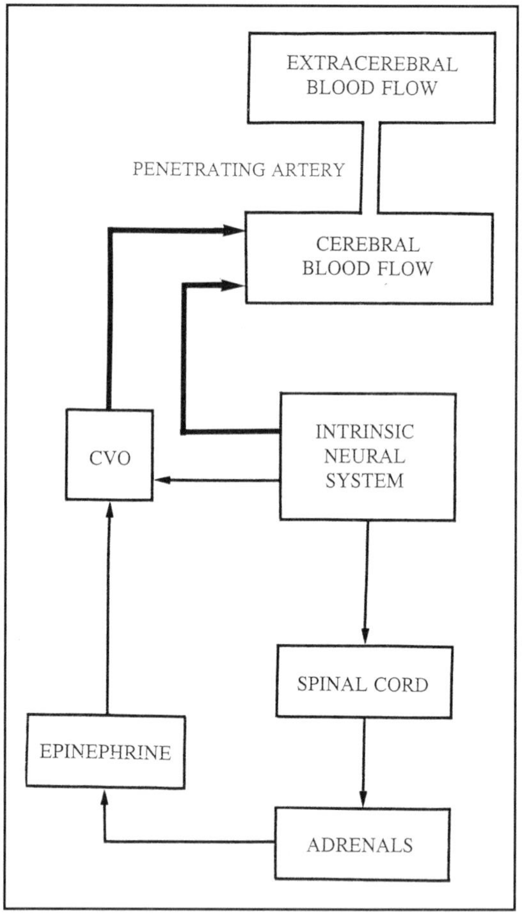

Usually, the changes in lCBF and lCGU throughout the brain have been either grouped or analysed considering all regions as independent measures. Recently attempts are being made to extend this global analysis by taking into consideration anatomically and functionally coupled neural systems.

## Global analysis

In Chapter 9, W. Kuschinsky has described the basic principle of CBF regulation including the concept of coupling of blood flow and metabolism. In this chapter it was emphasized that intrinsic neural systems can also have a specific influence on the coupling of CBF and metabolism. As indicated in Chapter 8, the coupling of lCBF and lCGU indicates that cerebral structures are perfused in proportion to their metabolic demand. Using a linear regression analysis for

*Fig. 20. Block diagram illustrating possible neural–hormonal interaction upon CBF.*

all brain regions measured, a tight correlation with a correlation coefficient close to unity between lCBF and lCGU values is usually found. Uncoupling or dissociation between lCBF and lCGU is assumed when the changes in these variables do not change either proportionally or in the same direction. For example, blood flow may decrease or increase whereas glucose use remains unchanged.

In studying the relationship between lCBF and lCGU it was found that electrical stimulation of the FN elicits dif-

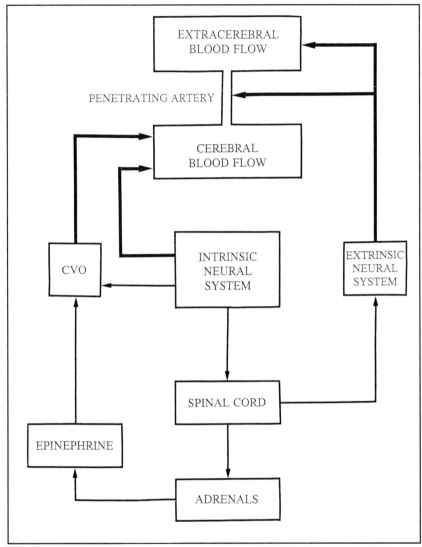

*Fig. 21. Summary of efferent mechanisms by which CBF is regulated.*

previously according to neuro-anatomical and/or functional criteria, the quantitative analysis of the relationship between flow and metabolism considered all measured regions as a whole. As we have seen, a tight correlation between lCBF and lCGU was invariably found, which was interpreted as a tight coupling between these two variables. In extending the global analysis to anatomically and functionally well-defined neural systems, Mraovitch *et al.* (1992) found that, under control conditions, there was a good correlation between lCBF and lCGU when the specific neural systems were analysed. For example, electrical stimulation of the CM-Pf resulted in a significant increase in the slope of the regression line between local blood flow and glucose utilization, analysed globally and within the reticular formation (RF) and the extrapyramidal motor system (EMS). This indicates that the coupling mechanisms for the RF and the EMS were not disturbed during stimulation, but merely reset. Although the coupling of lCBF and lCGU was preserved in the limbic system, CM-Pf stimulation did not significantly change their relationship from that of the unstimulated sham controls. The magnitudes of the increases in the lCBF/lCGU ratios for the three systems were also significantly different (RF being higher than the other two). It appears, thus, that CM-Pf stimulation can elicit a differential pattern of cerebrovascular and glucose metabolic response within well-defined neural networks. This finding provides, for the first time, evidence suggesting that anatomically and functionally defined neural networks have specific vascular and metabolic regulatory mechanisms.

ferential changes in lCBF and lCGU throughout the brain. Some regions showed only moderate increase in lCBF, whereas it was increased two- to threefold in others. Although the increase in lCBF was associated with a marked focal alteration in lCGU in some brain regions, the magnitude of the overall lCGU increase was small or even not significant.

Dissociation between lCBF and lCGU has also been observed during stimulation of the CM-Pf thalamic nucleus and the dorsal raphe nucleus (DR), and lesion of the substantia nigra pars compacta (SNc).

## Specific neural systems

Although the changes in lCBF and lCGU across the brain were grouped

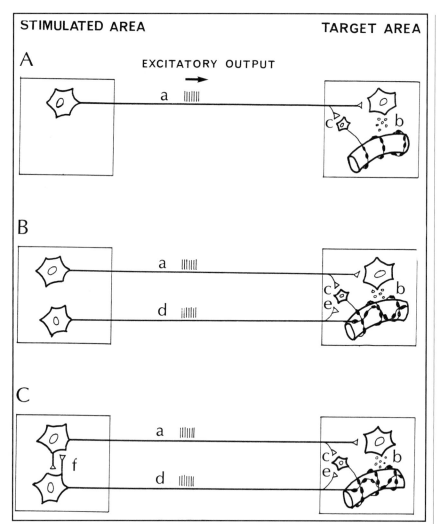

*Fig. 22. Summary of possible interactions leading to the cerebral blood flow-metabolism coupling. A. Metabolic (b) and neurogenic (via local neurons) coupling (c). B: Differentiated control of CBF and metabolism. The coupling implicates metabolic and independent neuronal (c and e) factors. C: Coupling occurs within the neurons at the stimulated region (f) implicated in the CBF regulation by appropriate adjustment of the cerebrovascular and 'metabolic' informations.*

## Where does coupling take place?

It is generally believed that coupling between neural activity and blood flow occurs locally at the target area (Fig. 22a) and involves metabolic (Fig. 22b) and neurogenic factors such as local neurons (Fig. 22c)

However, recent data demonstrating the dissociation of blood flow and metabolism in specific brain regions suggest that separate mechanisms may regulate these two variables in addition to that classically assumed to be in operation (Fig. 22b). It is also conceivable that differentiated information intended for cellular metabolic regulation and appropriate local vascular adjustment have their origins within the 'regulatory' (stimulated) regions implicated in the cerebrovascular and metabolic responses (Fig. 22cf).

## Are there cerebrovascular neurons?

Until recently, the results of research on intrinsic neural systems regulating CBF might be called descriptive. Several anatomically restricted cerebrovascular centres have been defined.

Two basic conclusions can be drawn from these studies. First, it is now clear that there is no single centre for cerebrovascular control. Second, the identified regions implicated in cerebrovascular regulation also participate in a variety of other functions. For example, a number of studies have demonstrated that the parabrachial nuclei (PB) participate in visceral, neuroendocrine and behavioural functions, including cardiovascular and cerebrovascular regulation (see Mraovitch, 1989). Activation of the FN is known to elicit a range of behaviours including feeding, drinking, predatory aggression, and cardiovascular and cerebrovascular responses (see Reis and Iadecola, 1989). The CM-Pf participates in neural mechanisms subserving cortical desynchronization and recruitment phenomena, conditioning, pain-related neuronal responses, blood pressure regulation and cerebrovascular control mechanisms (see Mraovitch, 1989).

Although some of the issues concerning the CBF regulation have been clearly established, many remained to be answered. For example, what are the pathways through which the effects on CBF are mediated and the efferent mechanisms through which neural events affect the blood vessels?

And (a challenging issue), are there cerebrovascular neurons that process 'cerebrovascular information' intended for intracerebral vascular smooth muscle?

At present, there is little sufficient morphological and electrophysiological data supporting the hypothesis of the existence of specific cerebrovascular neurons, nor is there data concerning the possible interaction between cardiovascular and cerebrovascular control mechanisms.

However, several lines of evidence indicate that some brain regions associated with cardiovascular and cerebrovascular regulation might operate via specific 'cardiovascular' and/or 'cerebrovascular' neurons and pathways. There is evidence that the FN may be functionally and anatomically differentiated with respect to cardiovascular and cerebrovascular control mechanisms (Mraovitch *et al.*, 1986) with neurons of the rostral FN mediating changes in arterial pressure and those along its rostrocaudal axis, changes in CBF. The frequency/response curves for neurons regulating the peripheral and the cerebral circulation are different. Further support for this hypothesis is the observation of Iadecola *et al.* (1990b) that the effects of alpha-chloralose anaesthesia on the elevation of arterial pressure and regional CBF elicited by FN stimulation are distinct, such that the CBF responses are more susceptible to the effects of the anaesthetic agent than are the arterial pressure responses, indicating that these responses are mediated by different neuronal populations and pathways. Finally, within the rostral ventral medulla, it is well known that caudally projecting PNMT neurons are implicated in the regulation of arterial pressure. It has been recently demonstrated (Iadecola *et al.*, 1993) that anatomically distinct nitric oxide synthase (NOS)-containing neurons projecting rostrally may be implicated in CBF control, suggesting specificity of efferent projections regulating CBF and metabolic responses.

**Acknowledgements:** The author thanks M. Underwood for helpful comments on the manuscript.

## Summary

Although the concept of the central nervous system regulating CBF and/or metabolism by activation of different brain regions is now widely accepted, the manner in which these regions and their pathways are organized is still unknown. The major factors impeding our understanding of how the CNS controls its own circulation have been a lack of information on the neuroanatomical organization of central cerebrovascular pathways and the variety of experimental approaches used to study cerebrovascular functions.

There is no doubt that the networks presented in this review are incomplete and other regions and neuronal circuits will likely appear. However, it is of interest and is perhaps more than just coincidence that the regions implicated in the regulation of CBF and metabolism are reciprocally interconnected and organized in horizontal and vertical loops. Another feature of this circuit is that two regions, the centromedian-parafascicular complex and the nucleus basalis, which are known to have diffuse cortical projections and cerebrodilatory effects upon the cortex, are the sites of convergence from all the regions believed to be implicated in CBF and/or metabolic regulation.

Our understanding of the complex interaction between different regions and their action upon the intracerebral vasculature will be greatly extended by electrophysiological and neuropharmacological studies on cerebrovascular and/or metabolic elicited responses, and detailed systematic anatomical studies on the axonal-vascular connections.

## Abbreviations

| | |
|---|---|
| AA, AAD, AAV, | anterior, dorsal, and ventral amygdaloid area |
| ACB | accumbens nucleus |
| APT | anterior pretectal area |
| B | basal nucleus of Meynert |
| BF | basal forebrain |
| CN | caudate nucleus |
| CPu | caudate putamen |
| CST | cortico-spinal tract |
| CXC24 | cingular cortex (area 24) |
| CXC32 | cingular cortex (area 32) |
| CXF | frontal cortex |
| CXP | parietal cortex |
| DN | dentate nucleus |
| DTg | dorsal tegmental nucleus |
| ECN | external cuneate nucleus |
| EP | entopeduncular nucleus |
| FCx | frontal cortex |
| FStr | fundus striati |
| GP | globus pallidus |
| HDB | horizontal nucleus of the diagonal band of Broca |
| HIPP | hippocampus |
| IC | internal capsule |
| ICP | inferior cerebellar peduncle |
| IN | nucleus interpopsitus |
| IO | inferior olivary nucleus |
| IP | interpeduncular nucleus |
| LH | lateral hypothalamic area |
| LHb | lateral habenular nucleus |
| LS | lateral septal nucleus |
| LVN | lateral vestibular nucleus |
| MB | mammillary complex |
| MCPO | magnocellular preoptic nucleus |
| MeA | medial anterior amygdaloid nucleus |

| | | | | | | |
|---|---|---|---|---|---|
| MnR | median raphe | PG | periventricular (pontin) grey | SI | substantia innominata |
| MLF | medial longitudinal fasciculus | PnC | pontin reticular nucleus caudal | SNc | substantia nigra compacta |
| MRF | mesencephalic reticular formation | PP | nucleus prepositus | SNr | substantia nigra reticulata |
| | | RD | raphe dorsalis | STN | spinal trigeminal nucleus |
| MS | medial septal nucleus | RMg | raphe magnus | STT | spinal trigeminal tract |
| MVN | medial vestibular nucleus | RO | raphe obscurus | V | IV ventricle |
| NA | nucleus ambiguus | RPn | raphe pontis | VL | ventrolateral thalamic nucleus |
| NSV | spinal trigeminal nucleus | RPa | raphe pallidus | VN | vestibular nucleus |
| NTS | nucleus tractus solitarii | RTh | reticular thalamic nucleus | VP | ventral pallidum |
| OCx | occipital cortex | RtTg | (RTN) reticular tegmental nucleus | VTA | ventral tegmental area |
| Par 1 | parietal cortex areas 1 | | | VTg | ventral tegmental nucleus |
| Par 2 | parietal cortex areas 2 | S | subiculum | ZI | zona incerta |
| PCx | parietal cortex | SCP | superior cerebellar peduncle | VII | facial nucleus |

## References for further reading (including references cited)

### General reviews

Iadecola, C. & Reis, D.J. (1990): Continuous monitoring of cerebrocortical blood flow during stimulation of the cerebellar fastigial nucleus: a study by laser-Doppler flowmetry. *J. Cereb. Blood Flow Metab.* **10**, 608–617.

Mraovitch, S. (1989): Functional organization of central cerebrovascular pathways with special reference to the thalamus. In: *Neurotransmission and cerebrovascular function II*, eds. J. Seylaz & R. Sercombe, pp. 343–368. Amsterdam: Elsevier Science.

Rakic, P. (1975): Local circuit neurons. In: *Neurosciences research program bulletin*, eds. F.O. Schmitt *et al.* Cambridge, MA.: The MIT Press.

Reis, D.J. & Iadecola, C. (1986a): Intrinsic neural control of cerebral circulation and metabolism. In: *Cerveau et hypertention arteriel*, eds. A. Bes & G. Gerand, pp. 76–89. Paris: Messon.

Reis, D.J. & Iadecola, C. (1986b): Regulation by the brain of its blood flow and metabolism. Role of intrinsic neuronal network and circulating catecholamines. In: *Neural regulation of brain circulation*, Vol. 8, Eric K. Fernnstrom Symposium, eds. C. Owman & J.E. Hardebo, pp. 129–145. Amsterdam: Elsevier Science.

Reis, D.J. & Iadecola, C. (1989): Central neurogenic regulation of cerebral blood flow. In: *Neurotransmission and cerebrovascular function II* eds. J. Seylaz & R. Sercombe, pp. 343–368. Amsterdam: Elsevier Science.

### Recent references

Adachi, T., Baramidze, D.G. & Sato, A. (1992): Stimulation of the nucleus basalis of Meynert increases cortical cerebral blood flow without influencing diameter of the pial artery in rats. *Neurosci. Lett.* **143**, 173–176.

Adachi, T., Inanami, O. & Sato, A. (1992b): Nitric oxide (NO) is involve in increased cerebral cortical blood flow following stimulation of the nucleus basalis of Meynert in anesthetized rats. *Neurosci. Lett.* **139**, 201–204.

Alheid, G.F. & Heimer, L. (1988): New perspectives in basal forebrain organization of special relevance for neuropsychiatric disorder: the striatopallidal, amygdaloid, and corticopetal components of substantia innominata. *Neurosci. Lett.* **27**, 1–37.

Arneric, S.P., Iadecola, C., Underwood, M.D. & Reis, D.J. (1987): Local cholinergic mechanisms participate in the increase in cortical cerebral blood flow elicited by electrical stimulation of the fastigial nucleus in the rat. *Brain Res.* **411**, 212–225.

Azmitia, E.C. (1978): The serotonin-producing neurons of the midbrain median and dorsal raphe nuclei. In: *Handbook of psychopharmacology*, Vol. 9, eds. L.L. Iversen, S.D. Iversen & S.H. Snyder, pp. 233–314. New York: Plenum.

Bernard, J.-F., Alden, M. & Besson, J.-M. (1993): The organization of the efferent projections from the pontine parabrachial area to the amygdaloid complex: a *Phaseolus vulgaris* leucoagglutinin (PHA-L) study in the rat. *J. Comp. Neurol.* **329**, 201–229.

Bonvento, G., Lacombe, P. & Seylaz, J. (1989): Effects of electrical stimulation of the dorsal raphe nucleus on local cerebral blood flow in the rat. *J. Cereb. Blood Flow Metab.* **9**, 251–255.

Bonvento, G., Lacombe, P., MacKenzie, E.T. & Seylaz, J. (1991): Effects of dorsal raphe stimulation on cerebral glucose utilization in

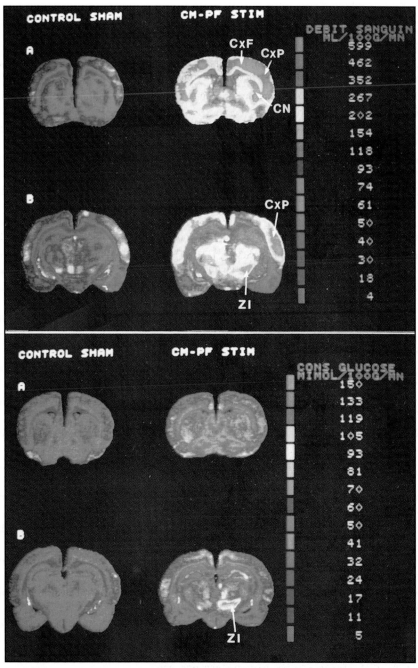

*Fig. 15. Representative colour-coded autoradiographs of coronal sections of the brain at the level of the caudate nucleus (A), posterior thalamus (B), mesencephalic reticular nucleus (C) and gigantocellular reticular nucleus (D) injected with $^{14}$C-IAP in sham-operated and CM-Pf-stimulated rats. Colour scales provide quantitative interpretation of lCBF expressed in ml/100 g per min. Note the marked increase in lCBF within the frontal and the parietal cortices, the caudate nucleus, the zona incerta, the mesencephalic reticular formation and the lower brainstem (from Mraovitch et al., 1992).(see colour section)*

the anaesthetized rat. *Brain Res.* **567,** 325–327.

Bonvento, G., Scatton, B., Claustre, Y. & Rouquier, L. (1992): Effects of local injection of 8-OH-DPAT into the dorsal or median raphe nuclei on extracellular levels of serotonergic projection areas in the rat brain. *Neurosci. Lett.* **137,** 101–104.

Bonvento, G., Charbonne, R., Correze, J.-L., Borredon, J., Seylaz, J. & Lacombe, P. (1994): Is a-chlorose plus holothan induction a suitable anesthetic regimen for cerebrovascular research? *Brain Res.* **665,** 2143–221.

Borowsky, I.W. & Collins, R.C. (1989a): Metabolic anatomy of brain: a comparison of regional capillary density, glucose metabolism, and enzyme activity. *J. Comp. Neurol.* **288,** 401–413.

Borowsky, I.W. & Collins, R.C. (1989b): Histochemical changes in enzymes of energy metabolism in the dentate gyrus accompany deafferentation and synaptic reorganization. *Neuroscience* **33,** 253–262.

Cao, W.H., Sato, A., Yusof, A.P.M. & Zhou, W. (1992): Regulation of regional blood flow in cerebral cortex by serotoninergic neurons originating in the dorsal raphe nucleus in the rat. *Biogenic Amines* **8,** 351–360.

Chida, K., Iadecola, C. & Reis, D.J. (1989): Global reduction in cerebral blood flow and metabolism elicited from intrinsic neurons of fastigial nucleus. *Brain Res.* **500,** 177–192.

Cudennec, A., Duverger, D., Serrand, A., Scatton, B. & Mackenzie E.T. (1988): Influence of ascending serotoninergic pathways on glucose use in the conscious rat brain. Effects of electrical stimulation of the rostral raphe nuclei. *Brain Res.* **244,** 227–246.

Dauphin, F., Lacombe, P., Sercombe, R., Hamel, E. & Seylaz, J. (1991): Hypercapnia and stimulation of the substantia inominata increase rat frontal cortical blood flow by different cholinergic mechanisms. *Brain Res.* **553,** 75–83.

Dietrich, W.D., Durham, D., Lowry, O.H. & Woolsey, T.A. (1982): 'Increased' sensory stimulation lends to changes in energy-related enzymes in the brain. *J. Neurosci.* **2,** 1608–1613.

Dudley, R.E., Nelson, S.R. & Samson, F. (1982): Influence of chloralose on brain regional glucose utilization. *Brain Res.* **233,** 173–180.

Eckenstein, F. & Baughman, R.W. (1984): Two types of cholinergic innervation in cortex, one co-localized with vasoactive intestinal polypeptide. *Nature* **309,** 153–155.

Gally, J.A., Montague, P.R., Reeke, G.N. Jr. & Edelman, G.M. (1990): The NO hypothesis: possible effects of a short-lived, rapidly diffusible signal in the development and function of the nervous system. *Proc. Natl. Acad. Sci. USA* **87,** 3547–3551.

Gautier, P. (1981): Pressor responses and adrenomedullary catecholamine release during brain stimulation in the rat. *Can. J. Physiol Pharmacol.* **59,** 485–492.

Goadsby, P.J., Piper, R.D., Lambert, G.A. & Lance, J.W. (1985a): Effect of stimulation of nucleus raphe dorsalis on carotid blood flow. I. The monkey. *Am. J. Physiol.* **248,** R257–R262.

Goadsby, P.J., Piper, R.D., Lambert, G.A. & Lance, J.W. (1985b): Effect of stimulation of nucleus raphe dorsalis on carotid blood flow. II. The cat. *Am. J. Physiol.* **248,** R263–R269.

Goadsby, P.J., Seylaz, J. & Mraovitch, S. (1993): Noncholinergic, nonadrenergic cortical vasodilation elicited by thalamic centrimedian-parafascicular comlex. *Am. J. Physiol.* **264,** R1150–R1156.

Golanov, E.V., Yamamoto, S. & Reis, D.J. (1994): Spontaneous waves of cerebral blood flow associated with a pattern of electrical activity. *Am. J. Physiol.* **266,** R204–R214.

Hökfelt, T., Fuxe, K, Goldstein, M. & Johansson, O. (1974): Immunohistochemical evidence for the existence of adrenaline neurons in the rat brain. *Brain Res.* **66,** 235–251.

Iadecola, C., Nakai, M., Arbit, E. & Reis, D.J. (1983a): Global cerebral vasodilatation by focal electrical stimulation within the dorsal medullary reticular formation in anesthetized rat. *J. Cereb. Blood Flow Metab.* **3,** 270–279.

Iadecola, C., Nakai, M., Mraovitch, S., Ruggiero, D.A., Tucker, L.W. & Reis, D.J. (1983b): Global increase in cerebral metabolism and blood flow produced by focal electrical stimulation of dorsal medullary reticular formation in rat. *Brain Res.* **272,** 101–114.

Iadecola, C., Underwood, M.D. & Reis, D.J. (1986): Muscarinic cholinergic receptors mediate the cerebrovasculation elicited by stimulation of the cerebellar fastigial nucleus in rat. *Brain Res.* **368,** 375–379.

Iadecola, C., Arneric, S.P., Baker, H.D., Tucker, L.W. & Reis, D.J. (1987): Role of local neurons in cerebrocortical vasodilation elicited from cerebellum. *Am. J. Physiol.* **252,** R1082–1091.

Iadecola, C., Springton, M.E. & Reis, D.J. (1990): Dissociation by chloralose of the cardiovascular responses evoked from the cerebellar fastigial nucleus. *J. Cereb. Blood Flow Metab.* **10,** 375–382.

Iadecola, C. (1992a): Does nitric oxide mediate the increase in cerebral blood flow elicited by hypercapnia? *Proc. Natl. Acad. Sci. USA* **89**, 3913–3916.

Iadecola, C. (1992b): Nitric oxide participates in the cerebrovasodilation elicited from cerebellar fastigial nucleus. *Am. J. Physiol.* **263**, R1156–R1161.

Iadecola, C., Faris, P.L., Hartman, B.K. & Xu, X. (1993): Localization of NADPH diaphorase in neurons of the rostral ventral medulla: possible role of nitric oxide in central autonomic regulation and oxygen chemoreception. *Brain Res.* **603**, 173–179.

Ishitsuka, T., Iadecola, C., Underwood, M.D. & Reis, D.J. (1986): Lesion of nucleus tractus solitarii globally impare cerebrovascular autoregulation. *Am. J. Physiol.* **251**, H269–H281.

Jacobs, B.L. & Azmitia, E.C. (1992): Structure and function of the brain serotonin system. *Physiol. Rev.* **77**, 165–229.

Katayama, Y., Tsubokawa, T., Hirayama, T., Kido, G., Tsukiyama, T. & Lio, M. (1986): Response of regional cerebral blood flow and oxygen metabolism to thalamic stimulation in humans as revealed by positron emission tomography. *J. Cereb. Blood Flow Metab.* **6**, 637–641.

Kimura, A., Sato, A. & Tano, Y. (1990): Stimulation of the nucleus basalis of Meynert does not influence glucose utilization of the cerebral cortex in anesthetized rats. *Neurosci. Lett.* **119**, 101–104.

Lacombe, P., Sercombe, R., Verrecchia, C., Philipson, V., Mackenzie, E.T. & Seylaz, J. (1989): Cortical blood flow increases induced by stimulation of the substantia innominata in the unanesthetized rat. *Brain. Res.* **491**, 1–14.

Lacombe, P.M., Iadecola, C., Underwood, M.D., Sved, A.F. & Reis, D.J. (1990): Plasma epinephrie modulates the cerebrovasodilation evoked by electrical stimulation of dorsal medulla. *Brain Res.* **506**, 93–100.

Levey, A.I., Hallanger, A.E. & Wainer, B.H. (1987): Cholinergic nucleus basalis neurons may influence the cortex via the thalamus. *Neurosci. Lett.* **74**, 7–13.

Linville, D.G. & Arneric, S.P. (1991): Cortical cerebral blood flow governed by the basal forebrain: age-related impairments. *Neurobiol. Aging* **12**, 503–510.

Maeda, M., Krieger, A.J. & Sapru, H.N. (1991): Chemical stimulation of the ventrolateral medullary depressor area decreases ipsilateral cerebral blood flow in anesthetized rats. *Brain Res.* **543**, 61–68.

Magistretti, P.J., Morrison, J.H., Shoemaker, W.J., Sapin, V. & Bloom, F.E. (1981): Vasoactive intestinal polypeptide induces glycogenolysis in mouse cortical slices: a possible regulatory mechanism for the local control of energy metabolism. *Proc. Natl. Acad. Sci. USA* **78**, 6535–6539.

Magistretti, P.J. & Morrison, J.H. (1988): Noradrenaline and vasoactive intestinal peptide-containing neuronal systems in neocortex: functional convergence with contrasting morphology. *Neuroscience* **24**, 367–378.

Marshall, J.F., Critchfield, J.W. & Kozlowski, R. (1981): Altered succinate dehydrogenase activity of basal ganglia following damage to mesotelencephalic dopaminergic projection. *Brain Res.* **212**, 367–377.

Morrison, J.H., Magistretti, P.J., Benoit, R. & Bloom, F.E. (1984): The distribution and morphological characteristics of the intracortical VIP-positive cell: an immunohistochemical analysis. *Brain Res.* **292**, 269–282.

Mraovitch, S., Kumada, M. & Reis, D.J. (1982): Role of nucleus parabrachialis in cardiovascular regulation in cat. *Brain Res.* **232**, 57–75.

Mraovitch, S., Iadecola, C., Ruggiero, D.A. & Reis, D.J. (1985): Widespread reduction in cortical blood flow and metabolism elicited by electrical stimulation of the parabrachial nucleus in rat. *Brain Res.* **341**, 283–296.

Mraovitch, S., Lasbennes, S.F., Calando, Y. & Seylaz, J. (1986): Cerebrovascular changes elicited by electrical stimulation of the centromedian-parafascicular complex in rat. *Brain Res.* **380**, 42–53.

Mraovitch, S., Feger, J., Calando, Y. & Seylaz, J. (1989): Cardiovascular and cerebrovascular alterations elicited by cholinergic stimulation of the centromedian-parafascicular complex in the rat. In: *Neurotransmission and cerebrovascular function I*, eds. J. Seylaz & E.T. MacKenzie, pp. 397–400. Amsterdam: Elsevier Science.

Mraovitch, S., Calando, Y., Pinard, E., Pearce, W.J. & Seylaz, J. (1992): Differential cerebrovascular and metabolic responses in specific neural systems elicited from the centromedian-parafascicular complex. *Neuroscience* **49**, 451–466.

Mraovitch, S., Calando, Y., Onteniente, B., Peschanski, M. & Seylaz, J. (1993): Cerebrovascular and metabolic uncoupling in the caudate-putamen following unilateral lesion of the mesencephalic dopaminergic nucleus in the rat. *Neurosci. Lett.*, in press.

Nakai, M., Iadecola, C. & Reis, D.J. (1982): Global cerebral vasodilatation by stimulation of rat fastigial cerebellar nucleus. *Am. J. Physiol.* **243**, H226–H235.

Nakai, M., Iadecola, C., Ruggiero, D., Yucker, L. & Reis, D.J. (1983): Electrical stimulation of the cerebellar fastigial nucleus increases cerebral cortical blood flow without changes in focal metabolism: evidence for an intrinsic system in brain for primary vasodilation. *Brain Res.* **260,** 35–49.

Ngai, A.C., Ko, K.R., Morii, S. & Winn (1988): Effect of sciatic nerve stimulation on pial arterioles in rats. *Am. J. Physiol.* **254,** H133–H139.

Palmer, R.M.J., Ferrige, A.G. & Moncada, S. (1987): Nitric oxide accounts for the biological activity of endothelium-derived relaxing factor. *Nature (Lond)* **327,** 524–526.

Raszkiewicz, J.L., Linville, D.G., Kerwin, Jr, J.F., Wagenaar, F. & Arneric, S.P. (1992): Nitric oxide synthase is critical in mediating basal forebrain regulation of cortical cerebral circulation. *J. Neurosci. Res.* **33,** 129–135.

Reis, D.J. (1984): Central neural control of cerebral circulation and metabolism. In: *Neurotransmitters and the cerebral circulation*, Vol. 2, eds. E.T. Mackenzie, J. Seylaz & A. Bes, pp. 91–119. New York: Raven Press.

Ruggiero, D.A., Cravo, S.L., Arango, V. & Reis, D.J. (1989): Central control of the circulation by the rostral ventrolateral reticular nucleus: anatomical substrates. In: *Progress in brain research 1981*, eds. J. Ciriello, M.M. Coverson & C. Polosa, pp. 49–79. Amsterdam: Elsevier.

Ruggiero, D.A., Mraovitch, S., Granata, A.R. & Reis, D.J. (1987): A role of insular cortex in cardiovascular function. *J. Comp. Neurol.* **257,** 189–207.

Schmidt, H.H.H.W., Gagne, G.D., Nakane, M., Pollock, J.S., Miller, M.F. & Murad, F. (1992): Mapping of neural nitric oxide synthase in the rat suggests frequent co-localization with NADPH diaphorase but not with soluble guanylyl cyclase, and novel paraneural functions for nitrinergic signal transduction. *J. Histochem. Cytochem.* **40,** 1439–1456.

Soeki, Y., Sato, A., Sato, Y. & Trzebski, A. (1989): Stimulation of the rostral ventrolateral medullary neurons increases cortical cerebral blow flow via activation of the intracerebral neural pathway. *Neurosci. Lett.* **107,** 26–32.

Takahashi, S., Crane, A.M., Jehle, J., Cook, M., Kennedy, C. & Sokoloff, L. (1995): Role of the cerebellar fastigial nucleus in the physiological regulation of cerebral blood flow. *J. Cereb. Blood Flow Metab.* **15,** 128–142.

Talman, W.T., Dragon, D.M.N., Heistad, D.D. & Ohta, H. (1991): Cerebrovascular effects produced by electrical stimulation of fastigial nucleus. *Am. J. Physiol.* **261,** H707–H713.

Toda, N. & Okamura, T. (1990): Possible role of nitric oxide in transmitting information from vasodilator nerve to cerebroarterial muscle. *Biochem. Biophys. Res. Commun.* **170,** 308–313.

Tsubokawa, T., Katayama, Y., Ueno, Y. & Moriyasu, N. (1981): Evidence for involvement of the frontal cortex in pain-related cerebral events in cats: Increase in local cerebral blood flow by noxious stimuli. *Brain Res.* **217,** 179–185.

Ueki, M., Mies, G. & Hossman, K.A. (1992): Effect of alpha-chloralose, halothane, pentobarbital and nitric oxide anesthesic on metabolic coupling in somatosensory cortex of rat. *Acta Anaesthiol. Scand.* **36,** 318–322.

Underwood, M.D., Iadecola, C., Sued, A. & Reis, D.J. (1992a): Stimulation of C1 area neurons globally increases regional cerebral blood flow but not metabolism. *J. Cereb. Blood Flow Metab.* **12,** 844–855.

Underwood, M.D., Bakalian, M.J., Arango, V., Smith, R.W. & Mann, J.J. (1992b): Regulation of cortical blood flow by the dorsal raphe nucleus: topographic organization of cerebrovascular regulatory regions. *J. Cereb. Blood Flow Metab.* **12,** 664–673.

# Chapter 13

# Clinical implications (1): Stroke, subarachnoid haemorrhage and epilepsy

**Barbro B. Johansson**

*Department of Neurology, University Hospital, S-22185 Lund, Sweden*

## A. Stroke

### Definitions and diagnosis

### Types and origins of stroke

Stroke or apoplexy is defined as an abrupt onset of neurological symptoms and signs indicating a disturbed cerebral circulation and lasting for at least 24 h or leading to death earlier. If lasting for less than 24 h, the episode is called a transient ischaemic attack (TIA) and is not included in the stroke definition. In Europe and North America about 80 per cent of strokes are caused by cerebral infarction and 20 per cent by cerebral haemorrhages, the figures for haemorrhages being somewhat higher in East Asia.

Infarction can be caused by occlusion of large extracranial or small intracerebral arteries. Occlusions can be caused by locally formed thrombi, by emboli coming from the heart or from the carotid and vertebral arteries. Arterial dissection, induced by trauma or occurring spontaneously, may be an underdiagnosed cause of brain infarction, particularly in young patients. Arterial dissection is often associated with localized headache or facial pain at or before the onset of neurological signs. Cerebral infarcts can also occur in connection with hypotensive episodes.

Cardiac emboli have been reported to

account for between 15 and 50 per cent of cases; the wide range illustrates that the diagnosis often is uncertain. Atrial fibrillation is a well recognized risk factor for cerebral embolism and the most common causes are rheumatic and ischaemic heart disease. The frequency of stroke increases with the duration of fibrillation. Cardiac embolism as a cause of stroke increases with age and may constitute about half of the cases in patients > 75 years of age.

As an intrinsic intracranial source of strokes, lacunae reflect arterial disease of the small penetrating arteries supplying the internal capsule, basal ganglia, thalamus and paramedian regions of the brain stem. They are thought to account for about 15–20 per cent of strokes. It is currently debated whether some lacunae may be of embolic origin.

Cerebral haemorrhages can be located intracerebrally or subarachnoidally. Since subarachnoid haemorrhages pose a special problem they are dealt with separately below (section B). Most intracerebral haemorrhages occur in the supratentorial compartment, mostly involving the basal ganglia and the thalamus. The second most common location is the subcortical white matter of the cerebral lobes. Less than 15 per cent of the haemorrhages are located in the cerebellum or pons. The symptoms will depend on the location and the size of the haematoma. Although the onset usually is abrupt, both the focal deficit and the level of consciousness usually undergo a gradual worsening due to further bleeding and/or secondary swelling. Haematomas of moderate or large size are accompanied by decreased levels of alertness.

## Diagnosis

### Generalities

The neurological symptoms and signs will depend on the vascular territory involved. Most infarcts are supratentorial with the vascular territory of the middle cerebral artery being most often affected. For description of the various clinical stroke syndromes, see Barnett *et al.* (1992) and Adams (1993). The clinical features sometimes suffice to differentiate acute haemorrhage from infarction. Classification of the ischaemic stroke into subtypes can to some extent be performed on clinical grounds. However, it is not possible, clinically, to definitely separate haemorrhages from infarction, as revealed by brain computed tomography (CT), which is now routinely used at admittance in most stroke centres. Small haemorrhages can give comparatively minor symptoms or even transient symptoms. Whereas haemorrhages are seen on CT scan immediately after onset, brain infarcts may not be visible during the first days and, if small, may not be detected at all. A lumbar puncture, previously the investigation of choice to separate haemorrhage from brain infarction, is now less frequently performed in centres where CT is available, but might be of value in selective cases.

### Doppler techniques and ultrasound imaging techniques

The Doppler techniques depend upon the reflection of a beam of sound of very high frequency by moving red blood cells. Doppler techniques are widely used for the diagnosis of cerebrovascular occlusive disease. The most commonly used technique is that of plotting the Dop-

pler frequency shift against time. When a stenosis is present, blood flow velocity increases in the stenotic portion of the vessel and is detected by an increase in the frequency of the Doppler shift signal. Turbulence distal to the stenosis produces a characteristic visual pattern. To visualize the carotid arteries, a two-dimensional ultrasound imaging technique is used. Currently, the two techniques (Doppler flow and imaging) are combined in real time, defining structure and flow. This technique is called echo-Doppler.

Transcranial Doppler ultrasonography is a non-invasive procedure for assessment of the intracranial cerebral circulation, allowing measurement of blood velocity in cerebral arteries at the base of the brain. However, since the diameter of the arterial lumen is unknown the blood flow cannot be determined. Despite this limitation, the method can be useful in answering specific questions such as detection of haemodynamically significant intracranial arterial stenosis. It is particularly useful in following changes in patients with subarachnoid haemorrhage who develop spasm, in intensive care unit monitoring of brain-injured patients and in intraoperative and postoperative monitoring of neurosurgical patients (Petty *et al.*, 1990).

### Magnetic resonance imaging (MRI)

The underlying principle of MRI is that many nuclei respond to the application of strong magnetic fields by absorbing and re-emitting radio waves, that can be detected and analysed and used to generate spectra indicating the concentration of various chemical species of these nuclei. Protons are among the most sensitive and abundant nuclei in biological tissues and have been widely utilized for MRI. Bone is not visualized and areas normally obscured by bone on CT scans are easily imaged. The resolution of grey and white matter is superior to that of CT scanning and cerebral infarction is evident much earlier, usually within 2 to 6 h. However, CT is superior in early identification of brain haemorrhages and will probably continue to be the screening method at admittance of stroke patients. The recent development in MR angiography, MR diffusion weighted imaging, MR spectroscopy and functional MRI together with the higher time resolution and more widely accessibility of MRI than PET have made these techniques powerful tools in experimental and clinical stroke research (Baron, 1993; Neil, 1993).

## Cerebral angiography

Direct puncture of arteries has been largely replaced by femoral catheterization under local anaesthesia with selective introduction into the major arteries leading to the brain. It is used predominantly to detect and evaluate extracranial, particularly carotid, stenosis, to diagnose arterial dessection and arterial and arteriovenous aneurysms.

## Single photon emission computerized tomography (SPECT) and positron emission tomography (PET)

In acute stroke, the normal coupling between cerebral blood flow and metabolism is not upheld. Without a concomitant determination of the oxygen extraction or cerebral metabolism, it is not possible to determine the state of the tissue by cerebral

blood flow studies alone. This has been illustrated in studies using PET (Wise et al., 1983; Brooks, 1991). In the early stage, the blood flow and metabolism might be low but oxygen extraction increased, suggesting that the tissue is still viable (Fig. 1). Later, particularly if the blood flow is restored, a hyperaemia occurs while the oxygen extraction is low or nil indicating that the tissue is severely damaged. The rate of spontaneous reperfusion increases gradually with time and occurs within the first two weeks after stroke onset in 77 per cent of patients with cortical infarcts (Jørgensen et al., 1994).

Examples of remote metabolic depression after focal stroke have been observed with PET in man. Reduced metabolic activity can be seen in the cerebral hemisphere contralateral to cerebral infarcts, ipsilateral to thalamic and lenticulo-capsular lesions, the thalamus ipsilateral to a cortical infarct, visual cortex distal to lesions of the optic radiations and the cerebellar cortex contralateral to supratentorial infarcts (crossed cerebellar diaschisis). The exact mechanism underlying these metabolic changes that usually are unaccompanied by changes in CT or MRI is not completely understood (Pappata et al., 1987; Ginsberg, 1990).

## Risk factors

The stroke incidence increases markedly with age. In a Swedish unselected population only 20 per cent of first-ever stroke patients are below the age of 65 and half of the patients are more than 75 years of age (Johansson et al., 1992). With the current ageing of the world population, stroke is

likely to remain a major medical problem.

## Hypertension

Hypertension is the most important risk factor for stroke. Hypertension predisposes to different types of intracerebral and extracerebral arterial lesions which may cause cerebrovascular events by different mechanisms (see Johansson, 1992). Hypertension leads to three main types of vascular changes: compensatory structural adaptation, degeneration vascular changes and, in the presence of other risk factors, to atherosclerosis.

## Structural adaptations to hypertension

When the blood pressure is increased abruptly to high levels in a previously normotensive individual, the autoregulatory capacity of the cerebral resistance vessels might be overcome and the blood flow increase (Fig. 2). A stepwise increase in blood pressure will be better tolerated. In chronic hypertension the vascular bed will adjust functionally and structurally to the increased load. The smooth muscle hypertrophy/hyperplasia will help to sustain wall tension and maintain adequate contractile function. In the cerebrovascular bed, an increased media thickness and lumen reduction has been observed over a large range of arterial sizes in vivo and in vitro, starting at rather large arteries (Fig. 3). These changes will shift the autoregulatory curve to the right (Fig. 4) and protect the blood–brain barrier (Strandgaard, 1978; Johansson, 1989). Although the structural adaptations basically are beneficial in protecting the vessels and preventing haemorrhage and permeability

creased peripheral resistance and reduced collateral capacity. In connection with an abrupt decrease in blood pressure in hypertensive individuals, infarction may occur in the border zones between the territories of the main cerebral arteries. These 'watershed' infarcts constitute approximately 10 per cent of all human brain infarcts (Torvik, 1984).

## Degenerative changes in hypertension

Degenerative changes in the small intracerebral blood vessels will occur when the compensatory mechanisms are not sufficient to protect the smaller intracerebral vessels. When the vessels yield to the high pressure, extravasation of plasma constituents seems to be the first step in hypertensive degenerative lesions in the vascular wall and in the develolpment of brain lesions (Johansson, 1992). Degenerative changes in the intracerebral arteries can lead to focal brain oedema, small lacunar infarctions or intracerebral haemorrhages due to intracerebral microaneurysms (see Johansson, 1992).

## Hypertension and atherosclerosis

Hypertension is a risk factor for atherosclerosis. Vascular changes related to atherosclererosis or ageing is much more pronounced in hypertensive individuals. The predominant sites, and usually earliest localization of atherosclerotic changes, are in the extracranial arteries. The second most common site is in the circle of Willis and with time the changes may also occur in the smaller intracerebral arteries. This is reported to occur earlier in hypertensive than in normotensive individual.

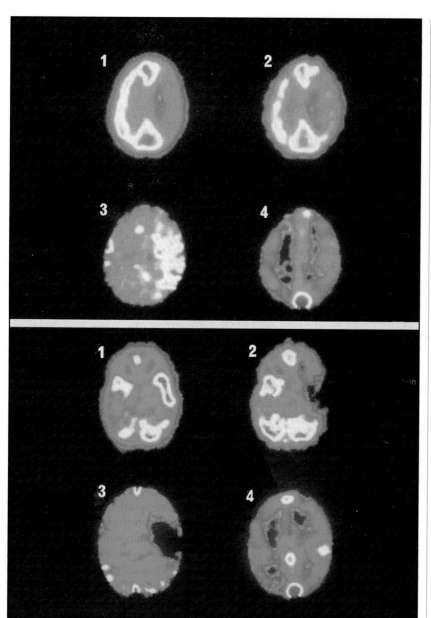

*Fig. 1. PET study of a patient with cerebral infarct 90 min (top panel) and 10 days (bottom panel) after stroke onset. 1.: CBF: 2: oxygen consumption; 3: oxygen extraction; 4: blood volume. At 90 min the blood flow and oxygen consumption is low in the affected area (right of picture) whereas the oxygen extraction is high indicating a viable tissue. At 10 days the blood flow is high in the infarcted area, 'luxury perfusion', whereas the lack of metabolims and oxygen extraction indicate that the tissue is dead. By courtesy of Professor R. Frackowiak, London.*

changes, they constitute a risk for ischaemia distal to any stenosis or occlusion or when the blood pressure is rapidly decreased because of the in-

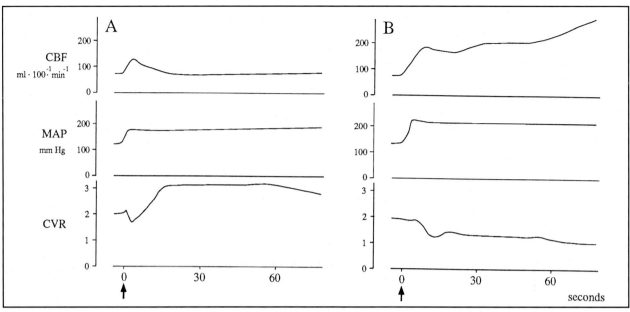

*Fig. 2. The vasomotor response to an abrupt but moderate increase in blood pressure in a previously normotensive rat is shown in (A). The vasoconstrictor response results in a perfect autoregulation within about 15 s. The lack of vasoconstrictor response after a large increase in blood pressure in (B) indicates that the upper limit of autoregulation is exceeded.*

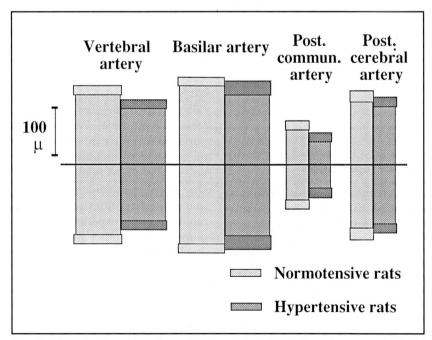

*Fig. 3. Schematic illustration of vessel lumen and media thickness in normotensive and hypertensive rats at 12 months of age. The media to radius ratio is increased by 27–50 per cent in the hypertensive rats. The increase is significant in all except the vertebral artery. Modified after Fredriksson et al. (1984).*

## Other risk factors

Other risk factors include diabetes, cigarette smoking, impaired cardiac function and high haematocrit and plasma fibrinogen levels. About 10 per cent of strokes are preceded by transitory ischaemic attacks. There

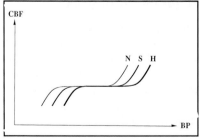

*Fig. 4. The normal autoregulatory curve (N) is shifted to the right chronic hypertension (H). The upper and lower limits are not fixed but vary with the rapidity of blood pressure increase, and with the sympathetic acitivity (S). A number of other factors can also influence the response. See Johansson (1990) by permission.*

are substantial racial differences with higher incidence in Japan and China, and in the US in the black population than in the white. There is an association between stroke and alcohol intake in men, which partly but not completely can attributed to coexistent hypertension. Although epidemiological studies are lacking, drug abuse, particularly amphetamine, heroin and 'crack', a cocaine preparation that is inhaled, have repeatedly been reported to be associated with stroke in young individuals.

Several studies suggest that family occurrence of stroke is an independent risk factor. Homocysteinaemia, long known to be a risk factor in homozygotes, has in several recent studies been reported to be a risk factor in heterozygotes also. For further discussions on risk factors, see Barnett *et al.*, 1992; Adams, 1993.

## Effects on brain tissue, mechanisms of neuronal death, post-ischaemic reperfusion

### Major effects of stroke

The high metabolic need and low energy reserve make the brain very vulnerable during ischaemia. When the perfusion pressure falls below critical levels, ischaemia develops and will progress to infarction if the flow is sufficiently reduced and the reduction persists long enough. In experimental studies on focal brain ischaemia, surprisingly extensive restoration of metabolism and function has been observed after ischaemic periods as long as 60 min or more. The length of time tolerated varies with factors such as

the degree of reflow after ischaemia, and plasma concentrations of glucose, corticosteroids and catecholamines.

The zone surrounding the core of the infarct where the flow is decreased and the neurons are metabolically lethargic and may be electrically silent but still viable has been called the penumbra. Currently the penumbra is looked upon as a dynamic process with impaired and unstable perfusion and metabolism which is the target for possible therapeutic intervention (Heiss & Graf, 1994). There is some disagreement as to the time during which the tissue may be rescued but this is likely to be related to the degree of perfusion deficit and may be up to a couple of hours. However, with long recovery periods, permanent selective neuronal damage is seen already after ischaemia of 5–15 min duration (Ito *et al.*, 1975). This 'maturation' of neuronal damage following transient global ischaemia, which has been verified in a number of studies, indicates an on-going process in the post-ischaemic period.

### Role of reperfusion

Clot resolution or recanalization of an occluded artery may occur spontaneously but the frequency is unknown. Animal as well as clinical studies have demonstrated that this will often lead to postischaemic hyperaemia. Whereas hyperaemia has been reported to be essential for the functional recovery in global ischaemia, hyperaemia has been associated with pronounced oedema and severe brain damage after focal ischaemia. Prevention of hyperaemia reduces the va-

sogenic oedema as well as the ischaemic brain tissue damage in focal ischaemia in experimental animals. If confirmed in man, these observations could be of importance for clinical trials with fibrinolytic agents and during operations on severely stenotic vessels where gradual rather than abrupt opening of an occluded vessel should be aimed at. However in the clinical situation it would be a difficult task to decide if and when attempts to reduce the blood flow should be tried. The marked hyperaemia following temporary occlusion is attenuated by chronic trigeminal ganglionectomy, indicating that neural factors are involved (Moskowitz *et al.*, 1989). An aggravating effect of postischaemic hyperemia could be related to any of the currently discussed mechanisms for nerve cell death ischaemia such as free radicals, lactacidosis, excitatory amino acids and possibly to some additonal intrinsic or extrinsic factors. However, so far clinical studies have not confirmed that hyperaemia aggravates the lesions, rather hyperaemia has been proposed to be a prognostically good sign (Jørgensen *et al.*, 1994; Marchal *et al.*, 1994). Preliminary reports from studies on thrombolytic therapy in stroke indicate that it might be a smaller problem than expected. One possible explanation could be that the reflow does not occur so rapidly as under experimental conditions. Further studies are needed to clarify the possible adverse and beneficial effects of hyperaemia in stroke.

### Neuronal damage

Energy is needed to uphold the ion gradients across nerve and glial cell membranes and energy failure will lead to a shift with efflux of $K^+$ from cells and influx of $Na^+$, $Cl^-$ and $Ca^{2+}$. This ion shift will lead to an accumulation of water within the cells, an intracellular oedema. Accumulation of metabolites within the cell will add to this oedema, which in the early stage is completely reversible. Ischaemia leads to a diffuse transmitter release since energy is needed to keep transmitters stored in their granulae. The electrical activity of the neurons stops when the blood flow is decreased to about one-third of the normal values (under normal temperature and blood glucose levels) but some basic cell function is still present and the cells can regain their function if the blood supply is restored. There are various hypotheses as to the triggering mechanisms for neuronal death and some will be presented below. Current research indicates that these mechanisms combine in the process that finally kills the neurons. The neuronal damage can be of two types: selective neuronal vulnerability affecting groups of neurons with a characteristic distribution within the brain, and infarction, affecting not only neurons but also blood vessels and glial cells.

## The calcium hypothesis

That cell death may be a specific consequence of disturbance in intracellular $Ca^{2+}$ homoeostasis has long been discussed (Schanne *et al.*, 1979). The concentration of calcium is 10,000 times higher in the extracellular fluid than in the cells and influx of calcium into the cells together with release of calcium from intracellular sources leads to uncontrolled activation of a number of calcium-dependent reactions (Siesjö, 1992; Morley *et al.*,

1994). The combination of energy failure and calcium influx/release can lead to an extensive breakdown of phospholipids and proteins, to proteolytic degradation of cytoskeletal components and to free radical formation (see below).

## The excitotoxic hypothesis

Glutamate and some other excitatory amino acids have a transmitter function in the brain but can also be toxic to the neurons. To prevent a high extracellular concentration of glutamate and other excitatory amino acids, they are taken up by efficient re-uptake mechanisms after release. The uptake mechanisms are energy-dependent and during ischaemia the excitatory amino acids may accumulate in too high concentrations. Current evidence indicates that ischaemic neuronal damage may be caused by enhanced release or diminished uptake of glutamate or other excitatory amino acids enabling an enhanced calcium influx through channels gated by excitatory amino acid receptors (Rothman & Olney, 1986; Hossmann, 1994). Antagonists which block NMDA and AMPA glutamate receptors can ameliorate neuronal damage in experimental ischaemia (see Choi, 1990; Graham *et al.*, 1993).

## Free radicals

The role of free radicals in the ischaemic nerve cell pathology has been debated for over a decade (see Siesjö, 1992; Chan, 1994). It has been suggested that free radicals are important particularly in the reperfusion period with good access to oxygen in an already damaged tissue. However, recent studies suggest a role also in permanent ischaemia and fee radical scavengers have been shown to re-

duce the infarct size in experimental brain infarction. Likewise, recent studies show that transgenic mice, overexpressing the enzyme CuZn-superoxide dismutase, develop smaller infarcts than control rats (Chan *et al.*, 1991). Calcium will enhance lipolysis and accumulation of arachidonic acid and interact with the free radical mechanisms in the degradation of lipids, proteins and DNA. Metals, including iron, increase the rate of lipid peroxidation and it has been suggested that free radical damage may also be related to alterations in iron-binding.

## Other factors

The role of acidosis for ischaemic brain damage is debated. Although several experimental and also some clinical studies indicate that pre-ischaemic hyperglycaemia can aggravate the damage, this seems to occur at very high tissue lactate levels (15–10 mmol/kg) corresponding to an intracellular pH of 6.0. The clinical implication of a slight to moderate hyperglycaemia is less clear (Siesjö, 1992).

Another factor thought to be of importance for ischaemic injury is platelet-activating factor which may at least in part act by generating free radicals (Lindsberg *et al.*, 1991). Since the brain metabolic needs increase with the temperature, lowering of the brain temperature will lead to better survival of neurons during ischaemia. Current evidence indicates a role for nitric oxide in the pathophysiology of focal cerebral ischaemia (Dalkara & Moskowitz, 1994).

## Animal models

The various stroke models available

and their advantages and disadvantages have been discussed by Ginsberg and Busto (1989). The animal models used can be divided into global and focal stroke models. Global ischaemia models are usually used to study selective neuronal vulnerability since they are easier to standardize than the focal models. The clinically most relevant stroke model is ligation of a middle cerebral artery. In most models of ischaemic stroke, young animals with no underlying chronic disease are used. Since hypertension is the largest risk factor for stroke and hypertensive blood vessels react differently to normotensive ones, it is relevant to use hypertensive rats in experimental stroke research. Atherosclerosis is not easily induced in rats but can be induced in rabbits and monkeys. For ethical and economic reasons studies on monkeys are few.

The relevance of animal models for human disease such as cerebrovascular disease has been questioned. Many drugs that have been shown to be useful in experimental models have not been of value in clinical trials. More attention should be directed towards the clinical outcome in animal research (see Johansson, 1995).

## Prevention of stroke. Acute treatment of stroke

The incidence of stroke has declined in many countries, paticularly in the United States and Japan, and it is believed that this to a large extent is related to treatment of hypertension (Whisnant, 1984). Since stroke increases so markedly with age and atherosclerotic mechanisms become more important, it is likely that intervention towards risk factors for athe-

rosclerosis may have some long term beneficial effect. Cigarette smoking about doubles the risk for stroke but there is no clear correlation between stroke and serum lipids.

The current concept of mechanisms behind ischaemic cell damage and possible therapeutic implications in the treatment of acute stroke have already been discussed. Although a number of experimental studies have shown a favourable effect of calcium blockers, clinical studies, with the possible exception of subarachnoid haemorrhage, have been disappointing. Whether the favourable effects of free radical scavengers and glutamate antagonists shown in animal studies are present also in clinical studies remains to be shown. Current data suggest that glucocorticoids may be harmful and should be avoided (Sapolsky and Pulsinelli, 1985). Presently, early mobilization and activation of stroke patients is probably of main importance and on-going animal studies in several laboratories indicate that physical exercise can influence the outcome of ischaemia. Likewise, pharmacological activation has been shown to increase the metabolic activity in alternative neuronal pathways after experimental vascular lesions. It is possible that release of trophic factors or inhibiton of negative factors may influence the recovery after stroke.

## Vascular dementia

The role of vascular disease as cause or contributor to the development of dementia syndromes is controversial (Tatemichi, 1990). According to Hachinski *et al.* (1974), when vascular diseases are responsible for dementia it

is through the occurrence of multiple small or large cerebral infarcts, multi-infarct dementia. Since both dementia and vascular lesions are common in an elderly population, it is inevitable that they must occur together in many instances. The Hachinski score developed to separate senile dementia from vascular dementia gives much weight to abrupt onset, history of stroke and focal neurological symptoms and signs in the ischaemic score. The role of hypertension in the development of dementia of insidious onset with no history of stroke has not been evaluated. However, recent data indicate that the prevalence of white matter lesions are high in hypertensive individuals in the absence of neurological symptoms or signs (Johansson, 1992). Whether such individuals with silent lesions are at a greater risk for developing vascular dementia is an important question that should be further studied. Binswanger's encephalopathy, a dementia syndrome marked by a progressive dementia and a step-wise neurological deterioration correlated with periventricular white matter attenuation, has been attributed to hypertension. The histopathological characteristics include white matter demyelination and, according to some reports, lacunae in the basal ganglia and the thalamus. Binswanger's encephalopathy is neither clinically nor pathologically a well-defined syndrome. It is currently debated whether ischaemia or plasma leakage with oedema is of primary pathogenetic significance. The current debate on possible long-term consequences of transient alterations of the blood–brain barrier as well as the recent finding of silent brain changes in hypertensive individuals suggest that further studies are needed in this field (for references, see Johansson, 1992, 1994).

## B. Subarachnoid haemorrhage

### Incidence and diagnosis

Subarachnoid haemorrhage constitutes about 5 per cent of stroke cases. In the majority of patients with acute, nontraumatic subarachnoid haemorrhage an arterial aneurysm can be found on arterial angiograms. The occurrence of intracranial aneurysms in the general population has been studied in various autopsy series. Depending on whether the aneurysms were prospectively searched for or merely noted as accidental findings, the incidences range from 0.2 per cent to 9 per cent. The mean incidence of 12 series comprising 87 700 post-mortem examinations was 1.6 per cent. The occurrence of aneurysms is higher in forensic series, since aneurysms are a relatively important cause of unexpected sudden death.

The diagnosis is suspected on the clinical history and verified by computer tomography where blood can be seen in the subarachnoid space. When performed within the first 3 days after ictus, CT is very sensitive in detecting subarachnoid blood. When performed later or if the amount of blood is low, CT might not give the diagnosis and lumbar puncture should be performed. In addition to counting the cells, the relative proportion of bilirubin and methaemoglobin can tell the age of the bleeding.

Because of the very high risk of rebleeding and development of cerebral vasospasm leading to ischaemia it is now generally though advisable to operate on the patients very early. Vasospasm is a serious complication that occurs in about 30 per cent of the patients 3–21 days after the initial haemorrhage (Barnett *et al.*, 1992). In spite of intense research the nature behind the vasospasm remains obscure. Some studies suggest that calcium entry blockers may have a favourable effect on the outcome and improve the cerebral circulation but without effect on the angiographically observed spastic portions of the arteries. The effect seems to be on smaller intracerebral vessels.

In 15–20 per cent of patients with subarachnoid haemorrhage, the cause of bleeding is undetermined. Patients with unknown aetiology have a better prognosis than patients with aneurysm. The pathophysiology may include bleeding abnormalities, rupture of an arteriographically occult vascular malformation and bleeding from an obliterated or partially thrombosed aneurysm.

### Causes and mechanisms

#### Hypertension and aneurysms

Hypertension predisposes not only to intracerebral microaneurysms, but also to saccular aneurysms. These macroscopically visible large aneurysms are formed in normally occurring gaps in the media at points of bifurcation, particularly at the circle of Willis. If an aneurysm ruptures, a subarachnoid haemorrhage will occur with or without concomitant bleeding into the brain parenchyma (see below). Clinically, about half of the patients with subarachnoid haemorrhages are hypertensive. The role of hypertension in the development of such lesions has been elegantly shown in experimental studies. Saccular aneurysms can be produced in rats and monkeys by combining experimental hypertension with ligation of one carotid artery to alter the haemodynamic stress. Administration of β-amino-propionitrile, a lathyrogen that inhibits cross-linking of collagen and elastin, increases the incidence of aneurysms. However, no aneurysms occur in the absence of hypertension and arterial ligation, indicating that the vascular tension and haemodynamic stress are the most important factors (Hazama and Hashimoto, 1987).

### Cerebral blood flow

A positive correlation between regional CBF and diameter of major supplying vessels has been observed with transcranial Doppler. However, in some cases with focal vasospasm the reduction in CBF is global and not restricted to the area of the spastic vessels. In such cases, the cerebral oxygen extraction was reduced but independent of the degree of vasospasm, speaking against vasospasm as a cause for the reduction of CBF. It has been proposed that the observed association between reduction in CBF and vasospasm could be caused by a common factor responsible for the devel-

Table 1. Spasmogens potentially responsible for vasospasm, evidence for and against their involvement, and their putative mechanism(s) of action

| Spasmogen | Evidence for | Evidence against | Mechanism of action |
|---|---|---|---|
| Oxyhaemoglobin | Time course of release into CSF Vasoactive Multiple sites of action High concentrations next to affected arteries Essential for development of vasoplasm Efficacy of U74006F (inhibitor of lipid peroxidation) | Not a potent vasoconstrictor | Direct free-radical-mediated effects on smooth muscle Stimulates lipid peroxidation Stimulates eicosanoid synthesis Metabolized to bilirubin Inhibits endothelium-dependent relaxation Neurogenic effects |
| Eicosanoids | Time course of release into CSF Vasoactive | Inadequate concentration next to affected arteries Contractions diminish with time Failure of antagonists to alleviate vasospasm | Receptor-mediated smooth muscle contraction |
| Bilirubin | Time course of release into CSF Vasoactive | Inadequate concentration next to affected arteries Elevated CSF levels in diseases not associated with vasospasm | Membrane perturbation with calcium influx Inhibits cell oxidative enzymes |
| Endothelin | Potent vasoconstrictor Increased synthesis following SAH | Contractions diminish with time | Receptor-mediated calcium influx, smooth muscle contraction |
| Lipid peroxides | Time course of release into CSF Vasoactive Efficacy of U74006F | Inadequate concentration next to affected arteries | Damage to cell membrane, permitting calcium influx Stimulation of eicosanoid synthesis |

CSF, cerebrospinal fluid; SAH, subarachnoid haemorrhage.

opment of both.

## Hypotheses of SAH-induced vasospasm

There can be no doubt that the subarachnoid blood clot is the essential cause of the delayed vasospasm. However, considerable debate has revolved around two hypotheses: either vasospasm is a prolonged and pronounced vasoconstriction (Findlay et al., 1991) – and as such leads to ultrastructural (and pharmacological) abnormalities – or ultrastructural changes induced by the clot actually cause the vasospasm and the luminal narrowing (Kassell et al., 1985).

The structural theory now proposes that inflammation and arterial wall fibrosis contribute to stiffening of the artery wall and prolongation of vasospasm. It has also been proposed that increased and excessive calcium entry, which begins very rapidly after haemorrhage, may be an important factor contributing to this pathological state (Bevan and Bevan, 1988). It is certain that many vasoconstrictor agents are released during the first hours and days, all of which induce increased cytosolic calcium, and recent evidence points to an early potentiation of calcium-induced constriction. The effects of several calcium entry blockers lend support to the general role of calcium, but do not dis-

tinguish between the two hypotheses.

Table 1 lists spasmogens potentially implicated in delayed vasospasm, according to Findlay et al. (1991). This list is headed by oxyhaemoglobin which has been shown to be formed in cerebrospinal fluid by erythrocyte haemolysis at a rate compatible with a role in the development of vasospasm, and to be an essential factor for it. It also leads to, or stimulates, the formation of other potent membrane-active agents such as lipid peroxides, free radicals, eicosanoids and bilirubin.

An interesting neurogenic hypothesis of vasospasm in the rat has been developed by Svendgaard et al. (1990).

According to this, sensory fibres on the major cerebral arteries (those originating in the trigeminal ganglion), excited by the periarterial blood clot, trigger via central catecholaminergic pathways a reflex release of hypothalamic spasmogenic peptides into the CSF (including vasopressin). Lesion of the central pathways or blockade of substance P receptors prevented the development of vasospasm. It is difficult at the present time to reconcile this hypothesis with the direct effects of blood spasmogens, but possibly such a mechanism could explain the generalized reduction of CBF seen in some cases during focal vasospasm.

## C. Epilepsy

## Background

Epilepsy is not a disease but a name used for a number of syndromes in which either a prior brain insult or – less frequently – hereditary factors have made a part of the brain electrically unstable. Epilepsy may occur after major head trauma, haemorrhage, brain infarction, infection, vascular malformations and brain tumours. A seizure can occur at the onset of these disorders but may also occur months or years after the initial insult. An epileptic seizure is a sudden, excessive and temporary nervous discharge and can occur in any individual under certain circumstances such as extreme fatigue, hypoglycaemia and excessive alcohol intake. An isolated seizure should not be called epilepsy.

## Clinical epilepsy

The pathophysiology of seizures is not well understood and seizures are usually classified according to their clinical picture. In the international classification now commonly used seizures are classified as to whether the onset is partial (focal) or generalized (Engel, 1989; Blum & Fisher, 1994). Simple partial seizures occur without loss of awareness and complex partial seizures even if accompanied by reduced alertness. The most common type of partial seizure in adults arises from limbic structures in the temporal lobe and was formerly called temporal lobe seizures or psychomotor seizures. In more than 50 per cent of cases such seizures are preceded by auras such as epigastral sensations, flushing, or cognitive distortions such as *déjà vu*. The patients are in a dreamlike state and automatic movements such as walking in circles, lip smacking and fumbling are common. In the motor area, a partial seizure often takes the form of Jacksonian attacks with rhythmical uncontrolled motor activity. Occipital seizures are associated with visual phenomena. Any partial seizure may spread to wide regions of the brain and become generalized. Partial seizures should be distinguished from generalized absence seizures (petit mal) which are characterized by a brief stare which can be accompanied by motor activities and if those are extensive it can be difficult to differentiate from complex partial seizures.

Generalized tonic–clonic (grand mal) seizures are characterized by stiffening and jerking accompanied by loss of consciousness and falling and sometimes bladder incontinence.

Epileptic discharges depend on excessive and abnormally synchronous firing of neurons (see Jefferys, 1994).

The electroencephalogram (EEG) remains the most useful test in diagnosis and classification of epilepsy. However, the clinical history is even more important because of the intermittent course of the disease and changes in EEG may not be present all the time. This is the reason for now commonly using long-term recording over several days or weeks. It is particularly important to elucidate whether or not the seizure starts from a single focus in patients with seizures resistant to antihypertensive drugs who might benefit from a surgical operation in case the single focus is verified. In order to correlate clinical and electroencephalic seizure activity, preoperative long-term video and electroencephalographic recordings are usually combined.

## Animal models of epilepsy

## Models

Epilepsy is a heterogenous disorder and a large number of experimental models have been used (Fisher, 1989; Jefferys, 1994). Basically, they can be divided into acute and chronic models for simple partial seizures, models for complex partial seizures, generalized tonic-clonic seizures and

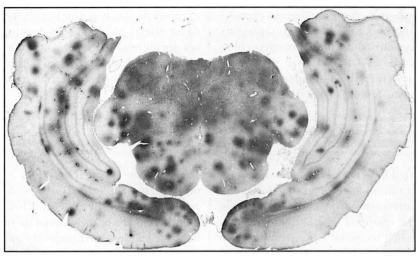

*Fig. 5. Immunohistochemical illustration of multifocal extravasation of endogenous serum albumin in the brain of a rat subjected to 5 min of epileptic seizures induced by bicuculline. The extravasation is most pronounced in the diencephalic structures. From Sokrab et al. (1990). With permission.*

generalized absence seizures. Acute models for simple partial seizures include topical application of penicillin or other focal chemical convulsants, acute electrical stimulation and GABA withdrawal.

Chronic simple partial seizures can be induced by cortically implanted alumina hydroxide, a model that has been rather extensively studied. Other metals such as cobalt, tungsten, zinc and iron can have similar effects and it has been discussed whether post-traumatic epilepsy can be related to the presence of blood and hence iron-containing haemoglobin in the cortex. Cryogenic injury and ganglioside antibody injection have also been used.

Models for complex partial seizures include kainic acid, tetanus toxin and kindling. In kindling, repeated shocks to various parts of the brain result in enhanced electrical excitability. Most regions of the forebrain can be kindled but kindling is usually initiated by electrical stimulation of the amygdala. After a few days of stimulation the shocks induce electrical afterdischarges, which become progressively more complex and prolonged with each kindling stimulus. Spontaneous epileptic seizures can be seen if the stimulation continues for a few weeks in rodents. Although the kindling phenomenon was described in 1958 and has been extensively used, the mechanism of kindling remains unknown.

For generalized tonic–clonic seizures some genetic models are available, such as photosensitive baboons, audiogenic seizures in mice, genetically epilepsy-prone rats and *Drosophila* shakers (a mutant fly). Other ways of inducing generalized seizures are by electroshock or administration of systemic convulsants such as pentylenetetrazol, systemic penicillin and GABA antagonists such as picrotoxin and bicuculline. For details on these and many other models see Fisher (1989).

## Relevance of models

An important question is how relevant the animal models are for human epilepsy. Although the relevance to clinical epilepsy remains uncertain, kindling studies are especially useful for investigations of changes that occur in the brain over time. It is hoped that elucidation of the mechanisms involved will advance our understanding of epilepsy and other plastic changes in the brain.

In addition to animal models, hippocampal slices and tissue cultures have been used for studying anticonvulsant mechanisms. Although such models are useful for studying some basic physiological and pharmacological principles they cannot be regarded as models of the complex behavioural manifestations of epileptic seizures.

## Local blood flow and glucose consumption

It has been known for decades that epileptic seizures markedly increase the cerebral blood flow and metabolism. The high blood flow is a combined result of metabolic vasodilatation and usually also an increase in blood pressure because of the sympathetic activation. However, the sympathetic activation probably also limits to some extent the CBF increase by acting on the cerebral arteries, as shown in animal studies (see Chapter 11). PET studies with $^{18}$F-deoxyglucose (FDG) have confirmed that generalized seizures cause a diffuse increase in glucose utilization. The increase in perfusion exceeds that of oxygen consumption and hence the oxygen extraction is reduced. In contrast, the relative changes in CBF and cerebral metabolic rate of glucose are similar. In the interictal period, 60–80 per cent of patients with lateralized foci on surface or depth EEG show ipsilateral glucose hypometabolism with the area of low FDG uptake being far more extensive than that of the EEG focus and the morphological lesions. The reason for the interictal hypometabolism round the focus has been ascribed to ischaemic penumbra surrounding a critical area of hypoperfusion, brain oedema, release of toxic cell components and release of neurotransmitters or false neuro-transmitters leading to disruption of surrounding metabolism. The same area shows hypermetabolism during seizures though changes may be more widespread. An interictal low flow and ictal high flow has also been observed with SPECT. Theoretical and practical aspects on the value of brain imaging in epilepsy have been reviewed by Brooks (1991), and Cook & Kilpatrick (1994).

Experimental as well as clonical seizures can increase the permeability of the blood–brain barrier (Fig. 6). If the concomitant increase in blood pressure is prevented no extravasation will occur, indicating that haemodynamic factors play a crucial role in the increased permeability. Although the blood–brain barrier function rapidly returns to normal after an epileptic seizure, extravasated plasma proteins remain in the tissue for some days and albumin accumulates in the Purkinje cells. It has been proposed that the loss of Purkinje cells in chronic epilepsy may partly be related to cumulative bouts of plasma extravasations (Sokrab *et al.*, 1990).

## D. Comparison of neuronal cell damaged in ischaemia, epileptic seizures and hypoglycaemia

### Differences and similarities

Although status epilepticus, ischaemia and hypoglycaemia all lead to neuronal necrosis, neurochemical events and the distribution of neuronal damage differ (Rothman and Olney, 1986; Olney *et al.*, 1986; Auer and Siesjö, 1988). In ischaemia there is an early dramatic loss of ion homoeostasis due to the deterioration of the tissue energy state and the acidosis is marked. In hypoglycaemic coma, oxygen consumption continues at a reduced rate and acidosis is absent. In epileptic seizures the metabolic rate is high, the ion homoeostasis minimal and acidosis moderate. Experimental studies suggest that the neuronal necrosis in epilepsy is unrelated to energy failure and occurs in spite of adequate oxygenation.

One common factor is that all three conditions lead to an enhanced release and accumulation of excitatory amino acids into the extracellular fluid. Furthermore, they are all accompanied by a calcium influx into the cells and there is evidence that they all can be accompanied by calcium-triggered proteolysis. Furthermore, lipolysis with accumulation of free fatty acids occurs in all three models. Nevertheless, the distribution of neuronal necrosis is different in the three conditions.

### Distribution of necrosis

In well oxygenated animals status epilepticus can cause moderate neuronal necrosis in the cerebral cortex, layers 3 and 4, in the hippocampus particularly CA 4 and CA 1 pyramidal cells, and in the ventromedial nucleus in the thalamus. Furthermore, infarction of the substantia nigra pars reticularis has been observed in a number of experimental models. In contrast, ischaemia tends to give rise to selective neuronal necrosis in all middle laminae of the cortex and in the same hippocampal regions as epilepsy whereas hypoglycaemia causes lesions predominantly in the dentate gyrus that is quite resistant to ischaemic injury. Cortical infarction is not seen in animals models of uncomplicated hypoglycaemia or epilepsy. High lactate levels and a drop in tissue pH has been thought to be important for the production of infarcts, i.e. necrosis not only of neurons but also of glial cells and vascular elements. It has been suggested that the vulnerability of specific cells is influenced by presynaptic inputs and by the type and state of the post-synaptic receptors (for references, see Auer & Siesjö, 1988).

## Summary

Stroke is a global problem with increasing significance because of the ageing population. Except for age, hypertension is by far the most important risk factor for stroke. Hypertension predisposes to a number of intracerebral and extracerebral vascular lesions which may cause cerebrovascular events by different mechanisms. The high metabolic need and low energy reserve make the brain very vulnerable to ischae-mia. During the last decade a number of experimental studies – supported by PET studies in man – suggest the presence of a therapeutic window, i.e. the time during which the neurons can be saved. The penumbra is the zone surrounding the core of the infarct where the flow is decreased and the neurons are metabolically lethargic and may be electrically silent but still viable. The presumed role of calcium, excitatory amino acids, free radicals, platelet-activating factor, acidosis and brain temperature in the process of neuronal death is briefly discussed. Epilepsy is not a disease but a name used for a number of syndromes in which either a prior brain insult or, less frequently, hereditary factors have made part of the brain electrically unstable. Epilepsy is a frequent and socially stigmatizing neurological problem. A number of animals models are used to elucidate the still poorly understood pathophysiology of seizures.

## References for further reading

Adams Jr., H.P. (1993): *Handbook of cerebrovascular diseases.* New York: Marcel Dekker.

Auer, R.N. & Siesjö, B.K. (1988) Biological differences between ischaemia, hypoglycemia, and epilepsy. *Ann. Neurol.* **24,** 699–707.

Barnett, H.J.M., Mohr, J.P., Stein, B.M. & Yatsu, F.M. (1992): *Stroke. Pathophysiology, diagnosis and management.* Vol. I and II. Churchill Livingstone, Edinburgh.

Baron, J.C. (1993): Neuroimaging procedures in acute ischemic stroke. *Curr. Opin. Neurol.* **6,** 900–904.

Bevan, J.A. & Bevan, R.D. (1988): Arterial wall changes in chronic cerebrovasospasm: *in vitro* and *in vivo* pharmacological evidence. *Am. Rev. Pharmacol. Toxicol.* **28,** 311–329.

Blum, D. & Fisher, R.S. (1994): Advances in epilepsy. *Curr. Opin. Neurol.* **7,** 96–101.

Brooks, D.J. (1991): PET: its clinical role in neurology. *J. Neurol. Neurosurg. Psychiatry* **54,** 1–5.

Chan, P.H. (1994): Oxygen radicals in focal cerebral ischemia. *Brain Pathol.* **4,** 59–64.

Chan, P.H., Yang, G.Y., Chen, S.F., Carlson, E. & Epstein, C.H. (1991): Cold-induced brain edema and infarction are reduced in transgenic mice overexpressing CuZn-superoxide dismutase. *Ann. Neurol.* **29,** 482–486.

Choi, D.W. (1990): Methods for antagonizing glutamate neurotoxicity. *Cerebrovasc. Brain Metab. Rev.* **2,** 105–147.

Cook, M.J. & Kilpatrick, C. (1994): Imaging in epilepsy. *Curr Opin. Neurol.* **7,** 123–130.

Dalkara, T. & Moskowitz, M. (1994): The complex role of nitric oxide in the pathophysiology of focal cerebral ischemia. *Brain Pathol.* **4,** 49–57.

Engel Jr.., J. (1989): *Seizures and epilepsy.* Philadelphia: F.A. Davis Company.

Findlay, J.M., MacDonald, R.L. & Weir, B.K.A. (1991): Current concepts of pathophysiology and management of cerebral vasospasm following aneurysmal subarachnoid haemorrhage. *Cerebrovasc. Brain Metab. Rev.* **3,** 336–361.

Fisher, R.S. (1989): Animal models of the epilepsies. *Brain Res. Reviews* **14,** 245–278.

Fredriksson, K., Nordborg, C. & Johansson, B.B. (1984): The haemodynamic effect of bilateral carotid artery ligation and the morphometry of the main communicating circuit in normotensive and spontaneously hypertensive rats. *Acta Physiol. Scand.* **121,** 241–147.

Ginsberg, M.D. (1990): Local metabolic responses to cerebral ischemia. *Cerebrovasc. Brain Metab. Rev.* **2,** 60–93.

Ginsberg, M.D. & Busto, R. (1989): Rodent models of cerebral ischaemia. *Stroke* **20,** 1627–1642.

Graham, S.H., Chen, J., Sharp, F.R. & Simon, R.P. (1993): Limiting ischemic injuury by inhibition of excitatory amino acid release. *J. Cereb. Blood Flow Metab.* **13,** 88–97.

Hachinski, V.C., Lassen, N.A. & Marshall, J. (1974): Multi-infarct dementia. A cause of mental deterioration in the elderly. *Lancet* **ii,** 207–210.

Hazama, F. & Hashimoto, N. (1987): An animal model of cerebral aneurysms. *Neuropathol. Appl. Neurobiol.* **3,** 77–90.

Heiss, W.-D. & Graf, R. (1994): The ischemic penumbra. *Curr Opin. Neurol.* **7,** 11–19.

Hossmann, K.-A. (1994): Glutamate-mediated injury in focal cerebral ischemia: the oxcitotoxin hypothesis revised. *Brain Pathol.* **4**, 23–26.

Ito, U., Spatz, M., Walker, J.T. Jr. & Klatzo, I. (1975): Experimental cerebral ischaemia in Mongolian gerbils: I. Light microscopic observations. *Acta Neuropathol. (Berl)* **32**, 209–223.

Jefferys, J.G.R. (1994): Experimental neurobiology of epilepsies. *Curr. Opin. Neurol.* **7**, 113–122.

Johansson, B.B. (1989): Hypertension and the blood–brain barrier. In: *Implications of the blood–brain barrier and its manipulation*, Vol. eds. J. Atkinson, C. Capdeville & F. Zannad, pp. 389–410. New York: Plenum Press.

Johansson, B.B. (1990): Hypertension and the cerebral circulation. In: *Coronary and cerebrovascular effects of antihypertensive drugs*, pp. 96–110. London: Transmedica.

Johansson, B.B. (1992): Vascular mechanisms in hypertensive cerebrovascular disease. *J. Cardiovasc. Pharmacol.* **19**,(Suppl. 3) S11–S15.

Johansson, B.B. (1994): Pathogenesis of vascular dementia – the possible role of hypertension. *Dementia* **5**, 174‘176.

Johansson, B.B. (1995): Functional recovery after brain infarction. A review of animal data. *Cerebrovasc. Dis.* (in press).

Johansson, B.B., Jadbäck, G., Norrving, B. & Widner, H. (1992): Evaluation of long-term functional status in first-ever stroke patients in a defined population. *Scand. J. Rehabil.* (Med. Suppl.) **26**, 105–114.

Jørgensen, H.S., Sperling, B., Nakayama, H., Raaschou, H.O. & Skyhøj Olsen, T. (1994): Spontaneous reperfusion of cerebral infarcts in patients with acute stroke. *Arch. Neurol.* **51**, 865–873.

Kassell, N.F., Sasaki, T., Colohan, A.R.T. & Nazar, G. (1985): Cerebral vasospasm following aneurysmal subarachnoid haemorrhage. *Stroke* **16**, 562–572.

Lindsberg, P.J., Hallenbeck, J.M. & Feuerstein, G. (1991): Platelet-activating factor in stroke and brain injury. *Ann. Neurol.* **30**, 117–129.

Marchal, G., Serrati, C., Rioux, P., Petit-Taboué, M.C., Viader, F., De La Sayette, V., Le Doze, F., Lochon, P., Derlon, J.M., Orgogozo, J.M. & Baron, J.C. (1993): PET imaging of cerebral perfusion and oxygen consumption in acute ischaemic stroke: relation to outcome. *Lancet* **341**, 925–926.

Morley, P., Hogan, M.J. & Hakim, A.M. (1994): Calcium-mediated mechanisms of ischemic injury and protection. *Brain Pathol.* **4**, 37–47.

Moskowitz, M.A., Sakas, D.E., Wei, E.P., Kano, M., Buzzi, M.G., Ogilvy, C. & Kontos, H.A. (1989): Postocclusive cerebral hyperemia is markedly attenuated by chronic trigeminal ganglionectomy. *Am J. Physiol.* **257**, H1736–H1739.

Neil, J.J. (1993): Functional imaging of the central nervous system using magnetic resonance imaging and positron emission tomography. *Curr. Opin. Neurol.* **6**, 927–933.

Norrving, B. & Lövenhielm, P. (1988): Epidemiology of stroke in Lund-Orup, Sweden 1983–1985. Incidence and age-related changes in subtypes. *Acta Neurol. Scand.* **78**, 408–413.

Olney, J.W., Collins, R.C. & Sloviter, R.S. (1986): Excitotoxic mechanisms of epileptic brain damage. *Adv. Neurol.* **44**, 857–877.

Pappata, S., Tran Dinh, S., Baron, J.C., Cambon, H. & Syrota, A. (1987): Remote metabolic effects of cerebrovascular lesions: magnetic resonance and positron tomography imaging. *Neuroradiology* **29**, 1–6.

Petty, G.W., Wiebers, D.O. & Meissner, I. (1990): Transcranial doppler ultrasonography: Clinical applications in cerebrovascular disease. *Mayo. Clin. Proc.* **65**, 1350–1364.

Rothman, S.M. & Olney, J.W. (1986): Glutamate and the pathophysiology of hypoxic-ischaemic brain damage. *Ann. Neurol.* **19**, 105–111.

Sapolsky, R.M. & Pulsinelli, W.A. (1985): Glucocorticoids potentiate ischaemic injury to neurons: therapeutic implications. *Science* **229**, 1397–1400.

Schanne, F.A.X., Kane, A.B., Young, E.E. & Farber, J. (1979): Calcium dependence of toxic cell death: a final common pathway. *Science* **206**, 700–702.

Siesjö, B. (1992): Pathophysiology and treatment of focal cerebral ischemia, Part II. Mechanisms of damage and treatment. *J. Neurosurg.* **77**, 337–354.

Skyhøj Olsen, T., Larsen, B., Herning, M., Bech Skriver, E. & Lassen, N.A. (1983): Blood flow and vascular reactivity in collaterally perfused brain tissue. *Stroke* **14**, 3, 332–341.

Sokrab, T-E.O., Kalimo, H. & Johansson, B.B. (1990): Parenchymal changes related to plasma protein extravasation in experimental seizures. *Epilepsia* **31,**(1), 1–8.

Strandgaard, S. (1978): Autoregulation of cerebral circulation in hypertension. *Acta Neurol. Scand.* **57,** (Suppl. 66), 1–81.

Svendgaard, N.A., Delgado-Zygmunt, T.J. & Arbab, M.A.R. (1990): Anatomical and physiological studies supporting the involvement of a reflex mechanism in cerebral vasospasm in the squirrel monkey. In: *Cerebral vasospasm*, eds. K. Sano, K. Takahura, N.F. Kassell & T. Sasaki, pp. 145–150, Tokyo: University of Tokyo Press.

Tatemichi, T.K. (1990): How acute brain failure becomes chronic: a view of the mechanisms of dementia related to stroke. *Neurology* **40,** 1652–1659.

Torvik, A. (1984): The pathogenesis of watershed infarcts in the brain. *Stroke* **15,** 221–223.

Wise, R.J.S., Bernardi, R.S.J., Frackowiak, N.J.L. & Jones, T. (1983): Serial observations on the pathophysiology of acute stroke. *Brain* **106,** 197–222.

Whisnant, J.P. (1984): The decline of stroke. *Stroke* **15,** 160–168.

# Chapter 14

# Migraine, autonomic dysfunction and the physiology of the cerebral vessels

Peter J. Goadsby

*Institute of Neurology, The National Hospital for Neurology and Neurosurgery, Queen Square, London WC1N, UK*

## Migraine

Traditionally migraine pathophysiology has been viewed primarily from a clinical perspective, limiting the consideration of the contribution of animal physiology. In this chapter laboratory data that characterize the cerebral circulation will be assessed with complementary results from careful clinical studies in an attempt to synthesize our current understanding of this complex problem. The main, and to an extent competing theories – the neural theory, that asserts that the fundamental defect is in the central nervous system (Lance *et al.*, 1983) or its connections with the vessels (Moskowitz *et al.*, 1979) and the vascular theory, that favours a disorder of the cranial vessels (Wolff, 1963) – will be reviewed in the light of the data presented.

## Definition and classification

The conventions of the Headache Classification Committee of the International Headache Society (Olesen, 1988) will be adopted throughout. Migraine is thus defined as headache lasting usually from 4 to 72 h which would be often unilateral, have a pulsating quality, be of moderate or severe intensity and be aggravated by physical activity. It is episodic and may often be associated with nausea, vomiting or photophobia. There should be no neurological abnormality of relevance on either clinical examination or investigation in the interictal period. In clinical practice two varieties of migraine attacks are commonly distinguished: those associated with an aura of neurological symptoms (previously called classical migraine), and a periodic headache without aura (common migraine) (Lance, 1982).

Migraine is predominantly an affliction of young people who are otherwise well. The prevalence of a family history and early onset of the disorder suggest that there is a strong genetic component. Taken together, these features suggest a subtle structural or functional defect that is not usually life-threatening. The female predominance and the association with menstruation does not assist in differentiating the site of the problem, since the hormonal changes described could affect either neural or vascular structures.

## Premonitory features

About 25 per cent of patients report symptoms of elation, irritability, depression, hunger, thirst or drowsiness during the 24 h preceding headache. Most of these manifestations can arise in the hypothalamus and this suggests a central site for their evolution. In addition the suprachiasmatic nucleus of the hypothalamus has been suggested to be one of two primary oscillators in the generation of circadian rhythms and thus could be easily implicated in the periodicity of migraine that is such an important clinical feature.

## Prodrome of migraine with aura

## Cerebral blood flow and the aura

Numerous studies over some years have confirmed that the aura phase of migraine is associated with a reduction in cerebral blood flow. Visual disturbances (such as the scintillating scotoma: flashing lights that move across the visual field), paraesthesiae or other focal neurological signs are associated with this reduction in cerebral blood flow. This flow change moves across the cortex as a 'spreading oligaemia' at 2–3 mm/min corresponding to the rate that Lashley estimated from plotting the progression of his own visual aura and the phenomenon of spreading cortical depression. This pattern reveals some common threads. First there is a focal reduction in flow that is usually posteriorly near Brodman area 7 and the superior part of area 19 (posterior parieto-visual association cortex) although focal frontal oligaemia without visual aura has been rarely reported. Secondly, this reduction enlarges and may involve the whole hemisphere. The changes first reported by Olesen (1991) and his colleagues noted above were produced by carotid angiography but similar changes have been seen in spontaneous attacks with single-photon emisson computed tomography (SPECT). The progression of the oligaemia across the cortex does not respect vascular territories and is thus unlikely to be primarily vasospastic and vasospasm is only rarely seen if patients have an angiogram during migraine. Vasoconstriction, however,

does occur as evidenced by retrograde flow from the carotid to vertebrobasilar circulation that has been seen in some patients. Furthermore there are reports that the oligaemia is preceded by a phase of focal hyperaemia. Such a change is again exactly what would be expected if a similar phenomenon to cortical spreading depression was involved. Following the passage of the oligaemia the cerebrovascular response to hypercapnia is blunted while autoregulation is intact. Again this pattern is repeated in spreading depression (see 'Neural phenomena associated with the aura' below). Usually the flow change is accompanied by a contralateral aura and the unilateral headache is homolateral with respect to the oligaemia, but patients have been reported to have unilateral headache with a homolateral aura suggesting a mismatch of the aura with the subsequent headache. Indeed the author has personally examined a patient with a scotoma homolateral to a unilateral headache. Headache may begin while cortical blood flow is still reduced thus making the likelihood that the pain arises from a primary, vascular abnormality less tenable.

Some methodological considerations have been suggested to account for the oligaemia and in particular to suggest that the flow changes observed reach an ischaemic level. Because of the errors involved only metabolic studies will be able to satisfactorily determine if ischaemia does occur. The best estimates of flow that take into account Compton scatter (an error generated by detectors picking up a signal from a well-perfused area when positioned over a poorly perfused area, thus overestimating flow) put the flow values during the aura at 20–25 ml/100 g/min. Electroencephalographic activity in humans undergoing carotid endarterectomy is reduced when flow drops below 23 ml/100 g/min, while at this level in the awake monkey a neurological deficit can be seen. Positron emission tomography (PET) studies of oxygen utilization have shown in a single patient during the aura that the flow reduction is balanced by an increase in oxygen extraction and thus oxygen metabolism is normal. This problem has been difficult to access because of the technological demands of the PET studies and for the moment it can be said that there is no hard evidence for cerebral ischaemia during the aura phase of migraine.

## Neural phenomena associated with the aura

### Spreading depression and the aura

What can be inferred from known cerebrovascular physiology concerning the aura? The cortical spreading depression of Leao (SD) was first reported to occur in the exposed rabbit cortex as a negative shift in DC potential that was measured and thus defined electrically. This shift corresponds ionically to a redistribution of $K^+$, $Na^+$, $Cl^-$, $Ca^{2+}$ and $H^+$ ions that has been carefully characterized by micropipette studies. Several features of SD have been used to characterize it and these include a rate of propagation of 2–6mm/min, limitation to one hemisphere and a refractory period for further SD of up to 3 min. SD has been observed in a number of species including rat, cat and in man. The changes can include subcortical structures, such as the caudate nucleus, which seem to be quantitatively less affected. In association with the ionic changes and shift in DC potential that are the electrophysiological markers of SD, distinct changes in cerebral blood flow have been described. Initially flow may increase after SD has been initiated and the increase has been reported to be up to 200 per cent and postulated to be related to the pre-SD level of blood flow. This is followed by a prolonged moderate reduction in cerebral blood flow that is associated with a marked blunting of cerebrovascular responses to hypercapnia with normal autoregulation. Recently it has also been shown that the reduction in cerebral blood flow seen after spreading depression may involve not only the cortex but may also include subcortical structures and even the brainstem. Furthermore although hypercapnic cortical vasodilatation is blocked by SD, neurogenic dilatation from either intrinsic (centromedian parafasicular complex, Chapter 12) or extrinsic (trigeminovascular, Chapter 11) neural sources is not. In addition, the suggested pathophysiological role of spreading depression in migraine and these findings reinforce the view that central mechanisms can still play a major role in the control of cerebrovascular tone even during the acute attack of migraine.

### Neurovascular changes

Experimental evidence suggests that the trigeminovascular system promotes vasodilatation (Chapter 11). Nerves that innervate the cerebral vessels through the trigeminovascular system contain almost exclusively vasodilator transmitters, such as calcitonin gene-related peptide (CGRP) and substance P (SP), release of which could not provoke spreading depression which is fundamentally vasoconstrictor in its expression. Available data suggests that lesions of the trigeminal ganglion do not affect resting cerebral blood flow or glucose utilization in the cat. They do, however, effect vasodilator protector mechanisms such as those seen during hyperaemia following ischaemia or epilepsy. In addition, it has recently been shown that in subarachnoid haemorrhage with threatened cerebrovascular compromise from vasospasm, venous calcitonin gene-related peptide levels are elevated in man. Electrical stimulation of the trigeminal ganglion in both man and the cat leads to increases in extracerebral blood flow and local release of both CGRP and SP (Goadsby et al., 1988). In the cat, trigeminal ganglion stimulation also increases cerebral blood flow by a pathway traversing the greater superficial petrosal branch of the facial nerve, again releasing a powerful vasodilator peptide, vasoactive intestinal polypeptide (VIP). Interestingly, the VIP-ergic innervation of the cerebral vessels is predominantly anterior rather than posterior and this may contribute to this region's vulnerability to spreading depression and in part explain why the aura is so very often seen to commence posteriorly. Stimulation of the more specifically vascular pain-sensitive superior sagittal sinus increases cerebral blood flow and jugular vein CGRP levels. Human evidence that CGRP is elevated in the headache phase of migraine (Goadsby et al., 1990) supports the view that the trigeminovascular system may be activated in a protective role in this condition.

Finally, it has been shown in the experimental animal that stimulation of a discrete nucleus in the brainstem, nucleus locus coeruleus (the main central noradrenergic nucleus), reduces cerebral blood flow in a frequency-dependent manner through an $\alpha_2$-adrenoceptor-linked mechanism. Importantly, while a 25 per cent overall reduction in cerebral blood flow is seen, extracerebral vasodilatation occurs in parallel. From the clinical standpoint, aura can exist in isolation from the pain as 'migraine equivalent'; it is thus possible that the aura originates in the central nervous system with the vascular changes being a secondary feature.

## The headache

## Pain processing

The two main clinical features of migraine most feared by the patient are pain and nausea. There is no strong evidence as to whether nausea is mediated by the area postrema or by gut mechanisms. Mild attacks of migraine without aura may be treated with an analgesic, such as aspirin, and an antiemetic, such as metoclopramide, which may act at either site since the area postrema is outside the blood–brain barrier and dopamine $D_2$ receptor antagonists clearly reduce nausea.

The pain of migraine may take many forms but it is usually identified by its severity, episodic nature and quality, the pulsatile or throbbing character. It has been shown that migraine headache was eased temporarily in one-third of patients by compression of the superficial temporal artery and in another third by compression of the common carotid artery. Vascular pain implicates the sensory innervation of the vessels peripherally or centrally. The cranial vessels are innervated by sensory fibres containing substance P and CGRP that arise from the trigeminal system and the upper cervical roots. The first (ophthalmic) division of the trigeminal nerve innervates predominantly the anterior cerebral circulation while the posterior circulation is innervated by the $C_{2-3}$ cervical roots (see Chapter 11).

Using the 2-deoxyglucose method in the cat, the first and second order neurons in the intracranial vascular pain pathway have been mapped. Electrical stimulation of the superior sagittal sinus, a pain-sensitive and C fibre innervated structure, increases both cerebral blood flow and glucose utilization in restricted and specific brain regions. Metabolic activation is seen in the second order neurons in the trigeminal nucleus caudalis, the dorsal horn of the $C_2$ region and in a curious group of cells in the dorsolateral white matter of the cervical cord again at the $C_2$ level (Goadsby and Zagami, 1991). These cells then project via the quintothalamic tract to decussate in the pons or midbrain and synapse on third order neurons at the level of the ventrobasal complex of the thalamus in the ventroposteromedial nucleus, medial nucleus of the posterior complex and intralaminar complex of the thalamus. These cells have the properties (latency and chemical nociceptive activation) of pain processing neurons in both the high cervical cord and the thalamus. These data compare with the distribution of metabolic activity seen in the hamster brain with vibrissal stimulation, a non-nociceptive trigeminal input, with increased activity in trigeminal nucleus principalis, interpolaris and caudalis, ventroposteromedial thalamus and $S_I$ somatosensory cortex. It is likely that these lower brainstem and cord areas represent the anatomical site of overlap between the trigeminal and cervical systems and thus explain the fronto-cervical pain that so often characterizes headache. Whether the pain arises from the vessels and their sensory innervation or whether there is central instability remains undecided.

Several well-conceived studies have implicated a role for the trigeminal neurons in pain generation. While it is almost self-evident that the pain is mediated or at least transduced in the trigeminal system, does it play a more direct role? Stimulation of the trigeminal ganglion in the rat leads to plasma extravasation and mast cell degranulation in the dura mater that can be blocked by dihydroergotamine and sumatriptan. Since these agents do not affect substance P or calcitonin gene-related peptide mediated changes in vascular permeability or mast cell degranulation in the dura mater when applied directly to cells, it has been suggested that their effect is prejunctional. This system is relatively non-specific as non-nociceptive trigeminal afferents, such as those concerned with a soft touch on the face, are also stimulated. In some of the studies the large currents employed (up to 3.0 mA) could have induced activity in neuronal systems distant to the trigeminal, such as the facial nerve dilator system, and thus the results may not be exclusively interpreted as being due to changes in the trigeminal system.

Careful clinical observations favour

the concept of central instability. Sudden pain in the head after eating ice-cream or swallowing a cold drink is a common experience that is more frequently felt by migraineurs. In addition about one-third of migrainous subjects locate the pain of ice-cream headache to the area habitually affected in their migraine headaches. Ice-pick pains, sudden jabs of pain in the head, are also commonly experienced by migrainous patients, 40 per cent of whom report that these are felt on the same side as their headaches. Taken together these observations suggest that there is hyperexcitability of trigeminal pathways that could discharge spontaneously to initiate a migraine attack. Furthermore the only operations which have produced lasting benefit in migraine are trigeminal rhizotomy and bulbar tractotomy while placement of electrodes into the region of the peri-acqueductal grey matter can induce migraine-like headaches which only resolve with their removal.

## Changes in cerebral blood flow

### Interictal studies

Using both $^{133}$Xe inhalation and HMPAO with SPECT, regional cerebral blood flow has been compared in migraine patients and age- and sex-matched controls. In 92 patients (60 with aura and 32 without aura) interictal flow asymmetries were found in about 40 per cent, considering both modalities. The largest asymmetries were seen in the group of patients with migraine with aura. Further studies to characterize whether these abnormalities correlate with the flow changes seen in the attack and whether interictal physiology (such as

hypercapnic vasodilatation) is normal are awaited.

## Studies during headache

### *Migraine with aura*

The headache phase of migraine may be accompanied by hyperaemia although in some studies the headache was not sufficiently characterized to determine its type. The pain of the headache may come during the oligaemic phase, so that it is unlikely that it is dilatation of the cerebral vessels that is alone responsible for the pain. Differences reported in various studies in the presence or absence of a flow change during the headache phase may be merely due to studying the patients at different phases of their attacks since patients studied serially may have either oligaemia, hyperaemia or no change during a headache dependent upon when they are studied.

### *Migraine without aura*

One early study has suggested that in migraine without aura there is hyperaemia in the headache phase but this has not been observed by others. Interestingly hypercapnic vasodilatation may also be blunted in migraine without aura although reports to date have demonstrated changes that, while reduced, were symmetrical.

### *Clinical observations*

Clinical studies have shown that in at least one-third of patients there is a significant extracerebral vascular component to the headache. The level of CGRP is elevated in the external jugular vein blood of migraineurs during headache clearly demonstrating some activation of trigeminovascular

neurons during migraine with or without aura. Whether the activity is peripherally generated is again uncertain, although it is clear that such changes can in part be seen in both man and cat with direct stimulation of the trigeminal ganglion. The release of these peptides offers the prospect of a marker for migraine that can be readily determined from a simple venous blood sample.

## Response to therapy and the neuropharmacology of migraine

The treatment of migraine is divided into two parts, prophylactic (interval) therapy and the management of the acute attack. Each has contributed different neuropharmacological data concerning the condition and its possible underlying mechanisms.

The vasoconstrictor action of ergotamine was once thought to explain the pathophysiology of migraine. The use of dihydroergotamine has made a vascular theory less tenable since dihydroergotamine is a very poor vasoconstrictor compared with ergotamine, affecting veins rather than arteries, but is equally effective as an antimigraine agent. Blood serotonin (5HT) levels drop at the onset of migraine and a low molecular weight serotonin-releasing factor is present during headache. Migraine may be precipitated by the injection of reserpine which releases serotonin from body stores while the intravenous injection of serotonin is effective in terminating migraine headache but has unacceptable side effects. Recent work has reconciled these two apparently disparate clinical observations by demonstrating that dihydroergotamine binds to α-adrenoceptors and serotonin receptors (5HT₁-like). The latter have recently been classified according to drug binding studies and three major types, $5HT_1$, $5HT_2$ and $5HT_3$, have been described (Bradley *et al.*, 1986). Parallel clinical experience with the newly synthesized 5HT₁-like agonist sumatriptan (GR43175) (Doenicke *et al.*, 1988) promises that such agents may provide a major new mode of acute attack therapy. These data suggest that it is the serotonergic agonist action of a drug that is an important determinant of its efficacy as an anti-migraine compound.

The interval therapy of migraine is dominated by drugs that interact with monoaminergic receptors. Both methysergide and pizotifen are $5HT_2$ receptor antagonists and are effective relatively safe long-term prophylactic drugs. They may also play a role in modifying the behaviour of central serotonergic neurons. The β-blockers lacking any agonist activity (propranolol, nadolol, atenolol, timolol and metoprolol) have been shown to be effective prophylactic drugs and may interact with central noradrenergic pathways from the locus coeruleus that also play a role in nociceptive control. Overall the pharmacological profile of interval therapy drugs is quite different to that of attack therapy and suggests a complex interaction with monoaminergic systems that is likely to be of central origin since it is these, and not the peripheral monoaminergic vascular fibres, that have an established role in the modulation of nociception.

For a neural theory to be plausible, these chemical agents must pass through the blood–brain barrier.

Many of the drugs active in the prophylaxis of migraine (propranolol, pizotifen and methysergide) do cross the blood–brain barrier and may cause central side-effects such as drowsiness. What of acute attack therapy? There is conflicting data for ergotamine derivatives: dihydroergotamine when studied by whole brain binding studies would appear not to cross the blood–brain barrier but when studied by more sensitive autoradiographic procedures can be clearly seen to bind in the area postrema, nucleus of the tractus solitarius, and very markedly in the dorsal raphe nucleus (Goadsby and Gundlach, 1991). Sumatriptan, however, only accesses the cerebrospinal fluid poorly and as measured in whole brain regions does not seem to enter the central nervous system significantly. It is possible that the termination of an acute attack may depend upon a peripheral action on the blood vessels or at a prejunctional serotonin receptor on the trigeminal nerves while prophylactic therapy may act mainly in the central nervous system.

## Hypothesis

Many of the premonitory symptoms of migraine suggest that some dysfunction involving the hypothalamus is taking place in the period before the headache. Areas responsible for biological rhythm, such as the hypothalamus, or external triggers (stress, either emotional or as excessive afferent stimulation) may overload brainstem mechanisms that normally 'gate' cranial nociception which, due to some inherited defect, cannot deal with such afferent volleys. An increased discharge from the locus coe-

ruleus, which would normally attempt to tune the signal-to-noise ratio for such inputs, would initially reduce cerebral blood flow and may thus initiate spreading depression and the prodromal phase of migraine with aura. As the firing increases in the locus coeruleus or as a result of a protective reflex firing of the central serotonergic system and trigeminovascular system or both, cerebral and extracerebral dilator nerves are excited. The combined firing and degree of excitation of these systems thus determines the extent and observability of extracerebral vascular changes and changes in neuropeptide levels. The depletion of brainstem monoaminergic nociceptive control mechanisms with continued activity and the antidromic activation of the trigeminal system releasing mediators of inflammation is thus perceived by the patient through the trigeminal innervation of the vessels as a throbbing and severe pain. The distribution of the pain in a fronto-occipital radiation results from the overlap of both trigeminal and cervical pain afferents in the high cervical spinal cord.

Such a scheme does not exclude humoral factors. Indeed there is experimental evidence that locus coeruleus stimulation can activate the adrenal gland and thus release noradrenaline that could, in turn, induce platelet aggregation and serotonin release. In addition, trigeminal system activation may itself be capable of causing platelet aggregation locally in the cerebral circulation. Cranial vessels being sensitized to nociceptive transmission and thus increasing activity in the trigeminovascular afferent system.

In conclusion, the pathophysiology of

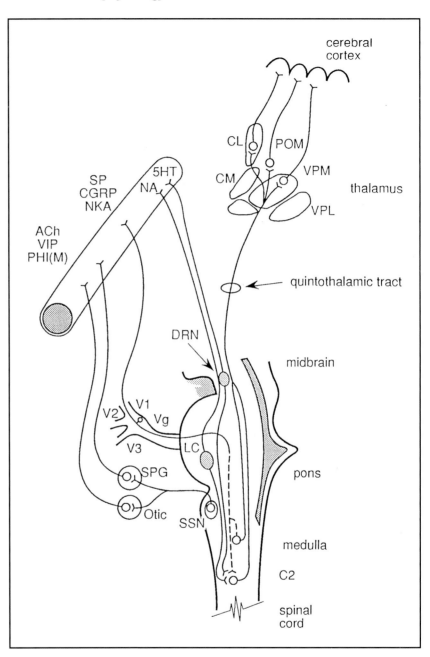

*Fig. 1. Line drawing presenting the basis of a neural schema for the pathophysiology of migraine. Migraine is presented as a central dysnociceptive condition which may be triggered by either centrally mediated events such as those from the hypothalamic circadian centres, or peripherally through the trigeminal system. The latter input is by way of the trigeminal ganglion (Vg) which contains the bipolar cell bodies for pain-sensitive innervation of the dura mater and blood vessels. The peripheral transmitters in this system are substance P (SP), calcitonin gene-related peptide (CGRP) and neurokinin A (NKA). The central first order synapse is in the trigeminal nucleus caudalis in the medulla and on neurons at the level of the second cervical segment of the spinal cord (C2). The second order neurons decussate and pass through the quintothalamic tract to the thalamus, specifically to the ventroposteromedial (VPM), centrolateral (CL) and Medial nucleus of the posterior group nuclei (POM). There is no apparent input to either the ventroposterolateral (VPL) or centromedian (CM) nuclei. A central reflex vasodilator response is also present and mediated largely by the facial (VIIth cranial nerve) with its outflow through the sphenopalatine (SPG) and otic ganglia. The post-ganglionic nerves contain the dilator transmitters vasoactive intestinal polypeptide (VIP), acetylcholine (ACh) and peptide histidine isoleucine [methionine in man, PHI(M)]. These systems are entrained or influenced by the central noradrenergic (locus coeruleus, LC) and serotoninergic (dorsal raphe nucleus, DRN) nuclei, which can normally function to gate incoming pain signals but in the setting of migraine are likely to be functioning abnormally.*

migraine has been examined in the light of recent physiological, neuropharmacological and clinical observations. Available information would suggest that migraine is caused by the response to afferent stimuli or biological rhythms of central nervous system regions that are defective in the control of craniovascular nociception, the inherited trait. Many of the recognized features of migraine can be explained in the light of recent observations made in patients and in the experimental animal. These provide a basis for an understanding of the condition, a rational approach to therapy and directions for basic research and pharmacological developments.

## B. Autonomic failure

Autonomic failure may be classified as either primary or secondary according to whether there is a definable cause for the disease process. Primary autonomic failure is of great interest in the study of the cerebral circulation as it comprises a group of subjects whose autonomic neurons degenerate while all other neurons are intact. Secondary failure is much less interesting because observations made concerning cerebral blood flow could be easily contaminated by the systemic and particularly non-neural effects of the initiating disease process. Primary autonomic failure may further be classified as either pure autonomic failure or multiple system atrophy (MSA). Pure autonomic failure implies involvement of exclusively autonomic neurons in the degenerative process while MSA may be operationally defined as autonomic degeneration plus degeneration of other neur-

onal systems. This latter group was in part that which has been known as the Shy–Drager syndrome (Shy & Drager, 1960). This is not a useful eponym and will not be further employed since there is considerable heterogeneity in the group. The group of MSA may be further subdivided into type I, autonomic failure with Parkinsonian features (also called striato-nigral degeneration), and type II, autonomic failure with cerebellar or pyramidal features (also called olivopontocerebellar atrophy). This division is based on the well-established clinical pictures that are seen and is determined by the predominant features of the individual case (Quinn, 1989; 1994). There have been no studies of cerebral blood flow in the group of primary autonomic failure although the question of cerebral blood flow looms when considering fainting as a manifestation of autonomic nervous sys-

tem disease (Hainsworth, 1988). There is some suggestion for pure autonomic failure that the primary pathology may be found in the sympathetic ganglia which have less mRNA for tyrosine hydroxylase than do patients with MSA (Foster and Lightman, 1990). This would make this condition attractive to study. Investigations of these rare and interesting patients are awaited with anticipation.

**Acknowledgements:** The work of the author included in this Chapter has been supported by the National Health and Medical Research Council of Australia and by grants from Warren and Cheryl Anderson, The J.A.Perini Family Trust, the Basser Trust and the Australian Brain Foundation. The author thanks Sandoz Australia Ltd for access to the data concerning dihydroergotamine. P.J.G. is a Wellcome Senior Research Fellow.

## Summary

Migraine is a common debilitating condition of young, otherwise well, individuals. The problem presents as a severe often pulsatile episodic headache that may be unilateral. Headache may be associated with a premonitory phase that has many characteristics to suggest involvement with the hypothalamus, such as thirst, hunger or mood changes. At the simplest level the primary classification is into headaches with and without associated neurological symptoms, the aura. If aura

is present there is often associated oligaemia and impaired hypercapnic vasodilatation but normal haemodynamic autoregulation. This triad of changes in cerebral blood flow is also seen in the spreading depression of Leao. The headache phase is very often, although not invariably, accompanied by hyperaemia. Analysis of the therapeutic measures employed to treat migraine yields important information. The development of a new potent and highly selective serotonin or $5HT_1$-like agonist has targeted this subgroup of receptors to be of crucial importance in migraine pathophysiology. It is

suggested that the fundamental defect in migraine is within the central nervous system and that various brainstem nuclei and the trigeminovascular system are intimately linked or marshalled into action by this defect. Understanding basic cerebrovascular physiology has produced much of the current knowledge of migraine and the pathogenesis of the condition cannot be understood completely without further studies to characterize the cerebral circulation and its neural innervation.

# References

Bradley, P.B., Engel, G., Feniuk, W., Fozard, J.R., Humphrey, P.P.A., Middlemiss, D.N., Mylecharane, E.J., Richardson, B.P. & Saxena, P.R. (1986): Proposals for the classification and nomenclature of functional receptors for 5-hydroxytryptamine. *Neuropharmacology* **25**, 563–576.

Doenicke, A., Brand, J. & Perrin, V.L. (1988): Possible benefit of GR43175, a novel 5-HT$_1$-like receptor agonist, for the acute treatment of severe migraine. *Lancet* **i**, 1309–1311.

Foster, O.J.F. & Lightman, S.L. (1990): *In situ* hybridization studies of sympathetic ganglia in multiple system atrophy and pure autonomic failure. *Proc. Soc. Neurosci.* **16**, 1345.

Goadsby, P.J. & Gundlach, A.L. (1991): Localization of [$^3$H]-dihydroergotamine binding sites in the cat central nervous system: relevance to migraine. *Ann. Neurol.* **29**, 91–94.

Goadsby, P.J. & Zagami, A.S. (1991): Stimulation of the superior sagittal sinus increases metabolic activity and blood flow in certain regions of the brainstem and upper cervical spinal cord of the cat. *Brain* **114**, 1001–1011.

Goadsby, P.J., Edvinsson, L. & Ekman, R. (1988): Release of vasoactive peptides in the extracerebral circulation of man and the cat during activation of the trigeminovascular system. *Ann. Neurol.* **23**, 193–196.

Goadsby, P.J., Edvinsson, L. & Ekman, R. (1990): Vasoactive peptide release in the extracerebral circulation of humans during migraine headache. *Ann. Neurol.* **28**, 183–187.

Hainsworth, R. (1988): Fainting. In: *Autonomic failure*, ed. R. Bannister, pp. 142–158. Oxford: Oxford University Press.

Lance, J.W., Lambert, G.A., Goadsby, P.J. & Duckworth, J.W. (1983): Brainstem influences on cephalic circulation: experimental data from cat and monkey of relevance to the mechanism of migraine. *Headache* **23**, 258–265.

Lance, J.W. (1982): *Mechanism and management of headache*, 4th edn. London: Butterworth Scientific.

Lashley, K.S. (1941): Patterns of cerebral integration indicated by the scotomas of migraine. *Arch. Neurol. Psychiatr.* **46**, 331–339.

Moskowitz, M.A., Reinhard, J.F., Romero, J., Melamed, E. & Pettibone, D.J. (1979): Neurotransmitters and the fifth cranial nerve: is there a relation to the headache phase of migraine? *Lancet* **ii**, 883–884.

Olesen, J. (1988): Classification and diagnostic criteria for headache disorders, cranial neuralgias and facial pain. *Cephalalgia* **8**, 1–96.

Oleson, J. (1991): Cerebral and extracranial circulatory disturbances in migraine pathophysiological implications. *Cereb. Brain Metab. Rev.* **3**, 1–28.

Quinn, N. (1989): Multiple system atrophy– the nature of the beast. *J. Neurol. Neurosurg. Psychiatr.* Suppl, 78–89.

Quinn, N. (1994): Multiple system atrophy. In *Movement disorders-3*, eds. C.D. Marsden & S. Fahn, pp. 262–281. New York: Raven Press.

Shy, G.M. & Drager, G.A. (1960): A neurological syndrome associated with orthostatic hypotension. *Arch. Neurol.* **2**, 511–527.

Wolff, H.G. (1963): *Headache and other head pain*. New York: Oxford University Press.

# Appendix

Principal receptor types with their specific agonists, antagonists and cellular actions. For more detailed information the reader is referred to the comprehensive table published by Watson & Abbot (1991): Receptor nomenclature supplement. *Trends in Pharmacol. Sci.* 11, Suppl. 1

| Receptor | Nomenclature | Potency order (endogenous ligands) | Selective agonists | Selective antagonists (pA2 in brackets) | Cellular actions | Notes |
|---|---|---|---|---|---|---|
| Adenosine (= $P_1$ purinoceptor) | $A_1$ | Adenosine > AMP ? | $N^6$-cyclopentyl-adenosine R-PIA 2-chloro-$N^6$-cyclo-pentyl-adenosine | DPCPX (8.3–9.3) 8 cyclopentyl-theophylline (7.4) | cAMP $\downarrow$ | |
| | $A_2$ | Adenosine > AMP ? | CGS 21680 CGS 22492 2-phenylamino-adenosine | CP 66713 (7.7) | cAMP $\uparrow$ | Subtypes $A_{2A}$, $A_{2B}$ proposed |
| $\alpha$-Adrenoceptor | $\alpha1$ | Noradrenaline $\geq$ adrenaline | Phenylephrine Methoxamine Ciralozine | Prazosine (8.5–10.5) Corynanthine (6.5–7.5) | $IP_3$/DG (stim $Ca^{2+}$ influx?) | Has been subdivided into $\alpha_{1A}$, $\alpha_{1B}$, $\alpha_{1C}$ |
| | $\alpha2$ | Adrenaline $\geq$ noradrenaline | UK 14304 BHT 920 Clonidine | Rauwolscine (7.5–9) Yohimbine (7.5–9) | cAMP $\downarrow$ $K^+$ channel $\uparrow$ (G) $Ca^{2+}$ channel $\downarrow$ (G) | Has been subdivided into $\alpha_{2A}$, $\alpha_{2B}$, $\alpha_{2C}$ |
| $\beta$-Adrenoceptor | $\beta1$ | Noradrenaline $\geq$ adrenaline | Noradrenaline Amoterol | CGP 20712A (8.5–9.3) Betaxolol (8.5) Atenolol (7) | cAMP $\uparrow$ | Also acts on Ca2+ channel $\uparrow$ (G) in heart |
| | $\beta2$ | Adrenaline > noradrenaline | Procaterol Terbutaline (?) | ICI 11855 (8.3–9.2) Butoxamine (6.2) $\alpha$-methylpropranolol (8.5) | cAMP $\uparrow$ | |
| | $\beta3$ | Noradrenaline > adrenaline | BRL 37344 | | | |

| Receptor | Nomenclature | Potency order (endogenous ligands) | Selective agonists | Selective antagonists (pA2 in brackets) | Cellular actions | Notes |
|---|---|---|---|---|---|---|
| Angiotensin | $AT_1$ | Angiotensin II > antiotensin III | | Losartan (8.1) | $IP_3/DG$ cAMP ↑ | $AT_2$ receptor in brain and periphery with unknown function |
| Atriatal natriuretic peptide (ANP) | $ANP_A$ | α-ANP > BNP (brain NP) | | | int cGMP ↑ | A non-selective antagonist exists |
| | $ANP_B$ | CNP (type-c NP) | CNP | | int cGMP ↑ | |
| Bradykinin (BK) | $B_1$ | $BK_{1-8}$ = Kallidin > BK | $BK_{1-8}$ Sar [DPhe$^8$)]-$BK_{1-8}$ | [Leu$^8$]$BK_{1-8}$ (6.7–7.3) | ? | |
| | $B_2$ | Kallidin > BK ≥ $BK_{1-8}$ | [Phe$^8$, ψ (CH$_2$-NH)-Arg$^9$]BK [Hyp$^3$, Tyr(Me)$^9$]BK | DArg[Hyp$^3$, Thi$^5$, DTic$^7$-Oic$^8$]BK (7.5–10.5) | $IP_3/DG$ | Other BK derivatives can be antagonists |
| Calcitonin gene-related peptide (CGRP) | $CGRP_1$ $CGRP_2$ | CGRPα CGRPβ | [Cys(ACM)$^{2,7}$]-CGRP (CGRP$_2$) | $CGRP_{8-37}$ (6–8) (CGRP$_1$) | cAMP ↑ | |
| Dopamine | $D_1$ | | SKF 38393 Fenoldopam | SCH 23390 SKF 83566 | cAMP ↑ | $D_3$, $D_4$ and $D_5$ receptors exist |
| | $D_2$ | | N-0437 Bromocriptine | (-)-sulpiride YM 091512 | cAMP ↓ K$^+$ channel ↑ (G) Ca$^{2+}$ channel ↓ (G) | |
| Endothelin (ET) | $ET_A$ | ET-1 = ET-2 > ET-3 | | BE 18257B (5.9) BQ-123 (7.4) | $IP_3/DG$ | A third subtype probably exists |
| | $ET_B$ | ET-1 = ET-2 = ET-3 | [Ala$^{1,3,11,15}$]ET-1 Sarafotoxin S6c | | $IP_3/DG$ | |
| GABA | GABA$_A$ (competitive) GABA$_A$ (benzodiazepine modulatory) | | Isoguvacine Muscimol Flunitrazepam Zoldipem | Bicuculline (6) SR 95531 Flumazenil | int. Cl$^-$ channel GABA-gated Cl$^-$ channel ↑ | Located on GABA$_A$ α subunit |
| | GABA$_B$ | | L-baclofen 3-aminopropyl-phosphinic acid | Saclofen (5.3) CGP 35348 (5) | cAMP ↓ K$^+$ channel ↑ (G) Ca$^{2+}$ channel ↓ (G) | |

| Receptor | Nomenclature | Potency order (endogenous ligands) | Selective agonists | Selective antagonists (pA2 in brackets) | Cellular actions | Notes |
|---|---|---|---|---|---|---|
| Glutamate (ionotropic) | NMDA | | NMDA | D-AP5 (5.2–5.9) <br> CGS 19755 (6) <br> MK 801 } channel <br> Phencyclidine } blockers | int. $Na^+/K^+/Ca^{2+}$ | Independent <u>metabotropic</u> receptors exist, acting via $IP_3$/DG or cAMP $\downarrow$ |
| | AMPA | | AMPA | NBQX (7.1) | int. $Na^+/K^+$ | |
| | Kainate | | Kainate <br> Domoate | | int. $Na^+/K^+$ | |
| Histamine | $H_1$ | | Pyridine-ethylamine | Mepyramine (9.1) <br> Triprolidine (9.9) | $IP_3$/DG | |
| | $H_2$ | | Dimaprit <br> Impromidine | Ranitidine (7.2) <br> Tiotidine (7.8) <br> Cimetidine | cAMP | |
| | $H_3$ | | R-α-methyl-histamine | Thioperamide (8.4) | ? | |
| 5-hydroxy tryptamine (5-HT) or serotonin | $5HT_{1A}$ | | 8-OH-DPAT <br> Ipsapirone | | cAMP $\downarrow$ <br> $K^+$ channel $\uparrow$ (G) | 5-$HT_1$ and 5-$HT_1$-like receptors (except 5-$HT_{1C}$) are selectively stimulated by 5-carboxamide-tryptamine and antagonized by methiothepin. The 5-$HT_{1B}$ receptor is the rodent homologue of the 5-$HT_{1D}$ receptor |
| | $5HT_{1C}$ | | α-methyl-5-HT | Ritanserin (8.7) <br> Pizotifen (8.1) | $IP_3$/DG | |
| | $5HT_{1D}$ | | Sumatriptan | | cAMP $\downarrow$ | |
| | $5HT_2$ | | α-methyl-5-HT | Ritanserin (9.5) <br> Pizotifen (8.9) <br> LY 53857 (8–9.5) <br> Ketanserin | $IP_3$/DG | |
| | $5HT_3$ | | 2-methyl-5-HT <br> 1-phenyl-biguanide | ICS 205930 (10–11) <br> Ondansetron (8–10) | int. cation channel | |
| | $5HT_4$ | | 5-methyl-5-HT <br> Renzapride | SD2 205557 (7.4) | cAMP $\uparrow$ | |
| Muscarinic receptors (acetylcholine) | $M_1$ | | | Pirenzepine (*) <br> Telenzepine (9.1) | $IP_3$/DG | Specificity of antagonists is limited. $M_4$ and $M_5$ subtypes also exist. |
| | $M_2$ | | | Methoctramine (7.9) <br> AFDX 116 (7.3) | cAMP $\downarrow$ <br> $K^+$ channel $\uparrow$ (G) | |
| | $M_3$ | | | Hexahydrosila-difenidol (8) <br> p-fluorohexahydro-siladifenidol (7.8) | $IP_3$/DG | |

| Receptor | Nomenclature | Potency order (endogenous ligands) | Selective agonists | Selective antagonists (pA2 in brackets) | Cellular actions | Notes |
|---|---|---|---|---|---|---|
| Neuropeptide Y (NPY) | $Y_1$ <br> $Y_2$ | NPY, Peptide YY | [Leu31, Pro34] NPY <br> [Pro 34] NPY ($Y_1$) | | | Possibility of $Y_3$ receptor |
| Nicotinic receptors (acetylcholine) | Muscle type | | | a-bungarotoxin <br> Gallamine  }channel <br> Decamethonium }blockers | int. $Na^+/K^+/Ca^{2+}$ | |
| | Neuronal type | | | Neuronal bungarotoxin (8.3) <br> Hexamethonium }channel <br> Mecamylamine  } blockers | int. $Na^+/K^+/Ca^{2+}$ | |
| Opoid receptors | $\mu$ | β-endorphin (β-end) > Dyndorphin A (dyn A) > [Met]enkephalin (met) > [Leu]enkephalin (leu) | DAMGO <br> Sufentanyl <br> PL 017 | CTAP (6.4–7.9) | cAMP ↓ <br> $K^+$ channel ↑ | |
| | $\delta$ | β-end = leu = met > dyn A | DPDPE <br> DSBULET <br> [DAla$^2$] deltorphin I or II | ICI 174864 (7.5) <br> Naltrindole (9.7) | cAMP ↓ <br> $K^+$ channel ↑ | |
| | $\kappa$ | dyn A >> β-end >> leu = met | U 69593 <br> CI 977 <br> ICI 197067 | Nor-binaltorphimine (10.3) | Ca channel ↓ | |
| Prostanoid receptors | DP | $PGD_2$ > $PGE_2$, $PGI_2$, $TXA_2$ | BW 245C <br> ZK 110841 | BWA 868C (9.3) | cAMP ↑ | |
| | EP1,EP2,EP3 | PGE2 > PGF2α, $PGI_2$ > $PGD_2$, $TXA_2$ | $EP_1$: 17-phenyl-w-trinor-PGE$_2$ <br> $EP_2$:Butaprost <br> $EP_3$: Enprostil | $EP_1$: SC 19220 (5.6) <br> AH 6809 (6.8) | $EP_1$: $IP_3$/DG <br> $EP_2$:cAMP <br> $EP_3$:$IP_3$/DG <br> cAMP ↓ | |
| | FP | $PGF_2$ > $PGD_2$ > $PGE_2$ > $PGI_2$, $TXA_2$ | Fluprostenol | | $IP_3$/DG | |
| | 1P (prostacyclin) | $PGI_2$ > $PGD_2$ > $PGE_2$ > $PGF_2$, $TXA_2$ | Cicaprost | | cAMP ↑ | |
| | TP (thromboxane) | TXA2 = $PGH_2$ >> $PGD_2$, $PGE_2$, PGF2α, $PGI_2$ | U 44619 <br> STA$_2$ <br> I-BOP | GR 32191 (8.8) <br> SQ 29548 (8.7) | $IP_3$/DG | |

| Receptor | Nomenclature | Potency order (endogenous ligands) | Selective agonists | Selective antagonists (pA2 in brackets) | Cellular actions | Notes |
|---|---|---|---|---|---|---|
| $P_2$ purinoceptors | $P_{2X}$ | ATP = ADP > AMP | $\alpha,\beta$ methylene-ATP $\beta.\gamma$ methylene-L-ATP | ANAPP3 (irrev.) Desensitization by $\alpha,\beta$ meATP $\beta.\gamma$-me-L-ATP | int. cation channel | |
| | $P_{2Y}$ | ATP = ADP > AMP | ADP$\beta$F HomoATP 2-methylthio-ATP | | IP$_3$/DG | |
| | $P_{2z}$ | ATP (ADP, AMP inactive) | 3'O (benzyol)-benzyol-ATP | 2-methylthio-L-ATP | int. cation channel | |
| | $P_{2T}$ | ADP (ATP, ADP antagonise) | 2-methylthio-ATP | 2-chloro-ATP (5.2) ATP (4.6) | int. cation channel | |
| Pyrimidine nucleotide receptors | Pyrimidine? | Uridine triphosphate = uridine diphosphate >>uridine monophosphate = cytidine triphosphate | | | Not determined | Characterizatio n incomplete |
| Tachykinin | NK$_1$ | Substance P (SP)> Neurokinin A (NKA) > NKB | SP methyl-ester [Sar[9], Met (O$_2$)[11]] 6 SP [Pro[9]] SP | CP 96345 (7–9) GR 71251 (7.7) L 688169 (7) | IP$_3$/DG | |
| | NK$_2$ | NKA > NKB ≥ SP | [$\beta$Ala[2]] NKA$_{4-10}$ GR 64349 [Lys[5], Meleu[9], – Nle[10]] NKA$_{4-10}$ | MEN 10207 (6.8–7.9) L 659877 (6.9–7.9) | IP$_3$/DG | |
| | NK$_3$ | NKB > NKA > SP | Senktide [MePhe[8]]NKB [Pro[7]]NKB | [Trp[7], $\beta$-Ala[8]] 6NKA$_{4-10}$ | IP$_3$/DG | |
| Vasopressin (VP) | $V_1$ ($V_{1A}$, $V_{1B}$) | VP > oxytocin (OT) | | d(CH$_2$)$_5$[Tyr(Me)[2]]–AVP (5.6–8.6) [DPen[1], Tyr(Me)[2]]–AVP (8.5) | IP$_3$/DG | $V_{1A}$ and $V_{1B}$ differentiated by antagonists |
| | $V_2$ | VP > OT | dDAVP d[Val[4]]AVP | d(CH$_2$)$_5$[DIle$_2$, Ile[4]]-AVP (8) OPC 31260 | cAMP ↑ | |
| Vasoactive intestinal polypeptide (VIP) | VIP | VIP > peptide histidine isoleucinamide (PHI) > peptide histidine metheonineamide (PHM) | | [4Cl-D Phe[6], Leu[17]]-VIP [AC-Tyr[1], DPhe[2]]-GRF$_{1-29}$ | cAMP ↑ | |

# INDEX

**A**

| | |
|---|---|
| α$_2$-receptors | 88, 265, 291, 297 |
| α-adrenergic | |
|   autoreceptors | 296 |
|   blockade | 290–293, 299 |
|   receptors | 34, 87, 264, 266–267, 273, 278, 289, 295, 297, 303 |
| α-blockers | 289 |
| α-kinase | 152, 153 |
| α-receptors | 88, 264–266 |
| accessory ganglia | 271 |
| acetazolamide test | 248 |
| acetoacetate | 179, 180, 186 |
| acetyl coenzyme A (CoA) | 69, 179, 182–184, 187–198 |
| acetylcholine esterase/AChE | 132, 133, 344 |
| acetylcholine/ACh | 11, 24, 33, 37, 58–59, 63, 80–82, 88, 91, 103, 132–133, 167, 170–172, 246, 263, 268–271, 273, 277, 279, 293, 297, 304, 306, 308, 313, 329, 337, 344, 350, 381, 387 |
|   biosynthesis | 69 |
|   -dependent potassium channels | 106 |
|   pathways | 70 |
|   release | 69–70 |
|   and VIP release | 76 |
| actin | 145, 148, 151 |
| action potential | 17, 53–57 |
| active transport | 49–50 |
| acute experimental preparation | 324–325 |

| | |
|---|---|
| adenohypophysis | 25, 31 |
| adenosine 5′-diphosphate/ADP | 63, 106–107, 160, 168, 170–172, 183–188, 194–195, 215, 250–251, 253–254, 278, 293, 298, 385 |
| adenosine 5′-monophosphate/AMP | 187–188, 195, 278 |
| adenosine 5′-triphosphate/ATP | 50, 63, 66, 70, 99, 107–108, 148, 151, 157, 168, 170, 178, 181–182, 184–185, 188, 193–194, 195, 228, 277–278, 296–299, 303 |
|   production | 79, 89, 106, 186–172, 227 |
|   -sensitive potassium channels | 106 |
| adenylate cyclase | 11, 64, 65, 152, 192, 273 |
| adrenal gland | 6, 17, 380 |
| adrenaline | 29, 33, 58, 135, 188, 266 |
| adrenergic receptors | 265 |
| α-adrenoceptors | 87, 385 |
| α$_2$-adrenoceptors | 88, 102, 378 |
| β-adrenoceptors | 87, 102, 106 |
| β$_2$-adrenoceptors | 88 |
| adventitia | 18, 126–127, 129, 131–133, 264, 268–269, 271–273, 302–303 |
| adventitial-medial border | 264, 266–267, 271, 280 |
| agonist-receptor interactions | 90–94, 149–150, 192 |
| agonists | 87, 92, 93, 96, 105, 167–169, 385 |
| alkaline phosphatase | 162 |
| allosteric interactions | 93–94 |
| amino acid activated chloride channels | 77 |

amino acids          23–24, 29, 33, 63, 65, 77–80, 88, 148, 160,
                     162–163, 186–188, 217, 228, 279, 280
  see also excitatory

4-Aminopyridine (4-AP)                         101, 154

AMPA receptor                            59, 104–105, 365

amphetamine                                        66, 67

anastomoses                        113, 114, 116, 117, 119
  arterio-venous                                      118
  inter-arterial                                      121
  interhemispheric                                    292

anaesthesia/anaesthetics/anaesthetized animal
                     99, 189, 201–205, 213, 224,
                     233, 249, 280, 289–92, 294, 308,
                     324, 331, 333–334, 339–340,
                     342, 345–347, 350, 353, 361

aneurysms                                          367

angiogenin                                         161

angiotensin converting enzyme /ACE                 275

angiotensin I/AI          63, 75, 88, 170, 275–276

angiotensin II/AII    63, 75, 167, 168, 275–276, 277, 385
  see also renin-angiotensin

animal preparation             199, 201–203, 365–366

antagonists                    87, 92–94, 105, 296, 385

anterior
  cerebellar arteries                               287
  cerebral arteries        287, 288, 292, 300, 302, 305
  communicating artery               269, 287, 300

aortic body                                         16

aortic depressor nerve                            18, 20

apomorphine                                        267

arachidonic acid          59, 65, 164, 169, 276, 277, 365

arachnoid                    120, 121, 134, 136, 298
  villi                                         137, 298

area postrema          11–13, 15, 17–18, 20, 24, 72, 75,
                       136, 164, 349, 378, 385

arginine-vasopressin/AVP                           274

aromatic l-amino decarboxylase                     162

arterial

blood pressure                         2, 9, 18, 20, 29, 32, 36,
                     203, 215–216, 290, 325, 336, 340, 353
  see also pial arterial blood pressure

supply                                         114–117

arterioles          2, 131, 135, 265, 267, 269, 288, 289

arteriovenous difference    179, 188–189, 206, 220–211

aspartate                                187, 280, 329
  and glutamate pathways                            63
  and glutamate synthesis/uptake/release          78–80

astrocytes        13, 126, 135, 158, 160, 161, 162, 193
  processes                        126, 162, 172, 268

atherosclerosis                        171, 362, 363, 366

ATPase                        107, 146, 150, 153, 192

atrial fibrillation                                360

atrial naturiuretic peptide/ANP    274–275, 279, 385

atropeptin                                         275

aura                               279, 369, 376–378

autonomic
  ganglia                                           5
  integration                                       5

autoradiographic studies    81, 189, 204, 209–210, 221–222,
                     224, 247, 250, 252, 265, 271, 324–325,
                     327, 330, 339–340, 346, 351, 380, 391

autoregulation                      255–258, 294, 303, 307,
                     311–312, 314, 330–331, 362–363, 377

autoregulatory plateau                             291

avian pancreatic polypeptide/APP                   288

axon reflex                                   271, 314


**B**

β-adrenergic receptors                    264–267, 289,
                     296, 297, 299, 323, 349

β-adrenoceptors                         89, 101, 385

β-blockers                                         380

β-hydroxybutyrate                              179 ff.

β-receptors                        265, 266, 275, 289

baroreceptor        16, 17–18, 32, 34–36, 290, 291, 331

basal forebrain/BF                29, 37, 136, 269, 329–330,
                                   332-3, 335–8, 340–1, 343–5, 350

basilar artery                   112, 114–118, 130, 133, 155,
165, 169–72, 275, 287, 289, 296–8, 305

bicuculline                                        295, 370

binding studies *see* ligand; radioligand

Binswanger's encephalopathy                        366

blood–brain barrier/BBB          78–79, 119, 136–137, 158,
                        160–164, 171–172, 200, 215–219,
                        222, 229, 232, 235, 256, 264–268, 270–272,
                        274–276, 278, 280, 289, 294, 303–304, 349,
                                        363, 366, 371, 378, 380

blood volume                                     214–219
   *see also* cerebral blood flow

bolus method                                 231–232, 234

bombesin                                          76, 275

bradykinin/BK          77, 167, 168, 171, 271, 275, 279, 386

brain

   activation

   dialysis                                  253–254, 257

   distribution of flow to                           118

   energy homeostasis                           194–195

   energy metabolism           177–180, 186–190, 192–194

      enzymes                          64, 189, 327–329

      heterogeneity                              192–'94

   microvessels *see* microvessels

   oedema              137, 164, 201, 275, 363, 371

   oxygen consumption/utilisation          177–178, 188

   parenchyma      121, 136, 161, 211, 218–219, 264, 266–267,
                              271–273, 293, 303, 367

   stem            20–21, 24–26, 28, 36, 68, 72, 78, 112,
                          324, 329, 332, 377–378, 380

   uptake index/BUI                             218, 219

   volume regulation                             36–37

bretylium                                       295, 296

**C**

C1 area                         23–25, 28, 34–36, 68, 80,
                        323, 331–333, 339, 340, 342

C2 and C3 roots           24–25, 29, 34, 68, 80, 309

C-fibres                                       74, 378
   *see also* trigeminal fibres

$Ca^{2+}$        47, 63–66, 69–71, 73, 78–79, 99, 103–104,
                        154–157, 192, 250, 365, 377

calcitonin gene-related peptide (CGRP)   133–134, 272–275,
                                   279, 300, 311–314,
                                   377–379, 381, 386

calcium                         57, 59–60, 98, 100, 106,
                        147–153, 156, 166, 257, 368

   activated postassium channels                98, 106

   ATPase                         50, 149, 152, 171

   channels              101–102, 149–150, 155
      *see also* potential-operated; receptor-operated;
         stretch-operated (calcium channels)

   entry blockers              150, 151, 366, 367

   extracellular                                    96

   hypothesis                                      365

   ionophore                                       167

   pump                              107–108, 190
   *see also* $Ca^{2+}$; sodium-calcium exchange

caldesmon                                         148

calmodulin          107, 147, 148, 150, 151, 153, 166

capacitance vessels                               265

capillary                                         269

   arrangement                                    121

   blood volume                                   216

   cycling                                   254–255

   density                                   254, 255

   innervation                                    135

   perfusion                                 254–255

   permeability                              216–219

   recruitment                                    255

capsaicin                                     271, 310

carbon dioxide          77, 179, 181–183, 185, 194,
                        204, 215, 222, 247–249
   *see also* $CO_2$; $PaCO_2$; $PCO_2$

carbonic anhydrase                            247–249

cardiac emboli | 359, 360
cardiopulmonary afferents | 331
cardiorespiratory reflex | 24
carnosine | 63
carotid
    arteries | 112, 114, 118, 157, 218, 359, 360, 367, 378
    body | 12, 16–17, 18, 24, 345
    rete | 278
    sinus | 12, 16, 17, 18, 20, 126, 249
    sinus nerve | 12, 16, 18
    *see also* external; internal (carotid)
catalase | 167
catecholamine | 17, 29, 63, 65–68, 127, 188, 264–267, 349, 364
    circulating | 65
cavernous sinuses | 18, 125
central
    neurogenic vasoconstriction | 345–347
    neurogenic vasodilation | 338–345
    pathways | 329–337
    serotonergic system
centromedian-parafascicular complex/CM-Pf | 28, 325, 333–338, 340–345, 348–353
cephalic
    arterial system supply | 111–118
    veins | 118–120
cerebellar
    arteries, anterior | 287
    arteries, posterior | 114, 269
    cortex | 35, 121–123, 362
    fastigial nucleus | 269
cerebellum | 2, 37, 68, 78, 112, 329, 330, 332, 360
cerebral
    angiography | 361
    arterial muscle | 155–158
    blood volume/CBV | 200, 214–219, 293
    circulation | 199–238, 263–281

cortex | 121–123, 294, 331, 335–337, 339, 342–343, 345–347, 377
    *see also* microcirculatory patterns
    energy consumption | 190–192
    energy metabolism | 194–195, 219–229
    energy metabolism homeostasis | 194–195
    haemorrhage | 268, 359, 360
    infarcts | 359, 360, 361, 362, 366
    metabolic rate/CMR | 189, 191, 199, 204, 219–222
    *see also* local cerebral metabolic rates
    metabolism | 180–186
    microvessels | 265, 268, 271, 272, 275, 276, 277
    oxygen consumption/utilization | 179–180
    vasoconstriction | 345–346
    vasodilatory responses | 342–343
    ventricles | 11
cerebrospinal fluid/CSF | 82, 120, 134, 163, 215, 245, 256, 266–267, 275, 303, 306, 314, 369, 380
    drainage | 137
    production | 137, 298–299
    *see also* coupling; uncoupling
cerebrovascular
    electrophysiology | 153–158
    function | 327–329
    neurons | 353–354
    resistance | 263–264
    response patterns | 337–351
cervical
    ganglion *see* inferior; middle; superior (cervical ganglion)
    sympathectomy | 348
    sympathetic ganglion/ganglia | 5
chemical regulation | 245–251
chemical stimulation | 32, 323, 324, 325–327, 338, 341–342, 345–347, 348
chemiosmotic theory | 185
chemoreceptors | 11, 13, 17, 20, 24, 32, 249–251, 291, 331, 345

chloride 96, 107, 108, 156
   *see also* Cl⁻
cholecystokinin/CCK 15, 63, 74, 76, 80, 81, 274, 311
choline acetyl transferase/ChAT 132, 133, 136, 162, 286, 304
cholinergic 69–70, 286, 325
   innervation 136, 304
   mechanisms 343–344
   nerve terminals 132
   pathways 69–70
   receptors 297
cholinesterase 58
chorda tympani 9, 12
choroid
   artery 131, 298
   epithelium 137, 298
   plexus 82, 114, 123, 124, 137, 170, 219, 269, 298, 303
   *see also* microcirculatory patterns
chronic experimental preparation 324–325
circle of Willis 112–118, 164, 167, 200, 286, 288–289, 302, 305–306, 363, 367
circumventricular organ 11, 349
citric acid cycle 77–8, 179, 327
Cl⁻ 47, 51–52, 65, 77–78, 154, 190, 365, 377
clearance techniques 201, 206–209
clinical epilepsy 369–370
$CO_2$ 223, 226, 230, 245, 246, 293
$[^{11}C]CO_2$ 235
coeliac ganglion 286
coenzyme Q 185
common anterior trunk 112
competitive antagonism 92
computer tomography/CT 360–362, 367
conduction 49, 53, 58, 59, 213
   *see also* saltatory conduction
conductivity (electrical or thermal) 218

confluens sinuum 118
conscious/awake animals 112, 189, 201, 204, 249, 252, 280, 289, 291, 324, 334, 339–340, 346
corrosion cast preparations 112
cortical spreading depression 201, 213, 273, 376, 377, 380
cotransmission 80–81
coupling 156, 251–254
   of cerebral blood flow and metabolism 185, 351–353
cranial nerve 7, 8–16, 24, 32, 33, 35, 36, 38, 246, 331
cranial–sacral system 1, 7
craniovascular circulation 312–313
curare/curarized animal 201, 202, 213
cyclic AMP/cAMP 59, 63, 65, 71, 89–90, 99, 102, 108, 149, 152, 164, 268, 299, 311, 329
cyclic GMP/cGMP 63, 65, 152, 164, 165
cyclooxygenase
   inhibitors 246–247
   pathway 247
   product 276
cytochrome
   a 226–227
   c 185
   oxidase 192, 193, 226, 254
$[^{11}C]$cytochrome oxidase/CO 233

**D**
D1/D2 receptors 378
dementia, multi-infarct 366
deoxyglucose/2DG/$[^{14}C]$2-deoxyglucose 189, 221–224, 234–238, 378
$[^{11}C]$deoxyglucose/$[^{11}C]$glucose 235
$[^{18}F]$deoxyglucose/$[^{18}F]$glucose (FDG) 223, 235, 238, 371
depolarization 52–53, 55–60, 65, 71, 73, 77, 99–101, 104, 106–107, 150, 154–157, 169–170, 264, 278, 295, 312, 325
diffusible tracer 205–211
dihydroergotamine 378–380
dihydropyridines 102, 150, 151, 155

3-4-dihydroxyl-L-phenylaline/DOPA                65, 71

DOPA decarboxylase/L-DOPA decarboxylase        65, 66, 67, 71, 80, 81, 162

dopamine β-hydroxylase/DBH               129, 134, 191, 286

dopamine/DA           28, 33, 63, 65, 66, 74, 88, 129, 188, 191, 266–267, 273, 279, 378, 386

  pathways                               67–68

  synthesis/uptake/release               65–67

Doppler techniques                            360

  *see also* laser

dorsal medullary reticular formation/DMRF        329, 331–333, 335, 342, 345, 348, 349

dorsal motor nucleus of the vagus        7–19, 21, 27, 34

dorsal raphe nucleus/DR            134–135, 308, 323, 333–335, 346–347, 352, 380, 381

dorsal root ganglion/ganglia              4, 27, 329

dura (mater)                    120, 136, 288, 299, 300, 303, 309–311, 313–314, 378, 381

dural sinuses                      119, 219, 288, 298

dural venous sinuses                        118, 137

dynorphin B                                273–274


**E**

Edinger–Westphal nucleus                    7, 8, 28

efferent mechanisms                       347–351

eicosapentaenoic acid                          276

electrical stimulation        9, 28, 31–32, 36–37, 295, 303, 310, 323–325, 330, 333–334, 336, 338–350, 352, 370, 377–8

electrochemical gradient               48, 49, 190–191

electrocorticogram                             31

electrogenic pump                           50–51

electron-dense vesicles                        132

electron-lucent vesicles              132, 135, 136

emissary veins                            118, 119

end-plate potential                          58–59

endopeptidase                                 273

endoperoxides                                 276

endorphin                        63, 74–75, 331

endothelial

  cell growth factor                     160–161

  cells        126, 135, 136, 257, 267, 268, 276, 278, 286

  factors                             159–160

  pathophysiology                      171–172

  receptors                           168, 270

endothelin        88, 169–170, 276, 277, 279, 286, 368

endothelium        136, 156–157, 161–163, 168–172, 250, 265, 267–273, 275

  derived contracting factor/EDCF      164, 170–171

  derived hyperpolarizing factor/EDHF        166

  derived relaxing factor/EDRF      164–171, 250, 278, 306

energy

  biosynthesis                        180–186

  consumption                         190–191

  metabolism                          199–238

  *see also* brain energy metabolism

enkephalin                       63, 74–75, 273

enteric nervous system                      14–15

ependyma                                      12

epidermal growth factor                       88

epilepsy                  327, 369–372, 377

  animal models                           370

  seizures and hypoglycemia                372

  *see also* clinical epilepsy

epinephrine        7, 23, 63, 65, 68, 80, 349, 350

  pathways                               67–68

  synthesis/uptake/release               65–67

  *see also* adrenaline

equilibrium

  method                    230–231, 233–234

  potential                              48–49

  studies                                94–96

escape phenomenon              292, 293, 294, 303

ethmoidal

  artery *see* external; internal (ethmoidal artery)

nerve                                                      305

excitatory amino acids          28, 36, 78, 104–105, 365, 372

excitatory post-synaptic potential/EPSP              59, 60

excitotoxic hypothesis                                   365

external

    carotid                                     16, 111, 286

    carotid plexus                                    16–18

    ethmoidal artery                                     116

    jugular venous system                                118

extracerebral supply by intracranial sources     117–118

extracranial

    contamination/counting                  206, 208–209

    supply                                        112–113

extradural cerebral

extravasation                                       363, 378

    *see also* protein


**F**

facial nerve                                              308

fastigial nucleus/FN             13, 36–37, 325–326, 329,
                                 331–333, 335, 337–345,
                                 347–348, 350–353

    *see also* cerebellar

feedback loops                              23–27, 33, 65

fibroblast growth factor                       89, 160–161

Fick principle                             205–206, 220, 230

filamin                                                   151

flow inhomogeneity                                       211

foetal reactivity                                    294–295

foetal stimulation                                   294–295

foramina of Luschka and Magendie                         137

forebrain          2, 15, 24, 27, 30–35, 324, 329, 335, 370

FP-receptors                                             276

free radicals                                            364


**G**

G proteins        59, 89–90, 92, 96, 98, 104, 106, 150

coupled receptors                                        149

g-kinase                                         152, 153, 165

galanin                                             80, 274, 279

gamma glutamyl transpeptidase                            162

gamma interferon                                         161

gamma-aminobutyric acid/GABA        24, 29, 32–33, 36,
                                 59, 74, 77–78, 103, 105, 162,
                                 187–188, 193, 228, 280,
                                 293, 297–298, 329, 370, 386

pathways                                                  78

    receptors                                            105

    synthesis/uptake/release             47–48, 77–78

    transaminase                                         280

gamma-aminocbutyric acid a/GABAa receptor       98, 105

ganglia, cells                                       79, 286

gap junctions                                             60

gastrin-releasing peptide                                275

general visceral efferent        4, 8, 10, 11, 13, 16, 18, 27

geniculate, ganglion                                      12

glia/glial

    cells          27, 63, 79, 136, 162, 193, 286, 364, 372

    endfeet                                          136, 158

glibenclamide                                            106

global analysis                                     342, 351–352

glossopharyngeal nerve                                12, 16

gluconeogenesis                                     180, 182

glucose          79, 163, 178, 186, 189, 220–224,
                 236, 237, 327, 364, 370–372

    metabolism                                       28, 254

    oxidation                                        186–187

    transport                                        159, 160

    utilization              192, 203–205, 222–223,
                             234–238, 251–252, 255
                             308, 324–325, 329–330, 339,
                             341–343, 345–347, 351, 378

glucose/[$^{14}$C]glucose technique        221, 223–224

glutamate        59, 63, 78–80, 103, 187, 188,
                 228, 280, 329, 365, 386

    receptor                                         104–105

glutamic acid decarboxylase 77, 162

glutamine 79, 104, 187, 188, 228

glycine 63, 77, 78, 103, 104, 188, 280

   pathways 280

   receptor 105–106

   synthesis/uptake/release 78

glycogenesis 180, 184, 185–186

glycogenolysis 180, 184, 185–186

glycolysis 151, 178, 180, 181–182, 327

greater superficial petrosal/GSP 304

guanethidine 295, 297


**H**

H+ 158, 248, 250, 253, 254, 257

haemorrhage *see* intracerebral; subarachnoid

head pain 271

   *see also* headache

headache 359, 375, 376, 377, 378–379

   *see also* migraine

heart rate 9, 18, 20, 34, 36, 203, 336, 340

hematomas 360

high-density lipoprotein/HDL 160

hill plot 96, 97

histamine 63, 161, 167–168, 188, 269–270, 279, 387

   receptors 168, 297

hydrogen *see* H+; sodium-hydrogen antiporter

5-hydroxydopamine 129, 131

6-hydroxydopamine 29, 346

5-hydroxyindole acetic acid/5-HIAA 71, 134

5-hydroxytrypamine/5HT *see* serotonin

hyperaemia 246, 248, 254, 255, 314, 362, 377, 379

   focal 376

   functional 246, 253

   postischaemic 364

hypercapnia 224, 246–248, 289, 291–292, 294, 303, 307, 350, 376, 377

   hypercapnic vasodilation 306, 314, 379

hyperpolarization 2, 4, 56, 60, 77, 100–101, 105, 107, 151–152, 155, 157, 166–167, 168

hypertension 18, 171, 256–257, 290–291, 294, 302–303, 331, 362–363, 366, 367

   non reflex-induced 291

   reflex-induced 291

   *see also* spontaneously hypertensive

hypertrophy, sympathetic-induced 302

hypocapnia 230, 245, 291–292, 311

hypoglycaemia 179, 193, 280, 369, 372

hypophyseal arteries 124

   portal vessels 112, 124, 125

hypophysis, venous drainage of 123–125

hypotension, haemorrhagic 290–291, 303

hypothalamus 28, 342, 346, 376, 380

hypoxia 20, 193, 228, 249, 250–251, 254, 278, 291–292, 294, 303, 307, 311, 314, 344, 345

PO$_2$ 157


**I**

immobilization stress 202

*in vitro* vessels/isolated vessels 78, 111, 134, 148, 152, 157, 160–162, 182, 187, 205, 215, 216227, 263–266, 268, 270–272, 274–277, 280, 291, 295–298, 303, 306–308, 310–312, 363

   stimulation 307–308

*in vivo* vessels/*in situ* vessels 67, 89, 111–113, 116, 124, 127, 146–148, 156, 162–163, 166, 180, 187, 189, 201, 205, 213, 221, 226–229, 257, 265–266, 268, 270–273, 275–276, 278–280, 288, 291, 297–298, 303, 306, 311–314, 363

indomethacin 246

inferior cervical ganglion 5

inferior petrosal nerve 9

inflammation 133, 201, 256, 268, 277, 368, 380

   *see also* neurogenic

inhibitory post-synaptic potential/IPSP    60, 190

inner ethmoidal artery, *see* internal ethmoidal artery

inner ophthalmic A, *see* internal aphthalmic artery

inositol 1,4,5trisphosphate/IP3    65, 90, 99, 149, 152, 153

insular cortex

insulin    63, 161

internal

    carotid artery    111, 113, 115–17, 125, 130, 208, 268,
        272, 274, 278, 286–267, 289,
        292, 300, 302, 304–306, 308, 348

      elastic lamina    126

      thhmoidal artery    116–117

      jugular bulbs    189

      jugular veins    189

      ophthalmic artery    117

intimal cushions    126–127

intracarotid injection    206, 266, 267–268, 270, 271, 273

intracerebral

    haemorrhages    360, 363

    pathways    342–343

intracisternal injection    276

intracranial

    circulation    112–113

    pressure/ICP    137, 214–215, 293

    sources    117–118

intrahypothalamic injection    268, 272, 280

intraparenchymal

    arteries    131, 293, 347, 348

    arterioles    131, 135, 293, 347, 348

    injection    266, 271, 273

    venules    288

intrathecal injection    272

intravenous infusion    222

intraventricular injection    248, 252, 266, 273, 379

intravertebral administration    268, 273

intrinsic innervation    134–136

intrinsic-extrinsic interaction    128, 324, 347–348

ionic

    channels    96–107

    flux    51–52

    pumps    107–108

    transport    190–191

irritant receptor    18

ischaemia    211, 226, 228, 232, 254, 256, 278,
    280, 309, 344–345, 360,
    363–365, 367, 372, 376–377

    focal    364, 365–366

    global    364, 366

    transitory    359

isolated ring segments    216

isometric changes in tone    215

isoprenaline    265

isopropyl-$[^{23}$I]iodoamphetamine    232

**J**

jugular

    bulbs, internal    189

    veins    118, 119, 377

    veins, internal    189

    venous system, external    118–119

**K**

$K^+$    47–48, 50–51, 59, 64, 78,
    103–104, 153–158, 190, 193,
    215, 250, 252–254, 257, 365, 377

kainate    104

    receptor    105, 280

kainic acid    325, 339, 342, 347, 348, 370

kallikrein    275

$K_d$    91–93, 95–96

ketone bodies    179–180, 193, 221

kindling    370

kinetic

    binding studies    94–96

rate constants                                      95–96
kininogen                                              275
Kölliker-fuse                          26, 27, 28, 29, 333


**L**
labyrinthine arteries                                  117
lactate                                            193, 221
lacunes                                                366
laser-Doppler flowmetry/LDF          201, 204, 205, 213–214,
                                      293, 308, 325, 346, 350
Latch hypothesis                                       151
lateral geniculate body                                  8
leptomeninges                                          120
leukotrienes                        169, 170, 275, 277, 279
ligand
    binding             87–89, 91–92, 94, 95, 161, 265, 271
    -gated channels                             102–107
    -gated potassium channels                        106
    interactions                                   93–94
    see also radioligand
lipolysis                                              179
lipoxygenases                                          276
local
    cerebral metabolic rates              192, 226–227
    cortical neurons                    336–337, 343–344
locus coeruleus/LC         29–30, 33–34, 37, 68, 73, 75–76, 135,
                           164, 246, 265, 308, 323, 329,
                           331–336, 345, 378, 380–381
low molecular mass neurofilament protein/NF-L     128
lumped constant/LC                         223, 235, 236
luteinizing hormone                            76–77, 106
lymphatic system                                        11
lysine-vasopressin                                     274


**M**
magnetic resonance imaging              360–361, 362
mandibular nerve                                         9

mast cells                          72, 127, 132, 161, 267,
                                    268–270, 272–273,
                                    275, 286, 300, 303, 313–314
    degranulation                              311, 378
    see also perivascular
medial cerebral artery see middle
median eminence                                125, 136
medulla                            26, 27, 30, 36, 78, 266, 294,
                                    304, 329–332, 346, 381
    oblongata            12, 13, 20–26, 37, 72, 76, 119
membrane
    conductance                                    153
    current                                         49
    electrical events                          155–158
membrane (continued)
    polarization                                   53
    potential                          99, 153–154, 156
    properties                                    157
    rectifying properties                         52
meningeal arteries                    133, 288, 309
mesencephalic reticular formation           21–22
mesencephalon                                        334
mesenteric                                           286
metabolic
    mechanisms                                    257
    pathways                                       63
    response patterns                        337–347
    theory                                        194
metabolites                                          188
microapplication
microcirculatory patterns
    cerebral and cerebellar cortex          121–125
    choroid plexus                              123
microganglia                                   304, 306
microglia                                       12, 126
microspheres                               211–212

microvessels 136, 158, 162–163, 264–265, 267–269, 274, 280, 343–344, 346

midbrain 2, 8, 26–30, 36, 78, 246, 378

middle cerebral artery/MCA 248, 257, 265, 267, 269, 275, 287–288, 292, 296, 298, 300, 302, 305, 309, 360, 366

middle cervical ganglion 5, 286, 287

migraine 34, 280, 285, 300, 309, 314, 375–381

  classification 375–376

  definition 375–376

  neuropharmacology 379–380

  varieties 375–376

miniature potentials 59

molecular microspheres 212

monoamine 33, 59, 80, 82, 279

  oxidase/MAO 66, 71, 81, 129, 131, 162, 286

morphometric observations 302–303

multiple system atrophy 382

multiple tracer 224

muscarinic receptors 88–89, 101, 106, 167, 246, 268, 297, 306, 343, 344, 387

muscimol 280

myoendothelial junctions 126

myofilaments 265

myogenic mechanisms 257–258

myosin 145, 146–147, 152

  light chain kinase 146

**N**

N-methyl-d-aspartate/NMDA 280, 365

  receptor 59, 104, 105, 202

Na+ 50, 64, 77, 103–104, 154, 190–191, 253, 365, 377

NAD/NADH 182

  redox state 182, 184, 194, 226–227

nasociliary nerve 272, 305

necrosis 372

neonatal subjects 294–295

Nernst equation 48–49, 52

neural

  mechanisms 246

  pathways 348, 349

  phenomena 377–378

  vascular plexus 5–6

neuroactive peptides 72–77

neuroendocrine system 26, 28

neuroexocrine function 313

neurogenic

  inflammation 275

  regulation 258, 285–315

  responses

neurohypophyseal hormones 24

neurohypophysis 248, 249

neurokinin A/NKA 133, 272, 279, 311, 381

  *see also* substance P/NKA-fibers

neuromuscular junction 58

neuronal

  cell damage 372

  death 364–366

  hormonal interaction 324, 348–350

  necrosis 372

  noradrenaline, release of

neuropeptide Y/NPY 11, 33, 76, 80, 82, 129–130, 266, 270, 273, 279, 297–298, 303, 313, 387

neuropeptides 23, 28, 33, 34, 72, 73, 270–275, 314

  biosynthesis 72–73

neurosecretory neurons 2, 64

neurotensin 75, 82, 275

neurotransmitter systems 63–82, 187–188, 191, 263–281

newborns 158–159, 179–180, 247, 294–295

nicotinic acetylcholine receptor 98, 103–104

nicotinic receptors 87, 286, 305, 387

nitric oxide/NO 165, 171, 255, 275, 324, 344, 347, 365

  intrinsic neuronal interaction 350

synthase-containing nerves | 133–134, 166–167
synthase/NOS | 134, 350, 353
nociceptive signals | 35, 75, 310, 378, 380
non-competitive antagonism | 93
non-diffusible tracer | 201, 206, 211–212
non-specific binding | 94
non-tracer techniques | 206, 213–214
noradrenaline/NA | 10–11, 21, 33, 37, 129, 135, 188, 191, 263–266, 269–270, 272–274, 277–279, 286, 293–294, 296–299, 303, 313, 380
neurally released | 265
noradrenergic
innervation | 135
pathways | 67–68
norepinephrine | 7, 11, 63, 80, 82, 246, 329, 349
pathways | 67
synthesis/uptake/release | 65–67
see also noradrenaline
nuclear magnetic resonance/NMR | 189–190, 220, 227–229, 248
nucleus parasolitarius | 13, 36
nucleus tractus solitarii/NTS | 9, 11–20, 22–24, 26–27, 29–36, 68, 329, 331–333, 335–336, 339, 380

**O**

occipital
cortex | 345, 347
sinuses | 118
oedema, vasogenic | 270
olfactory artery | 117
oligaemia | 376, 379
opiates | 74, 273–274, 279
receptors | 74, 75, 82
opioid
peptides | 26
receptors | 388

organ bath | 263
osmotic pressure | 47, 137
otic ganglion | 9, 27, 304, 305, 306, 308, 312, 313, 381
oxidative phosphorylation | 184–185
$[^{15}O]$oxygen | 231, 233, 234
oxygen
consumption/utilization | 16, 151, 177–178, 191, 226, 229, 233–234, 251, 254, 268, 274, 278, 361, 371
extraction | 233
polarography | 226
radicals | 275, 277
reactivity | 249–250
see also $PaO_2$; $PO_2$
oxyhaemoglobin | 170, 368
oxytocin | 28, 31, 63, 74, 75, 275

**P**

P1-receptors | 194, 278
P2x-receptors | 278, 388
P2y-receptors | 278, 388
pacchionian granulations | 137
$PaCO_2$ | 215, 292, 293, 295
pain | 4, 26, 34, 35, 72, 74, 75, 133, 272, 289, 330, 335, 345, 378–379
see also headache
$PaO_2$ | 234, 248, 292, 293
parabrachial
complex/afferents/efferents/pathways | 26–29, 33
nuclear complex/PB | 26, 29, 31, 331–334, 345–347, 353
parasympathetic
blockade | 306–307
innervation | 132–133, 327
nervous system | 5, 7–11, 258, 299, 304–309, 312
preganglionic motor neurons | 21, 27, 28
sacral/cranial nerves | 30, 285
stimulation | 308–309

parasynaptic transmission 81–82

paraventricular hypothalamic nucleus/nuclei 24, 28

parenchymal

  arteries 264, 266, 269

  arterioles 126, 266, 269

  microvessels 266, 267, 269, 270

partial agonists 92

partition coefficient/Lamda 211

patch clamp 57, 156, 157

  technique 55, 97

$PCO_2$ 226, 245–246, 293, 308

penetrating

  arteries 270, 271, 272, 280, 360

  arterioles 120, 121, 266, 270, 271, 272, 288, 348

pentose phosphate pathway 180, 183, 185, 193

penumbra zone 363

peptide histidine isoleucine/PHI 271, 304, 306

peptide histidine methionine/PHM–27
271, 304, 306, 313, 381

peptide YY/PYY 313

peptidergic

  innervation 133, 258, 270

  mechanisms 270

peptides 74–77, 279

  neuroactive 73–74, 76–77

  peptide-containing cell groups and pathways 72–77

perfusion pressure 255–258, 269, 271, 290, 364

pericytes 126, 135, 158, 160

peripherin 135

perivascular

  glia processes 136

  mast cells 299–301

  microapplication 245

  plexus 126, 288

periventricular

  fibre system 24–26

  organs 15, 27

permeability

  of capillaries, BBB 136

  selective 47–48

permeability-surface/PS product 218

petrosal nerve 305, 308, 312

  inferior 9

  *see also* superficial

PGE2 276–277

PGF2a 276–277

PGI2-sensitive receptor 276–277

  pH 157–158, 185, 215, 228, 245–247, 250, 253, 266, 372

pharmacological stimulation 246, 263, 307

phenyl-ethanolamine-n-methyl transferase/PNMT
66, 68, 353

phosphocreatine/PC4 187, 192, 195, 228

pia (mater) 269

pial

  arterial blood pressure 215–216, 294

  arteries 120, 122, 128–129, 132–134, 167–168, 170, 245, 250, 264–267, 278, 280, 289, 293, 302, 347, 348

  arterioles 133, 168, 170, 258, 264–265, 268, 270, 272–276, 278, 280, 294–295

picrotoxin 370

pituitary gland 72, 75, 76, 217

platelets 161, 267, 268, 277

  aggregation 380

  platelet derived growth factor 88, 159–160

  platelet-activating factor 365

$PO_2$ 157, 188, 226, 249–251, 250, 293

  hypoxia 157

polarization 53

polyamines 161

pons 2, 11, 25, 26–30, 34, 36, 78, 305, 308, 329–330, 332–334, 360, 378

portal vessels 123–125

positron emmission tomography/PET 209, 221, 223, 230–238, 361–362, 371, 377, 390, 392

post-ischaemic reperfusion 364–366

posterior

    cerebellar artery 114, 269

    cerebral artery 114–115, 287, 300, 302, 305

    communicating arteries 114–116, 118, 133, 287, 297, 302

potassium 152, 154, 163, 271

    channels 49, 50, 52–56, 59, 96, 99, 101

        delayed rectifier 154

    repolarization 53–54

    see also K+; sodium-potassium ATPase

potential-operated calcium channels 150, 152

precapillary

    arterioles 120

    sphincters 127

prejunctional inhibition 297–298

primary neurogenic

    cerebrovasodilation 338, 340–345, 348, 349

    vasoconstriction 345–346

probe technique 207–208, 226–227

prodrome 376–378

prostacylin/PGI2 276

prostaglandins 63, 170, 246–247, 276

prostanoids 158, 270, 271, 275, 276–277, 297, 388

protein

    contractile 146–148

    extravasation 311

pseudocholinesterase 162

pterygopalatine (ganglion) 308–309

pupillary

    constrictor reflex 8, 35

purines 161, 168, 278, 279, 296, 297

pyruvate 178, 192, 221

**Q**

quantal release 59

quinolinate 280

quisqualate 104, 105, 280

**R**

radioactive

    emitters (beta, gamma) 233

    microspheres 211–212

    tracers 200

radioactivity 188

radioligand binding and techniques 94, 96, 104, 268

Ramus communicants/Rami communicantes 5

Ranvier's node 54, 57

Raphe (nuclei) 26, 134

receptor 87–90

    -coupled channels 99

    -operated calcium channels 149, 150, 156, 166

    structures 2, 385

rectification 101

reflex/reflex arc 18, 20, 35

regional variation 292–293

reinnervation 302

relaxation mechanisms 151–153

renin–angiotensin system 257, 275–276

reperfusion see post-ischaemic reperfusion

repolarization 53–54

reproducibility 204–205

respiratory

    chain 184–185

    control 183

resting potential 47–52

rete mirabilis 112

reticular formation 352–353

reticulohypothalamic neurons 24

rhodopsin-type receptors 89–90

rostral ventrolateral medulla/RVL 11, 23–24, 32, 323, 325, 331, 338, 342–345, 348

**S**

saltatory conduction 57

sarcoplasmic reticulum 148–152

saturation technique 95, 96, 209–210

sausage string 256

Scatchard analysis 95

Schild plot 93

Schwann

   cell 131

   sheath 11, 130, 132, 133

second messengers 87

secondary neurogenic cerebral

   vasoconstriction 347, 348

   vasodilation 345, 348, 349

seizures 246, 254, 278, 295, 312, 327, 369

sensitivity (of a technique) 204

sensorial innervation 258

sensory

   fibres 309, 378

   inputs 329–330

   nerve 285

   root 4

serine hydromethyltransferase/SHMT 78

serotonergic

   fibres 71, 267

   innervation 72, 134–135, 268, 336

   neurons 135, 267, 334

   system, central 70, 72

serotonin/5-HT 33, 63, 74, 80, 82 106–107, 129, 134, 167–168, 188, 191, 267–268, 273, 279, 288, 297, 300, 303, 346, 379–380, 387

   pathways 70–72

   receptors 88, 106–107, 327

   synthesis/uptake/release 71–72

shunt 60, 117, 118, 123

Shy–Drager syndrome 382

sigmoid sinus 118, 119

sinus rectus 131

smooth muscle

   cerebral arterial 153–154

   contraction 145–153

   phenotype 302

   receptors 59, 74, 265

sodium 96

   -calcium exchange 107–108, 149, 150

   channels 51–57, 59, 99–100

     inactivation 55–56

   -hydrogen antiporter 250

   polarization 53

   -potassium-ATPase 50–51, 107, 151, 154, 160, 163, 171, 190–191, 193

   pump 52, 107

   *see also* $Na^+$

somatosensory

   neurons 12

   relays 14

somatostatin 63, 76, 82, 275

   as neurotransmitter 24

somatotopic organisation 313

spare receptors 92

spatial resolution 204

special visceral afferent 10, 11, 13, 16

species differences 200–201, 287–288

specific

   binding 95, 96

   neural systems 352–353

SPECT measurement 209, 230, 232–233, 233, 361-2, 371, 376, 379

spectrometry 227–229

spectroscopy 227–229

sphenopalatine 271, 288, 304–309, 312, 313, 381

sphincters 127

spinal cord 4, 24, 26–27, 30, 32, 34, 76, 78, 285, 323, 349, 380–381

spinal nerves 35–36

spontaneously hypertensive rats/SHRs    302, 303
stellate ganglion    5, 286, 287, 298, 347
stress    202, 266, 367, 380
  *see also* immobilization stress
stretch-induced tone    263
stretch-operated calcium channels    149, 150, 156
stroke    32, 233, 285, 312, 359–367, 392
  acute treatment    366
  diagnosis    360–362
  models    365–366
  origins    359–360
  prevention    366
subarachnoid
  haemorrhage    267, 269, 309, 314, 360, 367–369, 377
  space    11, 20
subendothelial arterial cushions    126–127
subependymal glial striatum    12
subfornical organ    28
substance P/SP    33, 63, 74, 80, 82, 106, 133, 167–168, 172, 271–275, 277, 279, 300, 310–313, 369, 377–378, 381
  fibers    133, 271
  NKA-fibers    272
sulcus limitans    2
sumatripan    378, 380
superficial petrosal nerve    9, 308, 312
superior
  cerebellar peduncle    26
  cervical ganglion    5, 264, 286–289, 292, 300, 302–303, 312–313, 347–348
  saggital sinuses    120, 121, 137, 288, 310, 313, 377, 378
  salivatory nucleus    305, 313
superoxide dismutase    166
sympathectomy    129, 131, 134–135, 286–290, 292, 299, 301–303, 330, 348
  *see also* cervical
sympathetic

activation    288–290, 292
chain ganglia    6
denervation    288–290
fibres    286–288
ganglion/ganglia, cervical    5
ganglionectomy    128–129, 300
innervation    128–132, 327
nerves    264–267, 270, 274, 277, 285, 298–303, 347–348
nervous system    1, 3–7, 11, 249, 257–258, 286–303, 312, 348
outflow    6–7
spinal nerves
stimulation    292–298
trunk ganglia    5, 6
synaptic
  plasticity    104, 255
  transmission    57–60, 64–65, 72, 101
  vesicles    59, 101, 192

**T**
taurine    63, 188, 228
  as neurotransmitter    77
$^{99m}$Tc    212, 232
$^{99m}$Tc HMPAO    212, 230, 232
telencephalic nuclei    26, 72
temporal resolution    204–205
tentorial nerve    313
tetraethylammonium/TEA    54, 55, 101, 106, 154
tetrodoxin/TTX    54, 55, 99, 154, 295, 296, 307
thalamus    25, 28, 329–330, 332, 334–336, 340, 342–343, 345, 349–350, 360, 362, 366, 372, 378, 381
thromboxanes/TXA2    170, 276, 277
thyrotropin-releasing hormone    76, 275
tight junctions    120, 123, 126, 136, 137, 158, 161, 162
tissue PCO$_2$    226

tissue perfusion technique          201, 205–211, 213–214

TP-receptors                        276

tracee                              235

tracer technique          210–211, 217–219, 234–235, 236, 342

  *see also* diffusible; multiple; non-diffusible;
    radioactive trans-cranial Doppler ultrasonography

transforming growth factor (TGF)

transitory ischaemic attack/TIA     359

transmembrane ionic gradient        56

transmitter-receptor interactions   90–94

transport

  selective                         163–164

transverse

  basal sinus                       119

  sinuses                           118–119, 121

tricarboxylic acid cycle            180–184, 193, 327

trigeminal

  fibers                            271, 312

  ganglia/ganglion                  133–134, 272,
                                    308–313, 369, 377–379, 381

  ganglionectomy                    311, 364

  innervation                       133, 309–314

  nerve          273, 309, 310, 311, 312, 313, 378, 380

  neurons              309, 310, 312, 314, 378

  nucleus caudalis                  12, 13

  pain fibres (C-fibres)            378

  system                 309–313, 378, 380

    stimulation                     312–314

trigeminovascular system            285, 314

trophic influences                  302–303

tropomyosin                         147, 148

tryptophan                          65, 71, 188

  hydroxylase/TPH                   71, 134, 191

tumor necrosis factor               161

tympanic nerve                      9

tyrosine                            65, 66, 89

  hydroxylase/TH            65, 71, 191, 286, 382

**U**

U44069                              276

U46619                              276

ultrasound                          360

uncoupling of cerebrospinal
  fluid and metabolism              351–353

uridine

  diphosphate/UDP                   168

  triphosphate/UTP                  168, 186

**V**

vagus nerve                    11, 12, 63, 68

vascular

  changes, degenerative             366

  contraction                       146–151

  dementia                          366–367

  flowmetry                         214–215

  growth and development            158–161

  innervation                       127–136

  -meningeal relations              120–121

  smooth muscle contraction         145–153

  tone, modulation of               164–171

vasoactive intestinal peptide/vasoactive intestinal
  polypeptide/VIP                   63, 74, 76,
                        81–82, 129, 132–134, 136, 263,
                   269–271, 273, 277, 279, 288, 297, 299,
                      304, 306–309, 312–313, 329,
                            336–338, 369, 377, 389

  -ergic innervation                306

  nerves                            133

vasomotor activity                  215–216, 362

vasopressin          28, 31, 75, 167, 168, 274, 279

  V1/V2-receptors                   63, 389

vasospasm                   367, 368–369, 377

  neurogenic hypothesis             369

v. cerebri magna of Galen           118

veins                  118, 119, 131, 265, 288

velocimetry measurements *see* laser-Doppler

venous drainage | 118–120, 123–125
  hypophysis | 123–125
ventral root | 4
ventrobasal thalamus | 27
venules | 121, 294
verapamil | 151, 154
vertebral
  arteries | 112, 114, 118, 133, 286, 287, 348, 359, 363
vessel
  calibre | 215
  wall | 126–127
vinculin | 148
Virchow–Robin space | 11, 119, 120, 126
visceral
  afferents | 13, 14–20, 26, 30–33
  motor neurons | 20
  reflex control | 33–37

voltage clamp | 54–55, 59
voltage-dependent channels | 57, 98–101
von-Willebrand factors | 158, 162

**W**

wall mass | 302
wall to lumen ratio/wall/lumen ratio | 302
$[^{15}O]$water/$H_2O$ | 231, 232, 233
water/$H_2O$ | 179, 181, 182
watershed infarcts | 363

**X**

$[^{133}]$Xenon | 209, 220, 233, 308, 379

**Z**

zona incerta | 29